The Art of Writing Reasonable Organic Reaction Mechanisms

Second Edition

Springer

New York
Berlin
Heidelberg
Hong Kong
London
Milan
Paris
Tokyo

Robert B. Grossman
University of Kentucky

The Art of Writing Reasonable Organic Reaction Mechanisms

Second Edition

 Springer

Robert B. Grossman
Department of Chemistry
University of Kentucky
Lexington, KY 40506-0055
USA
rbgros1@uky.edu

Library of Congress Cataloging-in-Publication Data
Grossman, Robert B., 1964–
    The art of writing reasonable organic reaction mechanisms / Robert B. Grossman—2nd ed.
        p.   cm.
    Includes bibliographical references and index.

    1. Organic reaction mechanisms.   1. Title
QD502.5.G76 2002
    547'.139—dc21   2002024189

This material is based on work supported by the National Science Foundation under Grant 9733201.
Any opinions, findings, and conclusions or recommendations expressed in this material are those of
the author and do not necessarily reflect the views of the National Science Foundation.

Printed in the United States of America.   (EB)

ISBN 978-1-4419-3016-3   e-ISBN 978-0-387-21545-7

Springer-Verlag is a part of *Springer Science+Business Media*

springeronline.com

Preface to the Student

The purpose of this book is to help you learn how to draw reasonable mechanisms for organic reactions. A *mechanism* is a story that we tell to explain how compound **A** is transformed into compound **B** under given reaction conditions. Imagine being asked to describe how you travelled from New York to Los Angeles (an overall reaction). You might tell how you traveled through New Jersey to Pennsylvania, across to St. Louis, over to Denver, then through the Southwest to the West Coast (the mechanism). You might include details about the mode of transportation you used (reaction conditions), cities where you stopped for a few days (intermediates), detours you took (side reactions), and your speed at various points along the route (rates). To carry the analogy further, there is more than one way to get from New York to Los Angeles; at the same time, not every story about how you traveled from New York to Los Angeles is believable. Likewise, more than one reasonable mechanism can often be drawn for a reaction, and one of the purposes of this book is to teach you how to distinguish a reasonable mechanism from a whopper.

It is important to learn how to draw reasonable mechanisms for organic reactions because mechanisms are the framework that makes organic chemistry make sense. Understanding and remembering the bewildering array of reactions known to organic chemists would be completely impossible were it not possible to organize them into just a few basic mechanistic types. The ability to formulate mechanistic hypotheses about how organic reactions proceed is also required for the discovery and optimization of new reactions.

The general approach of this book is to familiarize you with the classes and types of reaction mechanisms that are known and to give you the tools to learn how to draw mechanisms for reactions that you have never seen before. The body of each chapter discusses the more common mechanistic pathways and suggests practical tips for drawing them. The discussion of each type of mechanism contains both worked and unworked problems. You are urged to work the unsolved problems yourself. **Common error alerts** are scattered throughout the text to warn you about common pitfalls and misconceptions that bedevil students. Pay attention to these alerts, as failure to observe their strictures has caused many, many exam points to be lost over the years.

> Occasionally, you will see indented, tightly spaced paragraphs such as this one. The information in these paragraphs is usually of a parenthetical nature, either because it deals with formalisms, minor points, or exceptions to general rules, or because it deals with topics that extend beyond the scope of the textbook.

Extensive problem sets are found at the end of all chapters. The *only way* you will learn to draw reaction mechanisms is to *work the problems!* If you do not work problems, you will not learn the material. The problems vary in difficulty from relatively easy to very difficult. Many of the reactions covered in the problem sets are classical organic reactions, including many "name reactions." *All* examples are taken from the literature. Additional problems may be found in other textbooks. Ask your librarian, or consult some of the books discussed below.

Detailed answer keys are provided in a separate volume that is available for download from the Springer–Verlag web site (http://www.springer-ny.com/detail.tpl?isbn=0387985409) at no additional cost. The answer keys are formatted in PDF. You can view or print the document on any platform with Adobe's Acrobat Reader®, a program that is available for free from Adobe's web site (http://www.adobe.com). It is important for you to be able to work the problems *without* looking at the answers. Understanding what makes *Pride and Prejudice* a great novel is not the same as being able to write a great novel yourself. The same can be said of mechanisms. If you find you have to look at the answer to solve a problem, be sure that you work the problem again a few days later. Remember, you will have to work problems like these on exams. If you can't solve them at home without looking at the answers, how do you expect to solve them on exams when the answers are no longer available?

This book assumes you have studied (and retained) the material covered in two semesters of introductory organic chemistry. You should have a working familiarity with hybridization, stereochemistry, and ways of representing organic structures. You do not need to remember specific reactions from introductory organic chemistry, although it will certainly help. If you find that you are weak in certain aspects of introductory organic chemistry or that you don't remember some important concepts, you should go back and review that material. There is no shame in needing to refresh your memory occasionally. Pine's *Organic Chemistry*, 5th ed. (McGraw-Hill, 1987) and Scudder's *Electron Flow in Organic Chemistry* (John Wiley & Sons, 1992) provide basic information supplemental to the topics covered in this book.

This book definitely does not attempt to teach specific synthetic procedures, reactions, or strategies. Only rarely will you be asked to *predict* the products of a particular reaction. This book also does not attempt to teach physical organic chemistry (i.e., how mechanisms are proven or disproven in the laboratory). Before you can learn how to determine reaction mechanisms experimentally, you must learn what qualifies as a reasonable mechanism in the first place. Isotope effects, Hammett plots, kinetic analysis, and the like are all left to be learned from other textbooks.

Errors occasionally creep into any textbook, and this one is no exception. I have posted a page of errata at this book's Web site (http://www.chem.uky.edu/research/grossman/textbook.html). If you find an error that is not listed there, please contact me (rbgros1@uky.edu). In gratitude and as a reward, you will be immortalized on the Web page as an alert and critical reader.

Graduate students and advanced undergraduates in organic, biological, and medicinal chemistry will find the knowledge gained from a study of this book invaluable for both their graduate careers, especially cumulative exams, and their professional work. Chemists at the bachelor's or master's level who are working in industry will also find this book very useful.

Lexington, Kentucky   Robert B. Grossman
January 2002

Preface to the Instructor

Intermediate organic chemistry textbooks generally fall into two categories. Some textbooks survey organic chemistry rather broadly, providing some information on synthesis, some on drawing mechanisms, some on physical organic chemistry, and some on the literature. Other textbooks cover either physical organic chemistry or organic synthesis in great detail. There are many excellent textbooks in both of these categories, but as far as I am aware, there are only a handful of textbooks that teach students how to write a reasonable mechanism for an organic reaction. Carey and Sundberg, *Advanced Organic Chemistry, Part A*, 4th ed. (New York: Kluwer Academic/Plenum Publishers, 2000), Lowry and Richardson's *Mechanism and Theory in Organic Chemistry*, 3rd ed. (New York: Addison Wesley, 1987), and Carroll's *Perspectives on Structure and Mechanism in Organic Chemistry* (Monterey CA: Brooks/Cole Publishing Co., 1998), are all physical organic chemistry textbooks. They teach students the experimental basis for elucidating reaction mechanisms, not how to draw reasonable ones in the first place. Smith and March, *March's Advanced Organic Chemistry*, 5th ed. (John Wiley & Sons, 2001) provides a great deal of information on mechanism, but its emphasis is synthesis, and it is more a reference book than a textbook. Scudder's *Electron Flow in Organic Chemistry* (John Wiley & Sons, 1992) is an excellent textbook on mechanism, but it is suited more for introductory organic chemistry than for an intermediate course. Edenborough's *Writing Organic Reaction Mechanisms: A Practical Guide*, 2nd ed. (Bristol, PA: Taylor & Francis, 1997) is a good self-help book, but it does not lend itself to use in an American context. Miller and Solomon's *Writing Reaction Mechanisms in Organic Chemistry*, 2nd ed. (New York: Academic Press, 1999) is the textbook most closely allied in purpose and method to the present one. This book provides an alternative to Miller & Solomon and to Edenborough.

Existing textbooks usually fail to show how common mechanistic steps link seemingly disparate reactions, or how seemingly similar transformations often have wildly disparate mechanisms. For example, substitutions at carbonyls and nucleophilic aromatic substitutions are usually dealt with in separate chapters in other textbooks, despite the fact that the mechanisms are essentially identical. This textbook, by contrast, is organized according to mechanistic types, not according to overall transformations. This rather unusual organizational structure, borrowed from Miller and Solomon, is better suited to teaching students how to draw reasonable mechanisms than the more traditional structures, perhaps because the all-important first steps of mechanisms are usually more closely related to the conditions under which the reaction is executed than they are to the overall transformation. The first chapter of the book provides general information on such basic concepts as Lewis structures, resonance structures, aromaticity, hybridization, and acidity. It also shows how nucleophiles, electrophiles, and leaving groups can be recognized, and it provides practical techniques for determining the general mechanistic type of a reaction and the specific chemical transformations that need to be explained. The following five chapters examine polar mechanisms taking place under basic conditions, polar mechanisms taking place under acidic conditions, pericyclic reactions, free-radical reactions, and transition-metal-mediated and -catalyzed reactions, giving typical examples and general mechanistic patterns for each class of reaction along with practical advice for solving mechanism problems.

This textbook is *not* a physical organic chemistry textbook! The sole purpose of this textbook is to teach students how to come up with reasonable mechanisms for reactions that they have never seen before. As most chemists know, it is usually possible to draw more than one reasonable mechanism for any given reaction. For example, both an S_N2 and a single electron transfer mechanism can be drawn for many substitution reactions, and either a one-step concerted or a two-step radical mechanism can be drawn for [2 + 2] photocycloadditions. In cases like these, my philosophy is that the student should develop a good command of simple and generally sufficient reaction mechanisms before learning the modifications that are necessitated by detailed mechanistic analysis. I try to teach students how to draw reasonable mechanisms by *themselves*, not to teach them the "right" mechanisms for various reactions.

Another important difference between this textbook and others is the inclusion of a chapter on the mechanisms of transition-metal-mediated and -catalyzed reactions. Organometallic chemistry has pervaded organic chemistry in recent years, and a working knowledge of the mechanisms of such reactions as metal-catalyzed hydrogenation, the Stille and Suzuki couplings, and olefin metathesis is absolutely indispensable to any self-respecting organic chemist. Many organometallic chemistry textbooks discuss the mechanisms of these reactions, but the average organic chemistry student may not take a course on organometallic chemistry until fairly late in his or her studies, if at all. This textbook is the first on organic mechanisms to discuss these very important topics.

In all of the chapters, I have made a great effort to show the forest for the trees and to demonstrate how just a few concepts can unify disparate reactions. This philosophy has led to some unusual pedagogical decisions. For example, in the chapter on polar reactions under acidic conditions, protonated carbonyl compounds are depicted as carbocations in order to show how they undergo the same three fundamental reactions (addition of a nucleophile, fragmentation, and rearrangement) that other carbocations undergo. Radical anions are also drawn in an unusual manner to emphasize their reactivity in $S_{RN}1$ substitution reactions.

This philosophy has led to some unusual organizational decisions, too. $S_{RN}1$ reactions and carbene reactions are treated in the chapter on polar reactions under basic conditions. Most books on mechanism discuss $S_{RN}1$ reactions at the same time as other free-radical reactions, and carbenes are usually discussed at the same time as carbocations, to which they bear some similarities. I decided to locate these reactions in the chapter on polar reactions under basic conditions because of the book's emphasis on teaching practical methods for drawing reaction mechanisms. Students cannot be expected to look at a reaction and know immediately that its mechanism involves an electron-deficient intermediate. Rather, the mechanism should flow naturally from the starting materials and the reaction conditions. $S_{RN}1$ reactions usually proceed under strongly basic conditions, as do most reactions involving carbenes, so these classes of reactions are treated in the chapter on polar reactions under basic conditions. However, Favorskii rearrangements are treated in the chapter on pericyclic reactions, despite the basic conditions under which these reactions occur, to emphasize the pericyclic nature of the key ring contraction step.

Stereochemistry is not discussed in great detail, except in the context of the Woodward–Hoffmann rules. Molecular orbital theory is also given generally short shrift, again except in the context of the Woodward–Hoffmann rules. I have found that students must master the basic principles of drawing mechanisms before additional considerations such as stereochemistry and MO theory are loaded onto the edifice. Individual instructors might wish to put more emphasis on stereoelectronic effects and the like as their tastes and their students' abilities dictate.

I agonized a good deal over which basic topics should be covered in the first chapter. I finally decided to review a few important topics from introductory organic chemistry in a cursory fashion, reserving detailed discussions for common misconceptions. A basic familiarity with Lewis structures and electron-pushing is assumed. I rely on Weeks's excellent workbook, *Pushing Electrons: A Guide for Students of Organic Chemistry*, 3rd ed. (Saunders College Publishing, 1998), to refresh students' electron-pushing abilities. If Weeks fails to bring students up to speed, an introductory organic chemistry textbook such as Joseph M. Hornback's *Organic Chemistry* (Brooks/Cole, 1998) should probably be consulted.

I have written the book in a very informal style. The second person is used pervasively, and an occasional first-person pronoun creeps in, too. Atoms and molecules are anthropomorphized constantly. The style of the book is due partly to its evolution from a series of lecture notes, but I also feel strongly that anthropomorphization and exhortations addressed directly to the student aid greatly in pushing students to think for themselves. I vividly remember my graduate physical organic chemistry instructor asking, "What would you do if *you* were an electron?", and I remember also how much easier mechanisms were to solve after he asked that question. The third person and the passive tense certainly have their place in scientific writing, but if we want to encourage students to take intellectual control of the material themselves, then maybe we should stop talking about our theories and explanations as if they were phenomena that happened only "out there" and instead talk about them as what they are: our best attempts at rationalizing the bewildering array of phenomena that Nature presents to us.

I have not included references in this textbook for several reasons. The primary literature is full of reactions, but the mechanisms of these reactions are rarely drawn, and even when they are, it is usually in a cursory fashion, with crucial details omitted. Moreover, as stated previously, the purpose of this book is not to teach students the "correct" mechanisms, it is to teach them how to draw *reasonable* mechanisms using their own knowledge and some basic principles and mechanistic types. In my opinion, references in this textbook would serve little or no useful pedagogical purpose. However, some general guidance as to where to look for mechanistic information is provided at the end of the book.

All of the chapters in this book except for the one on transition-metal-mediated and -catalyzed reactions can be covered in a one-semester course.

The present second edition of this book corrects two major errors (the mechanisms of substitution of arenediazonium ions and why Wittig reactions proceed) and some minor ones in the first edition. Free-radical reactions in Chapter 5 are reorganized into chain and nonchain processes. The separate treatment of transition-metal-mediated and -catalyzed reactions in Chapter 6 is eliminated, and more in-text problems are added. Some material has been added to various chapters. Finally, the use of italics, especially in Common Error Alerts, has been curtailed.

I would like to thank my colleagues and students here at the University of Kentucky and at companies and universities across the country and around the world for their enthusiastic embrace of the first edition of this book. Their response was unexpected and overwhelming. I hope they find this new edition equally satisfactory.

Lexington, Kentucky   Robert B. Grossman
January 2002

Contents

1

The Basics

1.1 Structure and Stability of Organic Compounds

If science is a language that is used to describe the universe, then Lewis structures—the sticks, dots, and letters that are used to represent organic compounds—are the vocabulary of organic chemistry, and reaction mechanisms are the stories that are told with that vocabulary. As with any language, it is necessary to learn how to use the organic chemistry vocabulary properly in order to communicate one's ideas. The rules of the language of organic chemistry sometimes seem capricious or arbitrary; for example, you may find it difficult to understand why RCO_2Ph is shorthand for a structure with one terminal O atom, whereas RSO_2Ph is shorthand for a structure with two terminal O atoms, or why it is so important that $\longleftrightarrow$ and not $\rightleftarrows$ be used to indicate resonance. But organic chemistry is no different in this way from languages such as English, French, or Chinese, which all have their own capricious and arbitrary rules, too. (Have you ever wondered why I, you, we, and they walk, but he or she walks?) Moreover, just as you need to do if you want to make yourself understood in English, French, or Chinese, you must learn to use proper organic chemistry grammar and syntax, no matter how tedious or arbitrary it is, if you wish to make yourself clearly understood when you tell stories about (i.e., draw mechanisms for) organic reactions. The first section of this introductory chapter should reacquaint you with some of the rules and conventions that are used when organic chemistry is "spoken." Much of this material will be familiar to you from previous courses in organic chemistry, but it is worth reiterating.

1.1.1 Conventions of Drawing Structures; Grossman's Rule

When organic structures are drawn, the H atoms attached to C are usually omitted. (On the other hand, H atoms attached to heteroatoms are *always* shown.) It is extremely important for you not to forget that they are there!

* **Common error alert:** *Don't lose track of the undrawn H atoms.* There are big differences among isobutane, the *t*-butyl radical, and the *t*-butyl cation, but if you lose track of your H atoms you might confuse the two. For this reason, I have formulated what I modestly call Grossman's rule: **Always draw all bonds and**

all hydrogen atoms near the reactive centers. The small investment in time required to draw the H atoms will pay huge dividends in your ability to draw the mechanism.

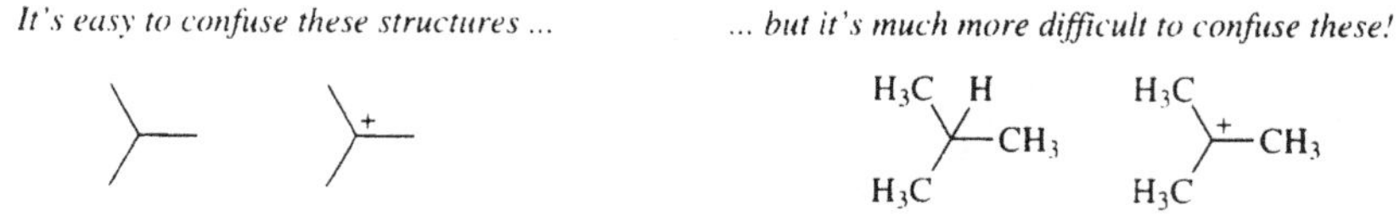

It's easy to confuse these structures ...   *... but it's much more difficult to confuse these!*

$$H_3C \quad H$$
$$\underset{H_3C}{>}\!\!-CH_3 \qquad \underset{H_3C}{\overset{H_3C}{>}}\!\!\overset{+}{-}CH_3$$

Abbreviations are often used for monovalent groups that commonly appear in organic compounds. Some of these abbreviations are shown in Table 1.1. *Aryl* may be phenyl, a substituted phenyl, or a heteroaromatic group like furyl, pyridyl, or pyrrolyl. *Tosyl* is shorthand for *p*-toluenesulfonyl, *mesyl* is shorthand for methanesulfonyl, and *triflyl* is shorthand for trifluoromethanesulfonyl. TsO⁻, MsO⁻, and TfO⁻ are abbreviations for the common leaving groups tosylate, mesylate, and triflate, respectively.

* **Common error alert:** *Don't confuse Ac (one O atom) with AcO (two O atoms), or Ts (two O atoms) with TsO (three O atoms). Also don't confuse Bz (benzoyl) with Bn (benzyl).* (One often sees Bz and Bn confused even in the literature.)

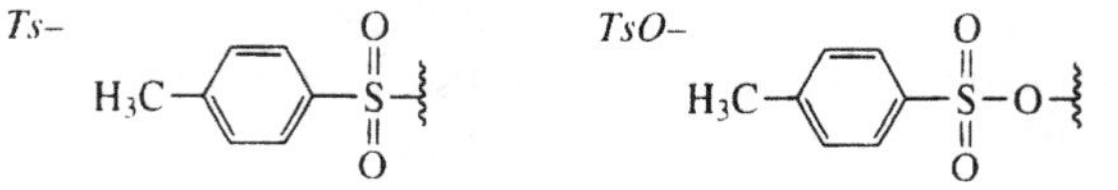

Sometimes the ways that formulas are written in texts confuse students. The more important textual representations are shown below.

* **Common error alert:** *It is especially easy to misconstrue the structure of a sulfone (RSO_2R) as being analogous to that of an ester (RCO_2R).*

RCOR	ketone	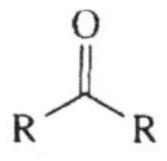	RSOR	sulfoxide	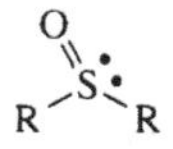
RCO$_2$R	ester	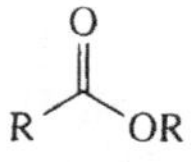	RSO$_2$R	sulfone	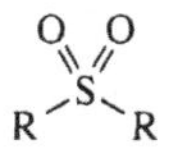
RCHO	aldehyde	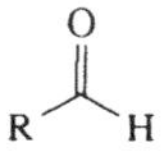	RSO$_3$R	sulfonate ester	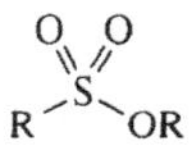

TABLE 1.1. Common abbreviations for organic substructures

Me	methyl	CH$_3$–	Ph	phenyl	C$_6$H$_5$–
Et	ethyl	CH$_3$CH$_2$–	Ar	aryl	(*see text*)
Pr	propyl	CH$_3$CH$_2$CH$_2$–	Ac	acetyl	CH$_3$C(=O)–
i-Pr	isopropyl	Me$_2$CH–	Bz	benzoyl	PhC(=O)–
Bu, *n*-Bu	butyl	CH$_3$CH$_2$CH$_2$CH$_2$–	Bn	benzyl	PhCH$_2$–
i-Bu	isobutyl	Me$_2$CHCH$_2$–	Ts	tosyl	4-Me(C$_6$H$_4$)SO$_2$–
s-Bu	*sec*-butyl	(Et)(Me)CH–	Ms	mesyl	CH$_3$SO$_2$–
t-Bu	*tert*-butyl	Me$_3$C–	Tf	triflyl	CF$_3$SO$_2$–

Conventions for the representation of stereochemistry are also worth noting. A *heavy* or *bold* bond indicates that a substituent is pointing *toward* you, out of the plane of the paper. A *hashed* bond indicates that a substituent is pointing *away* from you, behind the plane of the paper. Sometimes a *dashed* line is used for the same purpose as a hashed line, but the predominant convention is that a dashed line designates a partial bond (as in a transition state), not stereochemistry. A *squiggly* or wavy line indicates that there is a mixture of *both* stereochemistries at that stereocenter, i.e., that the substituent is pointing toward you in some fraction of the sample and away from you in the other fraction. A *plain* line is used when the stereochemistry is unknown or irrelevant.

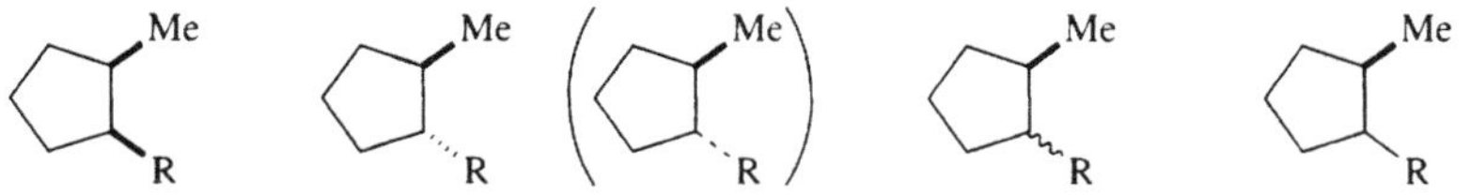

R pointing out of plane of paper — *R pointing into plane of paper* — *R pointing in both directions* — *Stereochemistry of R unknown*

Bold and hashed lines may be drawn either in tapered (wedged) or untapered form. The predominant convention is that tapered lines show *absolute* stereochemistry, whereas untapered lines show *relative* stereochemistry. European and U.S. chemists generally differ on whether the thick or thin end of the tapered hashed line should be at the substituent. Bear in mind that these conventions for showing stereochemistry are not universally followed! A particular author may use a dialect that is different from the standard.

trans, racemic — *trans, enantiopure (U.S.)* — *trans, enantiopure (European)*

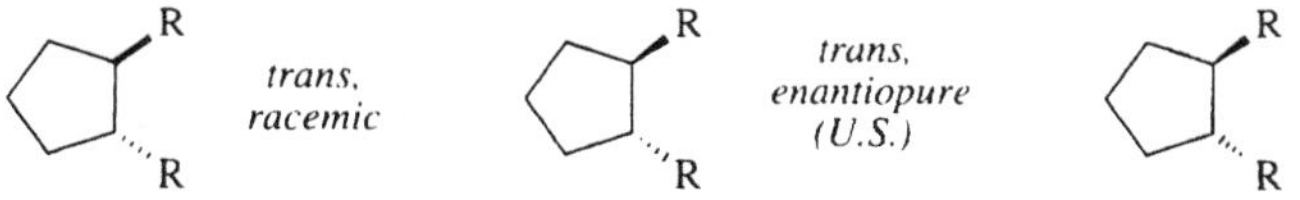

1.1.2 Lewis Structures; Resonance Structures

The concepts and conventions behind Lewis structures were covered in your previous courses, and there is no need to recapitulate them here. One aspect of drawing Lewis structures that often creates errors, however, is the proper assignment of formal charges. A formal charge on any atom is calculated as follows:

$$\begin{aligned}\text{formal charge} = {} & (\text{valence electrons of element}) \\ & - (\text{number of } \pi \text{ and } \sigma \text{ bonds}) \\ & - (\text{number of unshared valence electrons})\end{aligned}$$

This calculation always works, but it is a bit ponderous. In practice, correct formal charges can usually be assigned at a glance. Carbon atoms "normally" have four bonds, N three, O two, and halogens one, and atoms with the "normal" number of bonds do not carry a formal charge. Whenever you see an atom that has an "abnormal" number of bonds, you can immediately assign a formal charge. For example, a N atom with two bonds can immediately be given a formal charge of -1. Formal charges for the common elements are given in Tables 1.2 and 1.3. It is very rare to find a nonmetal with a formal charge of ± 2 or greater, although the S atom occasionally has a charge of $+2$.

TABLE 1.2. Formal charges of even-electron atoms

Atom	1 Bond	2 Bonds	3 Bonds	4 Bonds
C		0*	+1 (no lp)§ −1 (one lp)	0
N, P	0†	−1	0	+1
O, S	−1	0	+1	0 or +2‡
Halogen	0	+1		
B, Al			0§	−1

Note: lp = lone pair

*Carbene

†Nitrene

‡See extract following Table 1.2 for discussion of S.

§Has an empty orbital

The formal charges of quadruply bonded S can be confusing. A S atom with *two single bonds and one double bond* (e.g., DMSO, $Me_2S=O$) has one lone pair and no formal charge, but a S atom with *four single bonds* has no lone pairs and a formal charge of +2. A S atom with six bonds total has no formal charge and no lone pairs, as does a P atom with five bonds total. There is a more complete discussion of S and P Lewis structures later in this section.

Formal charges are called *formal* for a reason. They have more to do with the language that is used to describe organic compounds than they do with chemical reality. (Consider the fact that electronegative elements often have formal positive charges, as in $\overset{+}{N}H_4$, H_3O^+, and $Me\overset{+}{O}=CH_2$.) Formal charges are a very useful tool for ensuring that electrons are not gained or lost in the course of a reaction, but they are not a reliable guide to chemical reactivity. For example, both $\overset{+}{N}H_4$ and $\overset{+}{C}H_3$ have formal charges on the central atoms, but the reactivity of these two atoms is completely different.

To understand chemical reactivity, one must look away from formal charges and toward other properties of the atoms of an organic compound such as *electropositivity*, *electron-deficiency*, and *electrophilicity*.

• *Electropositivity* (or electronegativity) is a property of an element and is mostly independent of the bonding pattern of that element.

• An atom is *electron-deficient* if it lacks an octet of electrons in its valence shell (or, for H, a duet of electrons).

• An *electrophilic* atom is one that has an empty orbital that is relatively low in energy. (Electrophilicity is discussed in more detail later in this chapter.)

TABLE 1.3. Formal charges of odd-electron atoms

Atom	0 Bonds	1 Bond	2 Bonds	3 Bonds
C				0
N, P			0	+1
O, S		0	+1	
Hal	0	+1		

* **Common error alert:** *The properties of electropositivity, electron-deficiency, electrophilicity, and formal positive charge are independent of one another and must not be confused!* The C and N atoms in CH_3^+ and NH_4^+ both have formal positive charges, but the C atom is electron-deficient, and the N atom is not. The C and B atoms in $\cdot CH_3$ and BF_3 are both electron-deficient, but neither is formally charged. B is electropositive and N is electronegative, but BH_4^- and NH_4^+ are both stable ions, as the central atoms are electron-sufficient. The C atoms in CH_3^+, CH_3I, and $H_2C=O$ are all electrophilic, but only the C in CH_3^+ is electron-deficient. The O atom in $Me\overset{+}{O}=CH_2$ has a formal positive charge, but the C atoms are electrophilic, not O.

For each σ bonding pattern, there are often several ways in which π and non-bonding electrons can be distributed. These different ways are called *resonance structures*. Resonance structures are alternative descriptions of a single compound. Each resonance structure has some contribution to the real structure of the compound, but no one resonance structure is the true picture. Letters, lines, and dots are words in a language that has been developed to describe molecules, and, as in any language, sometimes one word is inadequate, and several different words must be used to give a complete picture of the structure of a molecule. The fact that resonance structures have to be used at all is an artifact of the language used to describe chemical compounds.

The true electronic picture of a compound is a weighted average of the different resonance structures that can be drawn (*resonance hybrid*). The weight assigned to each resonance structure is a measure of its importance to the description of the compound. The *dominant resonance structure* is the structure that is weighted most heavily. Two descriptions are shown to be resonance structures by separating them with a double-headed arrow ($\longleftrightarrow$).

* **Common error alert:** *The double-headed arrow is used* only *to denote resonance structures. It must not be confused with the symbol for a chemical equilibrium* ($\rightleftarrows$) *between two or more different species.* Again, resonance structures are alternative descriptions of a single compound. There is no going "back and forth" between resonance structures as if there were an equilibrium. Don't even think of it that way!

Diazomethane is neither this: *nor this:*

$$\overset{-}{\underset{..}{H_2C}}-\overset{+}{N}\equiv N: \qquad \longleftrightarrow \qquad H_2C=\overset{+}{N}=\overset{-}{\underset{..}{N:}}$$

but a weighted average of the two structures.

Low-energy resonance structures of a compound provide better descriptions of the compound's electronic nature than do high-energy resonance structures. The rules for evaluating the stability of resonance structures are the same as those for any other Lewis structure.

1. No first-row atom (B, C, N, O) can have more than eight electrons in its valence shell. (The octet rule is less sacred for heavier main group elements such as P and S, and it does not hold at all for transition metals.)

*   2. **Common error alert:** *Resonance structures in which all atoms are surrounded by an octet of electrons are almost always lower in energy than resonance structures in which one or more atoms are electron-deficient.* However, if there are electron-deficient atoms, they should be electropositive (C, B), not electronegative (N, O, halogen).

3. Resonance structures with charge separation are usually higher in energy than those in which charges can be neutralized.

4. If charge is separated, then electronegative atoms should gain the formal negative charge and electropositive ones should gain the formal positive charge.

These rules are listed in order of importance. For instance, consider $MeÖ\text{-}\overset{+}{C}H_2$ $\longleftrightarrow$ $MeO=CH_2$. The second resonance structure is more important to the description of the ground state of this compound, because it is more important that all atoms have an octet (rule 2) than that the more electropositive element C have the formal positive charge instead of O (rule 4). As another example consider $Me_2C=O \longleftrightarrow Me_2\overset{+}{C}\text{-}\overset{-}{O} \longleftrightarrow Me_2\overset{-}{C}\text{-}\overset{+}{O}$. The third structure is unimportant because an electronegative element is made electron-deficient. The second structure is less important than the first one because the second one has charge separation (rule 3) and an electron-deficient atom (rule 2). Nevertheless, the second structure does contribute *somewhat* toward the overall description of the ground state electronic structure of acetone.

Resonance structures are almost universally defined by organic chemists as structures differing only in the placement of π bonds and lone pairs. The σ network remains unchanged. If the σ networks of two structures differ, then the structures represent *isomers*, not alternative resonance descriptions.

How do you generate a resonance structure of a given Lewis structure?

• Look for an electron-deficient atom next to a lone-pair-bearing atom. The lone pair can be shared with the electron-deficient atom as a new π bond. Note the changes in formal charge when pairs of electrons are shared! Also note that the atom accepting the new bond must be *electron-deficient*.

* **Common error alert:** *A formal positive charge is irrelevant to whether an atom can accept a new bond.*

The curved-arrow convention is used to show how electrons in one resonance structure can be moved around to generate a new resonance structure. The curved arrows are entirely a formalism; electrons do not actually move from one location to another, because the real compound is a weighted average of the different resonance structures, not an equilibrium mixture of different resonance structures. The curved arrows help you not to lose or gain electrons as you draw different resonance structures.

• Look for an electron-deficient atom *adjacent to* a π bond. The electrons in the π bond can be moved to the electron-deficient atom to give a new π bond, and the distal atom of the former π bond then becomes electron-deficient. Again, note the changes in formal charges!

• Look for a radical *adjacent to* a π bond. The lone electron and one electron in the π bond can be used to make a new π bond. The other electron of the π bond goes to the distal atom to give a new radical. There are no changes in formal charges.

Half-headed arrows (fishhooks) are used to show the movement of single electrons.

• Look for a lone pair *adjacent to* a π bond. Push the lone pair toward the π bond, and push the π bond onto the farther atom to make a new lone pair. The atom with the lone pair may or may not have a formal negative charge.

When lone pairs of heteroatoms are omitted in structural drawings, a formal negative charge on a heteroatom can double as a lone pair. Thus, a curved arrow will often begin at a formal negative charge rather than at a lone pair.

• In aromatic compounds, π bonds can often be moved around to generate a new resonance structure that has no change in the total number of bonds, lone pairs or unpaired electrons, electron-deficient atoms, or formal charges, but that is, nevertheless, not the same structure.

• The two electrons of a π bond can be divided evenly or unevenly between the two atoms making up that bond: A=B ⟷ Å⁺–B̄ ⟷ Ā–B̊⁺ ⟷ Ȧ–Ḃ. The process usually generates a higher energy structure. In the case of a π bond between two different atoms, push the pair of electrons in the π bond toward the more electronegative of the two.

Two other important rules to remember when drawing resonance structures are the following:

• A lone pair or empty orbital cannot interact with a π bond to which it is orthogonal (perpendicular). The resonance structures in such cases often look hopelessly strained.

• Two resonance structures *must* have the same number of electrons (and atoms, for that matter). The formal charges in both structures must add up to the same number.

* Common error alerts:

• *Tetravalent C or N atoms (i.e., quaternary ammonium salts) have no lone pairs or π bonds, so they do not participate in resonance.*

• *Electronegative atoms like O and N must have their octet. Whether they have a formal positive charge is not an issue.* Like banks with money, electronegative atoms are willing to *share* their electrons, but they will not tolerate electrons' being taken away.

An electronegative atom is happy to share its electrons, even if it gains a formal positive charge ...

... and it can give up a pair of electrons if it gets another pair from another source ...

... but it will not give up a pair of electrons entirely, because then it would become elecron-deficient.

very high energy (bad) resonance structure!

• *If you donate one or two electrons to an atom that already has an octet, regardless of whether it has a formal positive charge, another bond to that atom must break.* For example, in nitrones (PhCH=NR–O) the N atom has its octet. A lone pair from O can be used to form a new N=O π bond only if the electrons in the C=N π bond leave N to go to C, i.e., PhCH=NR–O $\longleftrightarrow$ PhCH–NR=O. In the second resonance structure, N retains its octet and its formal positive charge.

• *In bridged bicyclic compounds, a π bond between a bridgehead atom and its neighbor is forbidden due to ring strain unless one of the rings of the bicyclic compound has more than eight or nine atoms* (Bredt's rule). Resonance structures in which such a π bond exists are very poor descriptions of the compound.

Problem 1.1. Which of the two resonance structures is a better description of the ground state of the following compound?

$$\text{Me}_2\overset{\cdot\cdot}{\text{N}}-\overset{\text{Me}}{\underset{\text{Me}}{\text{B}}} \quad\longleftrightarrow\quad \text{Me}_2\overset{+}{\text{N}}=\overset{\text{Me}}{\underset{\text{Me}}{\text{B}^-}}$$

Problem 1.2. Draw as many reasonable resonance structures for each of the following compounds as you can.

The second-best resonance structure often provides the key to understanding the chemical behavior of that compound. For example, the second-best resonance structure for acetone tells you that the carbonyl C is slightly electron-deficient and susceptible to attack by electron-rich species. This point will be revisited later.

In general, the more low-energy resonance structures a compound has, the lower its energy.

The ability to look at one structure and see its resonance structures is extremely important for drawing organic reaction mechanisms. If you require it, Chapters 1–3 of Daniel P. Weeks's *Pushing Electrons*, 3rd ed. (Saunders College Publishing, 1998), can help you acquire the necessary practice.

Compounds with a terminal O attached to S or P are fairly common in organic chemistry. Resonance structures in which S and P have extended the capacity of their valence shells (using relatively low energy 3d orbitals) to accommodate more than eight electrons are often written for these compounds. The extended-shell description can be very confusing; for example, DMSO (below) seems to be analogous to acetone, but the S in DMSO has a lone pair, whereas the C in acetone is moderately electron-deficient. The dipolar resonance structures are a better description of the ground state of these compounds, but old habits die hard among organic chemists. In any case, when you see S=O or P=O "π" bonds, be aware that the valence shell may have been extended beyond eight electrons and that you may not be looking at a conventional π bond.

1.1.3   Molecular Shape; Hybridization

Molecules are three-dimensional objects, and as such they have shapes. You must always keep the three-dimensional shapes of organic compounds in mind when you draw reaction mechanisms. Often something that seems reasonable in a flat drawing will manifest itself as totally unreasonable when the three-dimensional nature of the reaction is considered, and vice versa.

This tricyclic compound looks horribly strained ...

... until you look at its three-dimensional structure!

Organic chemists use the concept of atom *hybridization* to rationalize and understand molecular shape. The concept of hybridization is itself a strange hybrid of Lewis theory and molecular orbital (MO) theory, and there are serious questions about its basis in reality. Nevertheless, organic chemists use hybridization almost universally to rationalize structure and reactivity, because it is easy to understand and apply, and because it works!

Before hybridization is discussed, a brief review of the basics of MO theory is in order. The following discussion is meant to be a quick, qualitative recap, not a comprehensive treatment.

Electrons do not orbit nuclei like planets around a star, as one early theory of the nucleus proposed. A better analogy is that electrons around a nucleus are like a cloud of gnats buzzing around one's head on a summer day. To carry the analogy further, it's not possible to locate one gnat and define its location precisely; instead, one can only describe the likelihood of finding a gnat at a particular distance from one's mouth or nostrils. Likewise, the position of particular electrons cannot be defined; instead, a mathematical function called an *orbital* describes the *probability* of finding an electron of a certain energy in a particular region of space. The actual probability is given by the *square* of the value of the orbital at a particular point in space.

The atoms with which organic chemists are most concerned (C, N, O) have four valence *atomic orbitals* (AOs), one s and three p orbitals, each of which can contain no, one, or two electrons. Electrons in the valence s orbital of an atom are lower in energy than electrons in the valence p orbitals. The s orbital is spherical, whereas p orbitals are dumbbell-shaped and mutually perpendicular (*orthogonal*; i.e., they do not overlap). A p orbital has two *lobes*; in the mathematical function that defines these orbitals, one lobe has a value less than zero (negative), and the other has a value greater than zero (positive). (These arithmetic values should not be confused with charge.)

s orbital

spherical distribution of electron density; uniform arithmetical sign

p orbital

in this region of space, solution to wave equation has arithmetical value greater than zero

in this region of space, solution to wave equation has arithmetical value less than zero

Each p orbital describes a distribution of electrons centered around an *x*, *y*, or *z* axis, so the three p orbitals are mutually perpendicular, but when the three p orbitals are squared and added together, a spherical distribution of electrons is again described.

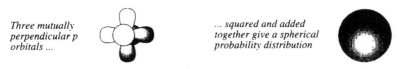

Three mutually perpendicular p orbitals ...

... squared and added together give a spherical probability distribution

Heavier elements may also have valence *d* and *f orbitals*. They need not concern you here.

When two atoms are close in space, the energies and probability distributions of the electrons on each atom change in response to the presence of the other nucleus. The AOs, which describe the electrons' probability distribution and energies, are simply mathematical functions, so the *interaction* of two spatially proximate AOs is expressed by arithmetically adding and subtracting the AO functions to generate two new functions, called *molecular orbitals* (MOs). The additive (in-phase) combination of AOs, a *bonding* MO, is lower in energy than either of the two starting AOs. The subtractive (out-of-phase) combination, an *antibonding* MO, is higher in energy than either of the two starting AOs. In fact, the destabilization of the antibonding MO is greater than the stabilization of the bonding MO.

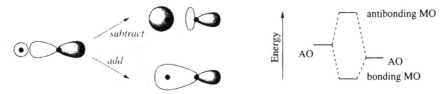

Why must two AOs interact in both a constructive and a destructive manner? The physical reality is that two AOs describe the distribution of four electrons in space. When two AOs interact, the resulting equations must still describe the distribution of four electrons in space. Two AOs, therefore, interact to give two MOs, three AOs interact to give three MOs, and so on.

When two AOs interact, if each AO has one electron, both electrons can go into the bonding MO. Because the total energy of the electrons is lower than it was in the separated system, a *chemical bond* is now present where there was none before. In contrast, if each AO is full, then two electrons go into the bonding MO and two into the antibonding MO; the total energy of the electrons is increased, the atoms repel one another, and no bond is formed.

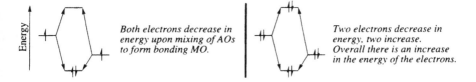

Both electrons decrease in energy upon mixing of AOs to form bonding MO.

Two electrons decrease in energy, two increase. Overall there is an increase in the energy of the electrons.

The valence electrons of any element in the main-group block reside in the four valence AOs. For example, a C atom has four valence electrons. One of these electrons can go into each valence orbital. The four half-filled AOs can then interact with four AOs from other atoms to form four bonds. Oxygen, by contrast, has six valence electrons. It has only two half-filled orbitals, so it makes only two bonds.

This simple picture is incomplete, though. Consider CH_4. If C used one s and three p AOs to make four bonds to H, one would expect that one of the C–H bonds would be different from the other three. This is not the case, though: most measures of molecular properties of CH_4 indicate that all four bonds are exactly equivalent. Why is this? Because all four bonding orbitals in CH_4 are equivalent, and the four AOs of C are simply mathematical functions, organic chemists hypothesize that the four AOs are "averaged," or *hybridized*, to make four new, equivalent AOs called *sp^3 hybrid orbitals* (because each one consists of one part s and three parts p). The four original AOs together describe a spherical distribution of electrons, so when this sphere is divided into four equal sp^3 orbitals, a *tetrahedral* array of four orbitals is created.

sp^3 hybrid orbital; large lobe is used in bonding

Tetrahedral array of sp^3 orbitals (back lobes omitted for clarity)

The AOs can be hybridized in other ways, too. One s and *two* p AOs can be averaged to give *three* new hybrid orbitals and *one unchanged* p orbital; this procedure is called *sp^2 hybridization*. Alternatively, one s and *one* p AO can be averaged to give *two* new hybrid orbitals and *two unchanged* p orbitals; this procedure is called *sp hybridization*. In summary, the characteristics of the three kinds of hybridization are as follows:

• sp^3 hybridization: The s and all three p orbitals are averaged to make four sp^3 orbitals of equal energy. The four orbitals point to the four corners of a tetrahedron and are 109° apart. The energy of each sp^3 orbital is $\frac{1}{4}$ of the way from the energy of the s AO to the energy of a p AO.

• sp^2 hybridization: The s and two p orbitals are averaged to make three sp^2 orbitals of equal energy, and one p orbital is left unchanged. The three hybrid orbitals point to the three corners of an equilateral triangle and are coplanar and 120° apart; the unhybridized p orbital is perpendicular to the plane of the hybrid orbitals. The energy of each sp^2 orbital is $\frac{2}{3}$ of the way from the energy of the s AO to the energy of a p AO.

• sp hybridization: The s and one p orbital are averaged to make two sp orbitals of equal energy, and two p orbitals are left unchanged. The sp orbitals point 180° apart from each other. The two unhybridized p orbitals are perpendicular to each other and to the line containing the sp orbitals. The energy of each sp orbital is halfway between the energy of the s AO and the energy of a p AO.

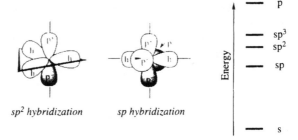

h = hybrid orbital.
p⁺ and p⁻ = lobes of p orbitals.
Back lobes of hybrid orbitals
omitted for clarity.

sp² hybridization   *sp hybridization*

The drawings of hybrid orbitals shown are simplistic. The sp³, sp², and sp orbitals do not actually have identical shapes. Better pictures of the actual shapes of these orbitals can be found in Lowry and Richardson's *Mechanism and Theory in Organic Chemistry*, 3rd ed. (Addison Wesley, 1987).

The hybridization of an atom is determined as follows. Hybrid orbitals are used to make σ bonds and to hold lone pairs *not* used in resonance; p orbitals are used to make π bonds and to hold lone pairs used in resonance, and they are used as empty orbitals. To determine the hybridization of an atom, add up the number of lone pairs not used in resonance and the number of σ bonds (i.e., atoms to which it is bound). If the sum is four, the atom is sp³-hybridized. If the sum is three, it is sp²-hybridized. If the sum is two, it is sp-hybridized.

Problem 1.3. Determine the hybridization of the C, N, and O atoms in each of the following compounds. (The black dot in the center of the final structure indicates a C atom.)

It is important to remember to think about the p orbitals as well as the hybrid orbitals when you think about the hybridization of an atom. It is also important to remember that the hybridization of an atom affects its properties and reactivity! This point will be illustrated many times in the future.

1.1.4 Aromaticity

An extra amount of stability or instability is associated with a compound that has a cyclic array of continuously overlapping p orbitals. Such a compound may have a ring with alternating single and multiple bonds, or the ring may contain both alternating π bonds and one atom with a lone pair or an empty orbital. If there is an odd number of electron pairs in the cyclic array of orbitals, then the compound is especially stable (as compared with the corresponding acyclic system with two additional H atoms), and it is said to be *aromatic*. If there is an even number of electron pairs, then the compound is especially unstable, and it is said to be *anti-aromatic*. If there is no cyclic array of continuously overlapping p orbitals, then the question of aromaticity doesn't apply, and the compound is *nonaromatic*.

The simplest case of an aromatic compound is benzene. Each of the C atoms in benzene is sp²-hybridized, so each has a p orbital pointing perpendicular to the plane of the ring. The six p orbitals make a cyclic array. Each C atom contributes one electron to its p orbital, so there is a total of three pairs of electrons in the system. Because three is odd, benzene is aromatic. In fact, benzene is about 30 kcal/mol lower in energy than 1,3,5-hexatriene, its acyclic analog.

There are many aromatic hydrocarbons other than benzene. Many are made up of fused benzene rings. All have an odd number of pairs of electrons in a cyclic array of orbitals.

Some aromatic hydrocarbons:

naphthalene   phenanthrene   benzo[*a*]pyrene   azulene

Furan, thiophene, pyrrole, and pyridine are all examples of heterocyclic aromatic compounds (heteroaromatic compounds). The heteroatoms in some of these compounds (furan, thiophene, pyrrole) contribute one lone pair to the aromatic system, whereas in others (pyridine) they contribute none. You can determine how many lone pairs a heteroatom contributes to the aromatic system by examining the effect of lone-pair donation on the hybridization of the heteroatom. For example, if the N atom of pyridine used its lone pair to participate in resonance, it would have to be sp-hybridized (one p orbital required for the N=C π bond, one for the lone pair used in resonance), but sp hybridization requires 180° bond angles, which are not possible in this compound. Therefore the N atom must be sp²-hybridized, and the N lone pair must be in a hybrid orbital that is orthogonal to the cyclic array of p orbitals. In pyrrole, by contrast, if the N atom uses its lone pair in resonance, the N atom must be sp²-hybridized, which is reasonable. Therefore, there is a cyclic array of p orbitals in pyrrole occupied by six electrons (two from each of the C=C π bonds and two from the N lone pair), and pyrrole is aromatic.

Some aromatic heterocycles:

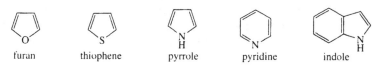

furan   thiophene   pyrrole   pyridine   indole

Problem 1.4. What is the hybridization of the O atom in furan? In what kind of orbitals do the lone pairs reside? How many lone pairs are used in resonance?

Certain charged compounds are aromatic also. The electron-deficient C atom in the tropylium and cyclopropenium ions is sp²-hybridized and has an empty p orbital. The tropylium ion has a cyclic array of seven p orbitals containing three pairs of electrons, and the cyclopropenium ion has a cyclic array of three p orbitals containing one pair of electrons; therefore, both ions are aromatic. (Note that cyclopropene itself is nonaromatic, because it doesn't have a cyclic array of p orbitals!) Similarly, the lone-pair-bearing C atom in the cyclopentadienide anion is sp²-hybridized so that the lone pair can be used in resonance; as a result, the cyclopentadienide anion has a cyclic array of five p orbitals containing three pairs of electrons, and it is aromatic, too.

Some aromatic ions:

| tropylium | cyclopropenium | cyclopentadienide | pyrylium |

Antiaromatic compounds are especially unstable compared with their acyclic analogs. Cyclobutadiene is isolable only in an inert matrix at very low temperatures. In dihydropyridazine, the two N-based lone pairs combine with the two C=C π bonds to create an eight-electron system that is particularly high in energy. The cyclopentadienyl cation is particularly high in energy too, as there are only two pairs of electrons in the cyclic array of five p orbitals (including the empty p orbital from the electron-deficient C). However, cyclooctatetraene, which at first glance appears to be antiaromatic, avoids antiaromaticity by bending into a tub shape so that its p orbitals don't overlap continuously.

Some antiaromatic compounds:

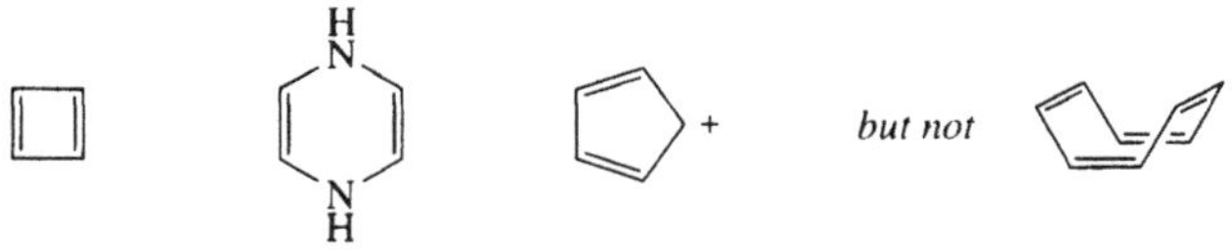

Some compounds have partial aromatic or antiaromatic character due to the presence of a minor aromatic or antiaromatic resonance structure. Tropolone (cycloheptadienone) is much more stable than one would expect from a highly unsaturated ketone because its C–O resonance structure is aromatic. On the other hand, cyclopentadienone is extremely unstable because its C–O resonance structure is antiaromatic.

aromatic   *antiaromatic*

To give you an idea of the amount of stabilization provided by aromaticity, consider 1,3-pentadiene and 1,3-cyclopentadiene. Both compounds are nonaromatic. Deprotonation of 1,3-pentadiene gives a nonaromatic compound, but deprotonation of 1,3-cyclopentadiene gives an aromatic compound. The acidity of cyclopentadiene ($pK_a = 15$) is about *20 orders of magnitude* greater than the acidity of 1,3-pentadiene and is about the same as the acidity of water. The establishment of an aromatic ring where there was none before provides an important driving force for many organic reactions.

> The amount of stabilization that aromaticity provides is greatest when the number of electrons is small. Naphthalene, a 10-electron aromatic system, is less stabilized than benzene, a 6-electron aromatic system. Also, all-carbon systems are more heavily stabilized than those with heteroatoms such as N, O, or S.

1.2   Brønsted Acidity and Basicity

An acid–base reaction involves the transfer of a proton H^+ from a Brønsted acid to a Brønsted base.

Some examples of some acid–base reactions follow:

$$CH_3CO_2H + CH_3NH_2 \rightleftarrows CH_3CO_2^- + CH_3NH_3^+$$
$$CH_3COCH_3 + t\text{-}BuO^- \rightleftarrows CH_3COCH_2^- + t\text{-}BuOH$$
$$H^+ + t\text{-}BuOH \rightleftarrows t\text{-}BuOH_2^+$$

A few points should be noted. (1) Bases may be anionic or neutral, and acids may be neutral or cationic. (2) The acid–base reaction is an equilibrium. The equilibrium may lie far to one side or the other, but it is still an equilibrium. (3) There is both an acid and a base on both sides of the equilibrium. (4) This equilibrium is not to be confused with resonance. (5) Proton transfer reactions are usually very fast, especially when the proton is transferred from one heteroatom to another.

* **Common error alert:** *The proton H^+ is not to be confused with the hydrogen atom (or radical) $H\cdot$ or the hydride ion H^-.*

1.2.1   pK_a Values

Acidities are quantified by pK_a values. The pK_a of an acid HX is defined as

$$pK_a = -\log \left(\frac{[H^+][X^-]}{[HX]} \right)$$

The larger the pK_a, the less acidic the compound. It is important that you develop a sense of the pK_a values of different classes of compounds and how variations in structure affect the pK_a. It is especially important for you to memorize the starred numbers in Table 1.4 in order to obtain a sense of relative acidities and basicities.

TABLE 1.4. Approximate pK_a values for some organic acids

$CCl_3CO_2\underline{H}$	0	Et$O\underline{H}$	*17
$CH_3CO_2\underline{H}$	*4.7	$CH_3CON\underline{H}_2$	17
pyr$\underline{H}^+$	*5	t-BuO$\underline{H}$	19
PhN$\underline{H}_3{}^+$	5	$C\underline{H}_3COCH_3$	*20
$HC{\equiv}N$	9	$C\underline{H}_3SO_2CH_3$	23
$N{\equiv}CC\underline{H}_2CO_2Et$	9	$HC{\equiv}C\underline{H}$	*25
$Et_3N\underline{H}^+$	*10	$C\underline{H}_3CO_2Et$	*25
Ph$O\underline{H}$	10	$C\underline{H}_3CN$	26
$C\underline{H}_3NO_2$	10	$C\underline{H}_3SOCH_3$	31
Et$S\underline{H}$	11	$N\underline{H}_3$	*35
MeCOC$\underline{H}_2CO_2Et$	11	$C_6\underline{H}_6$, $H_2C{=}C\underline{H}_2$	*37
EtO$_2$CC$\underline{H}_2CO_2Et$	*14	$C\underline{H}_3CH{=}CH_2$	37
$HO\underline{H}$	*15	alkanes	*40–44
cyclopentadiene	*15		

Note: pyr = pyridine.

You will sometimes see other pK_a values cited for certain compounds, especially alkanes. The pK_a of a compound changes dramatically with solvent, and it also depends on the temperature and the method of measurement. Approximate differences between acidities matter when organic reaction mechanisms are drawn, so the values given here suffice for the purposes of this text. For a more detailed discussion of acidity, see any physical organic chemistry textbook.

The following are some trends that can be discerned from the data:

• All else being equal, acidity *increases* as you move to the *right* in the periodic table (cf. $H_3C\underline{H}$, $H_2N\underline{H}$, $HO\underline{H}$), as electronegativity increases.

• All else being equal, acidity *increases* as you go *down* the periodic table (cf. Et$O\underline{H}$ with Et$S\underline{H}$) and size increases. This trend is opposite that for electronegativity. The trend is due to the increasingly poor overlap of the very small H(s) orbital with the increasingly large valence orbital of the atom to which it is bound.

* **Common error alert:** *Overlap effects come into play only when the acidic proton is* directly attached *to the heteroatom. Otherwise, inductive effects dominate.*

• All else being equal, a given atom is usually more acidic when it bears a formal positive charge than when it is neutral (cf. $NH_4{}^+$ with NH_3). However, it is not true that all positively charged acids are more acidic than all neutral acids (cf. R_3NH^+ with CH_3CO_2H). Conversely, an atom is usually more basic when it bears a formal negative charge than when it is neutral.

• The acidity of HA increases when inductively electron-withdrawing groups are attached to A and decreases when inductively electron-donating groups are attached to A (cf. CCl_3CO_2H with CH_3CO_2H, and HOH with EtOH).

• For uncharged acids, acidity *decreases* with *increased steric bulk* (cf. EtOH with t-BuOH). As the conjugate base becomes more hindered, the ability of the solvent to organize itself around the base to partly neutralize the charge by dipole effects or hydrogen bonds becomes increasingly compromised. As a result, the conjugate base becomes higher in energy, and the acid becomes weaker.

An inductive effect is often cited as the reason why *t*-BuOH is less acidic than EtOH. In fact, in the gas phase, where solvation plays no role, *t*-BuOH is more acidic than EtOH. Solvent effects play a very important role in determining acidity in the liquid phase, where most chemists work, but they are often ignored because they are difficult to quantify.

* **Common error alert:** *The rate of proton transfer from an acid to a base is not perceptibly slowed by steric hindrance.* The attenuation of acidity by steric bulk is a ground-state, thermodynamic effect.

• HA is much more acidic when the lone pair of the conjugate base can be stabilized by resonance (cf. PhOH with EtOH, $PhNH_3^+$ with Et_3NH^+, and $C\underline{H}_3CH=CH_2$ with alkanes). HA is especially acidic when the lone pair can be delocalized into a carbonyl group, and even more so when it can be delocalized into two carbonyl groups (cf. alkanes, CH_3COCH_3, and $EtO_2CCH_2CO_2Et$). The most common anion-stabilizing group is the C=O group, but nitro groups ($-NO_2$) and sulfonyl groups ($-SO_2R$) are also very good at stabilizing anions (cf. CH_3NO_2, CH_3COCH_3, and $CH_3SO_2CH_3$). Nitro groups are even more anion-stabilizing than carbonyl groups because of a greater inductive effect.

• For a given atom A, acidity *increases* with *increased s character* of the A–H bond; that is, A(sp)–H is more acidic than A(sp^2)–H, which is more acidic than A(sp^3)–H (cf. pyrH$^+$ with R_3NH^+, and HC≡CH, benzene, and alkanes). The lone pair of the conjugate base of an sp-hybridized atom is in a lower energy orbital than that of an sp^3-hybridized atom.

• Nonaromatic HA is much more acidic if its conjugate base is aromatic (cf. cyclopentadiene with propene). Conversely, a substance is a very poor base if protonation results in loss of aromaticity (e.g., pyrrole).

You can use these principles and the tabulated acidity constants to make an educated guess about the pK_a of a compound that you haven't seen before. It is important to know pK_a values because the first step in an organic reaction is often a proton transfer, and you need to know which proton in a compound is most likely to be removed.

Carbonyl compounds are perhaps the most important acidic organic compounds, so it is worth pointing out some of the factors that make them more or less acidic. The energy of a carbonyl compound is largely determined by the energy of its $R_2\overset{+}{C}-\overset{-}{O}$ resonance structure. The greater the ability of a group R to stabilize this resonance structure by lone pair donation, hyperconjugation, or inductive effects, the lower in energy the carbonyl compound is. The $\overset{+}{C}-\overset{-}{O}$ resonance structure, though, is much less important in the corresponding enolates, and as a result, most enolates have approximately the same energy. Because the acidity of a compound is determined by the difference in energy between its protonated and deprotonated forms, it turns out that acidities of carbonyl compounds correlate very well with their energies: *the lower in energy a carbonyl compound is, the less acidic it is.* (This correlation is not true of all compounds.) Thus, the order of increasing acidity is carboxylates < amides < esters < ketones < aldehydes < acyl anhydrides < acyl chlorides.

* **Common error alert:** α,β-*Unsaturated carbonyl compounds are* not *particularly acidic at the* α-*carbon atoms.* The C=O π bond prefers to be in conjugation and coplanar with the C=C π bond, so the C–H σ orbital does not overlap with the C=O π orbital. An unfavorable conformational change is required before deprotonation of the α-carbon can even begin. Note that the low acidity of α,β-unsaturated carbonyl compounds as compared with their saturated congeners contradicts the general rule that $C(sp^2)$ is more acidic than $C(sp^3)$, all else being equal.

lower energy conformation; deprotonation not possible

higher energy conformation; deprotonation possible

It is often more convenient to talk about basicities than acidities. In this textbook, the pK_b of a base is defined as the pK_a of its conjugate acid.* For example, according to this book's definition, NH_3 has a pK_b of 10 (because $NH_4{}^+$ has a pK_a of 10) and a pK_a of 35. The strength of a base correlates directly with the weakness of its conjugate acid. Factors that increase acidity decrease basicity, and factors that decrease acidity increase basicity. For example, EtS^- is less basic than EtO^-, just as EtSH is more acidic than EtOH.

1.2.2   Tautomerism

When acetone (CH_3COCH_3) is deprotonated, a compound is obtained that has a lone pair and a formal negative charge on C. A resonance structure in which the lone pair and formal negative charge are on O can be drawn. The true structure of the anion, of course, is a weighted average of these two resonance structures. If this anion reacts with H^+, the H atom may bind to either O or C. If the H atom binds to C, acetone is obtained again, but if it attaches to O, a compound (an *enol*) is obtained that differs from acetone only in the position of attachment of the H atom and the associated π system. Acetone and the enol are called *tautomers.*

Tautomers are isomers. They have different σ bond networks, which clearly distinguishes them from resonance structures. The most important kinds of tautomers are carbonyl–enol tautomers, as in the preceding example. Tautomerization is a chemical equilibrium that occurs very rapidly in acidic or basic media; it should not be confused with resonance, which is not an equilibrium at all.

---

*This definition differs from the standard one that $pK_b = 14 - pK_a$ (conjugate acid). (The 14 derives from the dissociation constant of H_2O.) The standard definition is much less convenient than the one used here because it requires that you learn two different numbering systems for what is essentially the same property. However, be careful not to use this book's definition in a different context; you are likely to be misunderstood.

1.3   Kinetics and Thermodynamics

Energies and rates are important aspects of reaction mechanisms. A reaction might be described as favorable or unfavorable, fast or slow, and reversible or irreversible. What does each of these terms mean?

A *favorable* reaction is one for which the *free energy* ($\Delta G°$) is less than zero (the free energy of the products is lower than the free energy of the starting materials). When $\Delta G° > 0$, the reaction is *unfavorable*. The free energy of a reaction is related to the *enthalpy* ($\Delta H°$) and *entropy* ($\Delta S°$) of that reaction by the equation $\Delta G° = \Delta H° - T\Delta S°$. In practice, enthalpies, not free energies, are usually used to determine whether a reaction is favorable or unfavorable, because $\Delta H°$ is easier to measure and because $T\Delta S°$ is small compared with $\Delta H°$ for most reactions at ordinary temperatures ($<100\ °C$). A reaction with $\Delta H° < 0$ is *exothermic*; one with $\Delta H° > 0$ is *endothermic*.

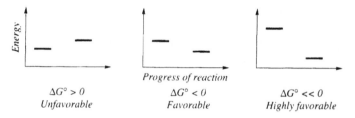

| $\Delta G° > 0$ | $\Delta G° < 0$ | $\Delta G° << 0$ |
| Unfavorable | Favorable | Highly favorable |

Of course, starting materials have to go through an energy barrier to become products; if there were no barrier, they couldn't exist! The energy required for the starting materials to reach the top of the barrier is called the *activation energy* ($\Delta G^{\ddagger}$). The arrangement of the reactants at the top of the barrier, where they can go either backward to starting materials or forward to products, is called the *transition state* (*TS*). The *rate* of a reaction is dependent on the size of the activation barrier, not on the energy difference between the starting materials and the products. A reaction is fast if the activation energy is low, and it is slow if the activation energy is high.

* **Common error alert:** *The rate of a reaction (depends on $\Delta G^{\ddagger}$) and the overall energetics of a reaction (depends on $\Delta G°$) are independent of one another.* It is possible to have a fast, unfavorable reaction or a slow, favorable reaction. An example of the former is the addition of water to the π bond of acetone to give the hydrate. An example of the latter is the reaction of gasoline with O_2 to give CO_2 and water at room temperature. The energy of the products does *not necessarily* influence the activation energy of a reaction.

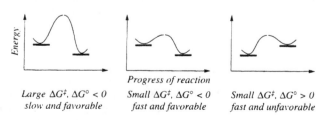

| Large $\Delta G^{\ddagger}$, $\Delta G° < 0$ | Small $\Delta G^{\ddagger}$, $\Delta G° < 0$ | Small $\Delta G^{\ddagger}$, $\Delta G° > 0$ |
| slow and favorable | fast and favorable | fast and unfavorable |

A reaction is in equilibrium when the rate of the forward reaction equals the rate of the reverse reaction. Such a reaction is *reversible*. In principle, all reactions are reversible, but in fact, some reactions have equilibria that lie so far to the right that no starting material can be detected at equilibrium. As a rule of thumb, if the equilibrium constant (K) is 10^3 or greater, then the reaction is irreversible. Reactions can also be made to proceed irreversibly in one direction by removing a gaseous, insoluble, or distillable product from the reaction mixture (LeChâtelier's principle). When starting materials and products or two different products are in equilibrium, their ratio is determined by the difference in free energy between them. However, if an equilibrium is not established, then their ratio may or may not be related to the difference in free energy.

* **Common error alert:** *If a reaction can give two products, the product that is obtained most quickly (the* kinetic *product) is not necessarily the product that is lowest in energy (the* thermodynamic *product).* For example, maleic anhydride reacts with furan to give a tricyclic product. If the progress of this reaction is monitored, it is seen that initially the more sterically crowded, higher energy endo product is obtained, but as time goes on this product is converted into the less crowded, lower energy exo product. The reason that the kinetic product goes away with time is that it is in equilibrium with the starting materials and with the thermodynamic product; the equilibrium, once established, favors the lower energy product. However, there are numerous cases in which a kinetic product is not in equilibrium with the thermodynamic product, and the latter is not observed. In other cases, the thermodynamic product is also the one that is obtained most quickly. One of the joys of organic chemistry is designing conditions under which only the kinetic or only the thermodynamic product is obtained.

kinetic, higher
energy product

lower
barrier

starting
materials

higher
barrier

thermodynamic,
lower energy product

Initially, equal amounts of kinetic and thermodynamic products are obtained. However, if the energy in the system is sufficiently high, the kinetic product can establish an equilibrium with the starting materials and eventually convert completely to thermodynamic product.

Many reactions proceed through unstable, high-energy intermediates with short lifetimes (e.g., carbocations). An *intermediate* is a valley in the reaction coordinate diagram. It is not to be confused with a TS, a peak in the reaction coordinate diagram. Transition states don't exist for longer than one molecular vibration and therefore can't be isolated, whereas intermediates may last anywhere from five molecular vibrations to milliseconds to minutes. Some reactions proceed through no intermediates, whereas others proceed through many.

It is very hard to get information about TSs, because they don't exist for more than about 10^{-14} s, but information about TSs is extremely important in thinking about relative rates and the like. The Hammond postulate states that a TS structurally resembles whichever of the two ground state species (starting materials, intermediates, or products) immediately preceding and following is higher in energy. The higher energy the species, the more it resembles the TS. For example, consider the reaction of isobutylene with HCl. This reaction proceeds through a high-energy intermediate, the carbocation. The rate-limiting step is formation of the carbocation. The Hammond postulate says that the energy of the TS leading to the carbocation is directly related to the energy of that carbocation; therefore, the rate of the reaction is related to the stability of the carbocation.

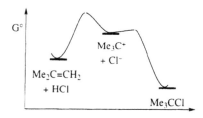

* **Common error alert:** *The Hammond postulate relates the structure and energy of the* higher energy *of the two ground state species immediately preceding and following the TS to the structure and energy of the TS.* For this reason, the TS of an exothermic reaction is not easily compared with the products of that reaction, because in an exothermic reaction the products are lower in energy than the starting materials.

> The term *stable* is ambiguous in organic chemistry parlance. When a compound is said to be "stable," it sometimes means that it has low energy ($\Delta G°$), i.e., it is thermodynamically stable, and it sometimes means that the barrier for its conversion to other species is high ($\Delta G^{\ddagger}$), i.e., it is kinetically stable. For example, both benzene and tetra-*t*-butyltetrahedrane are surprisingly stable. The former is both kinetically and thermodynamically stable, whereas the latter is kinetically stable and thermodynamically unstable. Certain kinds of compounds, like hemiacetals, are kinetically unstable and thermodynamically stable. In general, "stable" *usually* means "kinetically stable," but you should always assure yourself that that is what is meant. When in doubt, ask.

1.4   Getting Started in Drawing a Mechanism

Three features of the way organic reactions are written sometimes make it difficult for students to figure out what is going on. First, compounds written over or under the arrow are sometimes stoichiometric reagents, sometimes catalysts, and sometimes just solvents. Second, organic reactions are often not balanced. Little things like salts, water, or gaseous products are often omitted from the right side of the equation (but usually not the left). Balancing an unbalanced equation will often help you determine the overall reaction, and this may narrow your choice of mechanisms. Third, the product written on the right side of the equation is usually the product that is obtained *after aqueous workup*. Aqueous workup converts ionic products into neutral ones. Be aware of these conventions.

Solvent over arrow

Reagent over arrow; ionic product converted into neutral one by aqueous workup

Gaseous by-product omitted

By-product omitted; catalyst over arrow, solvent under arrow

When reagents are separated by a semicolon, it means that the first reagent is added and allowed to react, then the second reagent is added and allowed to react, and so on. When reagents are numbered sequentially, it may mean the same, or it may mean that the reaction mixture is worked up and the product is isolated after each individual step is executed.

Reagents added sequentially to reaction mixture

Reagents added sequentially or reaction mixture worked up after each step

A chemical reaction involves changes in bonding patterns, so probably the most important step when sitting down to draw a mechanism is to *determine which bonds are made and broken* in the course of the reaction. You can do this very easily as follows: Number the non-H atoms in the starting materials as sequentially as possible, then identify the same atoms in the products, using atom sequences and bonding patterns and minimizing the number of bonding changes. Remember to number carbonyl O atoms and both O atoms of esters, and *obey Grossman's rule*! Looking at the number of H atoms attached to certain C atoms in starting materials and products often helps you figure out how to number the atoms. Sometimes balancing the equation provides important numbering clues, too.

Many students are reluctant to take the time to number the atoms, but the importance of doing it cannot be overemphasized. If you don't number the atoms, you may not be able to determine which bonds are made and broken, and if you can't do that, you can't draw a mechanism! The time that you save by knowing which bonds are made and broken more than compensates for the investment of time in numbering the atoms.

Problem 1.4. Balance the following equations and number all atoms other than H in the starting materials and products.

(a)

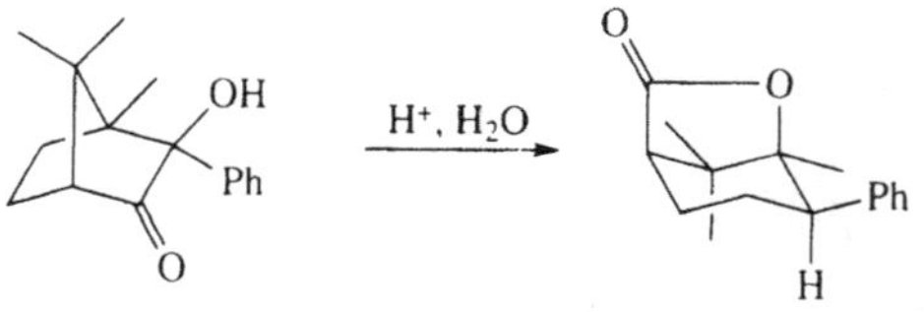

(b)

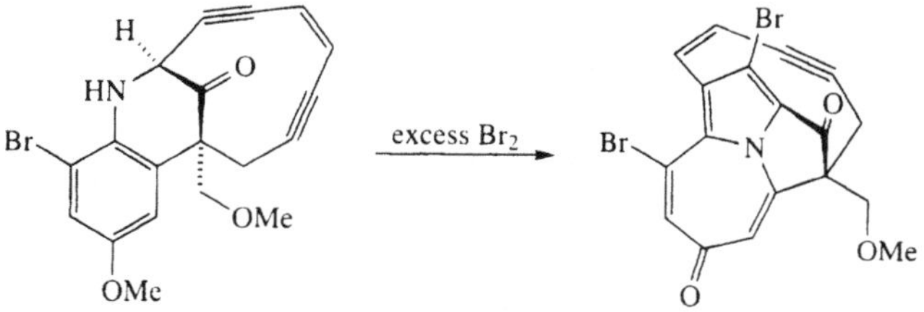

After you have numbered the atoms, make a list of σ *bonds between non-H atoms* that need to be made and σ *bonds between non-H atoms* that need to be broken. Do not list bonds to H that are made or broken, and do not list π bonds! This procedure will allow you to focus your attention only on what is important: the *changes* in bonding patterns that occur in going from starting material to product. Don't worry about the different appearances of starting material and product; all that matters are the changes in bonds between atoms. Products can sometimes look very different from starting materials even when only a few bonds have changed.

Problem 1.5. Make a list of all σ bonds made and broken between atoms other than H in the two preceding reactions.

Why should you not include π bonds in your list of bonds to make and break? The reason is that the location of the π bonds will follow naturally from where the σ bonds are made and broken. To illustrate the point, consider the tautomerization of acetone to the corresponding enol under basic conditions.

You might say that the important change in going from starting material to product is that the π bond migrates from C=O to C=C. It is true that the π bond migrates, but this fact does not tell you what mechanistic steps you need to execute. If you consider only the σ bonds, though, you see that you need to make an O–H σ bond and break a C–H σ bond. Because the conditions are basic, the first step is likely to be that a base deprotonates the acidic C to give an enolate, thus cleaving the C–H σ bond. Then the basic O atom deprotonates EtOH, forming the new O–H σ bond. Once the appropriate σ bonds have been made and broken, the π bond has naturally migrated to the appropriate position.

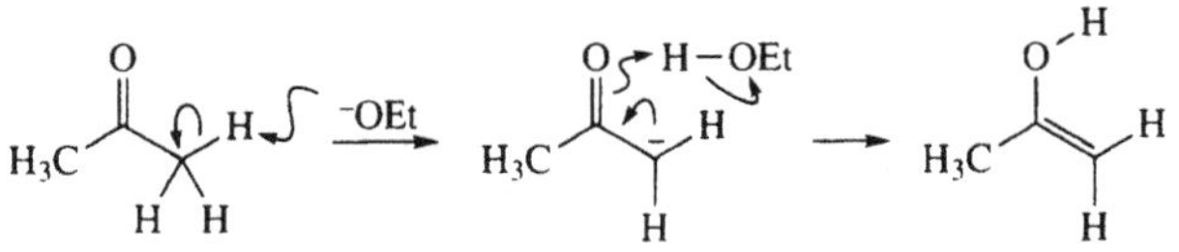

Note how obeying Grossman's rule makes much clearer which σ bonds are made and broken in the reaction.

Why should you not also include element–H σ bonds in your list of bonds to make and break? Because it is much easier to make and break element–H σ bonds than other kinds of σ bonds, especially in polar reactions, you will usually make and break the necessary element–H σ bonds naturally on the way to making and breaking σ bonds between heavy atoms. After you have made and broken the σ bonds between heavy atoms, if you have not yet arrived at the product, *then* look at the σ bonds involving H atoms.

1.5   Classes of Overall Transformations

Organic chemistry would just be a vast collection of often seemingly contradictory information if there were no way of classifying reactions. One way to classify reactions is by the overall tranformation, as defined by the relationships of the starting materials and the products. There are four basic kinds of overall transformations: *addition, elimination, substitution,* and *rearrangement.*

• In an *addition* reaction, two starting materials combine to give a single product. Usually a π bond in one starting material is replaced by two new σ bonds.

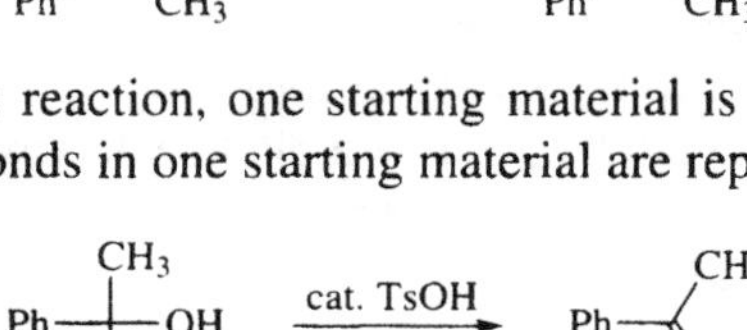

• In an *elimination* reaction, one starting material is divided into two products. Usually two σ bonds in one starting material are replaced by a new π bond.

• In a *substitution* reaction, an atom or group that is σ-bound to the rest of the starting material is *replaced* by another σ-bound atom or group.

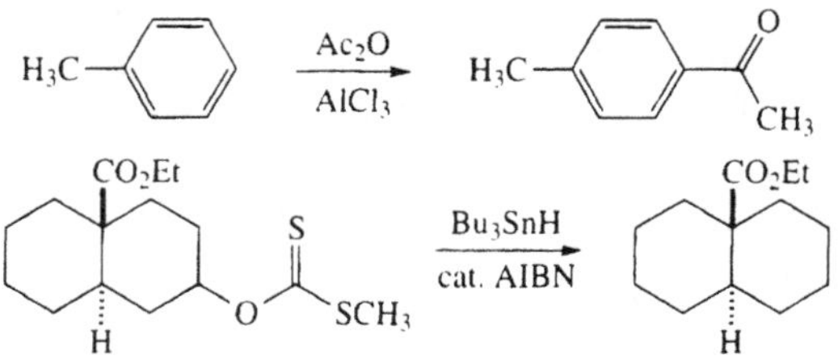

• In a *rearrangement*, one starting material gives one product with a different structure.

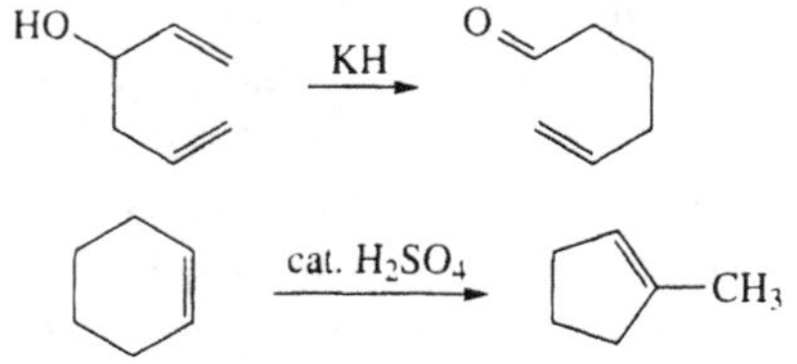

Some transformations constitute examples of more than one of these classes. For example, the reaction of esters with Grignard reagents gives alcohols in what is a substitution *and* an addition reaction. Remember that classification schemes are created by people. A compound does not stop and worry about whether a particular reaction fits within the boundaries of human classification schemes before it undergoes the reaction.

1.6   Classes of Mechanisms

A second way to classify reactions is according to the type of mechanism that is operative. The mechanism-based method of classification is used to organize the material in this text. However, it is important not to lose the forest (the overall transformation) for the trees (the mechanistic steps). Both classification methods have advantages, and it is useful to be able to move freely between them.

There are four basic types of organic reaction mechanisms—polar, free-radical, pericyclic, and metal-catalyzed or mediated.

• *Polar* reactions proceed by the movement of *pairs* of electrons from areas of high electron density (*nucleophiles*) to areas of low electron density (*electrophiles*), or from filled orbitals to empty orbitals. Polar mechanisms are further divided into those that proceed under basic conditions and those that proceed under acidic conditions.

• *Free-radical* reactions proceed by the movement of *single* electrons. New bonds are often formed using one electron from a half-filled orbital and another electron from a filled orbital. Free-radical reactions often proceed by a chain reaction, though not always.

• *Pericyclic* reactions are characterized by the movement of electrons in a cyclic formation.

• *Metal-mediated* and *metal-catalyzed* reactions require a transition metal. However, certain transition-metal complexes (e.g., $TiCl_4$, $FeCl_3$) act only as Lewis acids in organic reactions, and others ($TiCl_3$, SmI_2) act as one-electron reducing agents like Na and Li; reactions promoted by these compounds are classified in polar acidic, pericyclic, or free-radical classes.

The mechanism classification and the overall transformation classification are orthogonal to each other. For example, substitution reactions can occur by a polar acidic, polar basic, free-radical, pericyclic, or metal-catalyzed mechanism, and a reaction under polar basic conditions can produce an addition, a substitution, an elimination, or a rearrangement. Both classification schemes are important for determining the mechanism of a reaction, because knowing the class of mechanism and the overall transformation rules out certain mechanisms and suggests others. For example, under basic conditions, aromatic substitution reactions take place by one of three mechanisms: nucleophilic addition–elimination, elimination–addition, or $S_{RN}1$. If you know the class of the overall transformation *and* the class of mechanism, your choices are narrowed considerably.

1.6.1   Polar Mechanisms

In polar reactions, *nucleophiles react with electrophiles*. Furthermore, most polar reactions are carried out under either acidic or basic conditions.

The terms nucleophile and electrophile are used to refer both to whole compounds and to specific atoms or functional groups within compounds. This dual meaning can be confusing.

The concepts of nucleophilicity and electrophilicity are most useful in understanding polar reactions, but they apply to free-radical and pericyclic reactions, too.

1.6.1.1   Nucleophiles

A *nucleophile* is a compound that has a relatively high energy pair of electrons available to make a new bond. A nucleophilic atom may be neutral or negatively charged. There are three classes of nucleophiles: lone-pair nucleophiles, σ-bond nucleophiles, and π-bond nucleophiles.

The division of nucleophiles into three types is completely artificial. Nature does not stop to consider whether a nucleophile is a π-bond nucleophile or a lone-pair nucleophile before a reaction proceeds. The distinctions are intended only to train you to spot nucleophilic atoms or functional groups with ease.

• *Lone-pair nucleophiles* contain atoms with lone pairs. The lone pair is used to make a new bond to an electrophilic atom. Alcohols (ROH), alkoxides (RO^-), amines (R_3N), metal amides (R_2N^-), halides (X^-), thiols (RSH), sulfides (R_2S), and phosphines (R_3P) are all examples of lone-pair nucleophiles, as are the O atoms of carbonyl compounds ($X_2C=O$). When these compounds act as nucleophiles, the formal charge of the nucleophilic atom is *increased* by 1 in the product.

$$R_3N\colon \quad H_3C{-}I \quad \longrightarrow \quad R_3\overset{+}{N}{-}CH_3 \quad + \quad I^-$$

Curved arrows are used to show the flow of electron density *from* the nucleophile *to* the electrophile. The base of the arrow should be at the pair of electrons used to make the new bond. The head of the arrow may point either to the electrophilic atom or between the two atoms where the new bond is formed.

• *Sigma-bond nucleophiles* contain a bond between a nonmetal and a metal. The electrons of the element–metal bond are used to form a bond between the nonmetal and the electrophile. The formal charge on the nucleophilic atom does not change; the metal increases its formal charge by 1. The nucleophilic atom may be a heteroatom (as in $NaNH_2$ or KOH), carbon (as in Grignard reagents (RMgBr), organolithium reagents (RLi), and Gilman reagents (R_2CuLi), which have C–Mg, C–Li, and C–Cu bonds, respectively), or hydrogen (as in the complex metal hydrides $NaBH_4$ and $LiAlH_4$).

$$H{-}\overset{\overset{\displaystyle H}{|}}{\underset{\underset{\displaystyle H}{|}}{Al}}{-}H \quad \overset{Me}{\underset{Me}{\diagdown}}C{=}O \quad \longrightarrow \quad H{-}\overset{Me}{\underset{Me}{\diagup}}{-}O^-$$

The very polarized bond between a metal and a nonmetal, E–M, is often thought of as being like an ionic bond (E^- M^+); thus, PhC≡C–Li, H_3C–MgBr, and $LiAlH_4$ are sometimes drawn as PhC≡C$^-$, H_3C^-, and H^-, respectively. By this analogy, σ-bond nucleophiles are actually lone-pair nucleophiles.

• *Pi-bond nucleophiles* use the pair of electrons in a π bond, usually a C=C bond, to form a σ bond between one of the atoms in the π bond and the electrophilic atom. The formal charge and total electron count of the nucleophilic atom of the π bond do not change, but the other atom of the π bond is made *electron-deficient*, and its formal charge increases by 1. The π bonds of simple alkenes and arenes are weakly nucleophilic; π bonds that are directly attached to heteroatoms, such as in enolates (C=C–O$^-$), enols (C=C–OH), enol ethers (C=C–OR), and enamines (C=C–NR$_2$), are much better nucleophiles.

In principle, either of the C(sp^2) atoms in an alkene may form a new bond to an electrophile. If the alkene is differentially substituted, though, one is usually more nucleophilic than the other.

* **Common error alert:** *A nucleophilic alkene always reacts with an electrophile such that the C less able to be electron-deficient makes the new bond.* In alkenes directly substituted with lone-pair-bearing heteroatoms (enolates, enamines, enols, enol ethers), the β-carbon (not attached to heteroatom) is nucleophilic, and the α-carbon (attached to heteroatom) is not. In alkenes substituted only with alkyl groups,

the less substituted C is more nucleophilic, and the more substituted C becomes electron-deficient upon reaction of the π bond with an electrophile.

NOT

NOT

Different compounds have different nucleophilicities. The nucleophilicity of a compound is measured by determining how reactive it is toward CH_3Br in water at 25 °C. Nucleophilicity bears some relationship toward basicity, which measures reactivity toward H^+, but there are important differences. CH_3Br is uncharged, relatively large, and has a relatively high energy LUMO (lowest unoccupied molecular orbital). The H^+, on the other hand, is charged, tiny, and has a low-energy LUMO. Moreover, nucleophilicity (the rate of reaction with CH_3Br) is a *kinetic* property, whereas acidity (the equilbrium constant between protonated and unprotonated species) is a *thermodynamic* property. Because of these differences, some bases are poor nucleophiles, and vice versa. Increases in basicity generally parallel increases in nucleophilicity, except in the following ways:

• Nucleophilicity *increases* as you go down the periodic table, whereas basicity *decreases*. Thus, I^- is a great nucleophile, whereas Cl^- is fair, and Et_2S is a very good nucleophile, whereas Et_2O is very poor. In the case of anions (e.g., I^- versus Cl^-), the reason for the trend is that protic solvents such as MeOH (in which nucleophilicity is traditionally measured) can more strongly hydrogen-bond to smaller anions, making them less nucleophilic. The evidence for this rationale is that in aprotic solvents and in the gas phase, Cl^- is actually slightly more nucleophilic (*and more basic*) than I^-. In the case of neutral nucleophiles (e.g., Et_2S versus Et_2O), the lower electronegativity and the greater steric accessibility of the heavier atoms (because of the longer bond lengths) explain their greater nucleophilicity.

• Nucleophilicity *decreases dramatically* with increased crowding around the nucleophilic atom, while basicity *increases slightly*. So, although EtO^- is both a good base ($pK_b = 17$) and a good nucleophile, $t\text{-}BuO^-$ is a better base ($pK_b = 19$) and a really awful nucleophile.

• Delocalization of charge decreases basicity *a lot* and nucleophilicity *somewhat*. For example, EtO^- ($pK_b = 17$) reacts with $s\text{-}BuBr$ at lower temperatures than AcO^- ($pK_b = 4.7$), but EtO^- acts mostly as a base to give 2-butene by an E2 elimination (Chapter 2), whereas AcO^- acts mostly as a nucleophile to give $s\text{-}BuOAc$ by S_N2 substitution. AcO^- is both less basic and less nucleophilic than EtO^-, so it requires higher temperatures to react at all, but when it does react, the proportion of substitution product is greater, because AcO^- is much less basic and only somewhat less nucleophilic than EtO^-. Likewise, simple ester enolates ($pK_b = 25$) react with secondary alkyl halides such as $i\text{-}PrBr$ at lower temperatures to give elimination primarily, whereas malonate anion ($pK_b = 14$) requires higher temperatures to react with *i*-PrBr, but the substitution product is obtained primarily.

EtO⁻ is both more basic and more nucleophilic than AcO⁻ ...

... but it is a lot more basic and only somewhat more nucleophilic ...

... so the proportion of nucleophilicity to basicity is greater for AcO⁻ than for EtO⁻.

• The absence of hydrogen-bonding in *polar aprotic solvents* (particularly polar, liquid, nonviscous compounds that lack acidic protons) makes anions dissolved therein particularly reactive; as a result, both their basicity and nucleophilicity increase, but their nucleophilicity increases more. For example, F^- is extremely unreactive in H_2O because it is so strongly solvated, but in DMSO it is nucleophilic toward alkyl halides. Common polar aprotic solvents include DMSO, HMPA, DMF, DMA, NMP, DMPU, and pyridine, but not EtOH or H_2O. *Note:* For purposes of drawing mechanisms, it is much more important that you recognize that a particular solvent is polar aprotic than that you know its structure!

Some common polar aprotic solvents:

| DMSO | HMPA (or HMPT) | DMF | DMA | NMP | DMPU |

Nonnucleophilic bases play a special role in organic chemistry, and it is important to recognize some of the ones that are widely used: *t*-BuOK, LiN(*i*-Pr)$_2$ (LDA), LiN(SiMe$_3$)$_2$ (LiHMDS), KN(SiMe$_3$)$_2$ (KHMDS), NaH and KH, EtN(*i*-Pr)$_2$ (Hünig's base), and DBU (an amidine base). Most of these bases are quite hindered, except for DBU, which has some nucleophilicity toward 1° alkyl halides, and the metal hydrides, which are nonnucleophilic for kinetic reasons. Again, it is more important that you remember the properties of these compounds than that you memorize their structures.

DBU, or 1,8-diazabicyclo[5.4.0]undec-1(7)-ene, a widely used, relatively nonnucleophilic base.

Sometimes it is hard to identify the nucleophilic and electrophilic sites in a compound. In this case, obey Meier's rule: **When in doubt, draw in all the lone pairs, and draw resonance structures until the cows come home.** Often the second-best resonance structure for a compound shows where the nucleophilic sites are in that compound. For example, if resonance structures are drawn for anisole, it is easy to see why reaction with electrophiles always occurs at the ortho or para positions, never at the meta position. Incidentally, Meier's rule applies equally to identifying electrophilic sites. Thus, acetone (Me$_2$C=O) is easily seen to be electrophilic at C when its second-best resonance structure is drawn.

Different kinds of nucleophiles are typically seen under basic and acidic conditions.

• Lone-pair nucleophiles that are good bases (RO$^-$, RC≡C$^-$, RS$^-$, R$_2$N$^-$) can exist under basic conditions only, but weakly basic lone-pair nucleophiles (halides, RCO$_2{}^-$, RSO$_3{}^-$, R$_2$S, R$_3$P) can exist under either acidic or basic conditions. Water and alcohols can exist under either basic or acidic conditions, but under basic conditions they are deprotonated before they react with an electrophile, whereas under acidic conditions they are not. Under acidic conditions, amines (R$_3$N) exist as their ammonium salts, which lack any lone pairs, and so they must be deprotonated in an unfavorable equilibrium process before they can act as nucleophiles.

• Most σ-bond nucleophiles (PhMgBr, CH$_3$Li, LiAlH$_4$, NaBH$_4$) are quite basic, so they are generally used only under basic conditions. Only a handful of σ-bond nucleophiles are compatible with acidic conditions: Lewis acidic nucleophiles such as AlMe$_3$, ZnEt$_2$, and i-Bu$_2$AlH, and a few non–Lewis acidic H$^-$ sources such as NaBH(OAc)$_3$ and NaBH$_3$CN.

• Most π-bond nucleophiles (enols, enol ethers, enamines, as well as simple alkyl-substituted alkenes and arenes) can exist under either acidic or basic conditions, but enolates exist only under basic conditions. Among π-bond nucleophiles, however, only enolates and enamines are reactive enough to attack the electrophiles typically seen under basic conditions.

1.6.1.2   Electrophiles and Leaving Groups

An *electrophile* is a compound that has a relatively low energy empty orbital available for bond-making. An electrophile may be neutral or positively charged. There are three classes of electrophiles: Lewis acid electrophiles, π-bond electrophiles, and σ-bond electrophiles.

• *Lewis acid electrophiles* have an atom E that lacks an octet and has a low-energy nonbonding orbital, usually a p orbital. A pair of electrons from the nucleophile is used to form a new bond to E, giving it its octet. The formal charge of E *decreases* by 1. Carbocations, boron compounds, and aluminum compounds are common Lewis acid electrophiles.

• In π-*bond electrophiles*, the electrophilic atom E has an octet, but it is attached by a π bond to an atom or group of atoms that can accept a pair of electrons. Electrophiles of the π-bond type commonly have C=O, C=N, or C≡N bonds, in which the less electronegative atom is the electrophilic one. C=C and C≡C bonds are electrophilic if they are attached to electrophilic atoms (see below). Miscellaneous π-bond electrophiles include SO_3 and RN=O. Certain cationic electrophiles can be classified as either the Lewis acid type or the π-bond type depending on whether the best or second-best resonance description is used (e.g., $R_2C=\overset{+}{O}H \longleftrightarrow R_2\overset{+}{C}-OH$; also, $H_2C=\overset{+}{N}Me_2$, $RC\equiv\overset{+}{O}$, $O=\overset{+}{N}=O$, $N\equiv\overset{+}{O}$). When a nucleophile attacks a π-bond electrophile, the π bond breaks, and the electrons move to the *other* atom of the π bond, whose formal charge is decreased by 1.

• *Sigma-bond electrophiles* have the structure E–X. The electrophilic atom E has an octet, but it is attached by a σ bond to an atom or group of atoms X, called a *leaving group* (or *nucleofuge*), that wants to take the electrons in the E–X bond away from E and leave to form an independent molecule. A nucleophile reacts with a σ-bond electrophile by using its pair of electrons to form a bond to the electrophilic atom E, increasing its formal charge by 1. At the same time, X departs with the pair of electrons from the E–X σ bond and decreases its formal charge by 1. There is no change in the overall electron count at E.

Sigma-bond electrophiles can be further subdivided into three types: those where E is C, those where E is a heteroatom, and Brønsted acids, in which E is H. Sigma-bond electrophiles of the first class include alkyl halides, alkyl sulfonates and other pseudohalides, oxonium ions (e.g., $Me_3\overset{+}{C}OH_2$), and sulfonium ions (e.g., Me_3S^+). Sigma-bond electrophiles of the first class may also

undergo *spontaneous* cleavage of the C–X bond in the absence of a nucleophile
to give a carbocation.

Common σ-bond electrophiles of the second class include Br_2 and other elemental halogens, peracids (RCO_3H; the terminal O of the O–O bond is electrophilic), and RSX and RSeX (X is Br or Cl; S or Se is electrophilic). Sigma-
bond electrophiles of the second class do not undergo spontaneous cleavage
of the E–X bond because E is usually too electronegative to suffer electron-
deficiency.

Heavily substituted halocarbons such as CBr_4 and Cl_3CCCl_3 are electrophilic at the halogens, not at C, because the leaving groups Br_3C^- and Cl_3CCCl_2 are fairly weak bases—
all that electron-withdrawing power stabilizes them—and because the C atoms are so buried
underneath all those large halogen atoms that a nucleophile can't possibly reach them.

Leaving groups differ in their *leaving group ability*. The ability of a leaving group
to leave is quite closely related to the pK_b of the group: *The less basic a leaving
group is, the better is its leaving group ability*. In general, only excellent and good
leaving groups (as defined in Table 1.5) leave at reasonable rates in S_N2 substitution reactions. Fair and poor leaving groups can be expelled by a neighboring negative charge, as in addition–elimination reactions at carbonyl groups.

* **Common error alert:** *The groups H^-, O^{2-}, and unstabilized carbanions are
truly awful leaving groups and almost never leave!*

TABLE 1.5. Approximate order of leaving group ability

Excellent to good leaving groups	pK_b	Fair to poor leaving groups	pK_b
N_2	< -10	F^-	$+3$
$CF_3SO_3^-$ (TfO$^-$)	< -10	RCO_2^-	$+5$
I^-	-10	$^-C\equiv N$	$+9$
Br^-	-9	NR_3	$+10$
$ArSO_3^-$ (e.g., TsO$^-$)	-7	RS^-	$+11$
Cl^-	-7	stabilized enolates	$+9$ to $+14$
RCO_2H	-6	HO^-	$+15$
EtOH	-2.5	EtO^-	$+17$
H_2O	-1.5	simple enolates	$+20$ to 25
		R_2N^-	$+35$

Looking at pK_b values won't always give you the right order of leaving group ability. For example, one can hydrolyze carboxamides ($RCONR_2$) with strong aqueous base, but one can't hydrolyze alkynyl ketones ($RCOC\equiv CR$), even though the pK_b of $^-NR_2$ is 35, and the pK_b of $^-C\equiv CR$ is 25. The solution to this apparent contradiction is that the N of the carboxamide is probably protonated before it leaves, so that the leaving group is not really $^-NR_2$ at all but is HNR_2 ($pK_b = 10$). For the same reason, the ability of ^-CN to leave is considerably lower (relative to heteroatom leaving groups) than its pK_b would indicate. The electrophile structure, the presence of metal counterions, and the nature of the reaction medium are also crucial factors in leaving group ability.

The relief of strain can increase a group's leaving ability. For example, the reaction of epoxides with nucleophiles Nu^- to give $NuCH_2CHRO^-$ proceeds readily, even though alkoxides are normally very bad leaving groups.

You may have noticed that $C=C$ π bonds can act as either nucleophiles or electrophiles. How can you tell whether an alkene is nucleophilic or electrophilic? Alkenes are the chameleons of organic chemistry: how they react depends on what is attached to them. Alkenes or arenes that are attached to electrophilic groups such as $-CR_2$, $-COR$, $-CO_2R$, $-C\equiv N$, $-NO_2$, or $-CH_2X$ (where X is a leaving group) are electrophilic. Alkenes or arenes that are attached to nucleophilic groups like $RO-$, R_2N-, or $-CH_2MgBr$ are nucleophilic. Simple alkenes and arenes are nucleophilic also.

* **Common error alert:** *Alkenes and arenes that are substituted with electrophilic groups are electrophilic at the β-carbon (i.e., the C atom not attached to the electrophilic substituent), not at the α-carbon, because the electrons in the π bond need to end up on the α-carbon in order to interact with the electrophilic group.* This pattern parallels what is seen for alkenes substituted with electron-donating groups, which are nucleophilic at the β-carbon.

Alkenes that are normally nucleophilic can act as electrophiles toward potent-enough nucleophiles, and alkenes that are normally electrophilic can act as nucleophiles toward potent-enough electrophiles. For example, alkyllithium compounds (RLi) add to the "nucleophilic" alkene ethylene, and the "electrophilic" alkene 2-cyclohexen-1-one adds Br_2 across its $C=C$ π bond. These reactions provide additional support for describing alkenes as the chameleons of organic chemistry.

* **Common error alert:** *Do not confuse formal positive charge with electrophilicity!* Consider $CH_2=\overset{+}{O}CH_3$. The electronegative element O has a formal positive charge, but it is not electron-deficient, so there is no reason to think that it may be electrophilic. Now draw the resonance structure, $H_2\overset{+}{C}-O-CH_3$. O is neutral, the formal positive charge is on C, and C is electron-deficient. Considering the two resonance structures, which atom is more likely to be at-

tacked by a nucleophile, C or O? The same considerations apply to $Me_2\overset{+}{N}=CH_2$, $Me_3\overset{+}{O}$, and $PhCH_2=\overset{+}{N}(\overset{-}{O})Ph$. In general, *a heteroatom that has an octet and that bears a formal positive charge is* not *electrophilic; any atom to which it is bound that is more electropositive is electrophilic.* An archetypical example readily remembered by students is H_3O^+. It has an O with a formal positive charge, and it is electrophilic at H.

Carbenes, a class of transient organic species, are formally both nucleophilic *and* electrophilic, although their electrophilicity dominates their reactivity. Carbenes are divalent, six-electron carbon compounds (CR_2) with one unshared pair of electrons. You can think of them as $\pm CR_2$ or $:CR_2$. (No formal charge is implied by the $\pm$ symbol!) The best-known carbene is CCl_2, which is generated from $CHCl_3$ and strong base and is used to make dichlorocyclopropanes from alkenes. Carbon monoxide ($:\overset{..}{O}=C: \longleftrightarrow :\overset{..}{O}\equiv C:$) and isocyanides ($R-\overset{..}{N}\equiv C: \longleftrightarrow R-\overset{+}{N}\equiv \overset{-}{C}:$) can be described as especially stabilized carbenes. Carbenes are discussed in more detail in Chapters 2 and 5.

Different kinds of electrophiles are typically seen under basic and acidic conditions, too.

• Lewis acid electrophiles are Lewis acids, so they exist under acidic conditions only.

* **Common error alert:** *Free H^+ and carbocations R_3C^+ should not be drawn in mechanisms occurring under basic conditions.*

• Most π-bond electrophiles can exist under either acidic or basic conditions. Under acidic conditions, though, π-bond electrophiles are usually protonated or coordinated to a Lewis acid before they react with a nucleophile.

• Most σ-bond electrophiles, both of the C–X type (where X is a leaving group) and the heteroatom–heteroatom type, can exist under both basic and acidic conditions. Electrophiles of the C–X type, though, are usually unreactive toward the types of nucleophiles seen under acidic conditions. Under acidic conditions, they are usually converted into carbocations before they react with a nucleophile.

1.6.1.3   A Typical Polar Mechanism

The reaction of isobutylene (2-methylpropene) with HCl to give *t*-butyl chloride is a typical example of a polar reaction.

HCl is an acid; it dissociates into H^+ and Cl^-. A curved arrow is drawn to show the movement of electrons from the bond to the more electronegative element.

Three species are in the reaction mixture now. How would you characterize their reactivity? One (H$^+$) is an electrophile, and two (Cl$^-$, isobutylene) are nucleophiles. *Nucleophiles react with electrophiles!* Cl$^-$ can combine with H$^+$, but this reaction is simply the reverse of the first reaction. More productively, the electrons in the π bond in isobutylene can be used to form a new bond to H$^+$ to give a carbocation.

We note a few points about this step. First, there is a net positive charge on the left, so there *must* be a net positive charge on the right. Second, the number of electrons has not changed in going from the left to the right, nor has the number of hydrogen atoms. (Remember Grossman's rule!) Third, the H$^+$ can attach to either of two C atoms, but it attaches to the more nucleophilic, less substituted C. (Remember Markovnikov's rule!) Fourth, the number of electrons around the nucleophilic carbon (on the right in both starting material and product) does not change in the course of the reaction, but a pair of electrons is taken away from the nonnucleophilic carbon, so this atom is electron-deficient in the product.

The carbocation is an electrophile, and there is still a nucleophile (Cl$^-$) in the reaction mixture. In the final step of the reaction, a lone pair on Cl$^-$ is used to form a σ bond to the electrophilic C of the carbocation. The product is t-BuCl.

1.6.1.4   Acidic and Basic Conditions; The pK_a Rule

Polar reactions usually take place under acidic or basic conditions, and the characteristics of mechanisms that take place under these two sets of conditions are quite different. To identify a reaction that proceeds by a polar basic mechanism, look for bases such as amines, alkoxides, or Grignard or organolithium reagents. The presence of one of the typical nonnucleophilic bases discussed earlier is a sign of a polar basic mechanism, as is the presence of salts, especially ones where the cation is an alkali metal (e.g., Na, Li, or K) or is otherwise unreactive (e.g., Bu$_4$N$^+$). To identify a reaction that proceeds by a polar acidic mechanism, look for Brønsted or Lewis acids. Common Brønsted acids include carboxylic acids, TsOH and other sulfonic acids, mineral acids such as H$_2$SO$_4$ and HCl, and even mildly acidic ammonium salts such as NH$_4^+$Cl$^-$ or pyrH$^+$ TsO$^-$ (also known as PPTS). Common Lewis acids include BF$_3$, AlCl$_3$, TiCl$_4$, ZnCl$_2$, SnCl$_4$, FeCl$_3$, Ag(I) salts, and lanthanide salts such as Sc(OTf)$_3$. The reactions of 3° alkyl halides in protic solvents usually proceed by a polar acidic mechanism.

* **Common error alert:** *If a reaction is executed under acidic conditions, no strong bases can be present! Any negatively charged species* must *be weak bases (e.g.,*

Cl^-). *If a reaction is executed under basic conditions, no strong acids can be present! Any positively charged species* must *be weak acids.*

For example, transesterification can occur under either basic or acidic conditions.

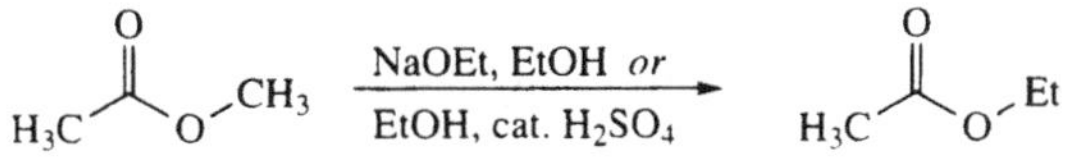

Under basic conditions, all species present are poor acids. Note the lack of any $R_2\overset{+}{O}H$ species in this mechanism.

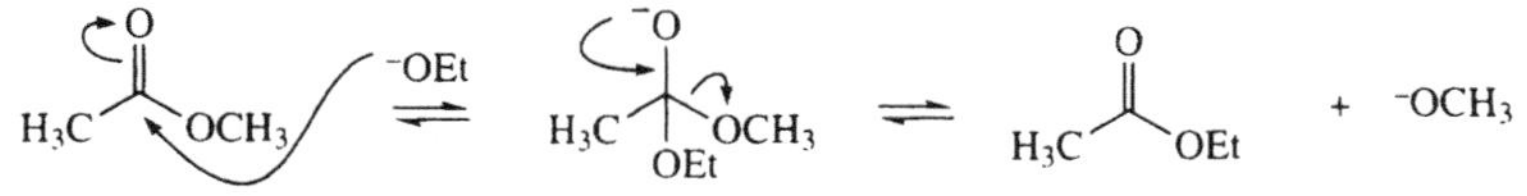

Under acidic conditions, all species present are poor bases. Note the lack of any RO^- species in this mechanism.

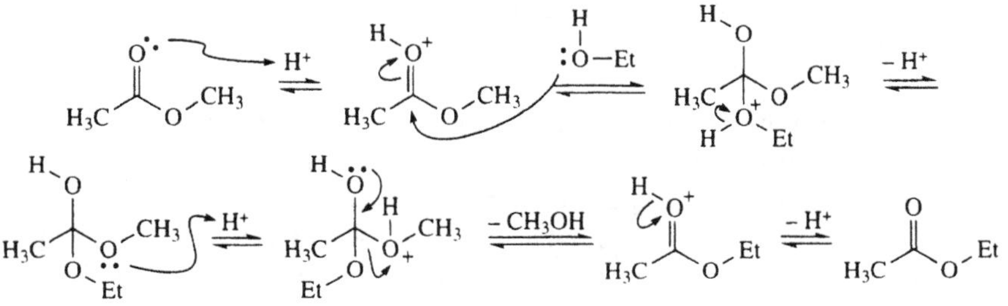

Under either set of conditions, direct attack of ethanol on the ester is not reasonable. The immediate product of such a direct attack would have both RO^- and $R_2\overset{+}{O}H$ groups, making it both a good acid and a good base. The presence of both a good acid and a good base in a single species creates a situation as untenable as the simultaneous existence of NaOH and H_2SO_4 in a reaction mixture.

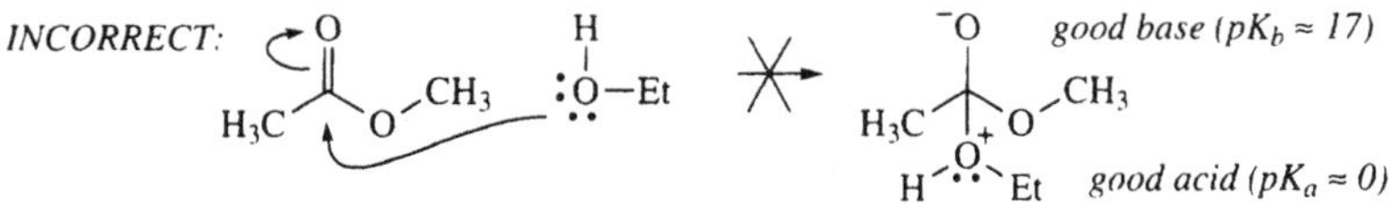

INCORRECT:

good base $(pK_b \approx 17)$

good acid $(pK_a \approx 0)$

Reactions that proceed by polar mechanisms often have a protonation or a deprotonation as the first step. If a reaction is run under basic conditions, look for acidic protons in the substrate to remove as a first step. If a reaction is run under acidic conditions, look for basic sites to protonate as a first step.

A weak base can deprotonate a weak acid to generate a stronger base in a rapid, reversible fashion, even though the deprotonation is thermodynamically unfavorable, *as long as the pK$_b$ of the base is no more than 8 units less than the pK$_a$ of the acid.* The "pK$_a$ rule" is a rule of thumb for determining whether a thermodynamically unfavorable deprotonation step in a mechanism is permissible. For example, it is permissible for ⁻OH to deprotonate acetone, because the pK_b of ⁻OH is only 5 units less than the pK_a of acetone, but it is not permissible for ⁻OH to deprotonate BuC≡CH, because the pK_b of ⁻OH is 10 units less than the pK_a of BuC≡CH. The same rule applies to protonations, of course, although the rule is not as important for reactions carried out under acidic conditions, because very strong acids are usually used in these reactions.

pK_a = 20 pK_b = 15   *unfavorable, but sufficiently rapid to be proposed in a mechanism*

pK_a = 25 pK_b = 15   *unfavorable, and not sufficiently rapid to be proposed in a mechanism*

Problem 1.6. PhC≡CH can be deprotonated by HO⁻, even though BuC≡CH cannot. What does this tell you about the relative acidities of PhC≡CH and BuC≡CH?

Often there is more than one basic or acidic site in a molecule. Remember that protonation–deprotonation reactions are very rapid and reversible reactions, so just because a substrate has an acidic or basic site, it doesn't mean that protonation or deprotonation of that site is the first step in your mechanism. It's up to you to figure out which site must be deprotonated for the reaction in question to proceed.

Problem 1.7. Which is the most acidic site in the following compound? Which site needs to be deprotonated for the reaction to proceed?

Sometimes basic conditions are specified, but there aren't any acidic protons. You might want to look for protons that are three bonds away from a leaving group X, e.g., H–C–C–X. If there are good nucleophiles present, then you may also need to look for electrophilic atoms. Don't forget that heteroatoms that have their octet and that bear a formal positive charge are not electrophilic!

1.6.2 Free-Radical Mechanisms

In free-radical reactions, odd-electron species abound. Not all reactions involving free radicals are chain reactions, and not all chain reactions involve free radicals, but the set of free-radical reactions and the set of chain reactions overlap enough that the two subjects are almost always discussed together.

A chain reaction consist of three parts: initiation, propagation, and termination.

Overall:

$$\underset{\underset{\text{Ph}-\overset{\displaystyle H}{\overset{|}{C}}\text{HCH}_3}{}}{} + \text{Br}_2 \xrightarrow{h\nu} \underset{\underset{\text{Ph}-\overset{\displaystyle Br}{\overset{|}{C}}\text{HCH}_3}{}}{} + \text{HBr}$$

In the *initiation* part of free-radical chain reactions, a small amount of one of the stoichiometric starting materials (i.e., a starting material that is required to balance the equation) is converted into a free radical in one or more steps. An *initiator* is sometimes added to the reaction mixture to promote radical formation. In the example, though, no initiator is necessary; light suffices to convert Br_2, one of the stoichiometric starting materials, into a free radical by σ-bond homolysis.

Initiation:

$$\text{Br}-\text{Br} \xrightarrow{h\nu} \text{Br}\cdot + \cdot\text{Br}$$

In the *propagation* part, the stoichiometric starting materials are converted into the products in one or more steps.

Propagation:

$$\text{Br}\cdot \quad \underset{\underset{H}{|}}{H-\overset{\overset{\displaystyle H_3C}{\diagdown}}{\underset{|}{C}}-\text{Ph}} \longrightarrow \text{Br}-\text{H} \quad \underset{\underset{H}{|}}{\cdot\overset{\overset{\displaystyle H_3C}{\diagdown}}{C}-\text{Ph}}$$

$$\underset{\underset{H}{|}}{\text{Ph}-\overset{\overset{\displaystyle CH_3}{/}}{C}\cdot} \quad \text{Br}-\text{Br} \longrightarrow \underset{\underset{H}{|}}{\text{Ph}-\overset{\overset{\displaystyle CH_3}{|}}{C}-\text{Br}} \quad \cdot\text{Br}$$

In the *termination* part, two radicals react to give one or two closed-shell species by *radical–radical combination* or *disproportionation*.

Termination:

$$\underset{\underset{H}{|}}{\text{Ph}-\overset{\overset{\displaystyle CH_3}{/}}{C}\cdot} \quad \cdot\text{Br} \longrightarrow \underset{\underset{H}{|}}{\text{Ph}-\overset{\overset{\displaystyle CH_3}{/}}{C}-\text{Br}}$$

$$\underset{\underset{H}{|}}{\text{Ph}-\overset{\overset{\displaystyle H\ \ H}{\diagdown/}}{C}-H} \quad \cdot\overset{\overset{\displaystyle H}{|}}{\underset{H_3C}{C}}-\text{Ph} \longrightarrow \underset{\underset{H}{|}}{\text{Ph}-\overset{\overset{\displaystyle H}{\diagup}}{C}=H} + \quad H-\overset{\overset{\displaystyle H}{|}}{\underset{H_3C}{C}}-\text{Ph}$$

etc.

The heart of free-radical chain reactions is the propagation part, and it is important to be able to draw one properly. Propagation parts have the following characteristics.

• Every step in the propagation part has *exactly one* odd-electron species on *each* side of the arrow. (Exception: O_2, a 1,2-diradical, can combine with a radical to give a new radical in a propagation step. Even in this case, though, there is still an *odd number* of electrons on each side of the arrow.)

* **Common error alert:** *Propagation steps do not involve the reaction of two radicals with each other. The only time two free radicals react with each*

other in chain reactions is in the termination part. (Exception: Reactions involving stoichiometric O_2.)

• The product radical from the initiation part of the chain mechanism is used as the starting radical in the first step of the propagation. The product radical in each step of the propagation is used as the starting radical in the next step of the propagation. The product radical in the last step of the propagation is the same as the starting radical in the first step of the propagation.

• If all the steps in the propagation part are added up, and like species on either side are canceled, an equation describing the overall transformation is obtained.

• Only stoichiometric starting materials or fragments thereof appear in the propagation.

  * **Common error alert:** *Although it is possible for an initiator to participate in the propagation part of a chain reaction, it is bad practice to write a mechanism in this way.* The concentration of initiator is usually very small, and the probability that a radical will encounter it is considerably smaller than the probability that the radical will encounter a stoichiometric starting material.

• Every step in the propagation part must be exothermic or nearly thermoneutral. If a particular step is endothermic, then radicals will accumulate at that point, and they will react with one another to terminate the chain.

All of these characteristics of propagation parts except the first are true of all chain reactions, not just chain reactions involving free radicals.

How do you recognize that a free-radical chain mechanism is operative? One key is the initiation step. Most free radicals are very unstable species, so when they are used in synthesis, they must be generated in the presence of their reaction partners. There are a limited number of ways of generating free radicals.

• Alkyl and acyl peroxides and AIBN (Chapter 5) are common initiators. These compounds readily undergo homolysis under the influence of light ($h\nu$) or heating to give free radicals.

  * **Common error alert:** *Hydrogen peroxide (H_2O_2) and alkyl hydroperoxides (ROOH) are* not *free-radical initiators.* The ·OH radical is too high in energy to be produced by homolytic cleavage of these compounds at ordinary temperatures.

• O_2, a stable 1,2-diradical, can abstract H· from organic compounds to generate radicals, or it can react with Et_3B to give $Et_2BOO\cdot$ and Et·. Remember that *air* contains about 20% O_2!

• Visible light ($h\nu$) has sufficient energy to cause weak σ bonds, such as the Br−Br bond to cleave, or to promote an electron from the highest occupied MO of a compound (HOMO), usually a π bond, to the LUMO, generating a 1,2-diradical. Even ambient light may suffice to initiate a free-radical chain reaction if there is a sufficiently weak bond in the substrate (e.g., C−I).

• Weak σ bonds, especially heteroatom–heteroatom σ bonds (O–O, N–O, etc.) and strained σ bonds, can undergo homolytic cleavage simply upon heating.

Some pericyclic reactions require light, so do not assume that a reaction requiring light must proceed by a free-radical mechanism. Also, a few free-radical chain reactions do not require added initiators. These reactions will be discussed as they arise.

Not all free-radical mechanisms are chain reactions, and those that are not do not require initiators. Reductions with metals such as Li, Na, or SmI_2 (often in liquid NH_3) and light-promoted rearrangements of carbonyl compounds proceed by *nonchain* free-radical mechanisms. Compounds containing weak σ bonds (typically either heteroatom–heteroatom bonds or very strained bonds) can undergo intramolecular rearrangements by nonchain free-radical mechanisms upon heating. Techniques with which you can distinguish chain and nonchain mechanisms are discussed in Chapter 5.

1.6.3 Pericyclic Mechanisms

Identifying pericyclic reactions takes care. They may be executed under acidic, basic, or neutral conditions, just like polar reactions. Many reactions involve several polar steps as well as one pericyclic step. Also, sometimes it's quite difficult to figure out the relationship between the starting materials and the products because of the extensive changes in bonding patterns that often occur with pericyclic reactions.

Some of the key features of pericyclic mechanisms are as follows:

• Pericyclic reactions involve the formation or cleavage of at least one π bond. Often more than one π bond is involved, and often either the starting material or product has two conjugated π bonds.

• Pericyclic reactions are *stereospecific*. When you start with a trans double bond, you get exactly one diastereomeric product; if you start with the cis double bond, you get the other diastereomer.

• There are no charged or electron-deficient intermediates in pericyclic reactions. They used to be called, somewhat facetiously, "no-mechanism" reactions. However, a mechanism may involve several polar steps as well as one or more pericyclic steps, and in such cases, charged and electron-deficient intermediates may abound.

Another reason why it is sometimes difficult to identify conclusively whether a reaction proceeds by a pericyclic mechanism is that it is usually possible to write very reasonable alternative free-radical or polar mechanisms for the same reaction. In this text, when both pericyclic and nonpericyclic mechanisms can be drawn for a reaction, the pericyclic mechanism will be assumed to be correct unless there is theoretical or experimental evidence to the contrary.

Some pericyclic reactions require light to proceed or give a particular stereochemical result only in the presence of light. Thus, the use of light may indicate either a free-radical or a pericyclic mechanism. If the bonding changes can be represented by electrons moving around in a circle, then a light-catalyzed reaction usually proceeds by a pericyclic mechanism.

1.6.4   Transition-Metal-Catalyzed and Mediated Mechanisms

Some very widely used organic reactions are catalyzed or mediated by transition metals. For example, catalytic hydrogenation of alkenes, dihydroxylation of alkenes, and the Pauson–Khand reaction require Pd, Os, and Co complexes, respectively. The d orbitals of the transition metals allow the metals to undergo all sorts of reactions that have no equivalents among main-group elements. This doesn't mean that the mechanisms of transition-metal-mediated reactions are difficult to understand. In fact, in some ways they are easier to understand than standard organic reactions. A transition-metal-catalyzed or -mediated reaction is identified by the presence of a transition metal in the reaction mixture.

Some transition-metal complexes are used simply as Lewis acids in organic reactions. Such reactions should be classified as polar acidic, not as transition-metal-mediated. Common Lewis acidic transition-metal complexes include $TiCl_4$, $FeCl_3$, $AlCl_3$, AgOTf, $ZnCl_2$, and $CeCl_3$. Also, a handful of transition-metal complexes are used as one-electron reducing or oxidizing agents including $FeCl_2$, $TiCl_3$, SmI_2, and $(NH_4)_2Ce(NO_3)_6$ (CAN). Reactions that use these compounds are usually best classified as free-radical reactions.

1.7   Summary

Getting started is usually the most difficult part of drawing a mechanism. Follow these simple steps, and you will be well on your way.

• *Label* the heavy atoms in starting material and product.

• *Make a list of which σ bonds* between non-H atoms are made or broken in going from starting material to product. *Do not list π bonds or bonds to H,* as they are easier to make and break at will.

• *Balance* the reaction.

• *Classify the overall reaction* by looking at the starting materials and products, including by-products. Is the reaction an addition, an elimination, a substitution, or a rearrangement? Some reactions may combine two or more of these types.

• *Classify the mechanism* by looking at the reaction conditions. Is it polar under basic conditions, polar under acidic conditions, free-radical, pericyclic, or metal-mediated? Some reactions, especially pericyclic ones, may combine two or more of these types.

Once you have classified the overall reaction and the mechanism, you may have a short series of mechanistic choices. For example, substitution of an aromatic ring under basic conditions generally occurs by one of three mechanisms.

• If the mechanism is polar, determine the nucleophilicity, electrophilicity, and acidity of the atoms to which σ bonds are made and broken. Under basic conditions, the first step is often deprotonation of an acidic atom, rendering it nucleophilic. Under acidic conditions, the first step may be protonation of a leaving group or a π bond.

When faced with a mechanism problem, students often ask, How does one *know* that a particular nucleophile attacks a particular electrophile to give a particular product? The answer to this question is relatively straightforward. A mechanism is a story that you tell about how compound **A** is transformed into compound **B**. To tell the story, you need to know what the product is! If you do not know what the product is, your story will sound something like the game in which 15 children in turn add one paragraph to a continuing story, and the final version wanders all over the known universe. Sometimes the product will be a minor or unexpected product, but a mechanism problem will always give you the product. Organic chemists do need to learn how to predict the product of a reaction, but this skill is deemphasized in this text, whose purpose is to help you to learn how to write mechanisms. Only occasionally will you be asked to predict the course of a reaction.

PROBLEMS

Chapters 1–3 in Daniel P. Weeks's *Pushing Electrons: A Guide for Students of Organic Chemistry*, 3rd ed. (Saunders College Publishing, 1998) are *strongly* recommended for refreshing your skills in drawing Lewis structures, drawing resonance structures, and using curved arrows in simple mechanism problems.

1. Explain each of the following observations.

 (a) Amides (R_2NCOR) are much more nucleophilic on O than they are on N.

 (b) Esters are much less electrophilic at C than ketones.

 (c) Acyl chlorides (RCOCl) are much more acidic than esters.

 (d) Compound **1** has a much larger dipole moment than its isomer **2**.

 (e) Compound **3** is much more acidic than **4**.

 (f) Imidazole (**5**) is considerably more basic than pyridine (**6**).

(g) Fulvene (**7**) is electrophilic at the exocyclic C atom.

(h) Cyclohexadienone (**8**) is much more prone to tautomerize than most carbonyl compounds. (*Note*: Tautomerizations of carbonyl compounds are almost always very fast, so this question is about thermodynamic propensities, not kinetic ones.)

(i) Cyclopentadienone (**9**) is extremely unstable.

(j) The difference between the pK_a values of PhSH and EtSH is much smaller than the difference between those of PhOH and EtOH.

(k) Furan (**10**) attacks electrophiles exclusively at C2, not at C3.

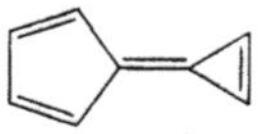

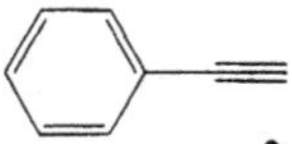

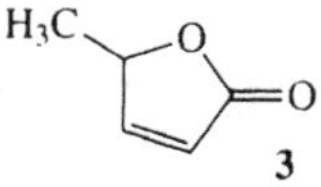

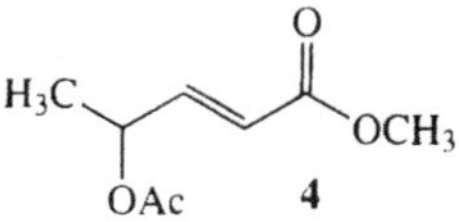

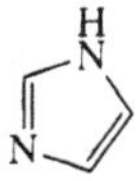

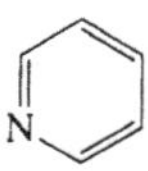

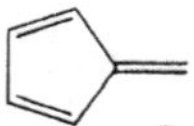

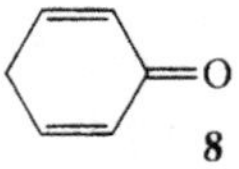

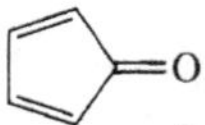

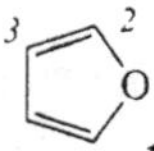

2. Indicate which of each pair of compounds is likely to be more acidic and why.

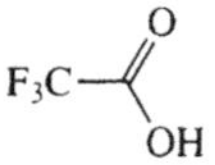

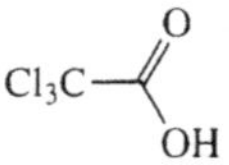

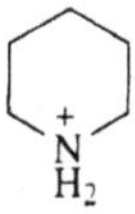

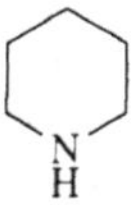

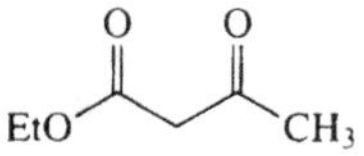

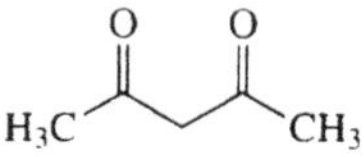

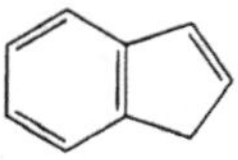

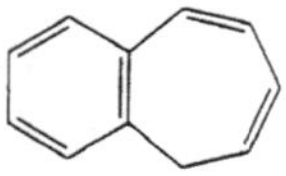

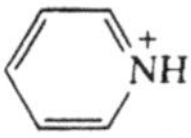

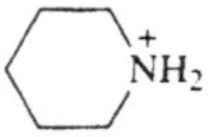

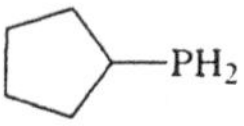

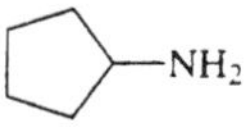

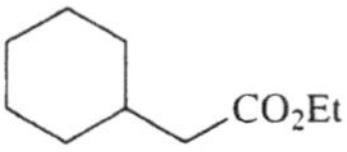

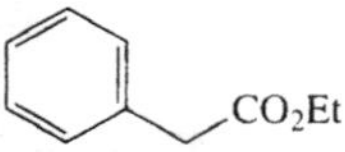

(h) 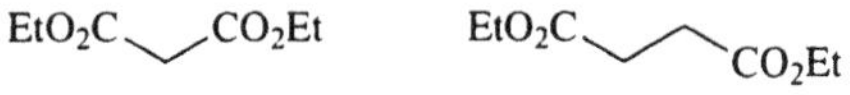
EtO2C
CO2Et
EtO2C
CO2Et

(i) 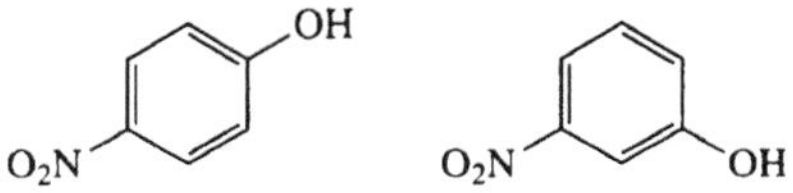
OH
O2N
O2N
OH

(j)
O
H3C
OH
O
H3C
NH2

(k)
Ph
CH3
Ph
H

(l)
O
O

(m) 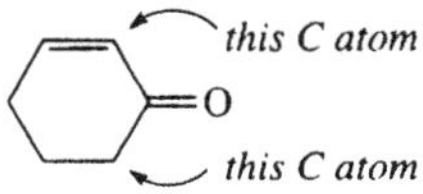
this C atom
O
this C atom

3. Classify each of the following reactions as polar, free-radical, pericyclic, or transition-metal-catalyzed or -mediated. For the polar reactions, determine whether the conditions are basic or acidic.

(a) 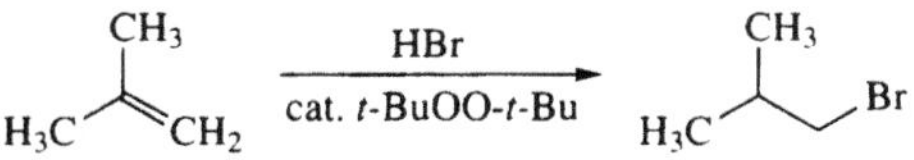
CH3
H3C
CH2
HBr
cat. t-BuOO-t-Bu
CH3
H3C
Br

(b) 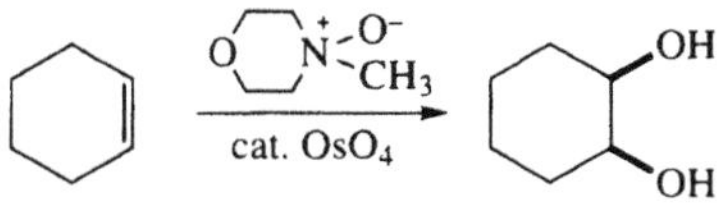
O
N
O−
CH3
cat. OsO4
OH
OH

(c) 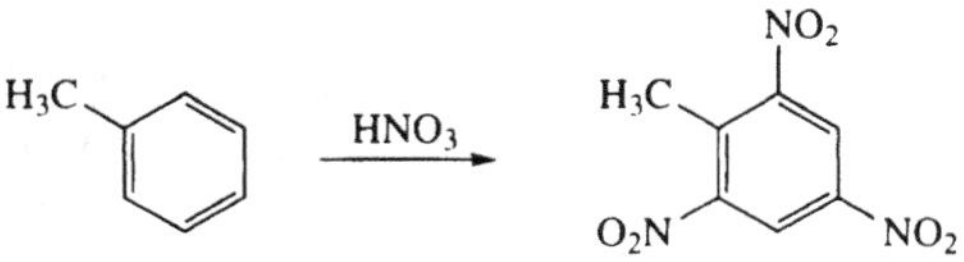
H3C
HNO3
NO2
H3C
O2N
NO2
(TNT)

(d) 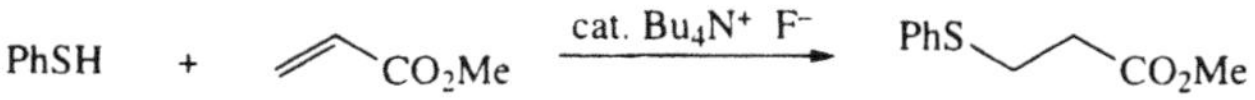
PhSH
+
CO2Me
cat. Bu4N+ F−
PhS
CO2Me

(e) 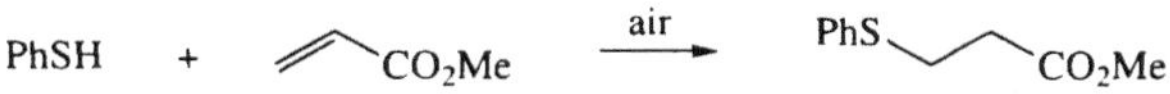
PhSH
+
CO2Me
air
PhS
CO2Me

(f) 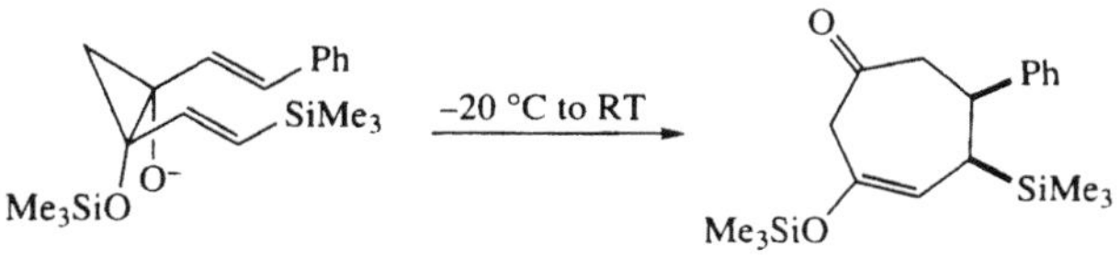
Ph
SiMe3
O−
Me3SiO
−20 °C to RT
O
Ph
SiMe3
Me3SiO

(g)

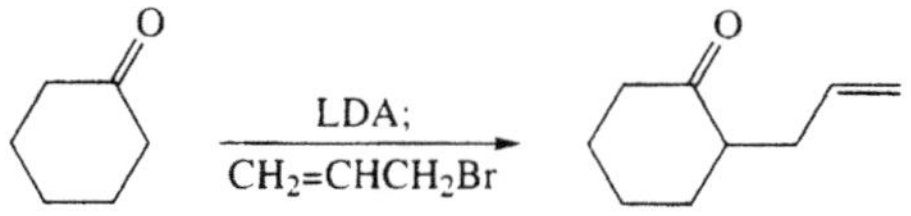

$$\xrightarrow[\text{CH}_2\text{=CHCH}_2\text{Br}]{\text{LDA;}}$$

(h)

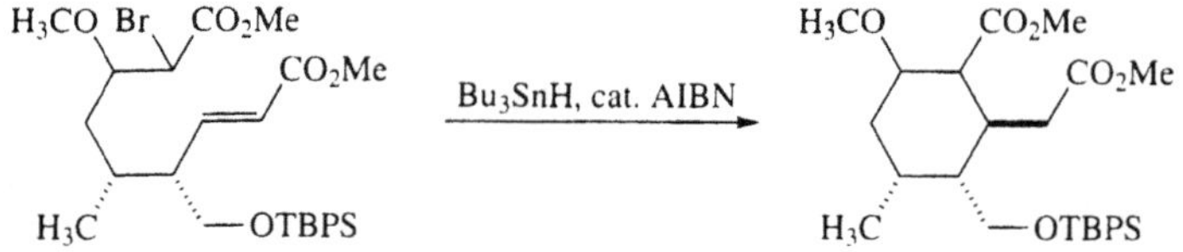

$$\xrightarrow{\text{Bu}_3\text{SnH, cat. AIBN}}$$

(i)

(j)

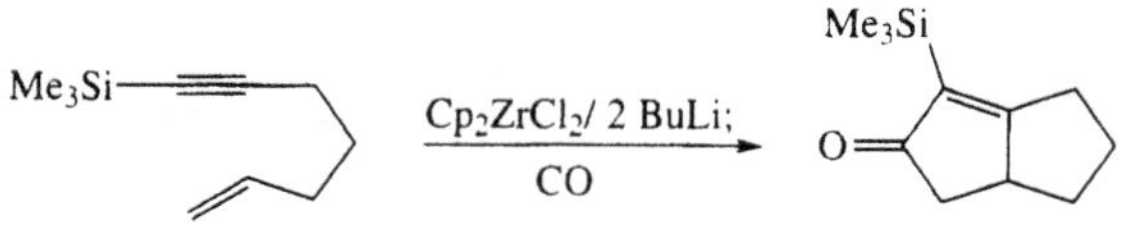

$$\xrightarrow[\text{CO}]{\text{Cp}_2\text{ZrCl}_2/\ 2\ \text{BuLi;}}$$

(k)

$$2\ \text{CH}_2\text{=CHCHO} \xrightarrow{\Delta}$$

(l)

$$\xrightarrow{\text{NaOEt}}$$

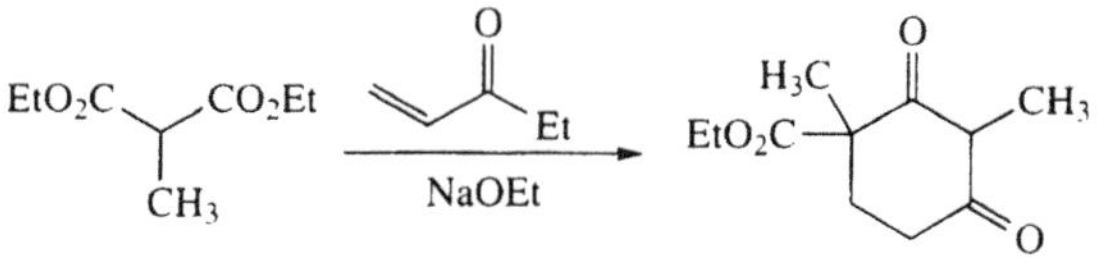

(m)

$$\xrightarrow{\Delta}$$

4. Most of the heavy atoms in the starting material(s) in each of the following reactions are numbered. Classify each reaction as polar acidic, polar basic, pericyclic, or free-radical. Then, number the atoms in the product(s) appropriately, and make a list of bonds made and broken between heavy atoms. Assume aqueous workup in all cases.

(a)

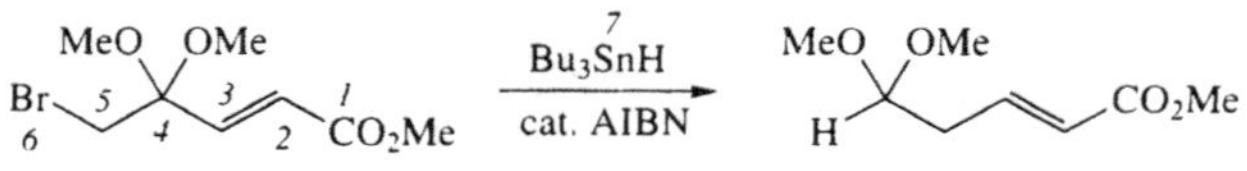

$$\xrightarrow[\text{cat. AIBN}]{\text{Bu}_3\text{SnH}}$$

(b)

$$\xrightarrow{\text{AgBF}_4}$$

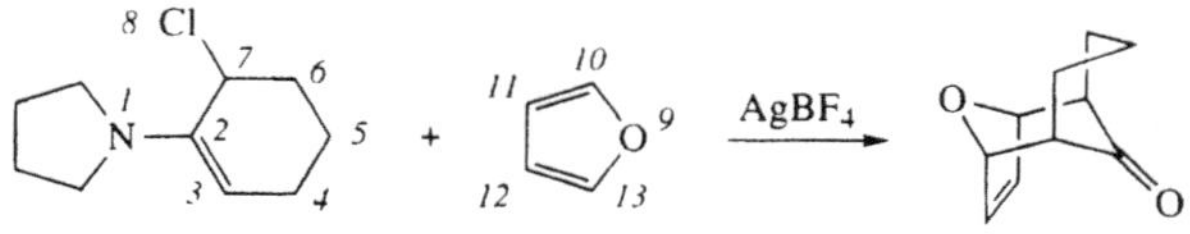

(c)

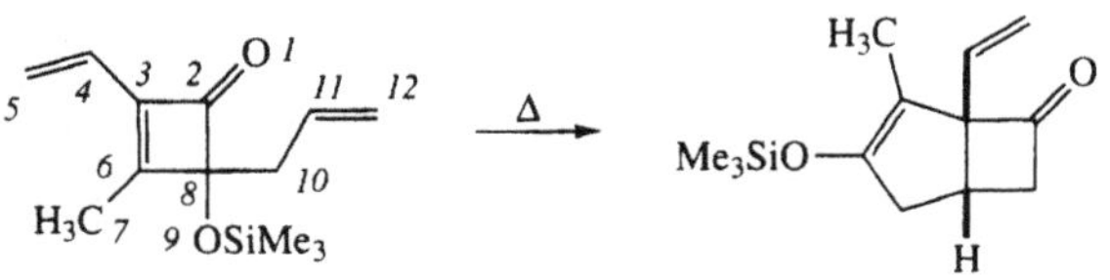

(d)

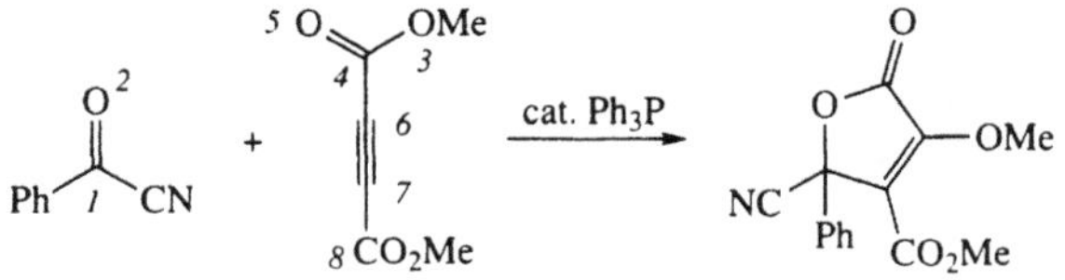

(e)

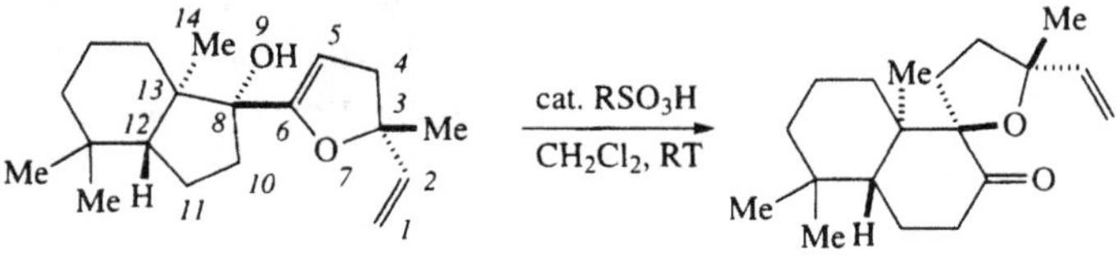

(f)

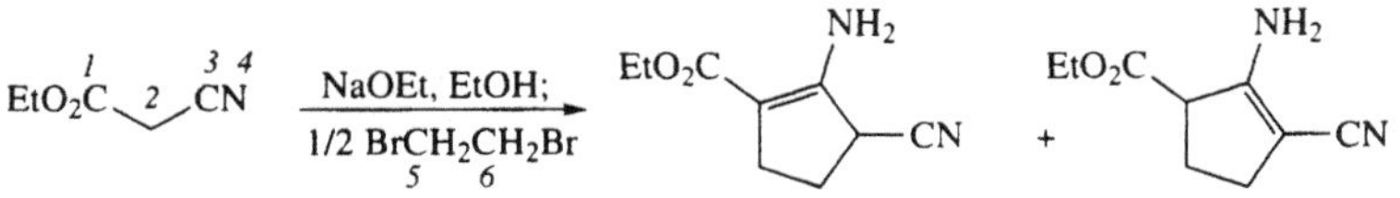

(g)

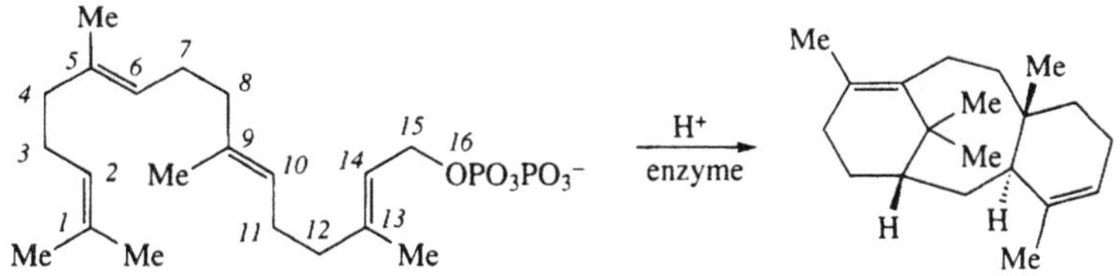

Geranylgeranyl pyrophosphate   *A Taxane*

(h)

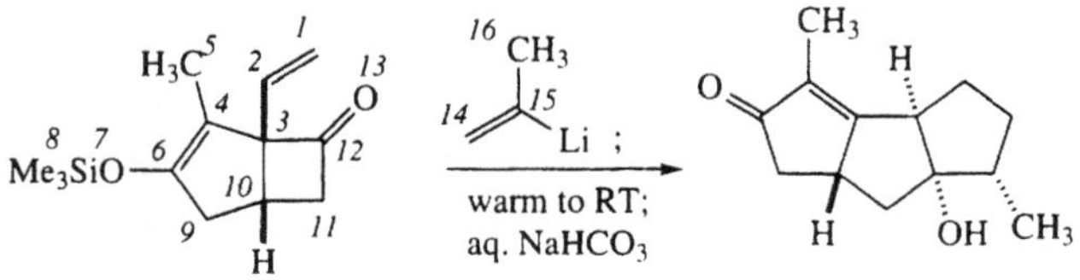

(i)

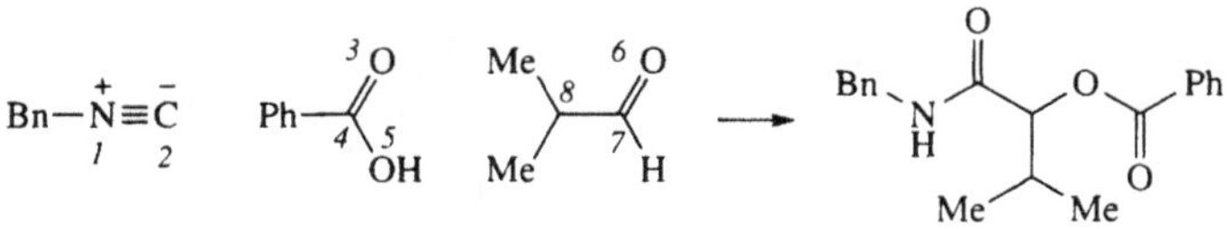

(j)

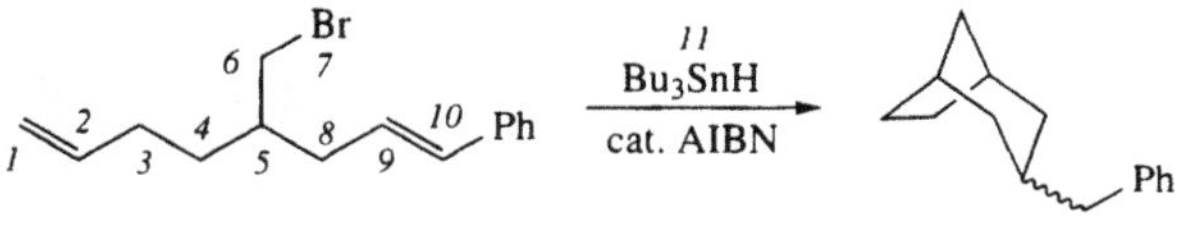

(k)

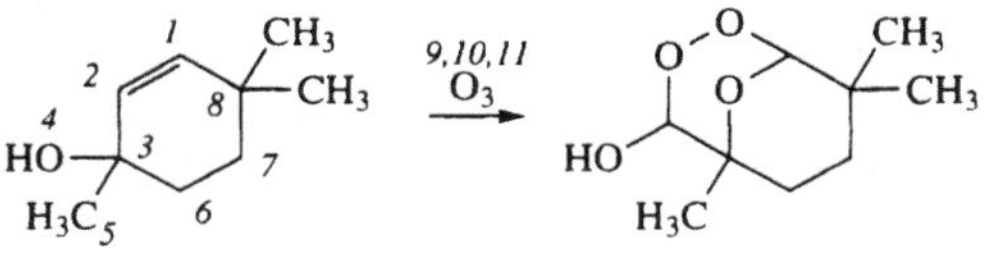

48   1. The Basics

(l)

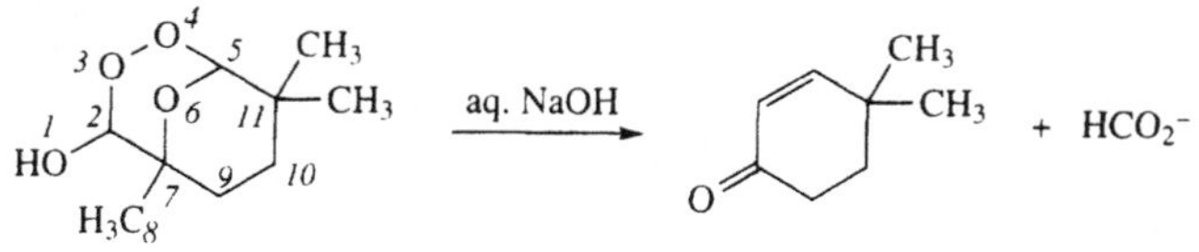

aq. NaOH →

(m)

Br₂ →

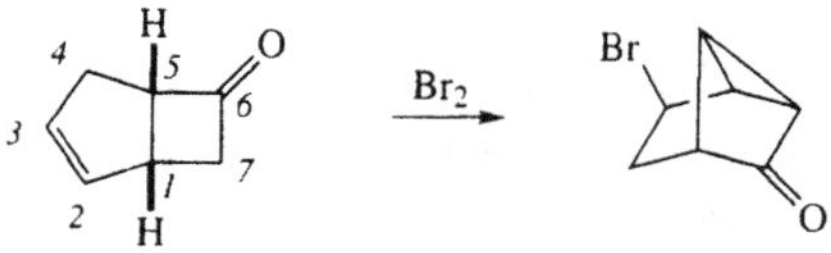

5. In each of the following compounds, a particular atom is indicated with an arrow. Determine whether this atom is nucleophilic, electrophilic, acidic. Some atoms may have none or more than one of these properties. For the purposes of this problem, "acidic" is defined as $pK_a \leq 25$.

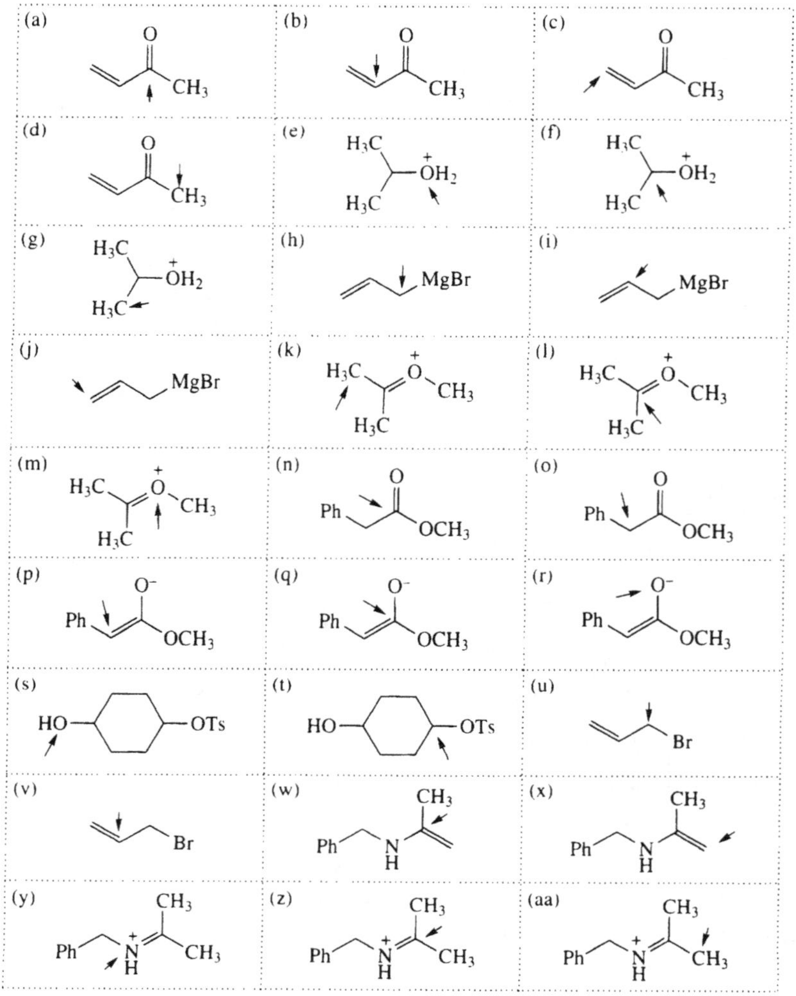

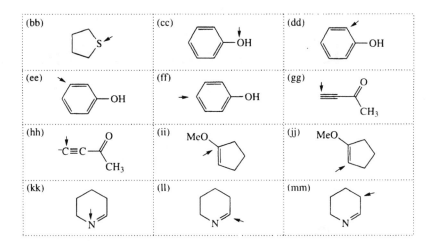

2

Polar Reactions under Basic Conditions

Now that you have learned how to identify the mechanistic class of a particular reaction, you are ready to learn how to write the mechanism itself. This chapter is devoted to polar mechanisms that occur under basic conditions. Archetypical mechanisms will be presented and discussed so that you become familiar with the patterns of reactivity of organic compounds under basic conditions.

Many of the reactions that will be discussed in this and later chapters are "name reactions." Name reactions are used as examples in this book simply because they are widely used reactions, not because you need to know the name of any reaction to be able to draw a mechanism for it. In fact, one of the goals of this text is to show you that you can draw a mechanism for any reaction *whether or not you have seen it previously.*

2.1 Substitution and Elimination at C(sp³)–X σ Bonds, Part I

Electrophiles that have leaving groups (X) attached to $C(sp^3)$ usually undergo *substitution* or *elimination* reactions. In a *nucleophilic substitution* reaction, a nucleophile–electrophile σ bond replaces the electrophile–X σ bond.

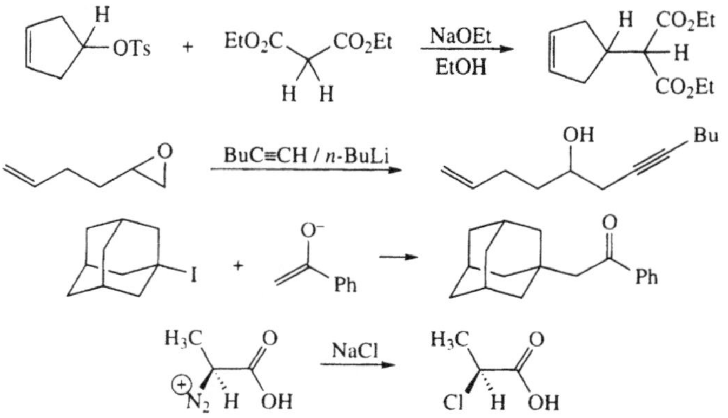

In an *elimination* reaction, the leaving group X (with its associated pair of electrons) and an electropositive group E (usually H) are removed from a compound. The most common kind of elimination is β-elimination, in which E and X are attached to adjacent atoms and a new π bond is formed between those atoms in the product.

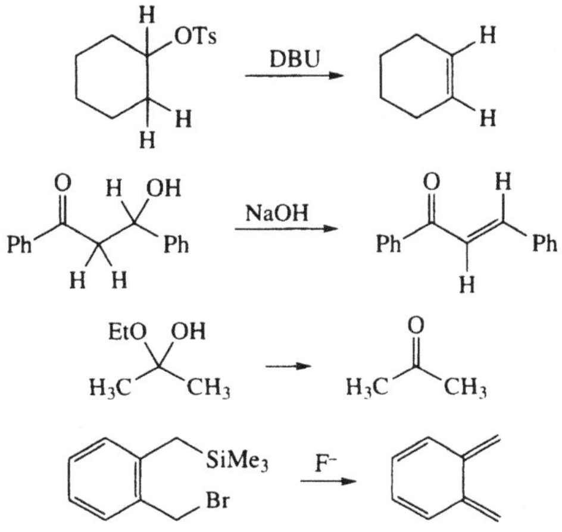

2.1.1 Substitution by the S_N2 Mechanism

Substitution reactions at 1° and 2° (but not 3°) C(sp³) usually proceed by the S_N2 *mechanism* under basic or neutral conditions. In the S_N2 mechanism, the nucleophile Nu⁻ approaches the electrophilic center opposite X and in line with the C–X bond. The lone pair on Nu⁻ is used to form the C–Nu bond, and the pair of electrons in the C–X bond simultaneously leaves with X as the bond breaks. The other three groups on C move away from Nu and toward X as the reaction proceeds, so that when the reaction is complete, the stereochemistry of C is inverted. The charges do not have to be as shown: the nucleophile may be anionic or neutral, and the electrophile may be neutral or cationic.

The curved arrow shows how the pair of electrons moves from the nucleophile to the electron-deficient center. The nucleophilic atom increases its formal charge by 1, and the leaving group decreases its formal charge by 1. The minus sign on Nu⁻ indicates both a formal charge and a pair of electrons. When the nucleophile is uncharged, the lone pair is usually drawn.

* **Common error alert:** *Intermolecular substitution by the S_N2 mechanism occurs only at 1° and 2° C(sp³).* Under basic conditions, intramolecular substitution at 3° C(sp³) may occur by the S_N2 mechanism, but substitution at C(sp²) never occurs by the S_N2 mechanism at all. Substitution at an electrophilic C(sp²) or intermolecular substitution at 3° C(sp³) *must* proceed by a mechanism other than S_N2 (see below).

* **Common error alert:** *F^-, HO^-, RO^- (except epoxides), H^-, and carbanions are almost never leaving groups in S_N2 reactions under basic conditions.*

Lone-pair nucleophiles are by far the most enthusiastic participants in S_N2 substitution reactions. Sigma-bond nucleophiles may also participate in S_N2 reactions, but they do not do so as often as lone-pair nucleophiles. By contrast, π-bond nucleophiles do not usually have sufficiently high energy to react with an atom that already has an octet. The major exceptions to this rule are the enamines ($R_2N-CR=CR_2 \longleftrightarrow R_2N=CR-\bar{C}R_2$), which are sufficiently nucleophilic at the β position to attack particularly reactive alkyl halides such as CH_3I and allylic and benzylic bromides, and enolates ($\bar{O}-CR=CR_2 \longleftrightarrow O=CR-\bar{C}R_2$), which react with many alkyl halides.

Electrophiles with allylic leaving groups can undergo either S_N2 or S_N2' substitution. In S_N2' substitution, the lone pair on Nu^- moves to form a bond to the γ-carbon of the allylic system. (The α-carbon is the C atom attached to the leaving group.) The electrons in the π bond between the β- and γ-carbon atoms move to form a new π bond between the β- and α-carbon atoms, and the leaving group is concomitantly expelled. The double-bond *transposition* is key to identifying an S_N2' substitution mechanism. It is very difficult to predict whether a particular allylic halide or pseudohalide will undergo S_N2 or S_N2' substitution with a particular nucleophile.

S_N2 substitution can occur at elements other than C. For example, substitution at a stereogenic S atom leads to inversion of configuration, suggesting an S_N2 mechanism for this process.

In contrast with first-row main-group elements, second-row and heavier atoms such as P and S can extend their octet, so substitution at these atoms can occur either by an S_N2 mechanism or by a two-step *addition–elimination* mechanism. In the first step, the nucleophile adds to the electrophilic heavy atom to give a hypervalent, 10-electron intermediate. In the second step, the nucleofuge leaves to give the substitution product. The lifetime of the 10-electron intermediate can vary from extremely short to very long, so both the S_N2 and the addition–elimination mechanisms are usually reasonable for S, P, or heavier elements.

addition–elimination mechanism for substitution at heavy elements

Not all substitution reactions under basic conditions occur with simple inversion. Sometimes, nucleophilic substitution at stereogenic C proceeds with *retention* of configuration. In such a reaction, two sequential nucleophilic substitutions have usually occurred. For example, the reaction of α-diazoniocarboxylic acids with nucleophiles such as Cl^- occurs with retention. In the first S_N2 substitution, the O^- of the carboxylate acts as a nucleophile, displacing N_2 with inversion to give a three-membered ring (an α-lactone). In the second S_N2 substitution, Cl^- displaces O with inversion to give the product with overall retention.

Sometimes, nucleophilic substitution at stereogenic C proceeds with *loss* of configurational purity. Such a reaction may be rationalized mechanistically in one of four ways.

1. The stereocenter in the starting material loses configurational purity before the S_N2 reaction occurs. For example, stereochemically pure 2-bromocyclohexanone may lose its configurational purity by deprotonation and reprotonation of the acidic stereocenter.

2. The stereocenter in the product loses configurational purity after the S_N2 reaction occurs.

3. The nucleophile can also act as a leaving group, or vice versa. This problem is especially severe with reactions involving I^-, which is both an excellent nucleophile and an excellent leaving group. The preparation of alkyl iodides such as 2-iodobutane in enantiopure form is difficult for this very reason.

4. The substitution may proceed by a non-S_N2 mechanism, e.g., $S_{RN}1$ or elimination–addition. These substitution mechanisms are discussed later in this chapter.

2.1.2 β-Elimination by the E2 and E1cb Mechanisms

$C(sp^3)−X$ electrophiles can undergo β-elimination reactions as well as substitutions. β-Elimination reactions proceed by the *E2* or *E1cb* mechanism under basic conditions. The concerted E2 mechanism is more common. The lone pair of the base moves to form a bond to a H atom on a C atom *adjacent to the electrophilic C atom*. The electrons in the H−C bond simultaneously move to form a π bond to the electrophilic C atom. Because this C atom already has its octet and is acquiring a new bond, the bond to the leaving group must break. The electrons in that bond leave with the leaving group to become a new lone pair.

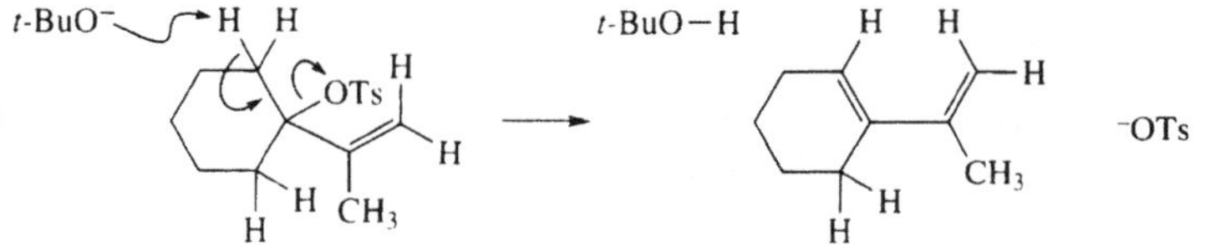

Because the H–C and the C–X bonds break simultaneously in the E2 mechanism, there is a *stereoelectronic requirement* that the orbitals making up these two bonds be *periplanar*, (i.e., parallel to each other), in the transition state of the elimination. Acyclic compounds can orient the two bonds in a periplanar fashion in two ways, *synperiplanar* (by eclipsing them at a 0° dihedral angle) and *antiperiplanar* (by staggering them at a 180° angle). The staggered arrangement is much lower in energy than the eclipsed arrangement, so antiperiplanar elimination is far more common than synperiplanar elimination.

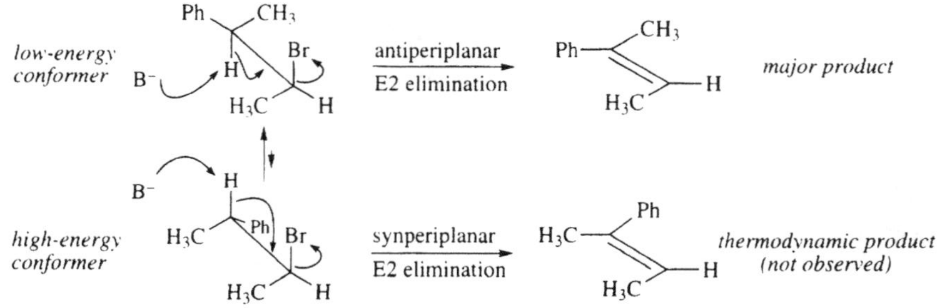

In cyclic compounds, there are much greater restrictions on conformational flexibility. In six-membered rings, the antiperiplanar requirement for E2 elimination is satisfied when both the leaving group and the adjacent H atom are *axial*. Compounds in which such a conformation is readily achievable undergo E2 elimination much more readily than those in which it is not. For example, in menthyl chloride, the C–Cl bond is not antiperiplanar to any adjacent C–H bond in the lowest energy conformation, and E2 elimination is therefore much slower than it is in the diastereomer neomenthyl chloride, in which the C–Cl bond is antiperiplanar to two C–H bonds in the lowest energy conformation. Moreover, two C–H bonds are antiperiplanar to the C–Cl bond in the reactive conformation of neomenthyl chloride, so two products are obtained upon E2 elimination, whereas only one C–H bond is antiperiplanar to the C–Cl bond in the reactive conformation of menthyl chloride, so only one product is obtained.

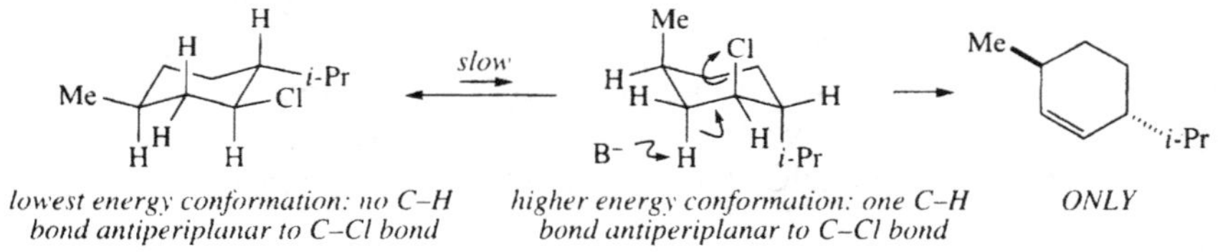

lowest energy conformation: no C–H bond antiperiplanar to C–Cl bond   *higher energy conformation: one C–H bond antiperiplanar to C–Cl bond*   *ONLY*

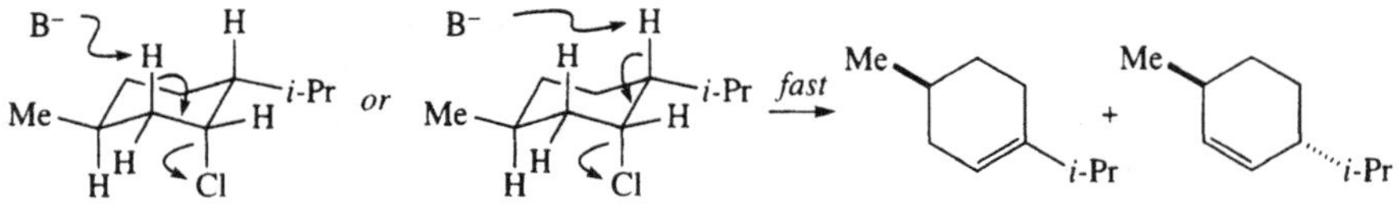

*lowest energy conformation: two C–H
bonds antiperiplanar to C–Cl bond*

Not only can 1° and 2° C(sp³)–X undergo β-elimination by the E2 mechanism under basic conditions, but so can 3° C(sp³)–X systems. Alkenyl halides
also undergo β-elimination readily. When there are H atoms on either side of an
alkenyl halide, either an alkyne or an allene may be obtained. Even alkenyl ethers
(enol ethers) can undergo β-elimination to give an alkyne.

Example

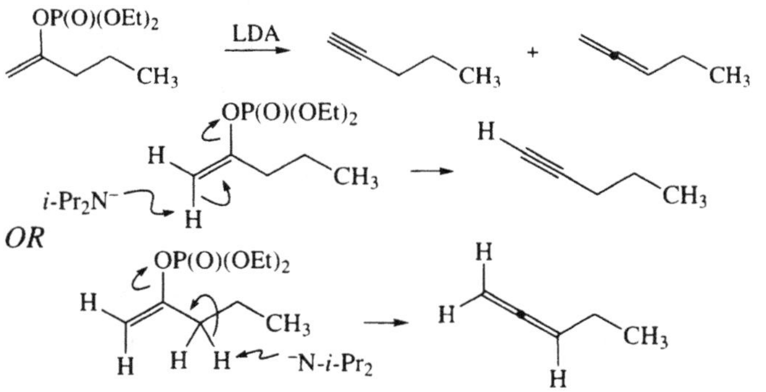

Problem 2.1. Draw a mechanism for the following reaction.

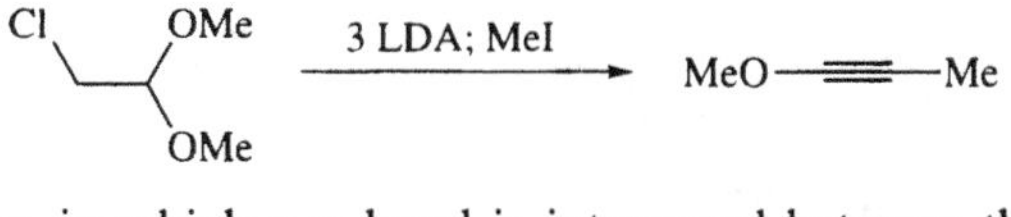

E2′ eliminations, in which a π bond is interposed between the two C atoms
at which bond breaking occurs, are also seen. In the following example, the base
is F⁻, and a Me₃Si group replaces the usual H.

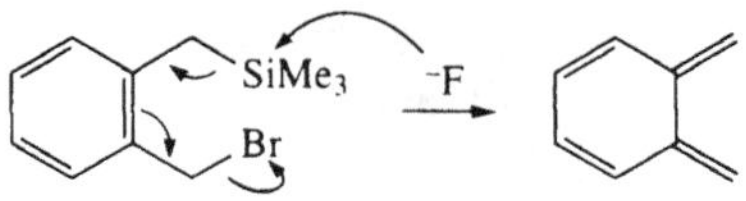

β-Elimination sometimes gives high-energy species. Under very strongly basic
conditions, halobenzenes undergo β-elimination to give *benzynes*, compounds that
are highly strained and reactive. β-Elimination from acyl chlorides occurs under
mildly basic conditions to give *ketenes*. Both of these reactions will be revisited later.

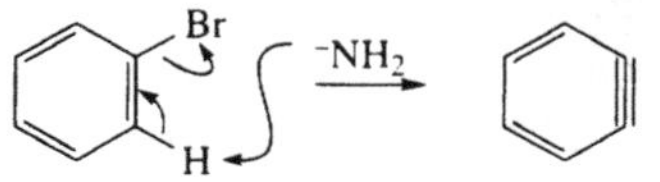

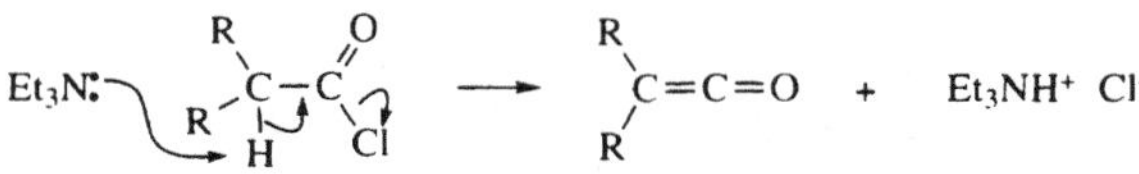

When the H is particularly acidic (usually because it is adjacent to a carbonyl) and the leaving group is particularly poor (especially $^-$OH and $^-$OR), a two-step mechanism called *Elcb* operates. In this mechanism, the acidic proton is removed first to make a stabilized carbanion. Then the lone pair on C moves to make a π bond to the neighboring electrophilic C atom, expelling the leaving group. The dehydration of an aldol (β-hydroxycarbonyl compound) is the most common example of an elimination reaction that proceeds by the Elcb mechanism.

The Elcb elimination is the usual mechanism by which a hemiacetal is converted to a carbonyl compound under basic conditions.

2.1.3 Predicting Substitution vs. Elimination

In principle, any compound with a lone pair can act as either a base or a nucleophile toward $C(sp^3)-X$, causing either E2 elimination or S_N2 substitution to occur. It is possible to predict whether substitution or elimination will occur with moderate accuracy (Table 2.1). Two factors largely determine the course of the reaction: (1) the nucleophilicity and basicity of the lone-pair-bearing compound, and (2) the identity of the substrate: Me or Bn, 1°, 2°, or 3° halide.

* **Common error alert:** *Whether substitution or elimination occurs is the outcome of a "negotiation" between the nucleophile and the electrophile. Neither the nucleophile nor the electrophile determines the outcome exclusively!*

The first factor that determines whether substitution or elimination occurs is the nucleophilicity and basicity of the lone-pair-bearing compound. The factors that influence nucleophilicity and basicity have already been discussed (Chapter 1). Nucleophiles–bases can be classified very simplistically into good nucleophiles–poor bases, good nucleophiles–good bases, and poor nucleophiles–good bases toward $C(sp^3)$ halides.

TABLE 2.1. Predicting substitution vs. elimination at $C(sp^3)-X$ under basic conditions

	Poor base, good nucleophile	Good base, good nucleophile	Good base, poor nucleophile
Me or Bn	S_N2	S_N2	S_N2 or N.R.
1°	S_N2	S_N2	E2
2°	S_N2	$S_N2 < E2$	E2
3°	E2, $S_{RN}1$, or N.R.	E2	E2

• *Second-row or heavier species and very stable carbanions are good nucleophiles and poor bases.* Br^-, I^-, R_2S, RS^-, and R_3P are in this class, as are ^-CN, stabilized enolates such as malonate anions and their derivatives, and certain organometallic compounds such as cuprates (R_2CuLi).

• *First-row, unhindered species and moderately stable carbanions are both good nucleophiles and good bases.* Unhindered RO^-, R_2N^-, and R_3N are in this class, as are enamines, simple enolates, and alkynyl anions ($RC{\equiv}C^-$). Cl^- (an honorary member of the first row) is also in this class.

• *First-row, hindered species and unstabilized carbanions are poor nucleophiles and good bases:* t-BuO^-, i-Pr_2NLi (LDA), $(Me_3Si)_2NK$ (KHMDS), i-Pr_2NEt (Hünig's base), and amidines and guanidines such as DBN, DBU, and TMG are in this class, as are $C(sp^2)$ and $C(sp^3)$ Grignard and organolithium reagents. NaH and KH are also in this class. F^- is a good nucleophile in aprotic solvents and a poor nucleophile in protic solvents, but it is always a good base.

Some common amidine and guanidine bases that are relatively nonnucleophilic:

DBU   DBN   TMG

Obviously the division of nucleophiles and bases into "good" and "bad" is a gross oversimplification. There is a continuum of nucleophilicity and basicity that runs from "excellent" through "good," "mediocre," "poor," and "awful." The simplistic classification given above merely provides a *guideline* for prediction of reactivity.

* **Common error alert:** *Don't forget that "good nucleophile" is defined with respect to reactivity toward CH_3Br at 25 °C in MeOH as solvent.* Some species, e.g., CH_3MgBr, that are not good nucleophiles toward σ-bond electrophiles are quite good nucleophiles toward π-bond electrophiles. For a discussion of how the properties of the electrophile influence the nucleophilicity of the nucleophile (hard/soft acid/base theory), see Fleming, Ian, *Frontier Orbitals and Organic Chemical Reactions* (New York: John Wiley & Sons, 1976), pp. 34ff.

The second factor that determines whether substitution or elimination occurs is the structure of the electrophile.

• MeX and BnX cannot undergo elimination (no β-hydrogens) and are good substrates for S_N2 (unhindered; the Bn group also stabilizes the TS for substitution), so these substrates undergo substitution only.

• 1° Alkyl halides love to do S_N2 (relatively unhindered) and don't like to do E2 (relatively unsubstituted alkene product, hence not very low in energy), so they tend to undergo substitution if the nucleophile is a good one, regardless of its basicity. If, however, the nucleophile is really poor but is a good base, they will consent to elimination.

• 3° Alkyl halides hate to do S_N2 (too hindered) and love to do E2 (highly substituted and low-energy alkene product, plus relief of steric congestion), so they tend to undergo elimination if the base is a good one at all, regardless of its nucleophilicity. If, however, the base is really poor but is a very good nucleophile, they might undergo substitution by the $S_{RN}1$ mechanism (see below), or there may be no reaction.

• 2° Alkyl halides are at the cusp. In general, they are perfectly happy to undergo either elimination or substitution, depending on whatever the base–nucleophile prefers.

If neither the electrophile (2° halide) nor the nucleophile (good base–good nucleophile) has a strong preference for substitution or elimination, a situation arises in which each asks the other, "I dunno, whadda you wanna do?" In this case, elimination tends to dominate over substitution, but the actual product ratio varies tremendously depending on substrate structure, solvent, and other factors. The proportion of S_N2 increases greatly for allylic, benzylic, and propargylic ($C\equiv C-C-X$) 2° halides. The portion of S_N2 can also be increased by executing the reaction in a polar aprotic solvent or by reducing the basicity of the nucleophile. E2 is more prone to occur for cyclic 2° halides than for acyclic 2° halides, but this tendency is ameliorated in polar aprotic solvents. Homoallylic substrates ($C=C-C-C-X$) and homopropargylic substrates ($C\equiv C-C-C-X$), regardless of their substitution pattern, are more prone to undergo elimination than other substrates due to the activation of the allylic or propargylic C–H bond toward deprotonation.

* **Common error alert:** *The leaving group ability of X⁻ affects the* overall rate *of both S_N2 and E2 reactions—the better the leaving group, the faster the rates— but it has relatively little influence on which of these pathways is the major one.* To determine whether S_N2 or E2 dominates, you must consider the factors enumerated above.

2.2   Addition of Nucleophiles to Electrophilic π Bonds

2.2.1   Addition to Carbonyl Compounds

Carbonyl compounds C=O have two major resonance structures, $R_2C=O \longleftrightarrow R_2\overset{-}{C}-\overset{}{O}$. In the second resonance structure, C is electron-deficient, so carbonyl compounds are good electrophiles. Carbonyl groups with α-hydrogen atoms are relatively acidic compounds, because the carbanions produced upon removal of the H are stabilized by resonance with the carbonyl group: $O=CR-\overset{-}{C}R_2 \longleftrightarrow \overset{-}{O}-CR=CR_2$. The enolate anions that are thereby obtained are nucleophilic at the α-carbon and on O. *Under basic conditions, then, carbonyl compounds are electrophilic at the carbonyl carbon and nucleophilic at the α-carbons (if they have H atoms attached).* All the chemistry of carbonyl compounds is dominated by this dichotomy.

The thermodynamic stabilities of carbonyl compounds are directly related to the stabilities of their $R_2\overset{-}{C}-\overset{}{O}$ resonance structures. The order of thermodynamic stabilities of the common types of carbonyl compounds is RCOCl (acyl chlorides) <

RCO_2COR (acid anhydrides) < RCHO (aldehydes) < R_2CO (ketones) < RCO_2R (acids, esters) < $RCONR_2$ (amides) < $ROCO_2R$ (carbonates) < $ROCONR_2$ (urethanes/carbamates) < R_2NCONR_2 (ureas) ≈ RCO_2^- (carboxylate salts). Addition of a nucleophile to a carbonyl compound gives an intermediate that has relatively little resonance stabilization, so the process costs most of the resonance stabilization of the carbonyl compound. As a result, the order of kinetic stabilities of carbonyl compounds is the same as the order of thermodynamic stabilities.

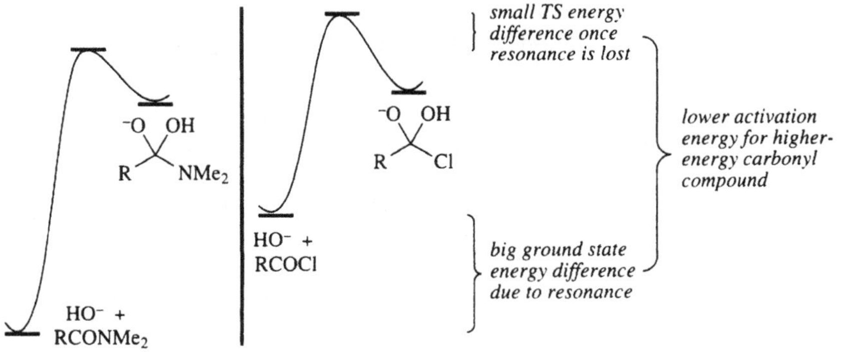

Sigma-bond nucleophiles such as Grignard reagents (RMgBr), organolithium compounds (RLi), and complex metal hydrides such as $NaBH_4$ and $LiAlH_4$ add to ketones and aldehydes to give alcohols. The mechanism of addition to ketones and aldehydes is complex and is still being debated, but a very simple picture suffices here. The electrons in the C–M or H–M σ bond move away from the metal M to form a bond between the nucleophilic atom and the electrophilic carbonyl C. At the same time, the electrons in the C=O π bond move to O to become a lone pair. After workup, an alcohol is obtained.

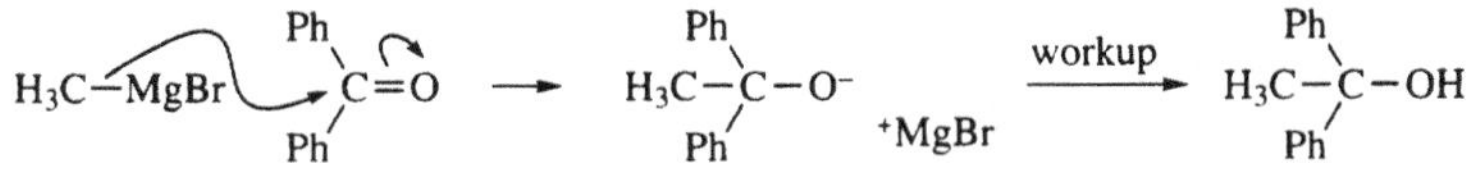

Grignard reagents and organolithium compounds are very strong bases, too, and you might ask how it can be predicted when they will act as bases toward carbonyl compounds and when as nucleophiles. 1,3-Dicarbonyl compounds are so acidic that they are always deprotonated by these compounds, but for simple ketones and aldehydes, the ratio of deprotonation to addition is substrate-dependent. In general, steric hindrance increases the amount of deprotonation that is seen. The reagent $CeCl_3$ greatly enhances the amount of addition, even for highly hindered substrates, for reasons that remain unclear.

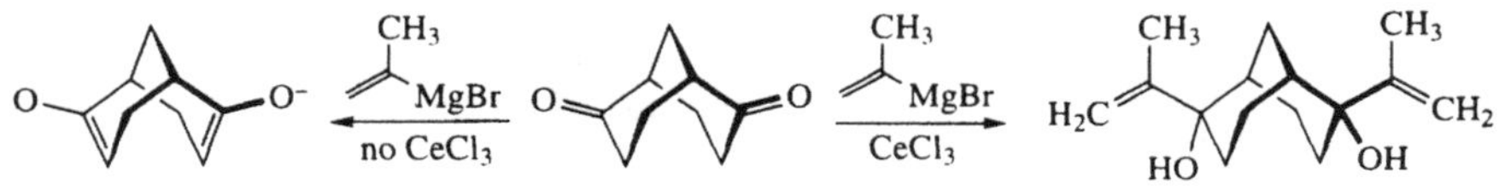

When the α-carbon of an aldehyde or ketone is a stereocenter, and addition of an organometallic nucleophile to the carbonyl group generates a new stereocenter, then selective formation of one diastereomer over the other is often observed. In many cases, the observed stereoselectivities are consistent with *Felkin–Anh* selectivity. The α-stereocenter of the carbonyl group has a large R_L group, a small R_S group, and H attached to it. Consider the different rotamers about the bond between the carbonyl carbon and the α-carbon. (The best way to understand this argument is to use Newman projections.) The nucleophile Nu^- will want to approach the carbonyl carbon from a trajectory that is as far away from R_L as possible. This condition is best satisfied if the $C–R_L$ and $C=O$ bonds are placed at a dihedral angle of 90° and Nu^- approaches the carbonyl group from opposite R_L. Two rotamers satisfy this condition: One has R_S nearly eclipsing the carbonyl substituent R and the other has H nearly eclipsing the carbonyl substituent R. The latter rotamer least hinders the approach of Nu^- to the carbonyl carbon. The predominant diastereomer of the product is then the one in which Nu is anti to R_L and OH is anti to H.

hindered trajectory   *preferred trajectory*

The diastereoselectivities of addition of organometallic compounds to ketones and aldehydes are often quite poor (2:1), and there are numerous cases in which anti-Felkin–Anh selectivity is observed, but the rule is widely invoked anyway.

Water and alcohols add reversibly to ketones and aldehydes under basic conditions to give *hydrates* or *hemiacetals*.

Example

R = H: *hydrate*
R = alkyl or aryl: *hemiacetal*

The equilibrium generally favors both ketones and aldehydes over the corresponding hydrates or hemiacetals, but it favors aldehydes less strongly than it does ketones. The equilibrium is pushed toward the hemiacetal by inductively electron-withdrawing groups on the α-carbon and when a five- or six-membered ring can be formed. Both of these factors are present in glucose, fructose, and other carbohydrates.

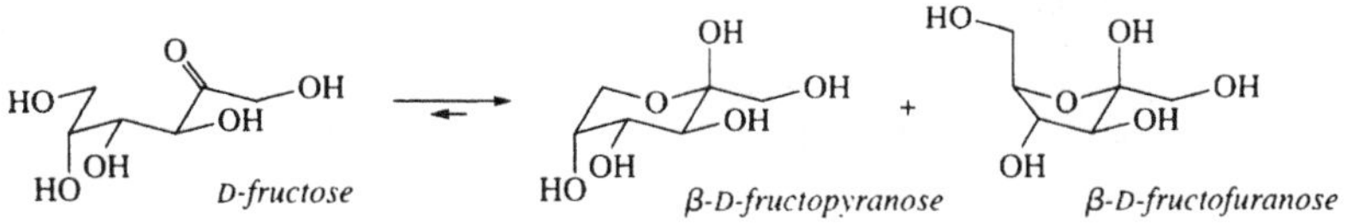

D-fructose   β-D-fructopyranose   β-D-fructofuranose

Primary amines (RNH$_2$) also add reversibly to ketones or aldehydes to give *imines* (Schiff bases) and related compounds via the intermediate *hemiaminals*. The position of the equilibrium depends on the structure of the amine and the carbonyl compound. With alkylamines, the equilibrium favors the carbonyl compound, but it can be driven to the imine by removal of H$_2$O. With hydrazines (R$_2$NNH$_2$) and hydroxyl- and alkoxylamines (RONH$_2$), the equilibrium greatly favors the *hydrazone, oxime*, or *oxime ether*, and it is difficult to drive the reaction in the reverse direction. Secondary amines (R$_2$NH) can form hemiaminals, but they cannot form imines.

Example

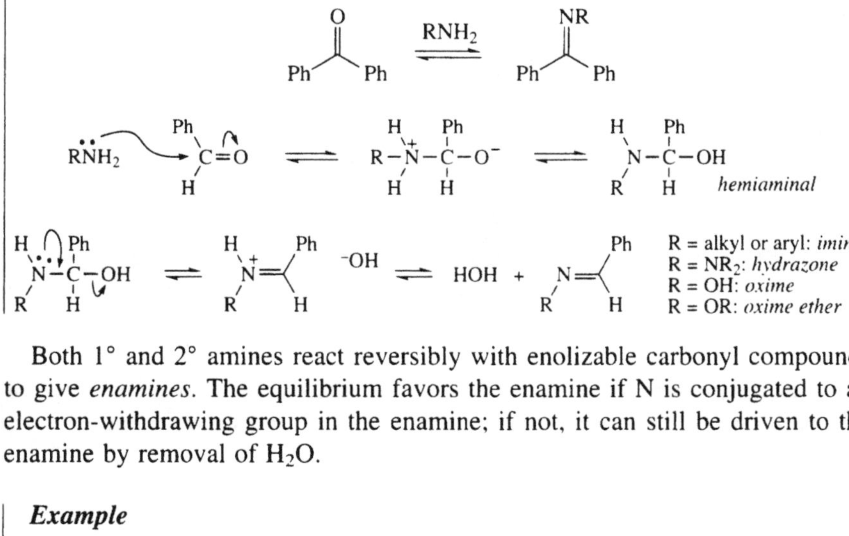

R = alkyl or aryl: *imine*
R = NR$_2$: *hydrazone*
R = OH: *oxime*
R = OR: *oxime ether*

Both 1° and 2° amines react reversibly with enolizable carbonyl compounds to give *enamines*. The equilibrium favors the enamine if N is conjugated to an electron-withdrawing group in the enamine; if not, it can still be driven to the enamine by removal of H$_2$O.

Example

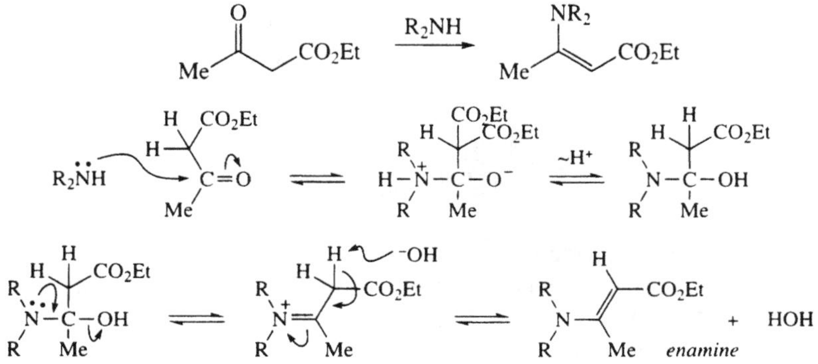

The reaction of cyanide ¯CN with carbonyl compounds to give a *cyanohydrin* proceeds similarly, with the position of the equilibrium dependent on the electrophilicity of the carbonyl compound.

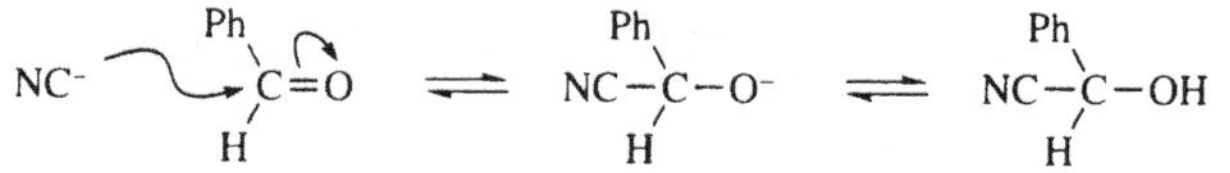

The conjugate bases of carbonyl compounds, *enolates*, react with ketones and aldehydes in the *aldol reaction*. The reaction is usually executed in two stages. First, the enolate is generated, usually by deprotonation of the carbonyl compound at low temperature with a strong base such as LDA, KHMDS, or LiHMDS, but not always. Then the electrophilic carbonyl compound is added to the reaction mixture. Under these conditions, the reaction usually stops at the β-hydroxycarbonyl stage. The nucleophilic component may be any carbonyl compound, or even a nitrile, a sulfonyl compound, or a nitro compound (honorary members of the carbonyl family).

Example

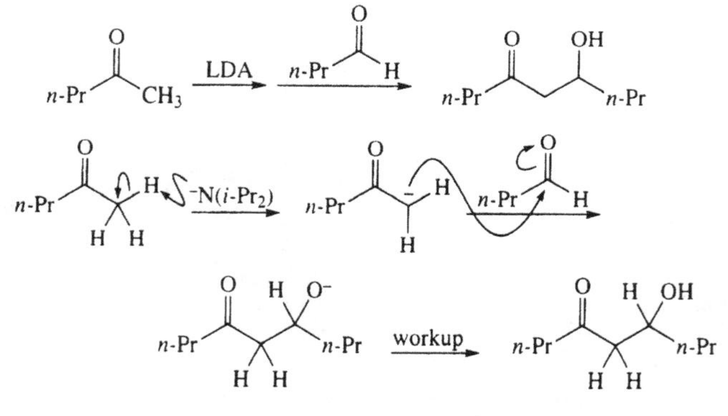

It matters *not one iota* whether you draw the $\bar{C}$ or $\bar{O}$ resonance structure of the enolate. The two resonance structures describe the same molecule.

Problem 2.2. Draw mechanisms for the following aldol reactions.

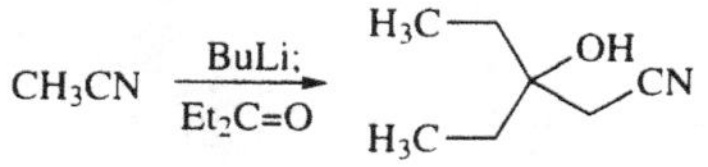

The enolate that is required for the aldol reaction can be generated in other ways, too. For example, trimethylsilyl enol ethers react with TBAF (tetrabutylammonium fluoride, $Bu_4N^+ F^-$) to give the corresponding enolates. Enolates may also be prepared by the dissolving metal reduction of α,β-unsaturated ketones (see Chapter 5).

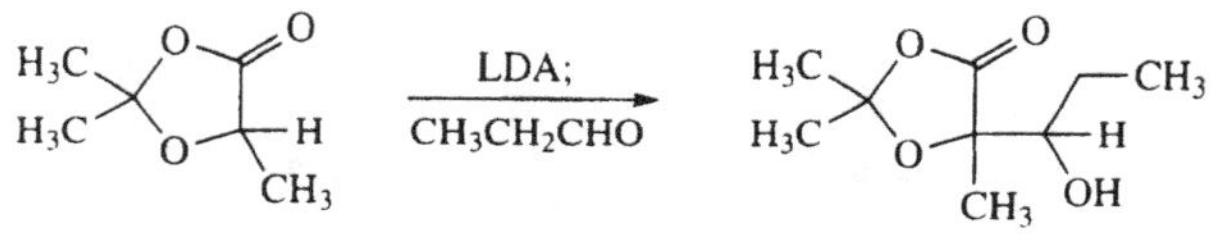

When a carbonyl compound with two α-hydrogen atoms (R^1CH_2COX) undergoes an aldol reaction with an aldehyde R^2CHO, two new stereocenters are created, and hence two diastereomers can be formed. The stereochemistry of these aldol reactions has been extensively studied since the 1970s, and it is now fairly predictable. When R^1CH_2COX is deprotonated with LDA, either the E- or the Z-diastereomer of the enolate $R^1CH=C(OLi)X$ may be generated in principle, but the enolate in which R^1 and OLi are Z (called a Z-enolate regardless of the nature of X) is lower in energy, so it usually predominates. The Z-enolate then undergoes the aldol reaction with an aldehyde via a transition state that features a six-membered, Li-containing ring. A six-membered ring is usually lowest in energy when it is in the chair conformation, and the aldol TS is no exception. When the enolate is Z, the R^1 substituent must be placed in a pseudoaxial position because it must remain cis to the enolate O, but the aldehyde substituent (R^2) can be placed in a pseudoequatorial position. After C–C bond formation and workup, the syn aldol is obtained selectively. (Whether an aldol product is syn or aldol is determined by drawing the main chain, including X and R^2, in zigzag fashion in the plane of the paper; it is syn when R^1 and OH reside on the same side of the plane of the paper, and it is anti when they reside on opposite sides.) If the E-enolate can be selectively generated, then both R^1 and R^2 can be placed in pseudoequatorial orientations in the aldol TS, and the anti aldol is obtained.

When the X group of the enolate also contains a stereocenter, two diastereomeric syn (or two anti) aldols can be formed. Depending on the nature of the X group, the diastereoselectivity may be very high. When an enantiopure, easily replaced X group is used to cause both new stereocenters in the aldol product to form with very high stereoselectivity, it is called a *chiral auxiliary*. The most widely used chiral auxiliaries are the oxazolidinones, which are derived from enantiopure α-amino acids such as (S)-valine. The aldol reaction of the enolate of an N-acyloxazolidinone (usually the Z-enolate) again proceeds through a chairlike TS. Two chairlike TSs are possible; the lower energy one has the oxazolidinone *i*-Pr substituent pointing away the aldehyde. After the aldol reaction, the oxazolidinone is hydrolyzed away and replaced with an OH group to give an enantiopure carboxylic acid.

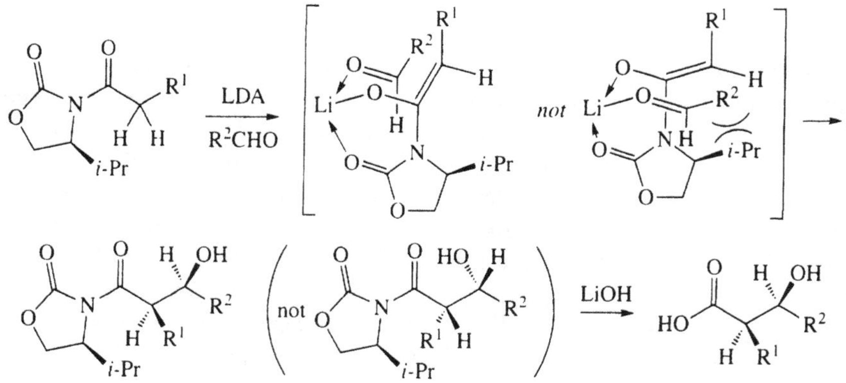

There are many, many variations of the aldol reaction involving metals other than Li (e.g., B, Ti, Sn) and various chiral auxiliaries and other X groups. Almost all of them proceed through selective formation of a Z- or E-enolate followed by a chair-like TS for the aldol reaction.

The aldol reaction of two different enolizable carbonyl compounds is not usually carried out simply by mixing them and adding a base, because either carbonyl compound can act as a nucleophile or as an electrophile, and a mixture of all four possible products (not counting stereoisomers) is usually obtained. However, if one of the carbonyl compounds is nonenolizable and is more electrophilic than the other (e.g., one compound is a ketone and the other is ArCHO, *t*-BuCHO, or EtO$_2$CCHO), then the two components can be combined and allowed to undergo an aldol reaction in the presence of a relatively weak base like NaOEt, KOH, or *t*-BuOK. Only one of the compounds can be converted into a nucleophile, and it attacks only the other component (and not another equivalent of itself) to give only one product. Under these conditions, the initial aldol product usually undergoes E1cb elimination to give an α,β-unsaturated carbonyl compound.

Example

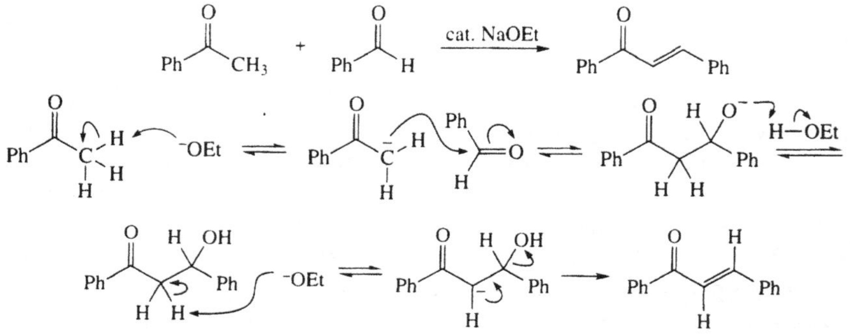

In principle, the protonation of O^- and subsequent deprotonation of C could be condensed into one intramolecular proton transfer step in which the O deprotonates C. However, such a transfer requires a very strained four-membered transition state. It is, therefore, considered poor practice to draw an intramolecular proton transfer, especially when the solvent can serve as a proton "shuttle" instead.

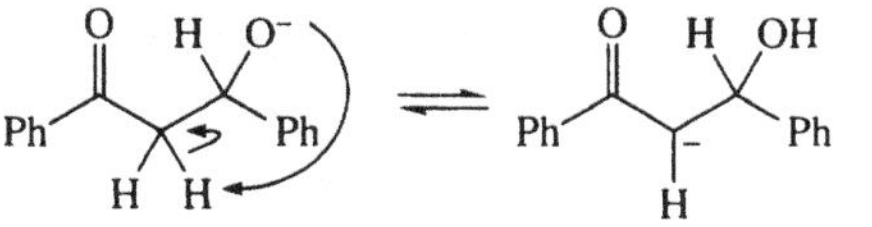

bad practice!

Problem 2.3. An aldol reaction is one of the key steps in the benzoin condensation, which requires a catalytic amount of ^-CN to proceed. Draw a reasonable mechanism.

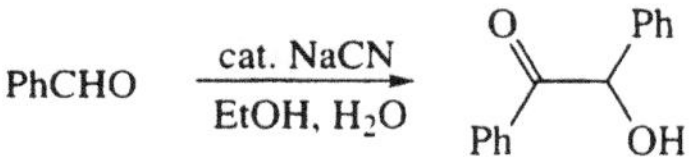

Aldol reactions are reversible, and you need to be able to recognize the reverse reaction, the *retro-aldol* reaction. Because a C–C σ bond is broken in the retro-aldol reaction, some driving force (e.g., relief of strain) is usually required for it to proceed to completion.

Example

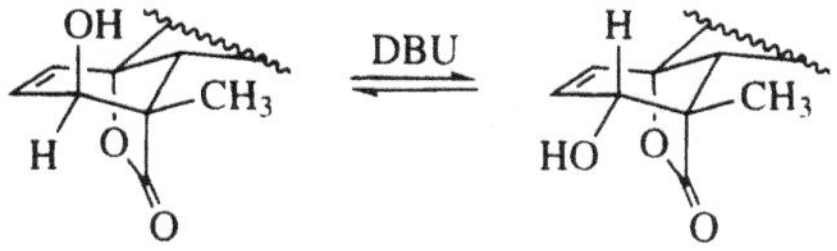

The mechanism of this reaction cannot involve inversion by an S_N2 substitution reaction because (1) no HO^- is present, and (2) HO^- is an awful leaving group in S_N2 reactions, anyway. The nonnucleophilic base DBU can deprotonate the OH group to give an alkoxide, which can undergo a retro-aldol reaction to give an aldehyde and an ester enolate. The aldehyde can then rotate about the single C–C bond to present its other face to the enolate for attack. An aldol reaction then gives the isomerized product.

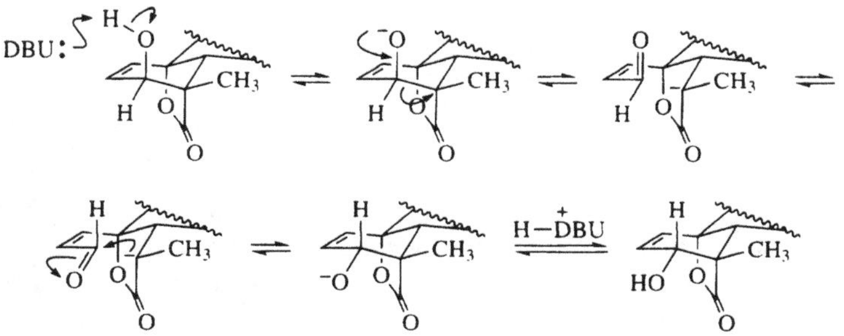

The Knoevenagel reaction, the condensation of 1,3-dicarbonyl compounds with aldehydes to give unsaturated compounds, is catalyzed by 2° amines. A perfectly reasonable mechanism involving deprotonation of the 1,3-dicarbonyl compound by base, aldol reaction with the ketone, and E1cb elimination of H_2O can be drawn. However, the Knoevenagel reaction does not proceed nearly so well using 3° amines, suggesting that the amine does not simply act as a base.

Example

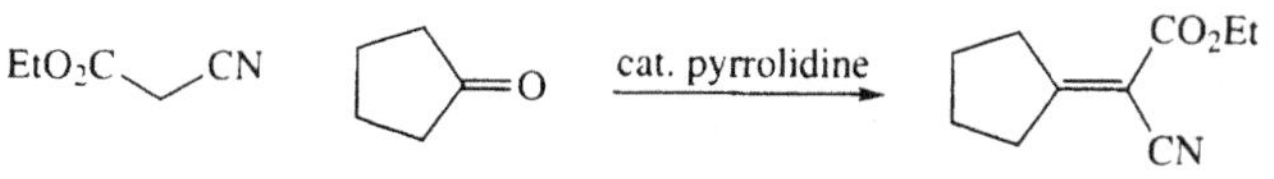

If the amine does not act as a base, it must act as a nucleophile. After addition to the electrophilic ketone to give a hemiaminal, ^-OH leaves, and an *iminium ion* is obtained. The iminium ion, the key intermediate, cannot be formed from a 3° amine.

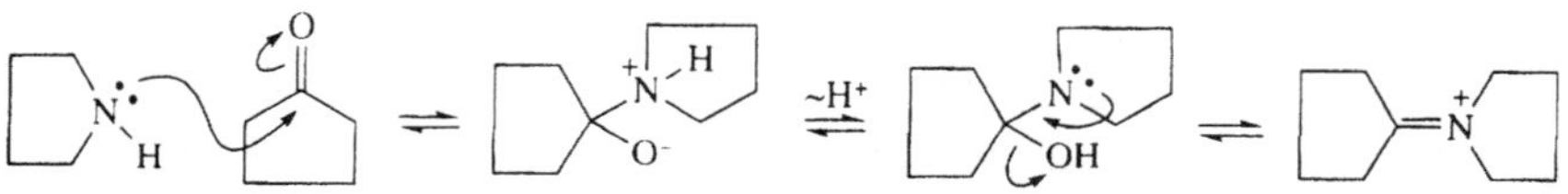

The iminium ion is more electrophilic than the ketone from which it is derived; it reacts with the deprotonated 1,3-dicarbonyl compound.

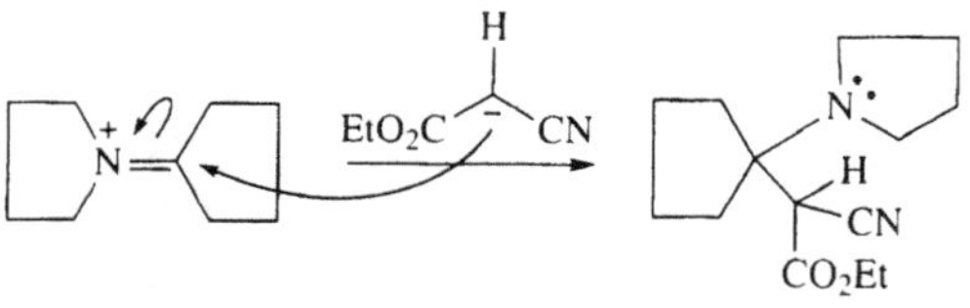

To arrive at the product, the elements of the amine need to be eliminated. After the amine is protonated to make it a better leaving group, E1cb elimination gives the product and regenerates the catalyst.

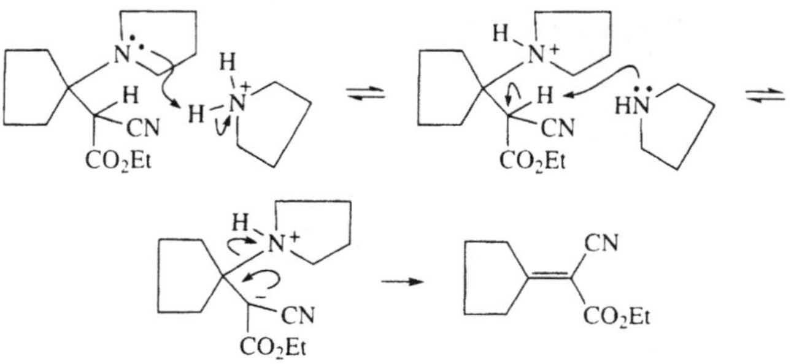

Grignard reagents and organolithium reagents add to nitriles to give imines, which are rapidly hydrolyzed to ketones upon addition of water. Only one addition to the C≡N triple bond takes place, because a second addition would produce a very high energy RN^{2-} species.

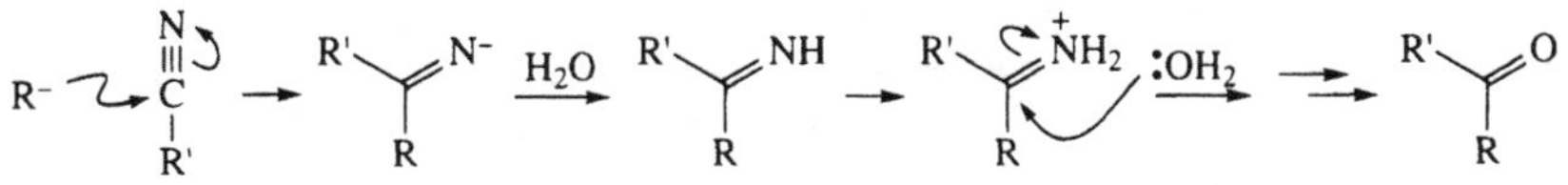

2.2.2 Conjugate Addition; The Michael Reaction

Alkenes and alkynes that are substituted with electron-withdrawing groups such as carbonyl, nitro, and sulfonyl groups are electrophilic. Many nucleophiles react with these alkenes in *conjugate* or *1,4-addition* reactions to give addition products. For example, alcohols, thiols, amines, and other heteroatomic nucleophiles react with electrophilic alkenes by an unremarkable mechanism. The nucleophile adds to the C atom that is β to the electron-withdrawing group, and the electrons in the π bond move to the C atom adjacent to the electron-withdrawing group to give a carbanion. The carbanion is low in energy because its lone pair can be delocalized into the electron-withdrawing group. Protonation of the carbanion completes the addition reaction.

Example

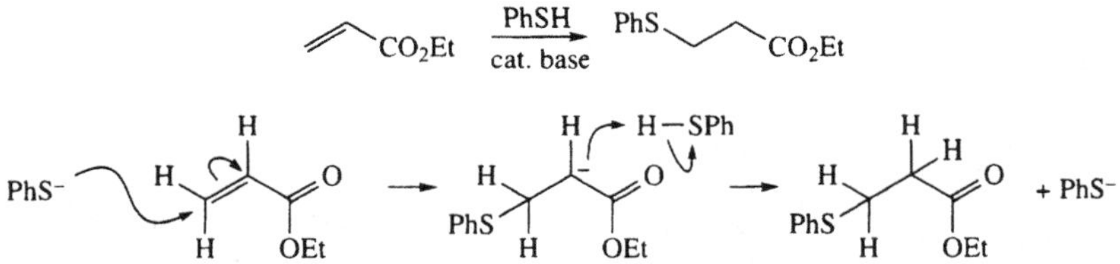

The most important kinds of conjugate addition reactions are Michael reactions, which involve the addition of C nucleophiles to C=C π bonds. The nucleophiles are often 1,3-dicarbonyl compounds such as malonates, cyanoacetates, β-ketoesters, and 1,3-diketones, but simple carbonyl compounds may also be used. Only catalytic amounts of base are usually required.

Example

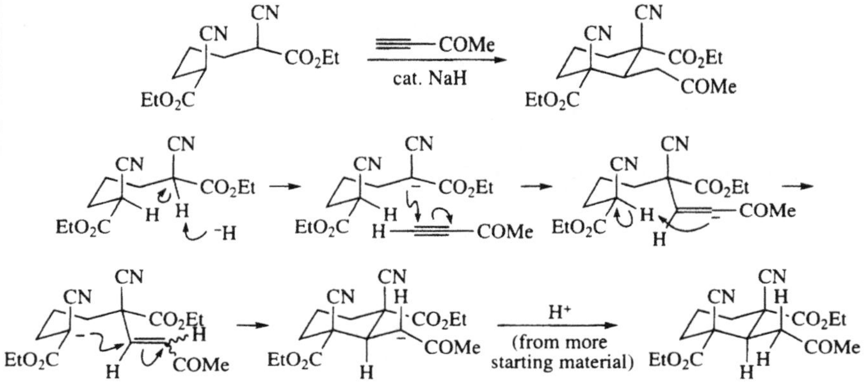

Often, the Michael reaction is followed by an aldol reaction, a substitution, or another Michael reaction. For example, the Robinson annulation consists of a Michael reaction, an aldol reaction, and a dehydration (β-elimination).

Example

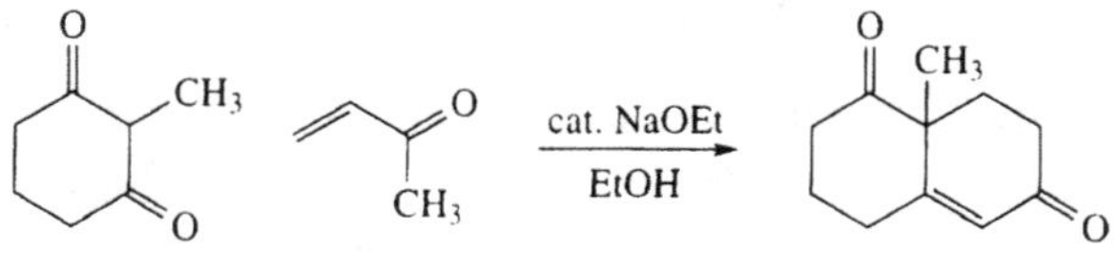

Numbering the atoms shows that bonds are made at C8−C9 and C13−C6, and the C6−O7 bond is broken.

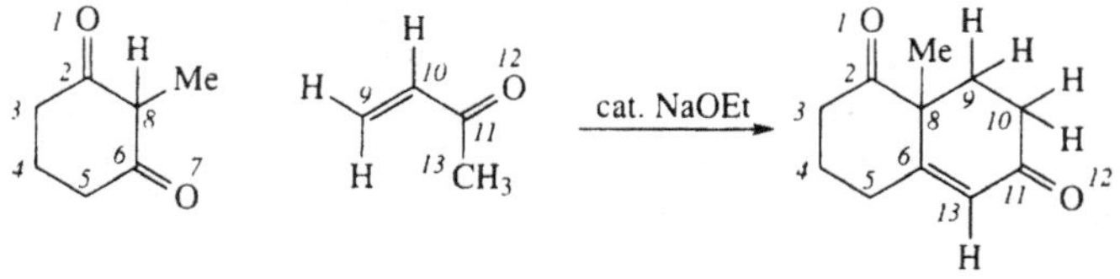

By far the most acidic site in both compounds is C8. Deprotonation of C8 turns it into a nucleophile, so it can add to C9 in a Michael reaction to give a C10 enolate. No more needs to be done with C10, so it is protonated.

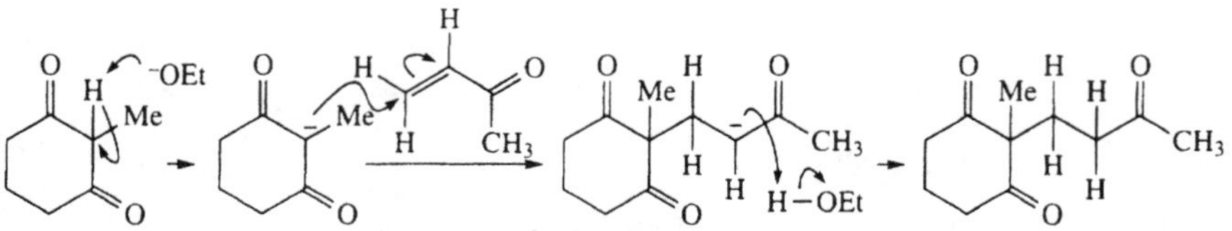

The C6−C13 bond still needs to be formed. C13 is made nucleophilic by deprotonation, then it attacks C6. The alkoxide product is protonated to give an aldol.

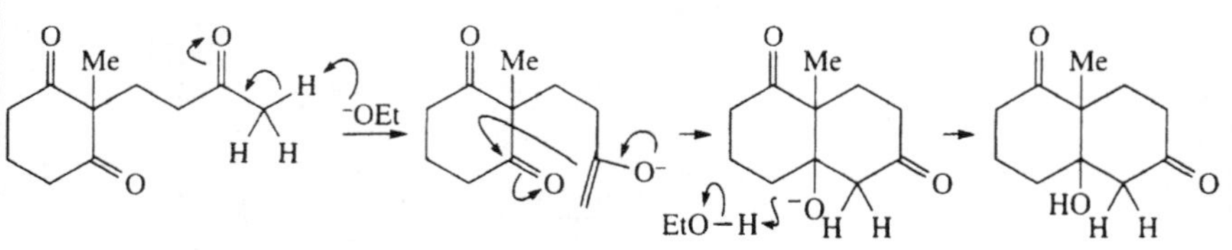

In the last step, the C6−O7 bond is broken, and the elements of H_2O are lost. This β-elimination reaction proceeds by the E1cb mechanism because of the poor leaving group ability of $^-$OH and the acidity of the C−H bond.

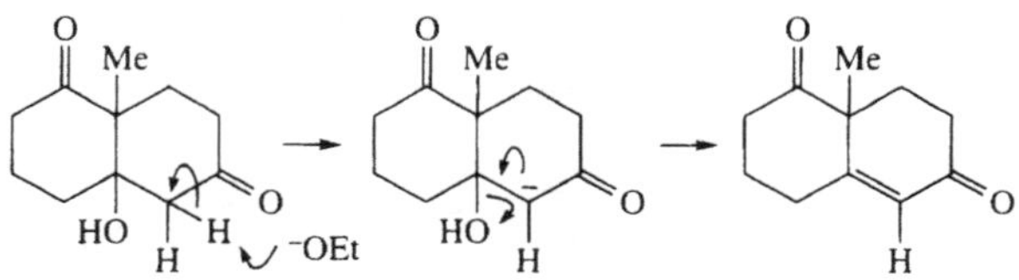

Problems 2.4. Each of the following reactions has a Michael reaction as a first step. Draw a reasonable mechanism for each reaction.

(a)

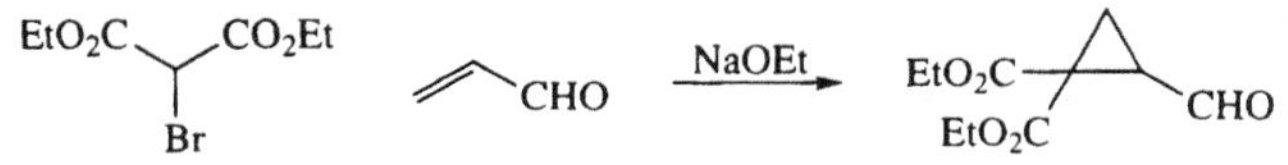

(b)

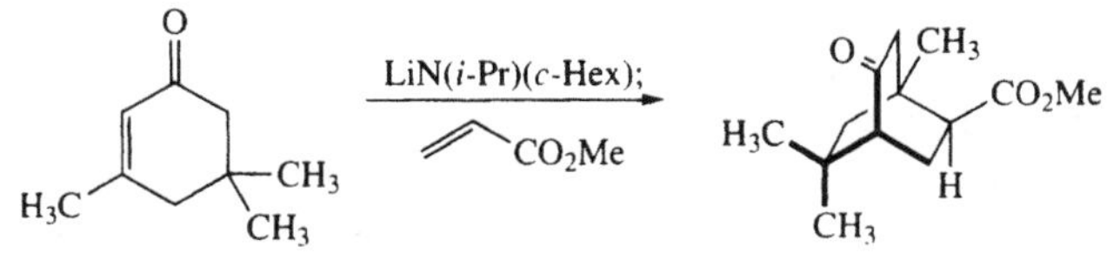

(c)

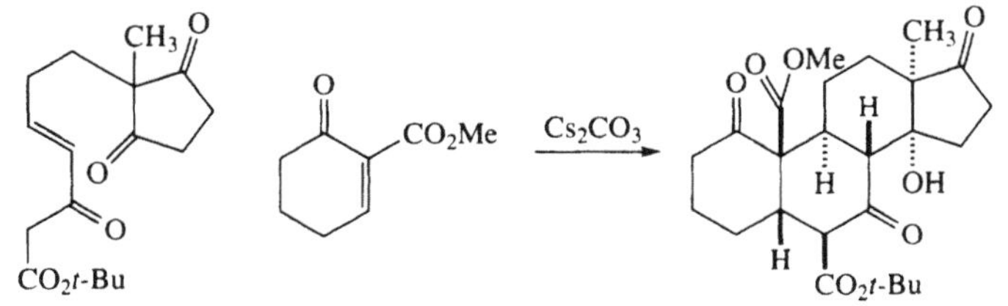

It is difficult to overemphasize the importance of the Michael reaction in synthesis. It is one of the mildest, most versatile, and most efficient methods for forming C–C bonds.

2.3   Substitution at C(sp²)–X σ Bonds

2.3.1   Substitution at Carbonyl C

Many carbonyl compounds, including esters, acyl chlorides, and acid anhydrides, have leaving groups attached to the carbonyl C, and many reactions proceed with substitution of this leaving group by a nucleophile. Substitutions at the carbonyl C usually occur by an *addition–elimination* mechanism. The nucleophile Nu⁻ adds to the electrophilic C of the carbonyl group to make a tetrahedral intermediate. The leaving group X⁻ then leaves to give a new carbonyl compound. Note that either X⁻ or Nu⁻ may be expelled from the tetrahedral intermediate. Which one is expelled depends on the nature of the two groups and the reaction conditions. Expulsion of Nu⁻, though, gives back the starting material, which is not very productive!

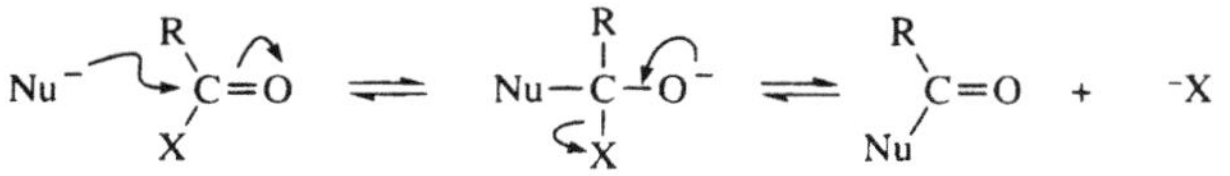

*   **Common error alert:** *C(sp²) electrophiles never, ever, ever undergo substitution by the S_N2 mechanism. The two-step mechanism for substitution at car-*

bonyl C is much more reasonable than a one-step, S_N2 mechanism for several reasons. The C=O π bond is higher in energy than the C(sp^2)–X bond, so it's easier to break. Moreover, addition of Nu$^-$ to the C=O π bond can occur along a trajectory that is out of the plane of the carbonyl group, whereas an S_N2 trajectory must be in the crowded plane of the carbonyl group. In addition to these theoretical considerations, plenty of experimental evidence suggests that the two-step mechanism is always operative. (See any physical organic chemistry textbook for details.)

Primary amines react with many esters just upon mixing to give amides. The amines are sufficiently nucleophilic to add to the ester carbonyl. After the nucleophilic N is deprotonated, the alkoxy group is a much better leaving group, so collapse of the tetrahedral intermediate occurs with expulsion of $^-$OR to give the amide as the product.

Example

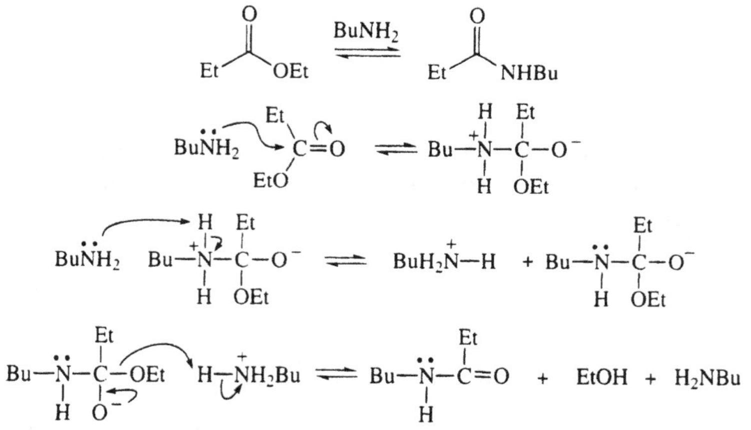

Transesterification of esters occurs by a mechanism very much like amide synthesis, but the reaction requires a catalytic amount of base (usually the Na salt of the alcohol). The nucleophile is the alkoxide. The reaction is driven in the forward direction by the use of a large excess of the starting alcohol.

$$RCO_2Et + MeOH \xrightarrow{\text{cat. NaOMe}} RCO_2Me + EtOH$$

Problem 2.5. Why can't carboxylic acids be similarly esterified?

$$RCO_2H + MeOH \xrightarrow{\text{cat. NaOMe}} \text{N.R.}$$

The reaction of alcohols with *enolizable acyl chlorides* or *anhydrides* can proceed by two different mechanisms. One is the addition–elimination mechanism that has already been discussed:

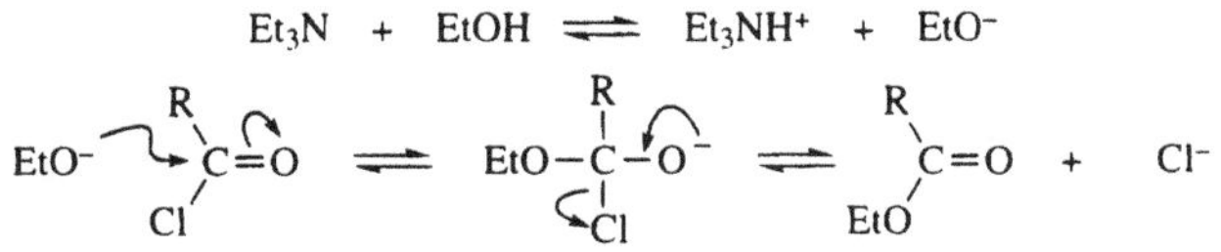

$$Et_3N + EtOH \rightleftharpoons Et_3NH^+ + EtO^-$$

The other is a two-step, *elimination–addition* mechanism. In the elimination step, β-elimination occurs by an E2 mechanism to give a *ketene*, a very reactive compound that is not usually isolable. In the addition step, the alkoxide adds to the electrophilic carbonyl C of the ketene to give the enolate of an ester. Acyl chlorides lacking α-hydrogens (*t*-BuCOCl, ArCOCl), of course, can react only by the addition–elimination mechanism.

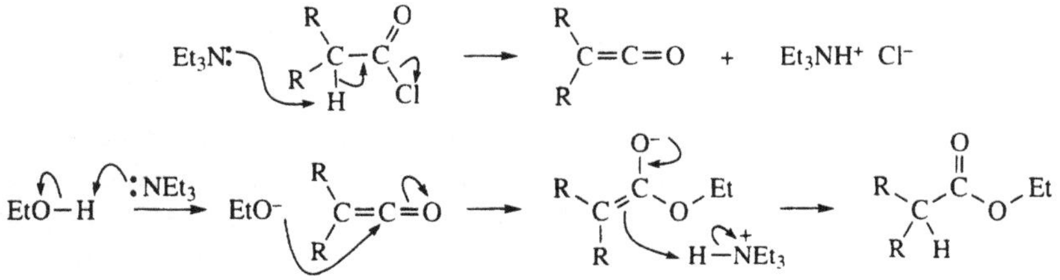

Often a nucleophilic catalyst such as DMAP (4-dimethylaminopyridine) is added to accelerate the acylation of alcohols ROH with acyl chlorides. The catalyst is a better nucleophile than RO$^-$, so it reacts more quickly with the acyl chloride than RO$^-$ to give an acylpyridinium ion by addition–elimination. The acylpyridinium ion, though, is more reactive than an acyl chloride, so RO$^-$ adds more quickly to it than to the acyl chloride. The two addition–eliminations involving DMAP are together faster than the single one involving RO$^-$. A rate acceleration is also observed with I$^-$, another great nucleophile and great leaving group.

without catalysis:

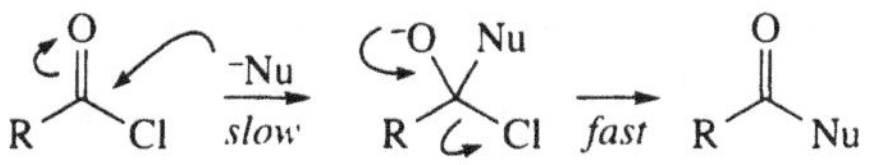

with catalysis:

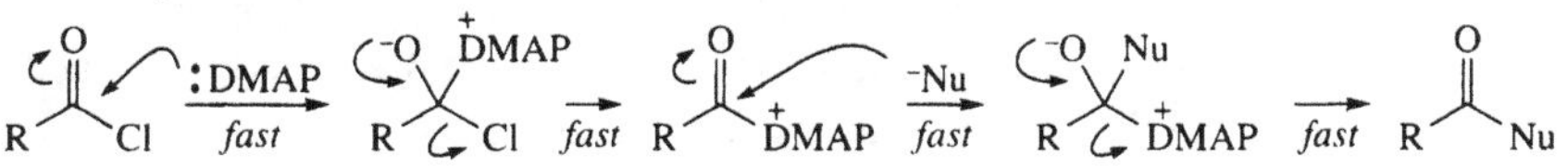

The *Claisen* and *Dieckmann condensations* are reactions in which an ester enolate acts as a nucleophile toward an ester. The Dieckmann condensation is simply the intramolecular variant of the Claisen condensation. In these reactions, the alkoxy part of the ester is substituted with an enolate to give a β-ketoester. A stoichiometric amount of base is required for this reaction, because the product is a very good acid, and it quenches the base catalyst. In fact, this quenching reaction is what drives the overall reaction to completion. The Claisen condensation is especially useful when one of the esters is nonenolizable (e.g., diethyl oxalate, ethyl formate, or diethyl carbonate).

Example

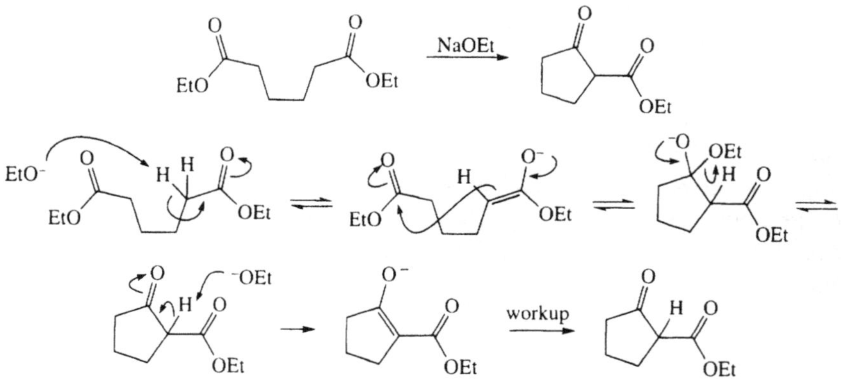

Ketones are acylated with esters in a similar manner. The product is a 1,3-diketone. Again, with intermolecular condensations, the reaction is especially important with nonenolizable esters such as diethyl carbonate, diethyl oxalate, and ethyl formate. And again, the reaction is driven to completion by deprotonation of the very acidic 1,3-dicarbonyl product.

Problem 2.6. Draw mechanisms for the following two acylation reactions.

(a)

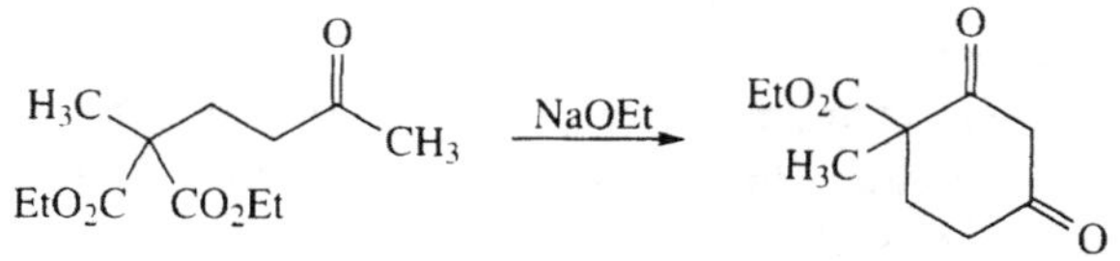

(b)

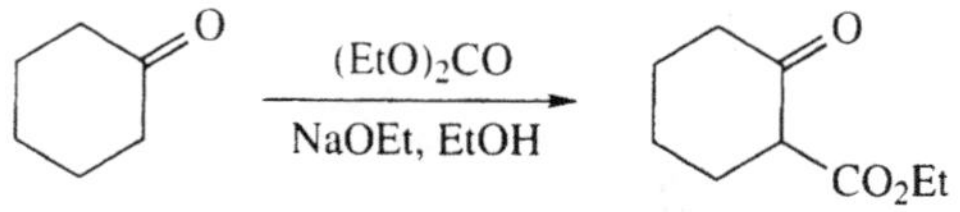

Like aldol reactions, the addition of enolates to esters is reversible. 1,3-Dicarbonyl compounds that cannot be deprotonated cleave readily to give two simple carbonyl compounds under basic conditions. The cleavage occurs by addition–elimination, and an enolate acts as a leaving group.

Example

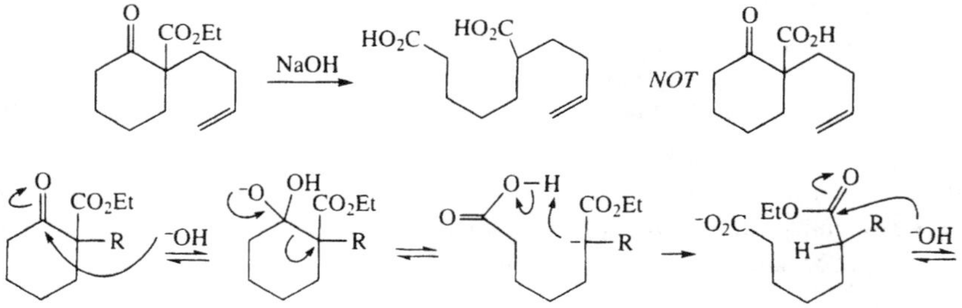

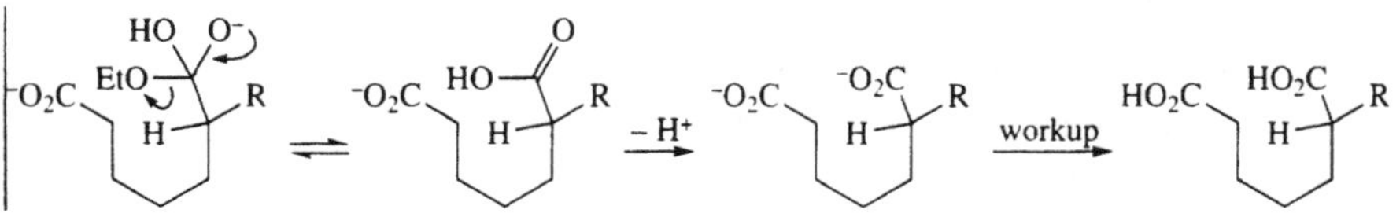

Grignard reagents, organolithium compounds, and complex metal hydrides react with esters to give alcohols. The nucleophiles first react by addition–elimination to give a transient ketone or aldehyde. The ketone or aldehyde is more reactive than the starting material toward the nucleophile, though, so another equivalent of nucleophile adds to it to give an alcohol after workup.

Example

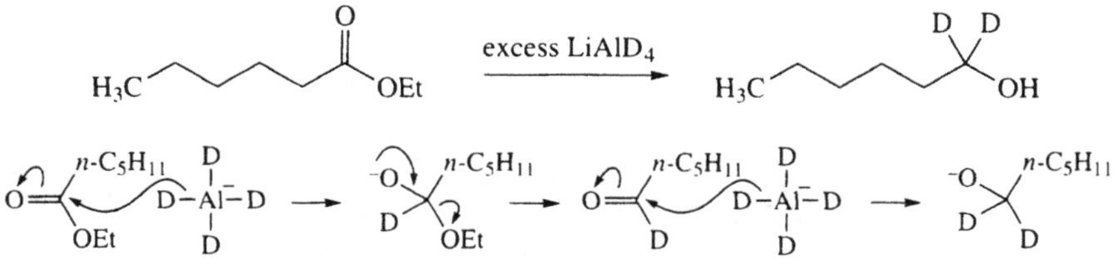

Nitriles are in the same oxidation state as esters, so you might expect the addition of Grignard reagents to nitriles to give primary amines, but, as noted earlier, addition to nitriles stops after a single addition because no leaving group can be expelled to give a newly electrophilic intermediate.

When Grignard reagents, organolithium compounds, or complex metal hydrides add to amides, the elimination step is slow at −78 °C, especially when the amine component is −N(Me)OMe (*Weinreb amides*). When the tetrahedral intermediate is sufficiently long lived, quenching of the reaction mixture with water at −78 °C gives the ketone or aldehyde rather than the alcohol.

Example

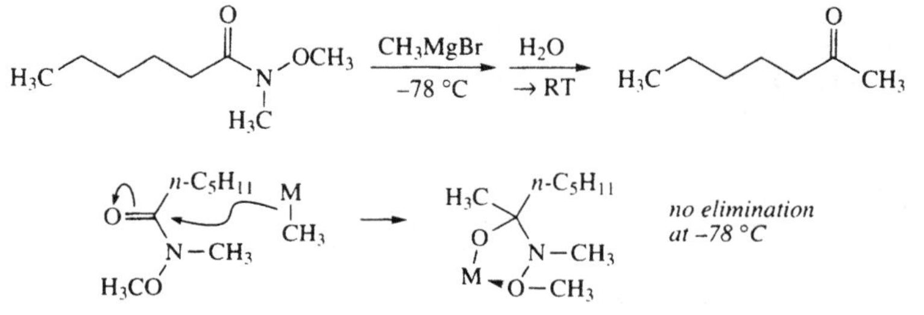

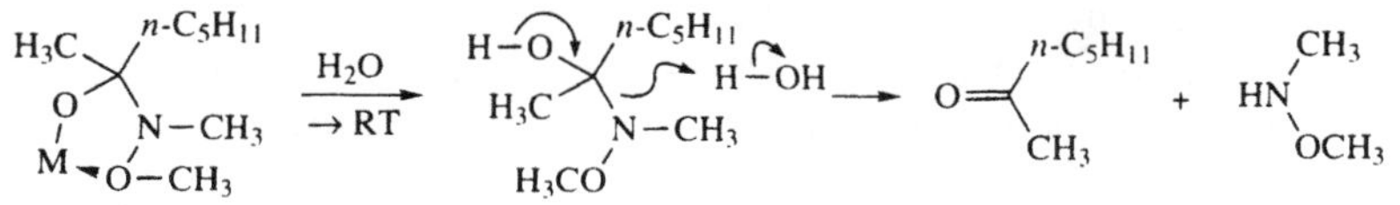

The O atom of sulfoxides is weakly nucleophilic, and it reacts with acyl anhydrides or acyl chlorides. In the *Pummerer rearrangement*, a sulfoxide and an acyl anhydride are converted to a mixed *O,S*-acetal, which can be hydrolyzed to provide an aldehyde and a thiol. The mechanism of the reaction begins with the sulfoxide O atom attacking Ac₂O (or some other acyl anhydride) in an addition–elimination reaction to give an acetoxysulfonium ion. E2 elimination of AcOH from this compound then provides a new sulfonium ion with a C=S⁺ π bond. The very electrophilic C atom is then attacked by acetate to give the *O,S*-acetal. Hydrolysis of the *O,S*-acetal proceeds by the usual addition–elimination mechanism. Later, we will see a reaction involving sulfoxides and an acyl chloride that begins the same way.

Example

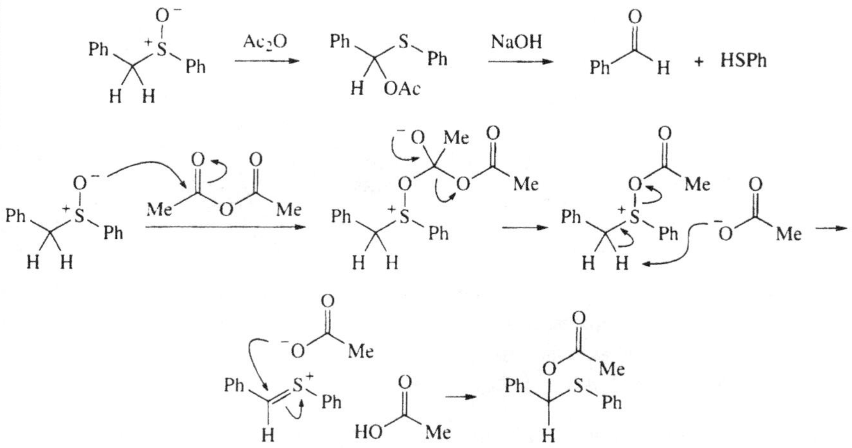

2.3.2 Substitution at Alkenyl and Aryl C

α,β-Unsaturated carbonyl compounds with a leaving group in the β position are susceptible to addition–elimination reactions just like ordinary carbonyl compounds. The functional group X–C=C–C=O is often described as a *vinylogous* form of X–C=O. For example, 3-chloro-2-cyclohexen-1-one reacts with NaCN to give 3-cyano-2-cyclohexen-1-one by an addition–elimination mechanism. The electrophile is a vinylogous acyl chloride, and its reactivity at the C attached to Cl is similar to the reactivity of an acyl chloride.

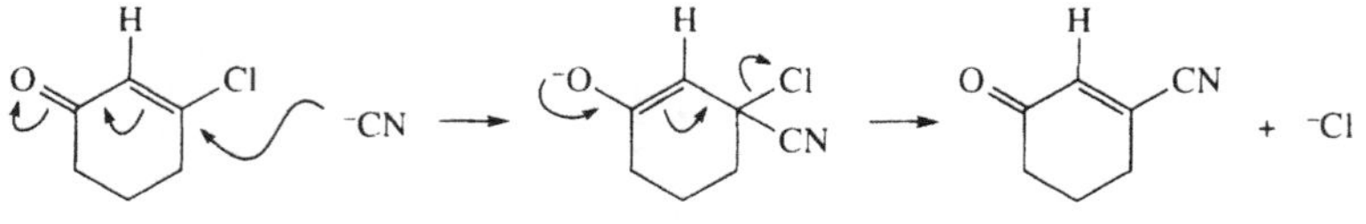

Aromatic compounds that are substituted with electron-withdrawing groups (usually nitro groups) undergo nucleophilic aromatic substitution reactions by this mechanism. Because addition of the nucleophile to the ring disrupts the aromaticity, two electron-withdrawing groups are usually required to make the aromatic ring sufficiently electron deficient that addition proceeds at a reasonable rate. Sanger's reagent, 2,4-dinitrofluorobenzene, reacts with amines in this fashion. Halopyridines undergo substitution reactions by an exactly analogous mechanism. Addition–elimination at aromatic rings is often called S$_N$Ar for no particularly good reason, and the anionic intermediate is called a *Meisenheimer complex*.

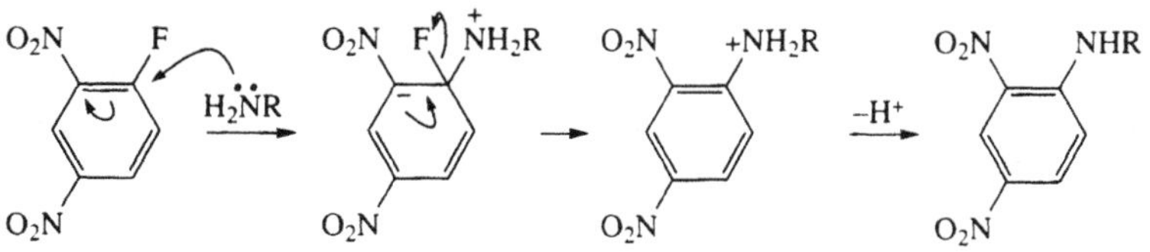

Problem 2.7. Draw mechanisms for the following two aromatic substitution reactions. *Note:* The second problem will be much easier to solve if you draw the by-product.

(a) ... NaOH ... *Smiles rearrangement*

(b) ... PBr$_3$ ...

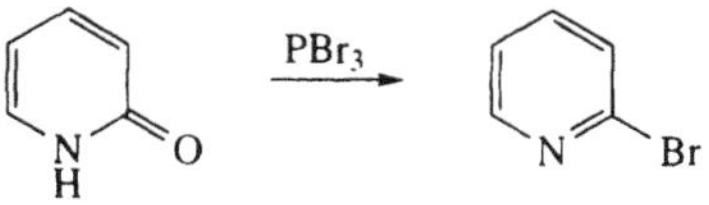

Nucleophilic substitution can also occur at aryl rings that are not substituted with electron-withdrawing groups, although it is not seen as often. Two mechanisms are possible: S$_{RN}$1 and elimination–addition. The *S$_{RN}$1 mechanism* is a radical chain mechanism. (The *R* in S$_{RN}$1 stands for *radical*.) As such, it has three parts: initiation, propagation, and termination. Consider the following reaction:

Overall:

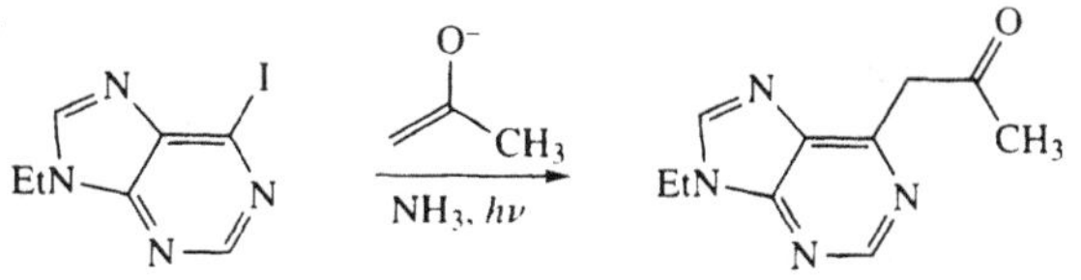

The reaction is *initiated* by electron transfer from the HOMO of the nucleophile (Nu$^-$) to the LUMO of the electrophile (R–X) to generate two radicals, Nu· and [R∶·X]$^-$ (a *radical anion*). (See Chapter 5 for a more complete discussion.) If the energies of the HOMO$_{nucleophile}$ and the LUMO$_{electrophile}$ are close enough, the transfer can occur spontaneously. Sometimes light (*hv*) is used to promote an electron in the HOMO of Nu$^-$ to a higher energy orbital so that electron transfer can occur more readily. If the electron transfer is particularly slow, catalytic amounts of one-electron reducing agents (e.g., $FeBr_2$) can be added to initiate the chain reaction by electron transfer to the electrophiles.

Initiation:

useful for $S_{RN}1$ problems — *lower-energy "toilet bowl" structure*

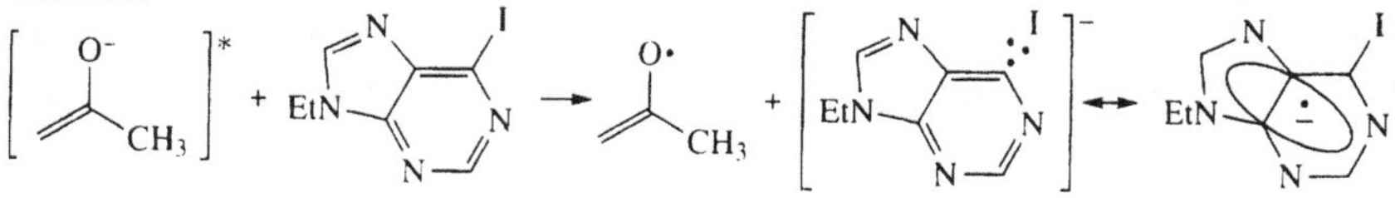

One can draw several resonance structures for the radical anion. This text uses an unusual resonance structure for the radical anion $[R{:}{\cdot}X]^-$ in which there is a *three-electron, two-center bond* between R and X. This resonance structure suggests that the odd electron is localized in the R–X σ^* orbital, making half a bond between R and X. The "toilet bowl" resonance structure is probably a better description of the ground state, reflecting the fact that the SOMO (singly occupied MO) of an aromatic radical anion is derived from the ring's p orbitals, but it is less useful for drawing the $S_{RN}1$ mechanism.

A curved arrow is not used to show an electron transfer.

The *propagation* part of an $S_{RN}1$ reaction has three steps. In the first step, $[R{:}{\cdot}X]^-$ dissociates into $:X^-$ and $\cdot R$. The departure of X^- is facilitated by the electron in the antibonding σ^* orbital.

Propagation:

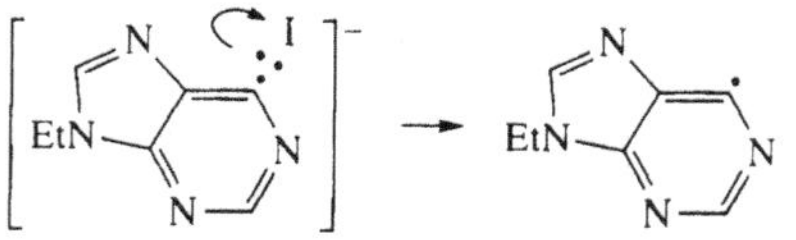

In the second propagation step, $R\cdot$ combines with Nu^- to give a new radical anion, $[R{:}{\cdot}Nu]^-$. This step is just the reverse of the first step (with a different nucleophile, of course)!

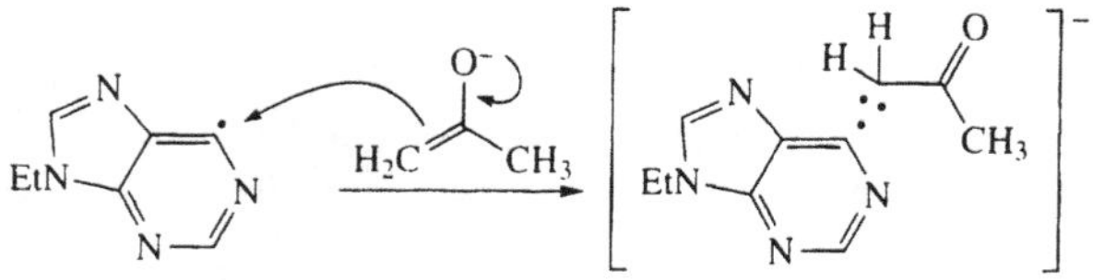

In the third propagation step, electron transfer from $[R{:}{\cdot}Nu]^-$ to R–X occurs to give R–Nu and $[R{:}{\cdot}X]^-$. The product radical of this step is the same as the starting radical of the first propagation step, so the chain is now complete.

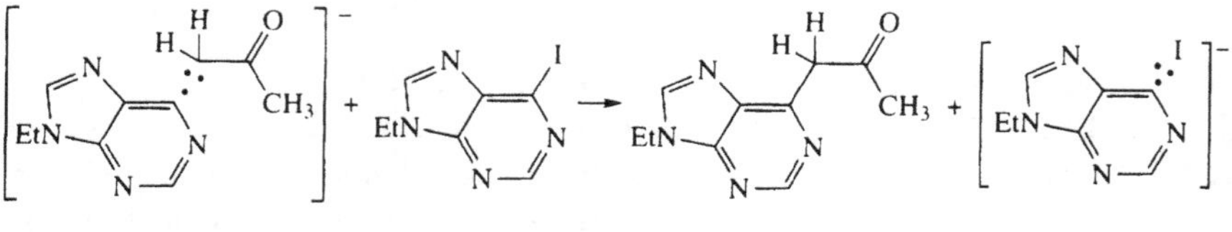

Note that the first two steps of the propagation part look a lot like the S_N1 substitution reaction (Chapter 3)—a leaving group leaves, then the nucleophile comes in—except that an aryl radical is formed as a transient intermediate instead of a carbocation.

The best nucleophiles for the S$_{RN}$1 mechanism can make a relatively stable radical in the initiation part, either by resonance (enolates) or by placing the radical on a heavy element (second-row main-group or heavier nucleophiles). The best electrophiles are aryl bromides and iodides. If light is required for substitution to occur, the mechanism is almost certainly S$_{RN}$1. Substitution at alkenyl C(sp^2)–X bonds can also occur by an S$_{RN}$1 mechanism.

Simple aryl halides undergo substitution reactions with very strong bases such as $^-$NH$_2$. Neither the addition–elimination mechanism nor the S$_{RN}$1 mechanism seems very likely for this pair of substrates.

The generally accepted mechanism for this reaction is a two-step *elimination–addition* mechanism. In the β-elimination step, NH$_2$$^-$ promotes the β-elimination of the elements of HI to give benzyne. In the addition step, NH$_2$$^-$ adds to the very strained triple bond to give an aryl anion, which undergoes proton transfer. Workup gives the neutral product.

One piece of evidence for this mechanism is that the reaction of strong bases with *o*- and *m*-substituted aryl halides gives a mixture of products resulting from substitution both at the C bearing the leaving group and at its next-door neighbor. Neither the addition–elimination mechanism nor the S$_{RN}$1 mechanism can account for this observation.

Problem 2.8. Draw a mechanism explaining why two products are obtained from the reaction of *o*-iodoanisole and sodamide.

Problem 2.9. Alkenyl halides such as CH$_3$CBr=CHCH$_3$ do not undergo substitution upon treatment with a strongly basic nucleophile such as $^-$NH$_2$. What reaction occurs instead, and why does substitution not occur?

Aromatic substitution reactions that proceed by the elimination–addition mechanism are not widely used synthetically, as a very strong base (p$K_b \geq 35$) is required to generate the aryne intermediate by β-elimination of HX. In addition, there is regiochemical ambiguity for unsymmetrical aryl halides. Even benzyne itself is much more readily prepared by several other methods.

In summary, aryl halides can undergo substitution by addition–elimination, $S_{RN}1$, or elimination–addition mechanisms under basic conditions. The addition–elimination mechanism is most reasonable when the arene is electron-poor. When the arene is not electron-poor, the $S_{RN}1$ mechanism is most reasonable either when the nucleophile is a heavy atom or is delocalized, when light is required, or when a catalytic amount of a one-electron reducing agent is required. The elimination–addition mechanism is most reasonable when the arene is not electron-poor and when very strong base ($pK_b \geq 35$) is used.

* **Common error alert:** *The S_N2 mechanism is not an option for aryl halides.*

> **Problem 2.10.** Hexachlorobenzene C_6Cl_6 undergoes six substitutions with PhS^- to give $C_6(SPh)_6$. Are the most reasonable mechanisms for the first and sixth substitutions the same or different? Draw the mechanisms for these two substitutions.

2.3.3 Metal Insertion; Halogen–Metal Exchange

Aryl and alkenyl halides undergo reactions with metals such as Zn, Mg, and Li to give products where the C–X bond is replaced with a C–metal bond. The best-known metal insertion reaction is the Grignard reaction, which uses Mg. Lithiation requires two equivalents of Li, because each Li supplies only one electron, but the Grignard and zinc insertion reactions require only one equivalent. The rate of insertion is strongly dependent on X, with $I > Br \gg Cl$, corresponding to the strength of the C–X bonds.

The mechanism of metal insertion is best understood as an electron transfer process. One-electron transfer from Li to PhI gives a radical anion that is best drawn with a three-electron, two-center bond, as in the $S_{RN}1$ mechanism. Departure of I^- gives Ph·, which reacts with another equivalent of Li by electron transfer to give the product.

When an aryl or alkenyl halide (RX) is treated with an alkyllithium compound (R'Li), *halogen–metal exchange* can take place. The Li and X swap places to give RLi and R'X. The conversion of the C(sp^3)–Li bond to the less basic C(sp^2)–Li bond provides the driving force for the halogen–metal exchange. Three reasonable mechanisms can be drawn for this reaction: (1) an S$_N$2 substitution at the halogen atom, with the organic anion R$^-$ acting as a leaving group; (2) an addition–elimination mechanism with a 10-electron Br intermediate; or (3) an electron transfer mechanism. The last mechanism, like the S$_{RN}$1 mechanism, has intermediates with three-electron, two-center bonds, but it is not a chain reaction.

S$_N$2 mechanism for halogen–metal exchange

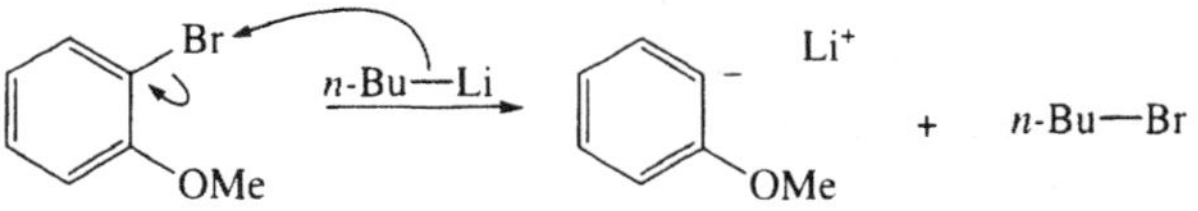

Addition–elimination mechanism for halogen–metal exchange

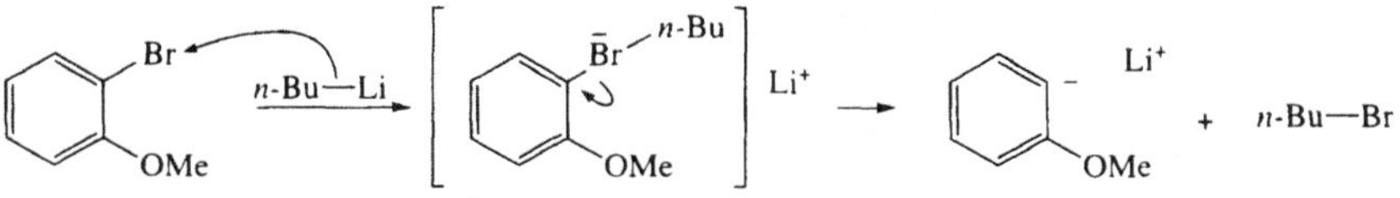

ten-electron Br intermediate

Electron transfer mechanism for halogen–metal exchange

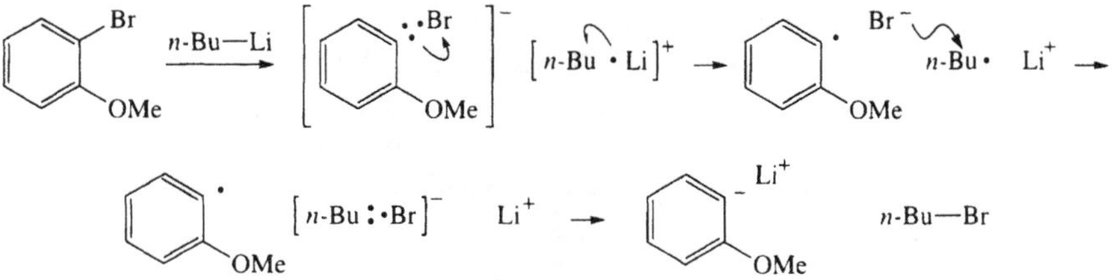

Halogen–metal exchanges are extremely fast reactions, occurring faster than mixing at −78 °C! Both aryl and alkenyl bromides and iodides are substrates for halogen–metal exchange, but tosylates and other pseudohalides do not undergo halogen-metal exchange, and the exchange with chlorides is very slow. The organolithium compound may be CH$_3$Li, n-BuLi, t-BuLi, or s-BuLi. When t-BuLi is used, two equivalents are required to drive the exchange to completion. The second equivalent of t-BuLi acts as a base toward the by-product t-BuX to give isobutane, isobutylene, and LiX by a β-elimination reaction.

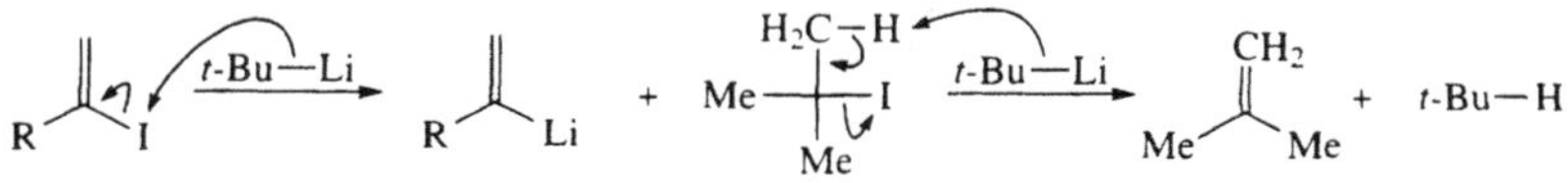

When CH$_3$Li or n-BuLi is used in halogen–metal exchange, a rather electrophilic MeX or n-BuX is obtained as a by-product. The alkyl halide can undergo S$_N$2 substitution with the organolithium compound as nucleophile to give the nucleophilic aromatic substitution product. However, S$_N$2 reactions of organolithium compounds with alkyl halides are rather slow at the low temperatures at which halogen–metal exchange is usually executed, so this side reaction is usually not a problem.

The aryl- or alkenyllithium compounds that are obtained from halogen–metal exchange are used as nucleophiles in subsequent reactions, usually toward π-bond electrophiles such as carbonyl compounds.

2.4   Substitution and Elimination at $C(sp^3)$–X σ Bonds, Part II

Substitution by the S_N2 mechanism and β-elimination by the E2 and E1cb mechanisms are not the only reactions that can occur at $C(sp^3)$–X. Substitution can also occur at $C(sp^3)$–X by the $S_{RN}1$ mechanism, the elimination–addition mechanism, a one-electron transfer mechanism, and metal insertion and halogen–metal exchange reactions. An alkyl halide can also undergo α-elimination to give a carbene.

2.4.1   Substitution by the $S_{RN}1$ Mechanism

Tertiary alkyl halides usually undergo E2 elimination under basic conditions, but sometimes nucleophilic substitution occurs. In these cases, the mechanism *cannot* be S_N2 (unless the reaction is intramolecular). However, the $S_{RN}1$ mechanism can operate at $C(sp^3)$ under basic conditions. The $S_{RN}1$ mechanism for $C(sp^3)$–X electrophiles is exactly the same as it is for $C(sp^2)$–X electrophiles. Initiation occurs by electron transfer from the nucleophile to the electrophile to give a radical anion. Sometimes light is used to photoexcite the nucleophile to encourage it to transfer its electron, and sometimes a one-electron reducing reagent (Chapter 5) is added to initiate the reaction instead. Propagation occurs by leaving-group dissociation to give the neutral radical, addition of the nucleophile to the radical to give a new radical anion, and electron transfer from the radical anion to another equivalent of starting electrophile.

Example

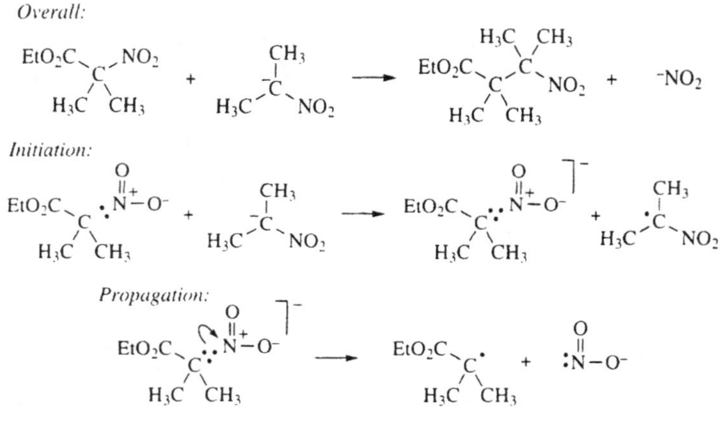

Overall:

Initiation:

Propagation:

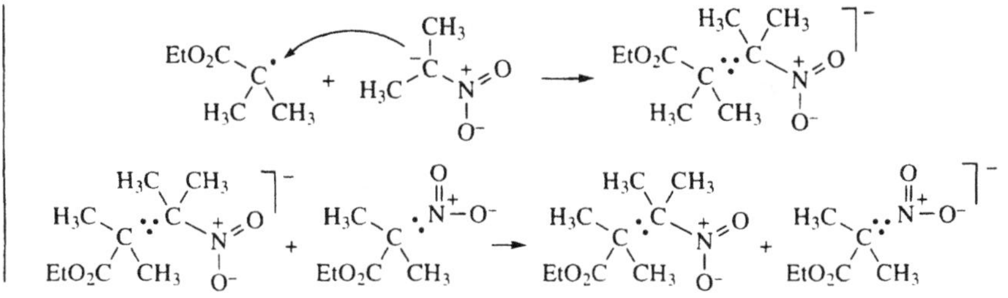

Again, the resonance structure for the radical anion $[R:\cdot X]^-$ in which there is a three-electron, two-center bond between C and X is most useful for drawing the $S_{RN}1$ mechanism. Of course, the LUMO need not be this particular orbital, and one may often draw better resonance structures in which the odd electron is localized elsewhere.

The best nucleophiles for the $S_{RN}1$ mechanism can make a relatively stable radical in the initiation part, either by resonance (enolates) or by placing the radical on a heavy element (second-row or heavier nucleophiles). The best electrophiles for the $S_{RN}1$ mechanism are able to delocalize the odd electron in the radical anion (aromatic leaving groups, carbonyl compounds), can make a stable radical (3° alkyl halides), and have a weak R–X (Br, I) bond. Tosylates and other pseudohalides are very poor $S_{RN}1$ electrophiles. If light is required for substitution to occur, the mechanism is almost certainly $S_{RN}1$.

2.4.2   Substitution by the Elimination–Addition Mechanism

A third mechanism for substitution at C(sp³)–X bonds under basic conditions, *elimination–addition*, is occasionally seen. The stereochemical outcome of the substitution reaction shown in the figure tells us that a direct S_N2 substitution is not occurring. Two sequential S_N2 reactions would explain the retention of stereochemistry, but the problem with this explanation is that backside attack of MeO⁻ on the extremely hindered top face of the bromide is simply not reasonable. The $S_{RN}1$ mechanism can also be ruled out, as the first-row, localized nucleophile MeO⁻ and the 2° alkyl halide are unlikely substrates for such a mechanism.

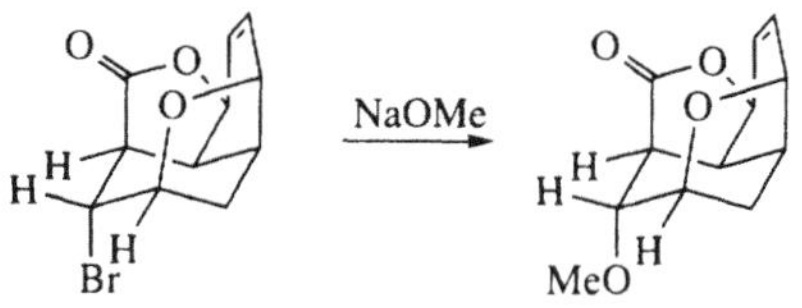

An *elimination–addition* mechanism can be proposed. MeO⁻ is both a good nucleophile and a good base. It can induce β-elimination of HBr (probably by an E1cb mechanism, because of the nonperiplanar relationship between the C–H and C–Br bonds) to give a compound *that is now π-electrophilic at the C atom that was formerly σ-electrophilic*. Another equivalent of MeO⁻ can now undergo *conjugate addition* to the electrophilic C atom; the addition occurs from the bottom face for steric reasons. Proton transfer then gives the observed product.

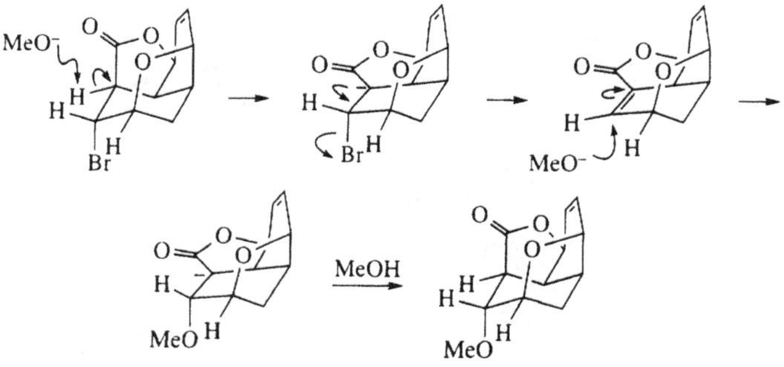

Sometimes both the S_N2 and the elimination–addition mechanisms are reasonable. The substitution reaction shown below proceeds more quickly than one would expect from a 2° alkyl chloride, suggesting that the elimination–addition mechanism is operative, but an S_N2 mechanism for the reaction is not unreasonable. Experimental evidence is required to determine unambiguously which mechanism is actually occurring.

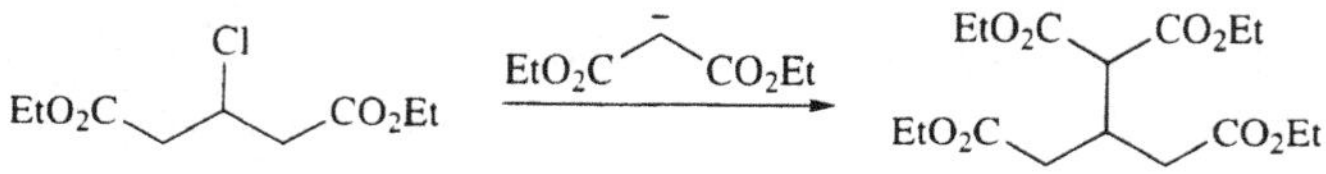

The elimination–addition mechanism for substitution is reasonable *only when elimination gives an alkene that is a π-bond electrophile at the C atom to which the leaving group was originally attached.* Most often the leaving group is β to a carbonyl, as in both the preceding examples, but not always. If a substitution reaction gives a stereochemical result other than inversion or involves a very hindered substrate, then a simple S_N2 mechanism can be ruled out, and elimination–addition (and, of course, $S_{RN}1$) should be considered.

Problem 2.11. Is an elimination–addition mechanism reasonable for the following reactions? Draw the most reasonable mechanism for each one.

(a)

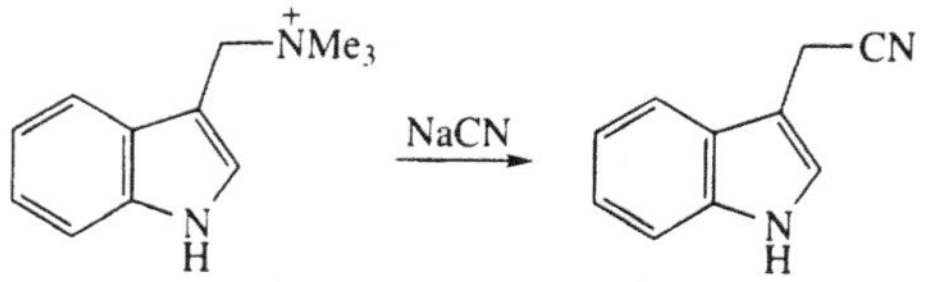

(b)

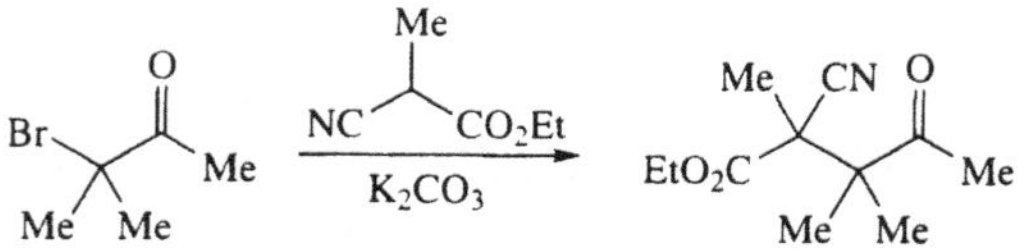

2.4.3 Substitution by the One-Electron Transfer Mechanism

One more mechanism for substitution at $C(sp^3)$–X bonds, the *one-electron transfer mechanism*, should be discussed. This mechanism is related to the $S_{RN}1$ mechanism, but it is not a chain mechanism. Instead, after one-electron transfer from

nucleophile to electrophile and departure of the leaving group, radical–
radical combination occurs to give the product.

$$\text{Nu:}^- \quad R\!-\!X \longrightarrow \text{Nu}\cdot \quad [R\!:\!X]^- \longrightarrow \text{Nu}\cdot\curvearrowright\cdot R \quad :X^- \longrightarrow \text{Nu}\!-\!R \quad :X^-$$

It is not possible to distinguish the electron transfer mechanism from the S_N2
or $S_{RN}1$ mechanism by cursory examination of the reaction conditions and substrates, so it is rarely appropriate to propose this mechanism. However, you should
be aware that some chemists believe that *all* substitution reactions are initiated
by electron transfer.

2.4.4   Metal Insertion; Halogen–Metal Exchange

Alkyl chlorides, bromides, and iodides undergo metal insertion reactions with Li,
Mg, and Zn, just as aryl and alkenyl halides do. The reaction is more facile for
heavier halogens.

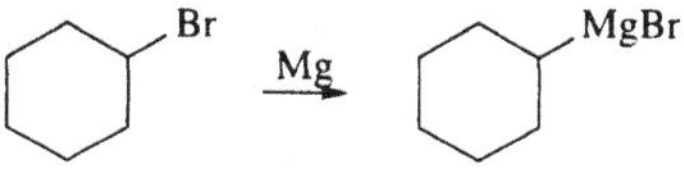

The *Reformatsky reaction*, in which an α-bromocarbonyl compound was treated
with Zn to give an enolate, was for a very long time the only way of quantitatively making enolates of weakly acidic carbonyl compounds. Nowadays it has
been superseded by strong nonnucleophilic bases like LDA and KHMDS.

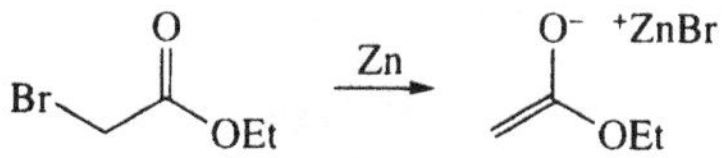

When metallation is carried out on an alkyl halide with a leaving group on the
adjacent atom, β-elimination occurs very rapidly to give an alkene. Zinc is usually used to execute this reaction. The adjacent leaving group is a halide or
pseudohalide, a carboxylate, or even an alkoxy group. The reaction accomplishes
the reverse of halogenation or cohalogenation of a double bond (Chapter 3).
Dibromoethane is often used to initiate Grignard reactions because it reacts
quickly with Mg to give the innocuous by-products ethylene and $MgBr_2$.

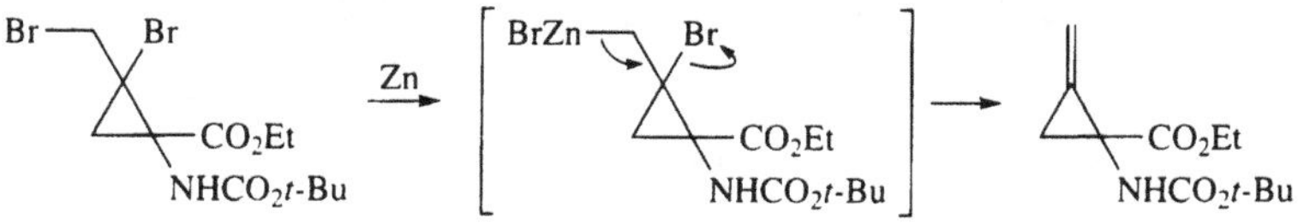

Problem 2.12. Draw a mechanism involving a metal insertion for the following reaction.

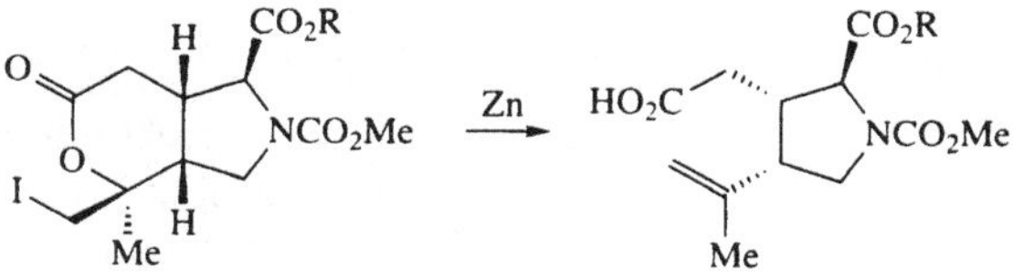

Halogen–metal exchange is much less generally useful for $C(sp^3)$ halides than it is for $C(sp^2)$ halides. The exchange is thermodynamically driven, so it proceeds at reasonable rates only with 1° alkyl iodides (RCH_2I) and two equivalents of *t*-BuLi (not MeLi, *n*-BuLi, or *s*-BuLi).

$$\underset{R}{\overset{H\ H}{\diagup}}\kern-0.5em I \quad \xrightarrow{\textit{t}\text{-Bu}-\text{Li}} \quad \underset{R}{\overset{H\ H}{\diagup}}\kern-0.5em Li \quad + \quad \textit{t}\text{-Bu}-I$$

However, bromo- and dibromocyclopropanes do undergo halogen–metal exchange at reasonable rates. The exocyclic bonds in cyclopropanes have a lot of s character, so the cyclopropyl anion is much lower in energy than an ordinary $C(sp^3)$ anion. Organolithium reagents less reactive than *t*-BuLi (e.g., *n*-BuLi and MeLi) can undergo halogen–metal exchange with cyclopropyl halides. In the case of dibromocyclopropanes, the bromolithium compound can react with an electrophile without undergoing α-elimination (see Section 2.4.5) if it is maintained at sufficiently low temperature.

2.4.5 α-Elimination; Generation and Reactions of Carbenes

Consider an alkyl halide like $CHCl_3$ that has no β-hydrogen and so cannot undergo β-elimination. The inductive withdrawing effect of the Cl atoms renders the H atom acidic enough to be removed by *t*-BuOK. However, once the H atom is removed, the lone pairs on the C and the Cl atoms repel one another, and the electropositive C atom is unhappy about having a negative charge, so a Cl atom can leave to give a *singlet carbene*, a neutral, divalent C with a lone pair and an empty orbital. The overall reaction is called α-*elimination*. It is convenient to depict a singlet carbene as a C atom with both a formal positive and a formal negative charge to emphasize the presence of both filled and empty orbitals, although of course the C atom has no formal charge at all.

The two unshared electrons in singlet carbenes have opposite spins. Carbenes generated by α-elimination are singlets because the two unshared electrons are derived from the same MO. *Triplet* carbenes, in which the two unshared electrons have parallel spins, are usually generated by photolysis of diazo compounds. Triplet carbenes are discussed in Chapter 5.

The most widely used substrates for α-elimination reactions are chloroform and bromoform, but other halides also can undergo α-elimination. For example, allyl chloride undergoes α-elimination when it is treated with a very strong base.

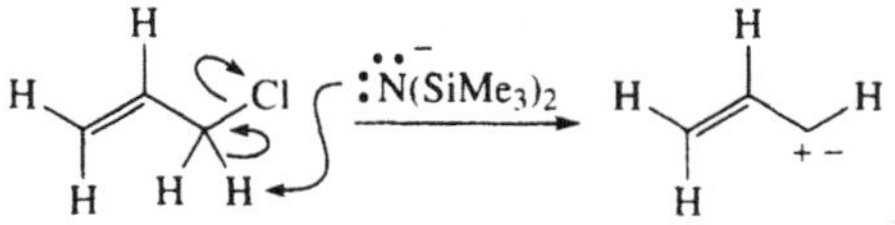

α-Elimination can also occur after halogen–metal exchange on a dibromocyclo-propane.

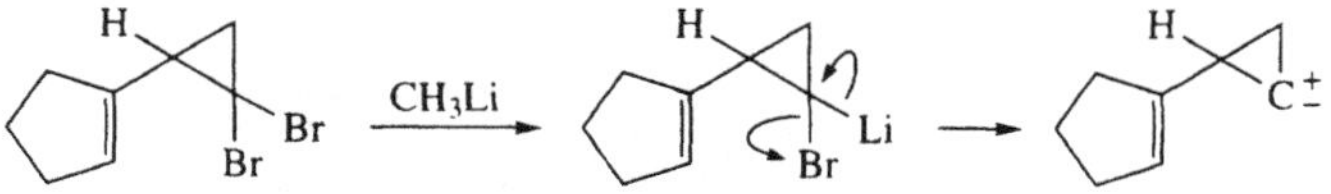

α-Elimination can occur even at N. For example. N-tosylhydrazones are converted into diazoalkanes by α-elimination of TsH (toluenesulfinic acid).

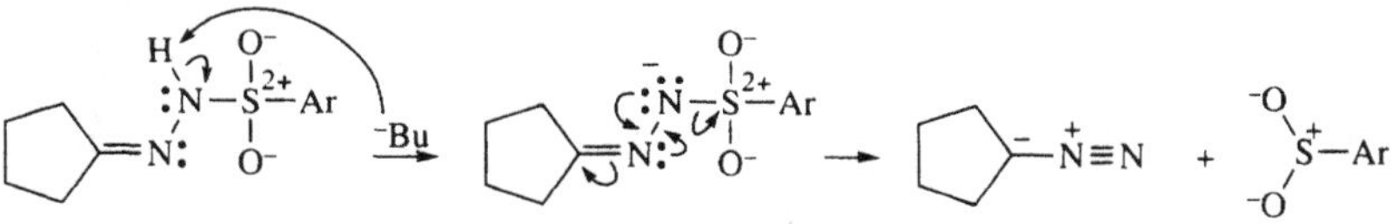

Carbenoids are compounds that react like carbenes but are not true divalent C species. The *Simmons–Smith reagent* (ICH_2ZnI) is a widely used carbenoid. The reagent is derived by insertion of Zn into one of the C–I bonds of CH_2I_2. (The rate of insertion of ordinary Zn dust into the C–I bond is very slow, so *zinc–copper couple*, or Zn(Cu), is used instead. The small amount of Cu on the Zn surface promotes the insertion reaction.) By examining the polarity of the C–I and C–Zn bonds, you can see how ICH_2ZnI might want to undergo α-elimination to give a carbene. In fact it doesn't, but it does undergo typical carbene reactions.

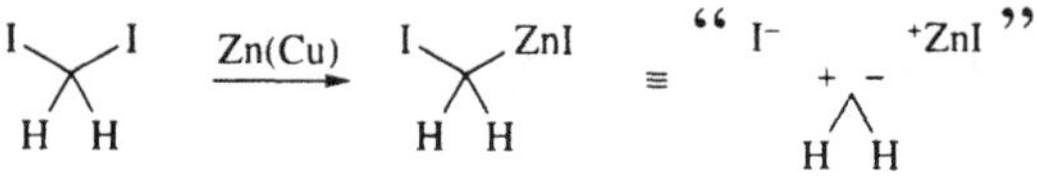

Carbenes and carbenoids can also be generated from diazo compounds ($R_2C{=}N_2 \longleftrightarrow R_2C{=}\overset{+}{N}{=}\overset{-}{N} \longleftrightarrow R_2\overset{-}{C}{-}\overset{+}{N}{\equiv}N$) by several nonbasic methods. They are discussed here because the reactivities of carbenes and carbenoids are the same no matter how they are generated. Diazo compounds are converted to singlet carbenes upon gentle warming and to carbenoids by treatment with a Rh(II) or Cu(II) salt such as $Rh_2(OAc)_4$ or $CuCl_2$. The transition-metal-derived carbenoids, which have a metal=C double bond, undergo the reactions typical of singlet carbenes. At this point you can think of them as free singlet carbenes, even though they're not. Metal carbenoids are discussed in more detail in Chapter 6.

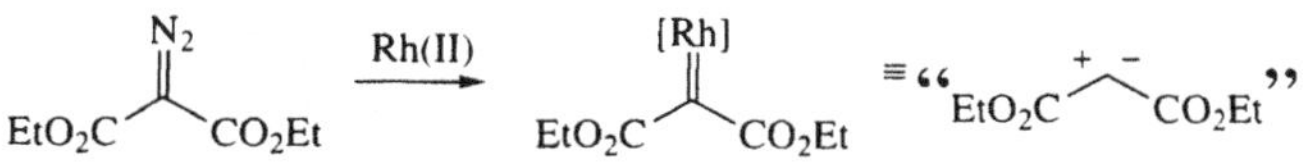

No matter how they are generated, carbenes and carbenoids undergo four typical reactions. The most widely used reaction is *cyclopropanation*, or *addition to*

a π bond. The mechanism is a concerted [2 + 1] cycloaddition (see Chapter 4). The carbenes derived from chloroform and bromoform can be used to add CX_2 to a π bond to give a dihalocyclopropane, whereas the Simmons–Smith reagent adds CH_2. Carbenoids generated from diazoalkanes with catalytic Rh(II) or Cu(II) also undergo cyclopropanations.

Example

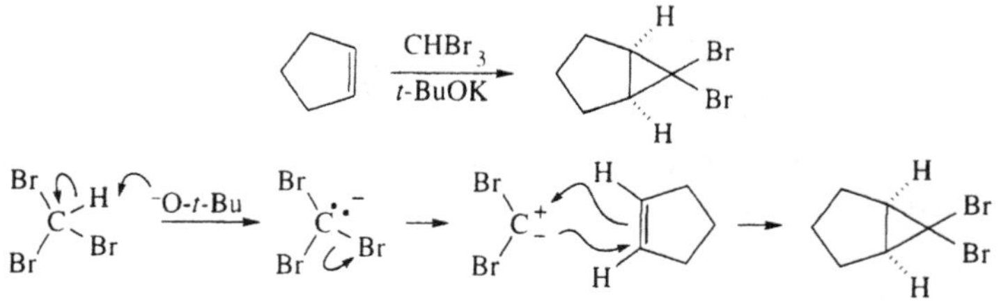

Problem 2.13. Draw mechanisms for the following two cyclopropanation reactions.

(a)

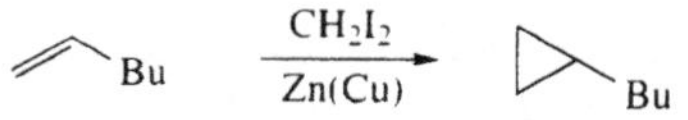

(b)

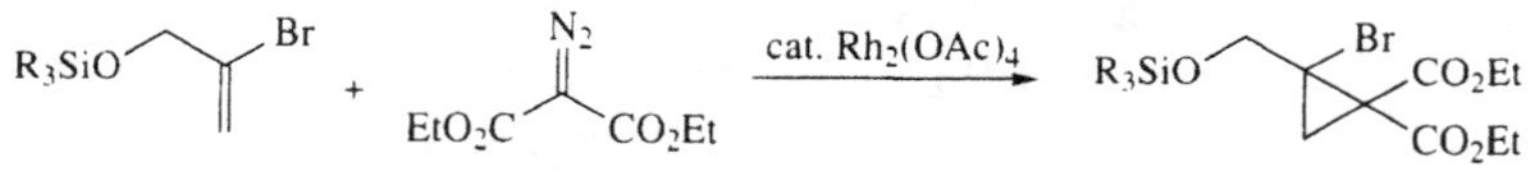

The second typical reaction of carbenes is *insertion into a C–H σ bond.* This three-centered reaction is similar to cyclopropanation, except that the carbene latches onto a σ bond instead of a π bond. Again, both carbenes and carbenoids can undergo this reaction. It is most useful when it occurs in intramolecular fashion. If the C of the C–H bond is stereogenic, the reaction proceeds with retention of configuration.

Example

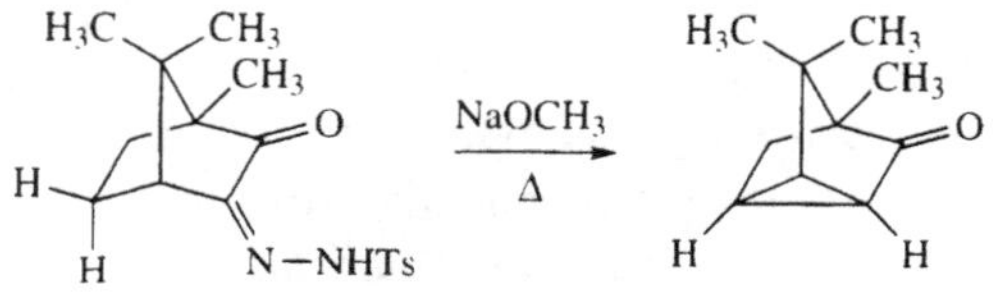

Base-promoted α-elimination of Ts–H from the hydrazone gives a diazo compound.

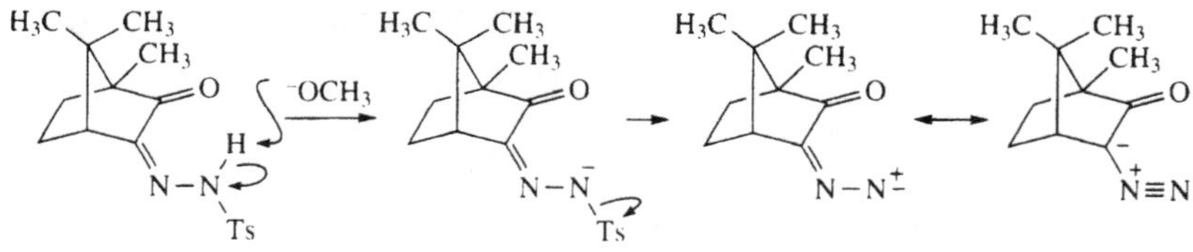

Under the reaction conditions the diazo compound decomposes to give a carbene, which undergoes a C–H insertion reaction to give the highly strained product.

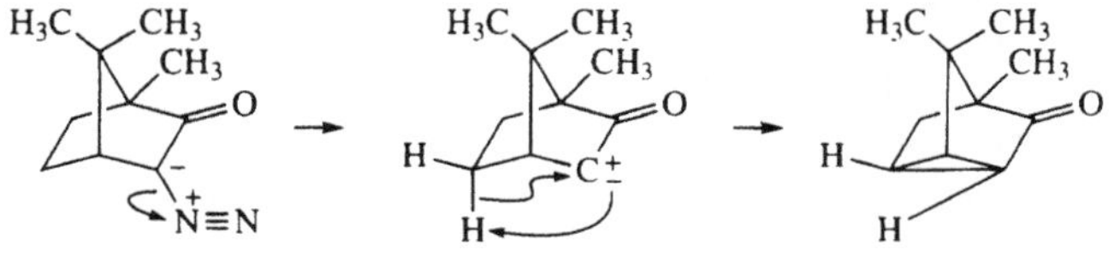

Problem 2.14. Draw a mechanism for the following C–H insertion reaction.

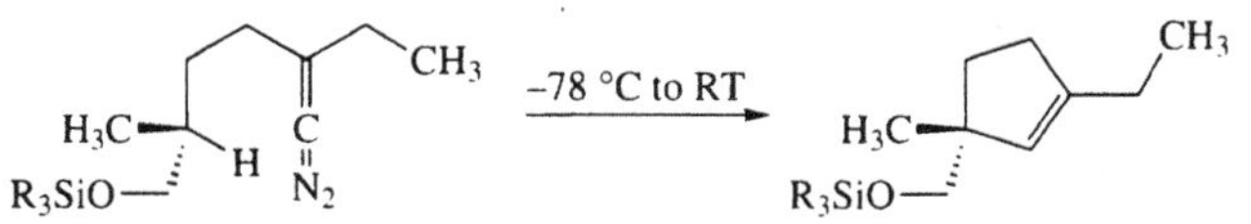

The third typical reaction of carbenes is *combination with a nucleophile*. Carbenes are electron-deficient species, so they combine with nucleophiles that have reactive lone pairs. Addition of a carbonyl O to a carbene gives a *carbonyl ylide*, a reactive compound useful for making furan rings by a 1,3-dipolar cycloaddition reaction (see Chapter 4).

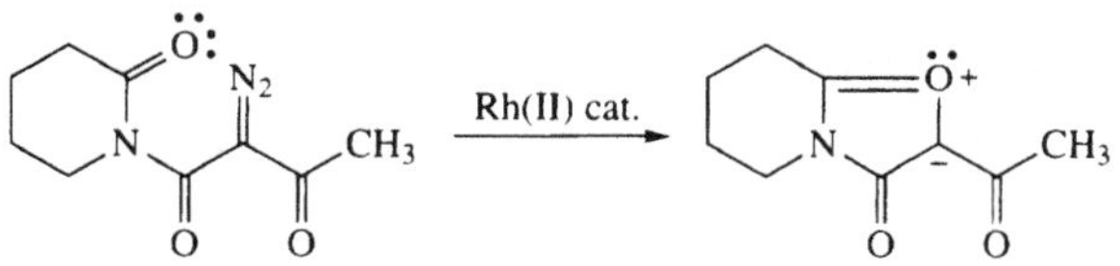

The fourth typical reaction of carbenes is a *1,2-shift*. A group on the adjacent C migrates to the carbene C with its pair of electrons, giving an alkene. The 1,2-shift severely limits the usefulness of many substituted carbenes. For example, when cyclohexene is treated with Zn/Cu and CH_2I_2, a cyclopropane is obtained in over 50% yield; with CH_3CHI_2, the yield is only 5%. Most of the $CH_3CH(I)ZnI$ is converted to $CH_2=CH_2$. The 1,2-shift can be made more useful, as in the Wolff rearrangement of α-diazoketones and Curtius rearrangement of acyl azides (see Sections 2.5.1 and 2.5.2), but these reactions generally don't proceed through free carbenes or nitrenes under thermal conditions.

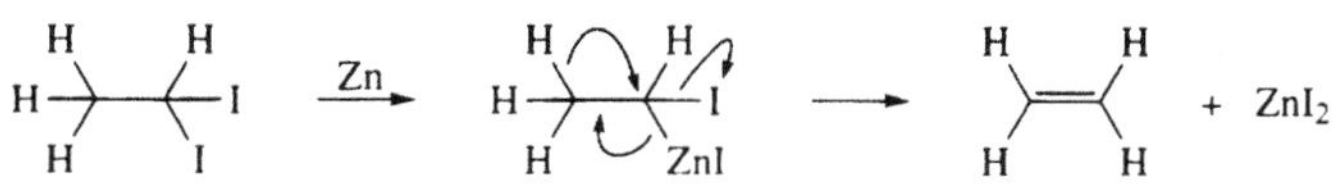

2.5 Base-Promoted Rearrangements

A rearrangement is a reaction in which C–C σ bonds have both broken and formed in the course of a reaction involving just a single substrate. It is often more difficult to draw a reasonable mechanism for a rearrangement than for any other kind of reaction. When you draw a mechanism for a rearrangement reaction, it is especially important to determine exactly which bonds are being broken and which are formed. Do this by numbering the atoms in the starting materials, identifying the atoms by the same numbers in the products, and making a list of bonds that are made and broken in the reaction.

Rearrangement reactions under basic conditions usually involve addition of a nucleophile to a carbonyl group followed by cleavage of one of the C–C σ bonds. The reaction is simply an addition–elimination reaction in which the leaving group is an alkyl group. The difference between this and most addition–elimination reactions of carbonyl compounds, of course, is that R^- is a horrid leaving group (unless of course it is somehow stabilized, e.g., as an enolate). Rearrangements occur when the atom α to the carbonyl C is electrophilic. R^- can then act as a leaving group because it *migrates* to the electrophilic atom in a 1,2-shift instead of leaving the compound entirely. The electrophilic atom may be either a σ-bond electrophile or a π-bond electrophile, and it may be C or a heteroatom.

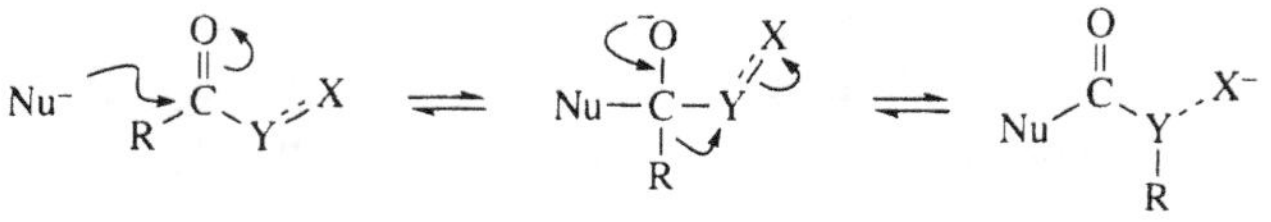

2.5.1   Migration from C to C

In the *benzylic acid rearrangement* of 1,2-diketones, HO^- adds to one ketone. The tetrahedral intermediate can collapse by expelling HO^-, giving back starting material, or it can expel Ph^-. The Ph^- group leaves because there is an adjacent electrophilic C to which it can migrate to give the product.

Example

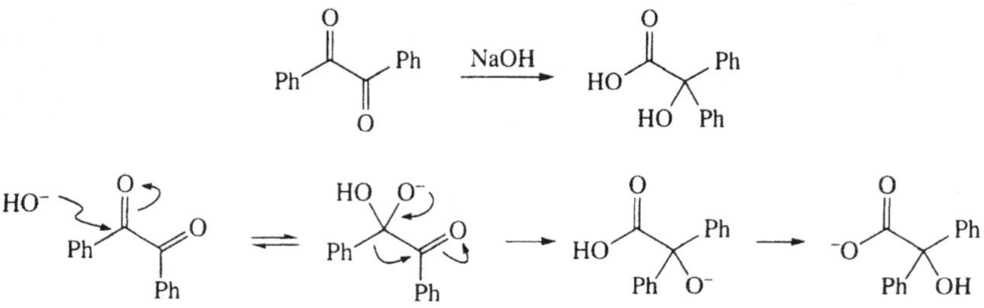

The *Favorskii rearrangement* differs from the benzylic acid rearrangement in that the α-carbon is a σ-bond electrophile rather than a π-bond electrophile.

Example

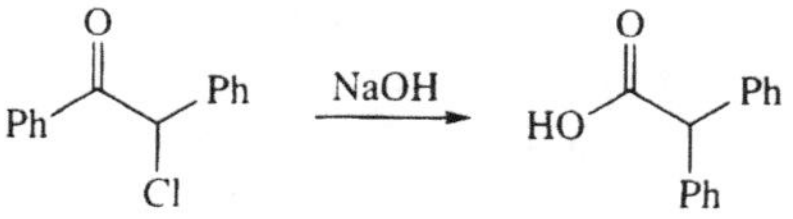

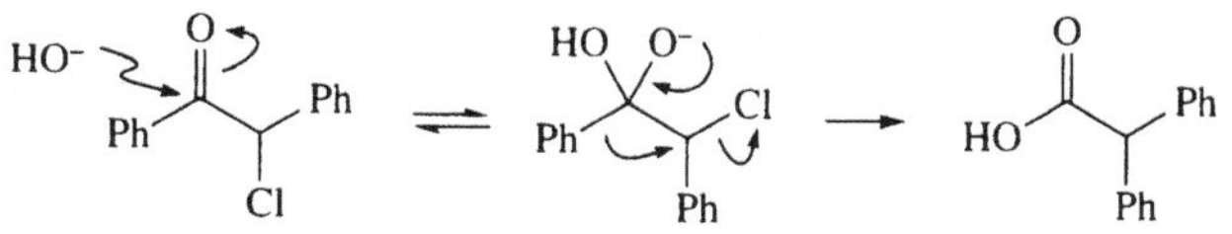

The mechanism shown, the *semibenzilic mechanism*, operates only when the ketone cannot enolize on the side opposite the leaving group, as in aryl ketones. When the ketone can enolize, a different mechanism is operative. This mechanism is discussed in Chapter 4.

Diazomethane (CH_2N_2) reacts with ketones to insert a CH_2 unit between the carbonyl and α-carbon atoms. The diazo compound acts as a nucleophile toward the electrophilic carbonyl C, and migration of R to the diazo C then occurs with expulsion of the great leaving group N_2. In principle the reaction can occur with other diazo compounds, but in practice other diazo compounds are rarely used for this purpose, as they are much more difficult to prepare than CH_2N_2. If the migrating R group is a stereocenter, retention of configuration occurs.

Example

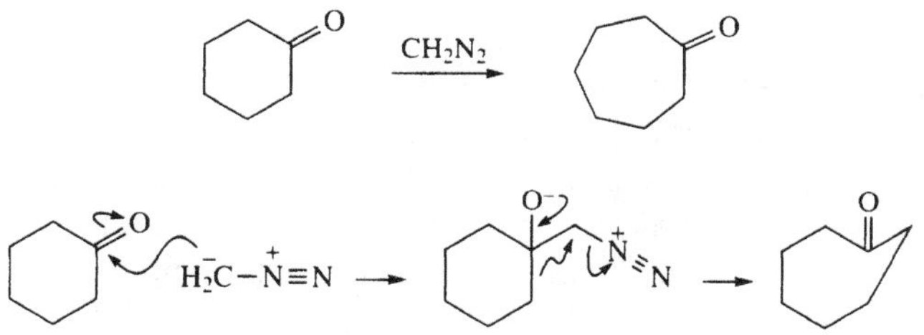

In the *Wolff rearrangement*, an α-diazoketone is heated to give a ketene. The mechanism of the Wolff rearrangement consists of one step: the carbonyl susstituent migrates to the diazo C and expels N_2. When the reaction is executed in H_2O or an alcohol, the ketene reacts with solvent to give a carboxylic acid or an ester as the ultimate product. (The mechanism of this reaction was discussed in Section 2.3.1.)

Example

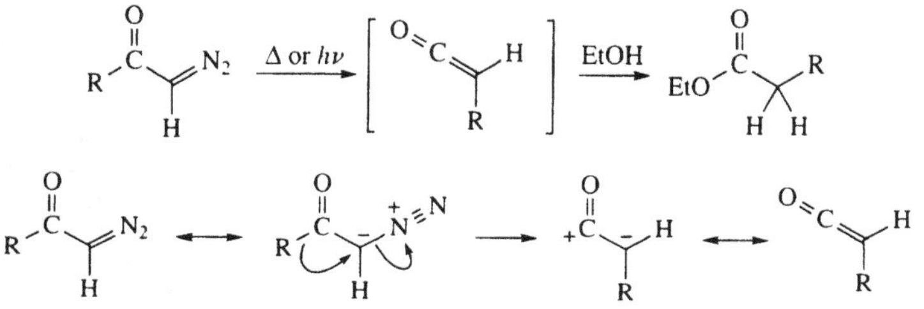

Light can be used to promote the Wolff rearrangement, too, and in this case the reaction is called the *photo-Wolff rearrangement*. The photo-Wolff rearrangement is probably a nonconcerted reaction. First, N_2 leaves to give a free carbene, and then the 1,2-alkyl shift occurs to give the ketene.

α-Diazoketones are often prepared by adding CH_2N_2 to an acyl chloride in a conventional addition–elimination reaction. In this case, the entire sequence from acid to acyl chloride to ketene to acid is called the *Arndt–Eistert homologation*. More complex α-diazocarbonyl compounds can be made by the reaction of unsubstituted or monosubstituted 1,3-dicarbonyl compounds with sulfonyl azides. If the nucleophilic carbon is monosubstituted, one of the carbonyl groups is lost.

Problem 2.15. Draw a reasonable mechanism for the diazo transfer reaction. Your task will be much easier if you draw the by-products.

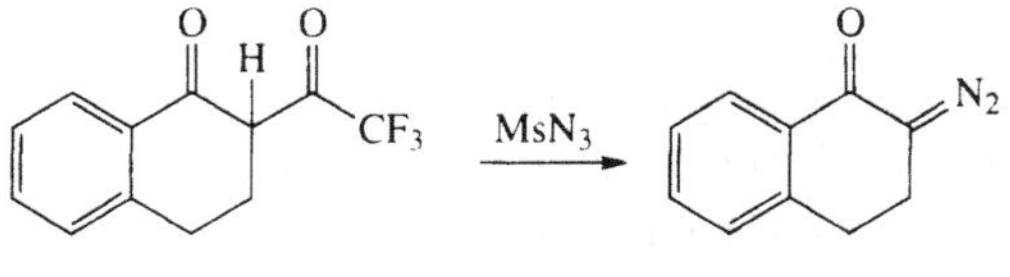

2.5.2 Migration from C to O or N

Ketones react with peracids (RCO_3H) to give esters in the *Baeyer–Villiger oxidation*. The peracid is usually *m*-chloroperbenzoic acid (mCPBA); peracetic acid and trifluoroperacetic acid are also commonly used. Peracids have an O–OH group attached to the carbonyl C. They are no more acidic than alcohols, because deprotonation of the terminal O does not give a stabilized anion, but they are usually contaminated with some of the corresponding acid. The Baeyer–Villiger reaction is thus usually conducted under mildly acidic conditions. Protonation of the ketone O occurs first to make the carbonyl C even more electrophilic than it normally is. The terminal O of the peracid then adds to the carbonyl C. The O–O bond is very weak, so migration of R from the carbonyl C to O occurs, displacing the good leaving group carboxylate. (Proton transfer occurs first to transform the carboxylate into an even better leaving group.) The product ester is obtained after final loss of H^+ from the carbonyl O.

Example

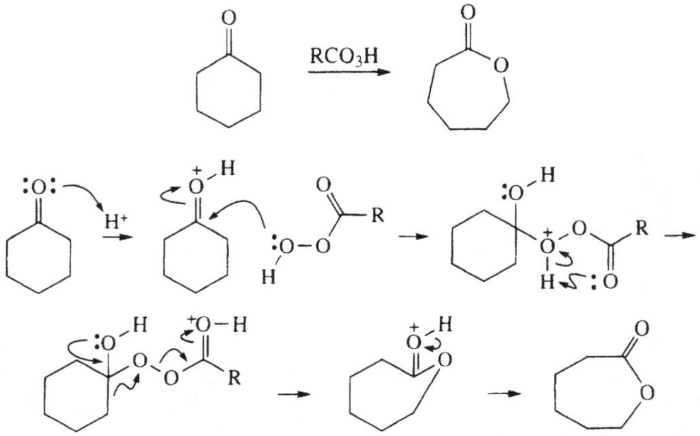

Problem 2.16. The Baeyer–Villiger reaction can also be carried out under basic conditions. Draw a reasonable mechanism.

In the *Curtius rearrangement*, an acyl azide undergoes rearrangement to an isocyanate upon gentle heating. The mechanism is exactly analogous to the Wolff rearrangement. The acyl azide may be prepared under basic conditions by reaction of an acyl chloride with NaN_3, or under acidic conditions by reaction of an acyl hydrazide ($RCONHNH_2$) with nitrous acid (HNO_2). (The latter reaction is discussed in Chapter 3.)

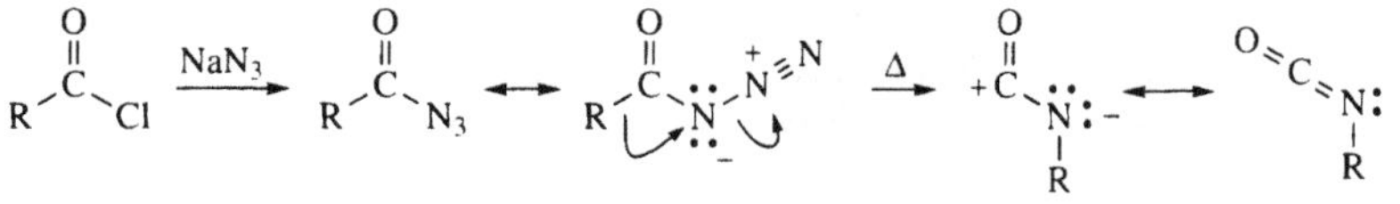

The *Hofmann rearrangement* can also be used to convert a carboxylic acid derivative to an isocyanate. A carboxamide is treated with Br_2 and aqueous base to give an isocyanate, which is usually hydrolyzed under the reaction conditions to give an amine with one fewer C atom than the starting material. The reaction proceeds by deprotonation of the amide and *N*-bromination, then a second deprotonation and rearrangement. The amide N is more reactive than the amide O in the bromination step because N is deprotonated under the reaction conditions.

Example

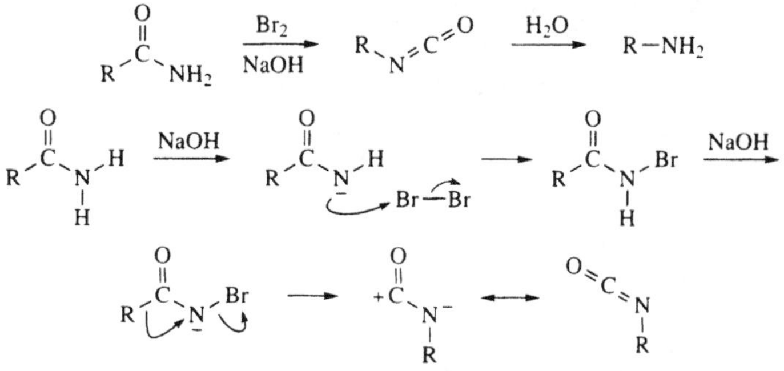

2.5.3   Migration from B to C or O

Migration of R to a neighboring atom occurs in situations other than when R is attached to a carbonyl C. Trialkylboranes also undergo migration reactions. For example, a trialkylborane (R_3B) will react with H_2O_2 under basic conditions to give a trialkyl borate (($RO)_3B$), which is hydrolyzed in situ to give ROH. The mechanism of this reaction is very similar to the mechanism of the Baeyer–Villiger reaction. In a trialkylborane, B is electron-deficient, so it wants an octet. Upon treatment with basic H_2O_2, HOO$^-$ adds to B to give an eight-electron complex. But B is electropositive, and it doesn't want a formal negative charge. So it throws off an alkyl group, which migrates to the adjacent O, displacing HO$^-$.

But now B is electron-deficient again, and it wants an octet again, so the process repeats itself until all the B–C bonds have been broken. The trialkylborate that is ultimately obtained undergoes hydrolysis to give the alcohol.

$$R-\underset{\underset{R}{|}}{B}-R \xrightarrow[\text{NaOH}]{\text{H}_2\text{O}_2} RO-\underset{\underset{OR}{|}}{B}-OR \longrightarrow ROH$$

$$R-\underset{\underset{R}{|}}{B}-R \xrightarrow{-O-OH} R-\underset{\underset{R}{|}}{\overset{O-OH}{B}}-R \longrightarrow \underset{\underset{R}{|}}{\overset{R-O}{B}}-R \quad {}^{-}OH \rightleftharpoons RO-\underset{\underset{OR}{|}}{B}-OR$$

Boron, the whiniest element in the periodic table, is never satisfied with its electron count. When it lacks its octet, it complains about being electron-deficient, but when a nucleophile comes along to give it its octet, it complains about its formal negative charge.

The alkyl groups in boranes can migrate to C, too. For example, addition of the carbenoid equivalent LiCHCl$_2$, obtained by deprotonation of CH$_2$Cl$_2$ with LDA at low temperatures, to an alkylboronate gives the one-carbon homologous product. Other carbenoid equivalents such as diazo compounds and CO can undergo similar reactions.

2.6   Two Multistep Reactions

A few well-known and widely used reactions use a cocktail of several different reagents and proceed by multistep mechanisms that are not easily discerned by the beginning student. Two of these reactions, the Swern oxidation and the Mitsunobu reaction, are discussed here. You will find that faculty members enjoy asking graduate students to draw mechanisms for these particular reactions, so you should learn them well!

2.6.1   The Swern Oxidation

When an alcohol is treated with a reaction mixture derived from oxalyl chloride, DMSO, and Et$_3$N, it is oxidized to the aldehyde or ketone. The order of addition is important. First, oxalyl chloride is added to DMSO; then, Et$_3$N and the alcohol are added, and the reaction mixture is allowed to warm to room temperature. The by-products are Me$_2$S (very smelly!), CO$_2$, CO, and Et$_3$NH$^+$ Cl$^-$.

The overall transformation with respect to the alcohol is an *elimination* of the elements of H_2. The problem is that H^- is not a leaving group. Either C or O of the alcohol must have an attached H replaced by a leaving group before an elimination reaction can take place. The O of the alcohol is nucleophilic under basic conditions, and therefore the elimination reaction may take place first by attachment of a leaving group to O of the alcohol, and then by an E2 elimination. The role of the DMSO and the oxalyl chloride, then, is to produce an electrophile that becomes a leaving group when attached to the alcohol O.

DMSO is nucleophilic on O. It reacts with the electrophilic oxalyl chloride by addition–elimination.

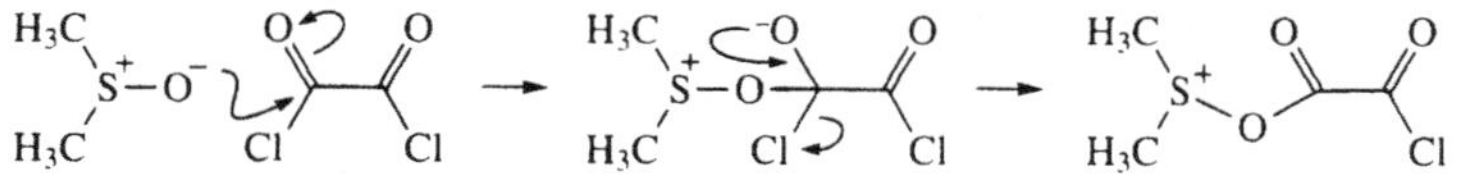

S now has a good leaving group attached. Cl^- can come back and attack S, displacing the oxalate, which decomposes to give CO_2, CO, and Cl^-. The S–O bond is thereby cleaved.

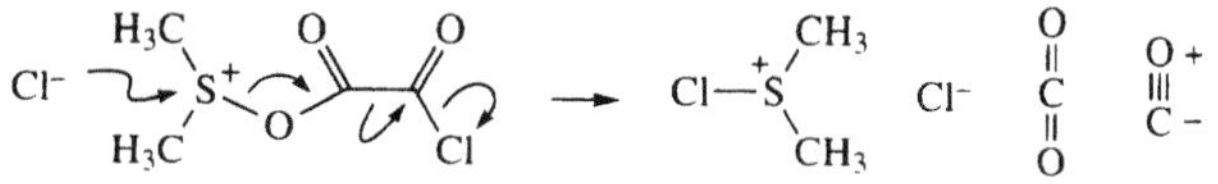

At this point S is a good electrophile. Now, the alcohol and Et_3N are added. The alcohol is deprotonated and attacks S, displacing Cl^-.

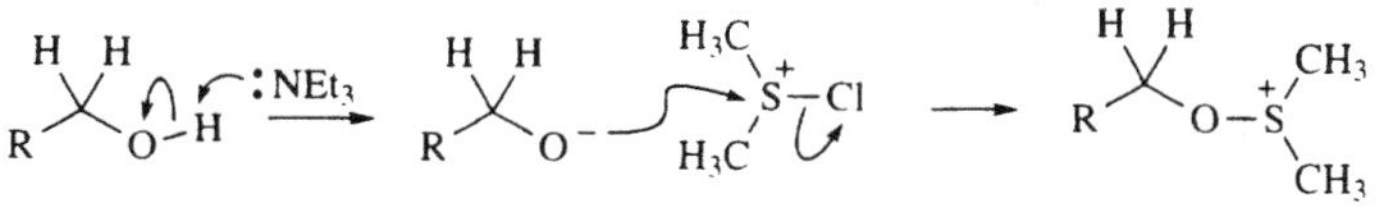

It is reasonable to draw the alkoxide attacking the S–oxalate compound directly, rather than having Cl^- intervene.

There is a good leaving group attached to the O of the former alcohol now, so an E2 elimination ensues, giving the aldehyde and Me_2S.

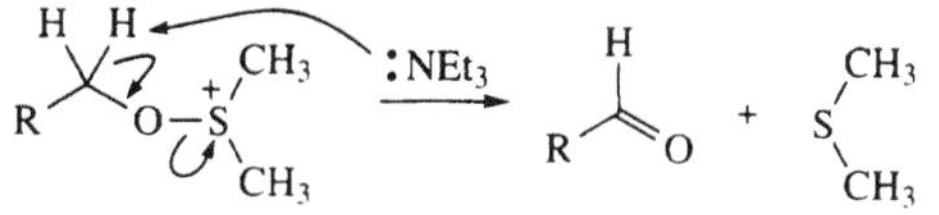

Although this last step is reasonable, it is not the way that the elimination actually takes place. In the currently accepted mechanism, the CH_3 group is deprotonated to give a *sulfur ylide*, and a *retro-hetero-ene* reaction ensues. Retro-hetero-ene reactions are discussed in Chapter 4.

Many variations on the Swern oxidation exist, usually involving replacing $(COCl_2)_2$ with another activating agent such as $(CF_3CO)_2O$ (trifluoroacetic anhydride, TFAA) or DCC (*N,N'*-dicyclohexylcarbodiimide, $C_6H_{11}N=C=NC_6H_{11}$).

Some of these reactions even have different names associated with them. In all of them, though the O of DMSO is converted into a good leaving group, the alcohol O displaces the leaving group from S, and an elimination of H^+ and Me_2S ensues to give the aldehyde (or ketone).

2.6.2 The Mitsunobu Reaction

In the Mitsunobu reaction, a chiral 2° alcohol and a carboxylic acid are converted to an ester with clean inversion at the electrophilic C. The reaction requires Ph_3P and $EtO_2CN=NCO_2Et$ (diethyl azodicarboxylate, DEAD). It is usually carried out by adding DEAD slowly to a mixture of the alcohol, Ph_3P, and the nucleophile in its protonated form.

The reaction is clearly not a simple S_N2 displacement of ^-OH from R by $PhCO_2{}^-$, because HO^- is a really awful leaving group in S_N2 reactions and will not leave under these mild conditions, and besides, Ph_3P and DEAD are required for the reaction. But equally clearly, S_N2 displacement must happen at some point, or clean inversion at R could not occur.

Balancing the equation shows that H_2O has been lost. Whenever Ph_3P is used in a reaction, it is almost always converted to Ph_3PO. The other by-product, then, must be $EtO_2CNHNHCO_2Et$. Because a single S_N2 displacement has occurred at the alcohol C, both O atoms in the product must come from benzoic acid, which means that the former alcohol O must end up attached to Ph_3P. Both the alcohol O and P are nucleophilic, so the role of the DEAD must be to convert one of them from a nucleophile into an electrophile. DEAD itself is a potent electrophile, with two electrophilic N atoms.

The first part of the mechanism of the Mitsunobu reaction involves addition of nucleophilic Ph_3P to the electrophilic N in DEAD. (The alcohol could add to DEAD instead, but P is far more nucleophilic.) The addition is preceded by protonation of DEAD by the carboxylic acid.

In the second part of the mechanism, the carboxylate displaces N from P in S_N2 fashion, giving an acyloxyphosphonium ion and a nitrogen anion. The latter then deprotonates the alcohol, generating an alkoxide, which displaces P from O to give an alkoxyphosphonium ion and to regenerate the carboxylate.

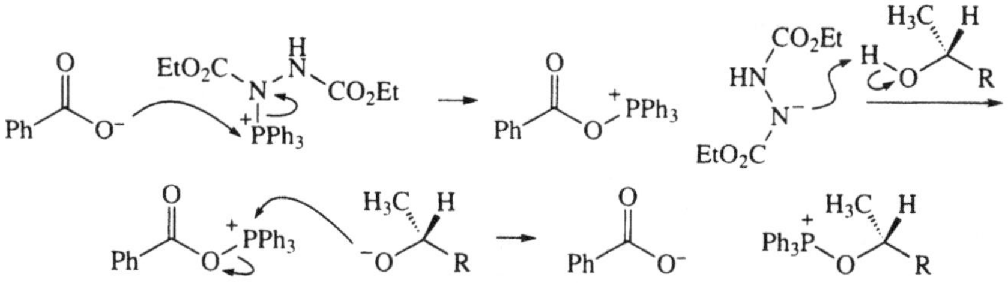

An alternative mechanism has the carboxylate deprotonate the alcohol, which then attacks the phosphoniohydrazine to displace N from P. Although this scheme is more direct, it violates the pK_a rule (pK_a alcohol ≈ 17, pK_b of carboxylate ≈ 5).

The nascent alkoxyphosphonium ion is a much better σ-bond electrophile than the alcohol, so in in the last part of the reaction mechanism, the carboxylate O attacks the electrophilic C of the oxyphosphonium salt in S_N2 fashion, displacing Ph$_3$PO to give the product.

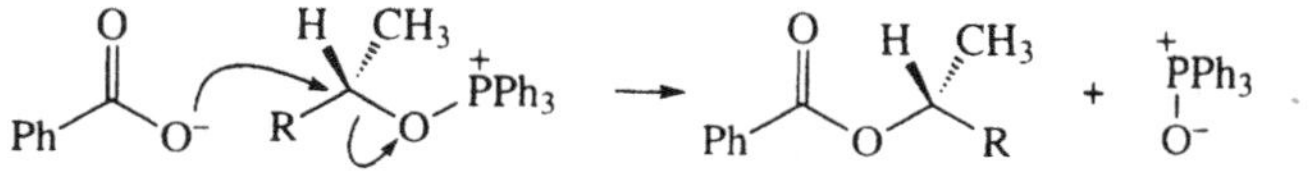

The mild conditions required for the Mitsunobu reaction make it one of the best methods for inverting the configuration of an alcohol or subjecting a 2° alcohol to nucleophilic substitution. Normally, E2 elimination competes with S_N2 substitution at 2° C(sp³), but little or no elimination occurs in the Mitsunobu reaction.

2.7 Summary

A saturated carbonyl compound is electrophilic at the carbonyl carbon. It is nucleophilic at the α-carbon (and on oxygen) *after it has been deprotonated* to make an enolate.

An α,β-unsaturated carbonyl compound is electrophilic at the β- and carbonyl carbons. It is nucleophilic at the α-carbon *after a nucleophile has added* to the β-carbon to make an enolate. It is *not* acidic at the α-carbon.

sp³-Hybridized atoms undergo substitution by

1. S_N2 (by far the most common: 1° or 2° halide, 3° halide only if intramolecular); or

2. $S_{RN}1$ (3° halide, delocalized or heavy nucleophile, sometimes $h\nu$ or a one-electron donor); or

3. elimination–addition (when β-elimination makes the carbon at which substitution occurs a π-bond electrophile); or

4. metal insertion (Mg, Li, Zn) and (occasionally) halogen–metal exchange (*t*-BuLi).

sp²-Hybridized atoms undergo substitution by

1. addition–elimination (by far the most common: carbonyls, electrophilic alkenes, aromatic compounds with electron-withdrawing groups);

2. elimination–addition (acyl chlorides; aromatic compounds when very strong base is used);

3. $S_{RN}1$ (delocalized or heavy-atom nucleophile, sometimes $h\nu$ or a one-electron donor); or

4. metal insertion (Mg, Li, Zn) and halogen–metal exchange (n-BuLi, t-BuLi, MeLi, s-BuLi).

Singlet carbenes undergo four typical reactions:

1. addition to a π bond (cyclopropanation);
2. insertion into a C–H bond;
3. addition to a lone-pair nucleophile; or
4. 1,2-shift.

Carbonyl compounds in which the α-carbon is electrophilic often undergo rearrangement reactions. Addition of a nucleophile to a carbonyl compound often prompts a rearrangement reaction.

Substitution reactions involve making a C–Nu bond and breaking a C–X bond. In principle, there are three modes by which the bond-making and bond-breaking can take place (Table 2.2). In the first mode, Nu⁻ adds, an intermediate is obtained, and then, in a second step, X⁻ leaves. In the second mode, Nu⁻ adds and X⁻ leaves simultaneously. In the third mode, X⁻ leaves, an intermediate is obtained, and then, in a second step, Nu⁻ adds. Examples of all three modes can be found in reactions under basic conditions. Addition–elimination takes place in the first mode; S_N2 takes place in the second mode; and elimination–addition and $S_{RN}1$ take place in the third mode. The first mode, in which Nu⁻ adds to give a discrete intermediate, occurs at C only when C can temporarily shed a pair of electrons, e.g., when it is π-bound to O. The third mode, in which X⁻ leaves first, seems to require a carbocationic intermediate. Because carbocations cannot exist under basic conditions, the third mode occurs under basic conditions only under special circumstances: either (1) when an electron is pumped into the system first so that a radical, not a carbocation, is obtained when X⁻ leaves ($S_{RN}1$), or (2) when a pair of electrons from an adjacent C–H bond can be used to satisfy the electron demand of the C at which substitution takes place (elimination–addition). Under acidic conditions, where carbocations can exist, the third mode is quite common; it is known as S_N1.

Elimination reactions involve breaking a C–H bond and a C–X bond. In principle, there are three modes by which the bond-breaking can take place (Table 2.3). In the first mode, the C–H bond breaks first to give an anionic intermediate, and then the C–X bond breaks. In the second mode, the two bonds break simultaneously. In the third mode, the C–X bond breaks to give a cationic intermediate, and then the

TABLE 2.2. Polar substitution mechanisms

Mode:	Nu⁻ adds first	Nu⁻ adds and X⁻ leaves simultaneously	X⁻ leaves first
Mechanism:	addition–elimination	S_N2	elimination–addition, $S_{RN}1$, S_N1

TABLE 2.3. Polar elimination mechanisms

Mode:	C–H breaks first	C–H breaks and X⁻ leaves simultaneously	X⁻ leaves first
Mechanism:	E1cb	E2	E1

C–H bond breaks. The E1cb and E2 mechanisms that occur under basic conditions take place by the first and second modes, respectively. The third mode generally does not take place under basic conditions, but under acidic conditions, where carbocations can exist, the third mode is quite common, and it is known as E1.

The final step in a reaction that occurs under basic conditions is often protonation of the conjugate base of the product. The requisite proton sometimes comes from an aqueous workup, sometimes from solvent, sometimes from the conjugate acid of a basic catalyst, and sometimes from more starting material. How do you know whence comes the proton? The answer depends on the amount of base and on the basicity of the conjugate base of the product.

• When a substoichiometric (catalytic) amount of base is used, the base must be regenerated so that it can go back and deprotonate more starting material; therefore, the proton must come from within the reaction mixture and not from aqueous workup. If the catalytic base is relatively weak (e.g., EtO^-, Et_3N), the proton can come from its conjugate acid. (The conjugate acid is generated in the very first step of the mechanism, when the base deprotonates the starting material.) If the catalytic base is very strong (e.g., NaH, LDA), the proton cannot come from its conjugate acid, so it must come from more starting material.

• When a stoichiometric or excess amount of base is used, all of the acidic protons of the starting material are consumed at the beginning of the reaction, and it is not necessary to regenerate the base at the end of the reaction. The proton of the final neutralization step must come from an aqueous workup.

• If the relative amount of base is not specified, you must use your judgment about pK_b's to judge whence the proton comes. Some reactions generate very acidic products that must be deprotonated and remain so until aqueous workup (e.g., hydrolysis of esters to carboxylic acids, Claisen condensation to give 1,3-dicarbonyl compounds). On the other hand, some reactions generate the conjugate bases of only moderately acidic products (e.g., aldol condensation). These bases may acquire a proton from solvent or from the conjugate acid of the added base if the conjugate acid is not too weak. Alkoxide bases are commonly used in the corresponding alcohol solvents (e.g., NaOEt in EtOH), and in these reactions, the solvent is often a proton source.

PROBLEMS

1. Determine and draw the mechanism (S_N2, $S_{RN}1$, addition–elimination, or elimination–addition) of each of the following substitution reactions. Some reactions might reasonably proceed by more than one mechanism; in these cases,

suggest an experiment that would allow you to determine the mechanism unambiguously.

(a)

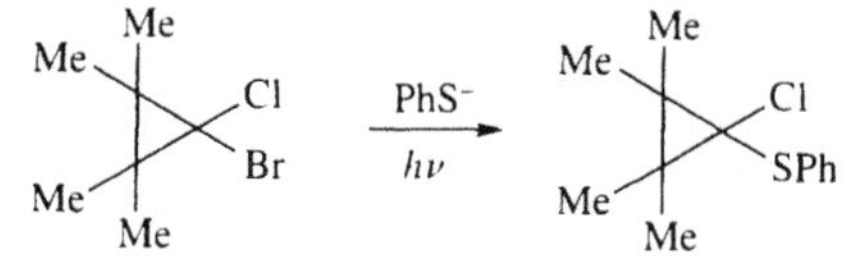

(b)

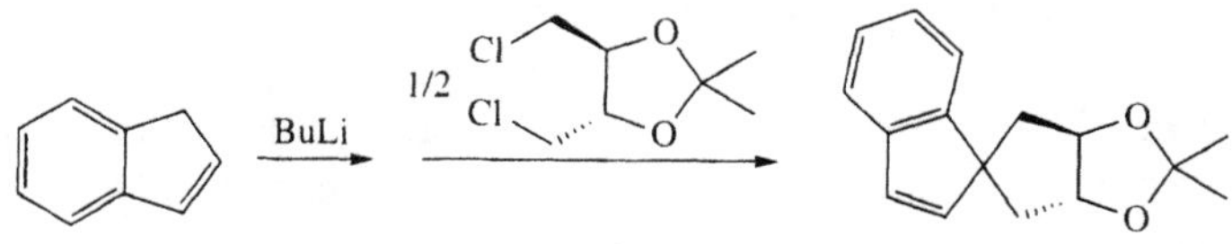

(c)

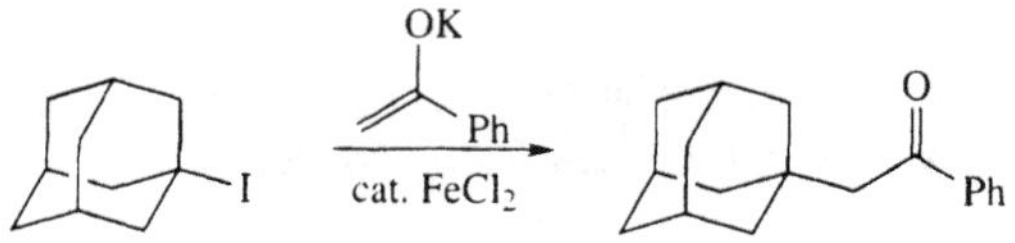

(d)

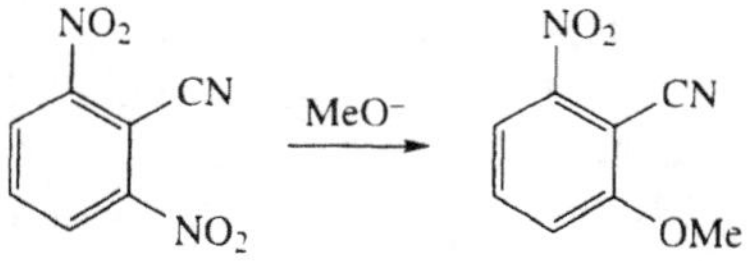

(e)

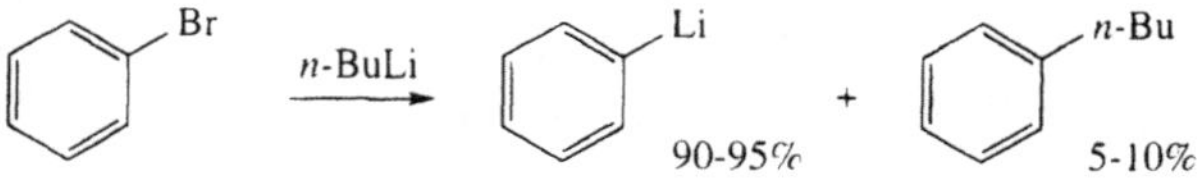

(f)

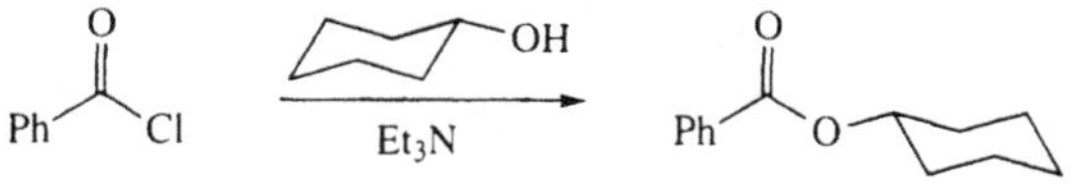

(g)

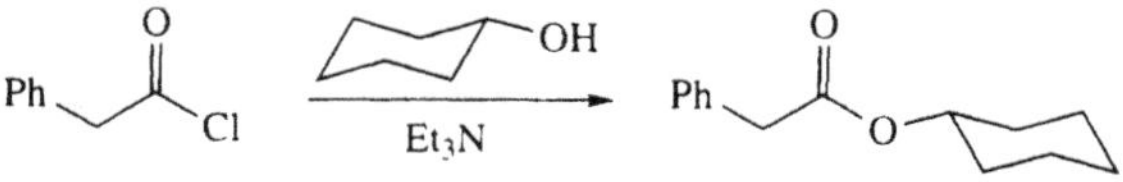

(h)   The product is a model for the antitumor agent duocarmycin, a potent
       alkylator of DNA.

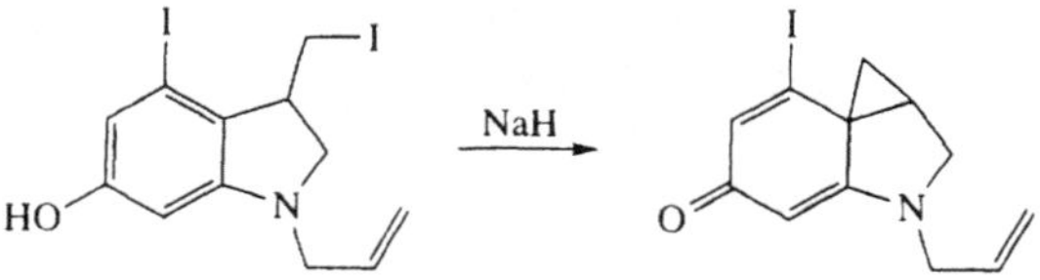

(i)

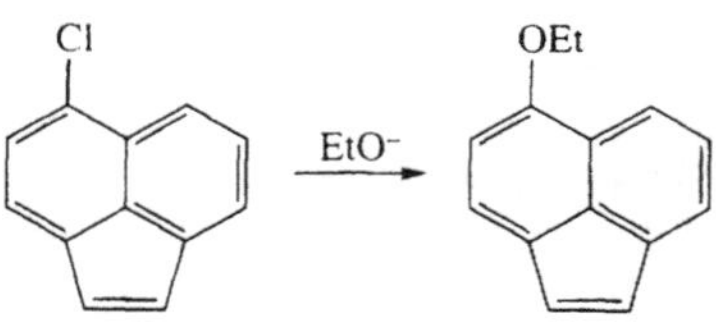

(j)

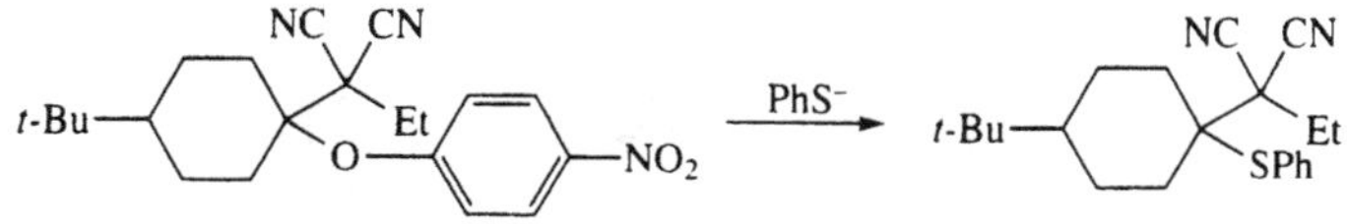

(k)

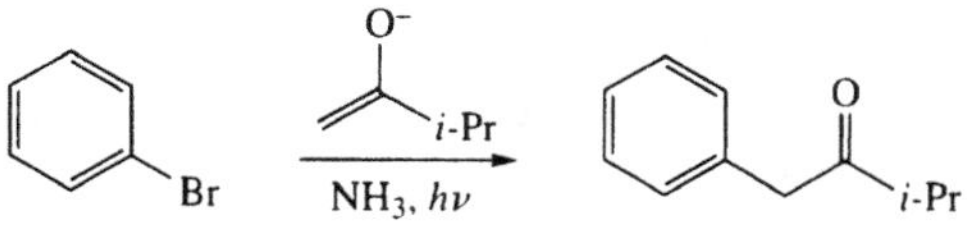

(l)   Note the stereochemistry!

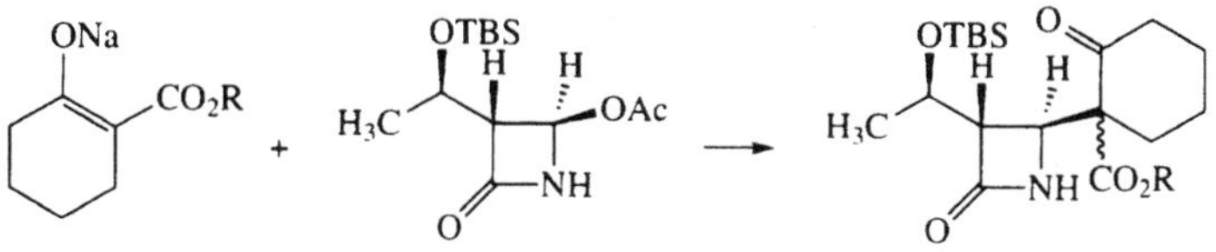

2. Draw mechanisms for the following reactions.

(a)

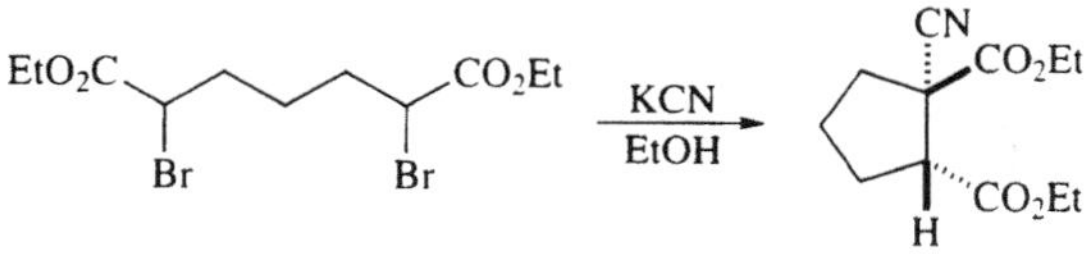

(b)

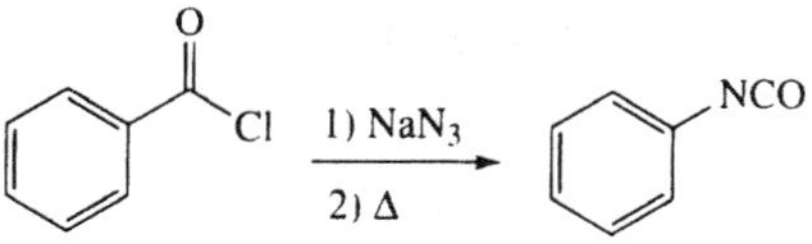

(c)

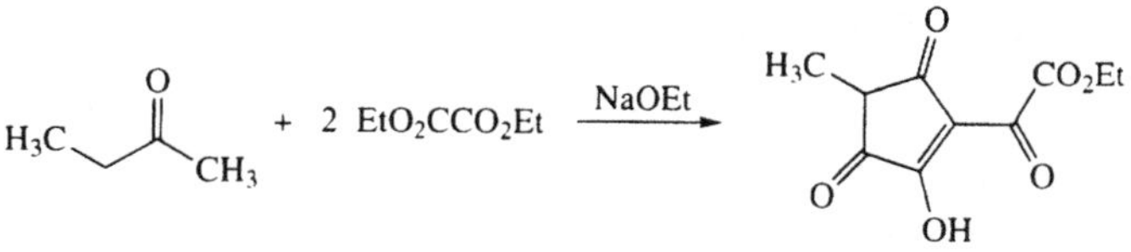

(d)

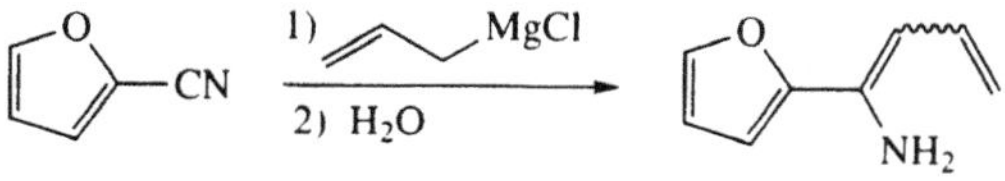

(e)   *The Darzens glycidic ester synthesis.*

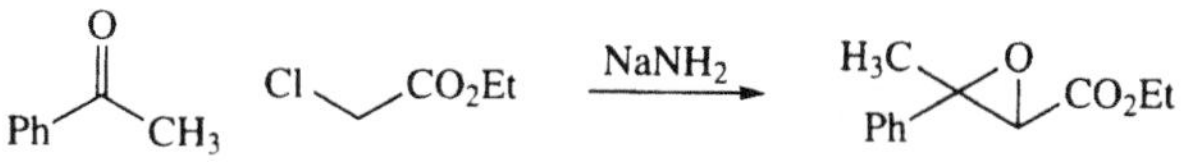

(f)

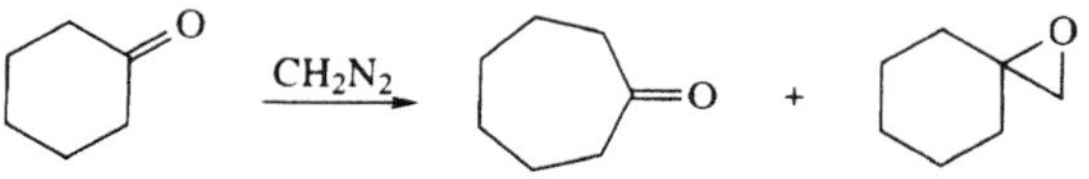

(g)

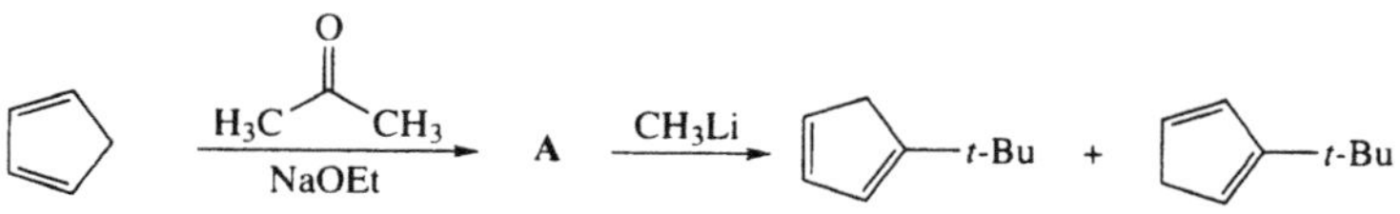

(h)

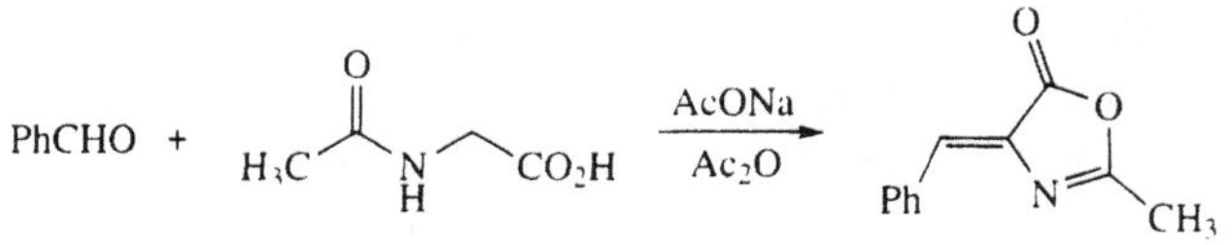

PhCHO + [H₃C–C(=O)–NH–CH₂–CO₂H] $\xrightarrow[\text{Ac}_2\text{O}]{\text{AcONa}}$ [oxazolone product with Ph and CH₃]

(i)

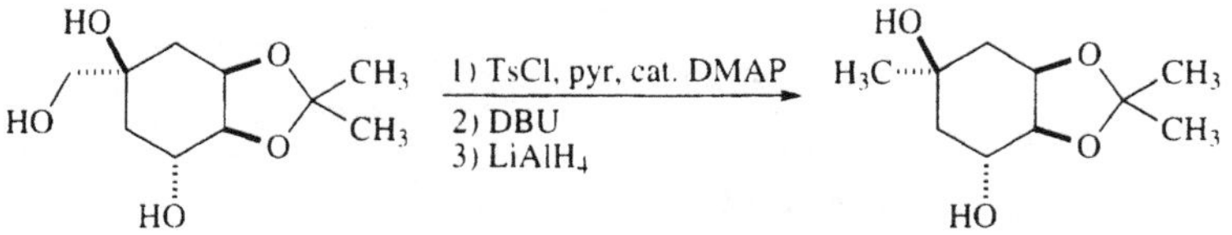

[bicyclic acetonide triol] $\xrightarrow[\substack{2)\ \text{DBU} \\ 3)\ \text{LiAlH}_4}]{1)\ \text{TsCl, pyr, cat. DMAP}}$ [bicyclic acetonide diol product]

(j)

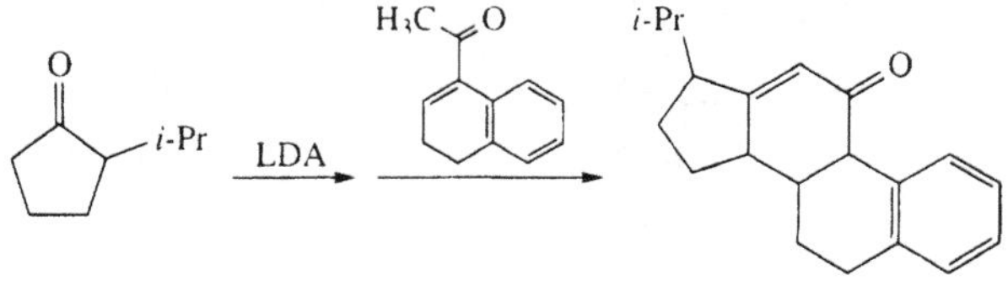

[cyclopentanone with i-Pr] $\xrightarrow{\text{LDA}}$ [enone intermediate with H₃C–C(=O)] $\longrightarrow$ [steroid-like tetracyclic ketone]

(k)

[fused bicyclic with Mg] $\xrightarrow[\text{CO}_2;\ \text{H}_3\text{O}^+]{\text{epoxide with Et}}$ [spirocyclic lactone product with Et and exocyclic methylene]

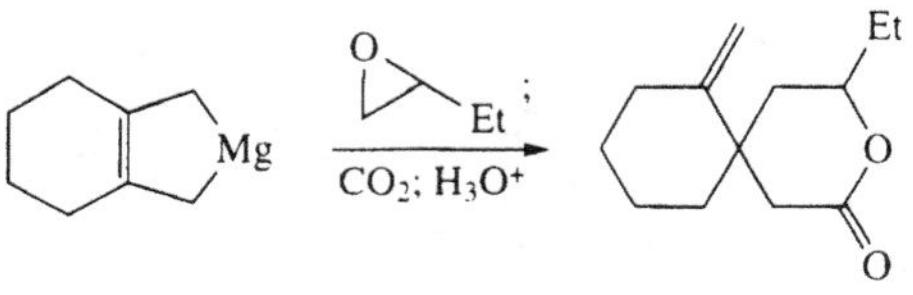

(l)   Two mechanisms can be drawn for the *Bayliss–Hillman* reaction. Draw one.

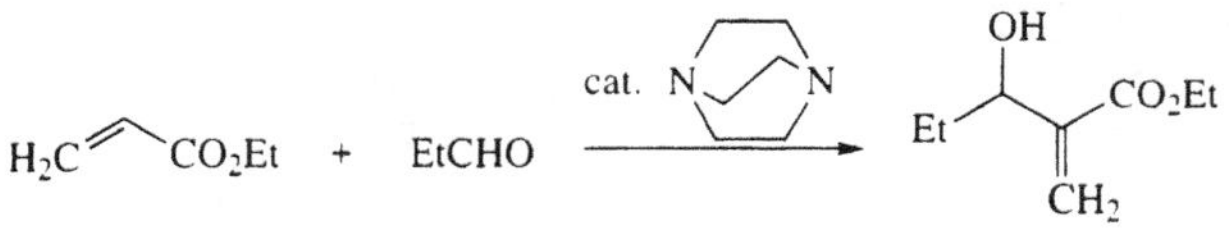

$H_2C{=}\,CO_2Et$ + EtCHO $\xrightarrow{\text{cat. N}\quad\text{N (DABCO)}}$ [Et–CH(OH)–C(=CH₂)–CO₂Et]

(m) The rate of the Bayliss–Hillman reaction is dramatically reduced when the steric bulk of the tertiary amine catalyst is increased. Is your mechanism consistent with this information? If not, draw another mechanism.

(n) TMEDA = $Me_2NCH_2CH_2NMe_2$; the reaction takes a different course in its absence, but it's not needed for the mechanism.

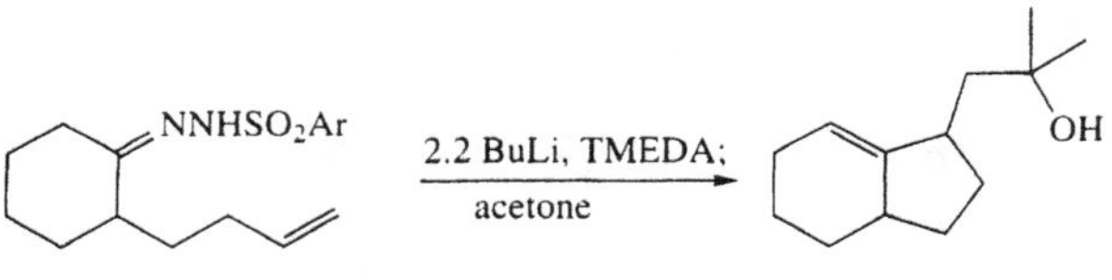

[cyclohexyl NNHSO₂Ar with butenyl chain] $\xrightarrow[\text{acetone}]{2.2\ \text{BuLi, TMEDA;}}$ [bicyclic product with C(CH₃)₂OH]

(o)

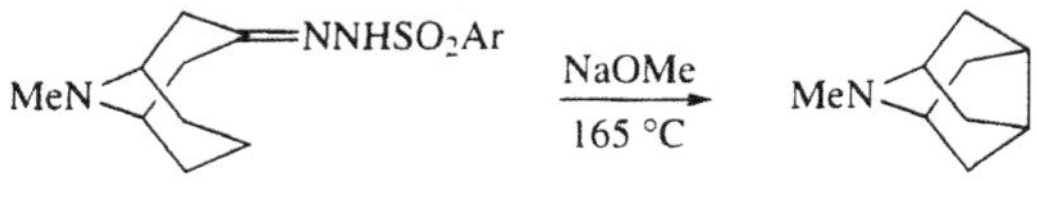

[MeN bicyclic =NNHSO₂Ar] $\xrightarrow[165\ °C]{\text{NaOMe}}$ [MeN bicyclic product]

(p)

$N_2{=}\,CO_2Et$ $\xrightarrow{\text{LDA; }i\text{-BuCHO}}$ A $\xrightarrow{\text{Rh}_2(\text{OAc})_4}$ [i-Bu–C(=O)–CH₂–CO₂Et]

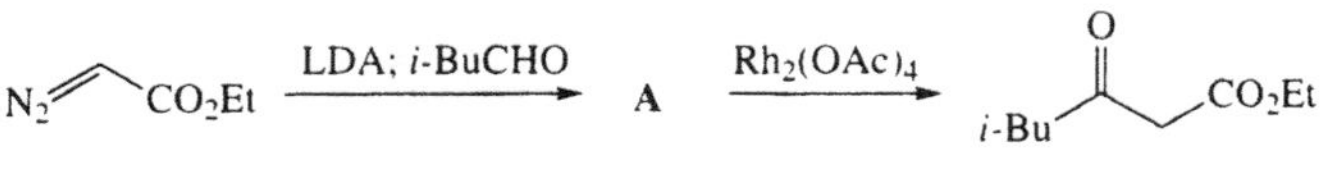

(q)

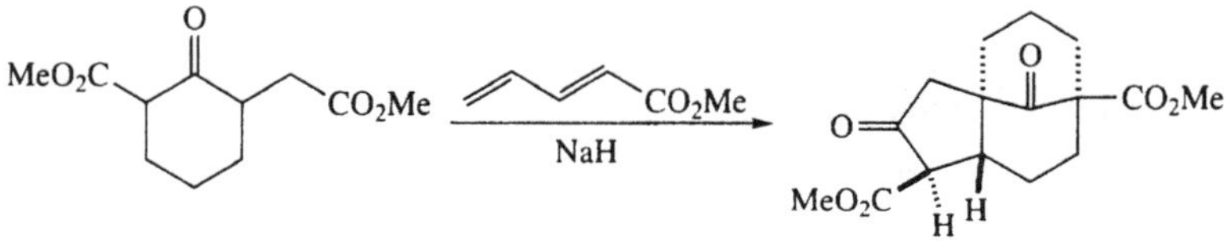

(r)   The *Arbuzov reaction.*

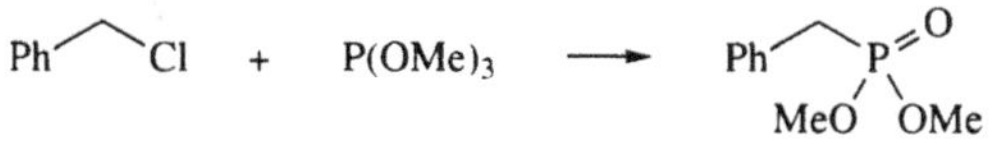

(s)   No ^{18}O is incorporated into the product!

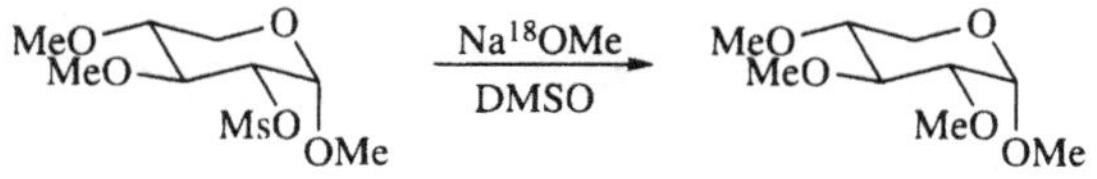

(t)

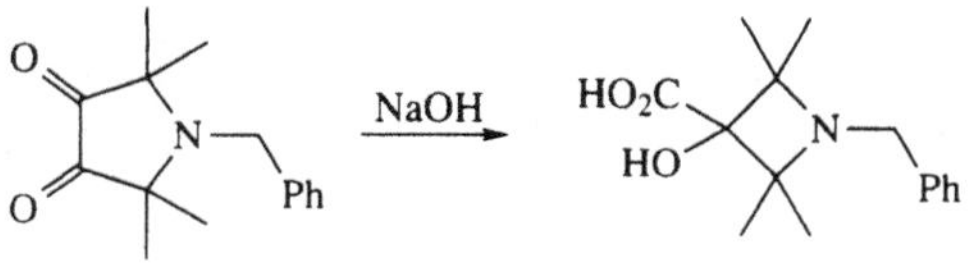

(u)

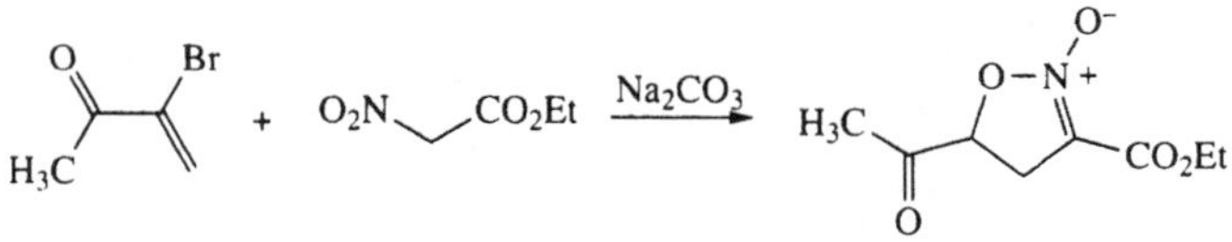

(v)   This is a very difficult problem. Here's a tip: No reaction between *p*-toluenesulfonylmethyl isocyanide (TosMIC) and cyclopentanone occurs until base is added to the reaction mixture. Draw a deprotonation and then the first bond-forming step, and *then* number the atoms.

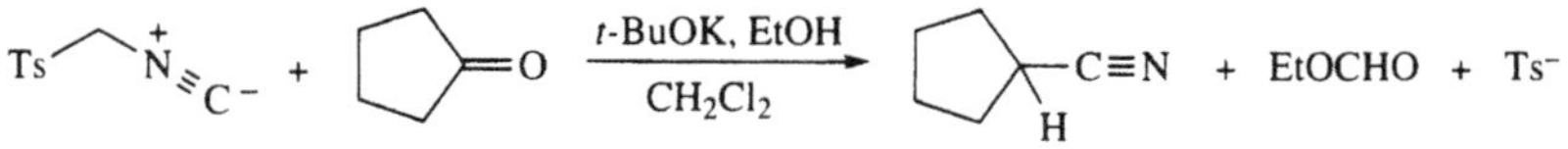

(w)   The *Potier–Polonovski rearrangement.* Pyr = pyridine.

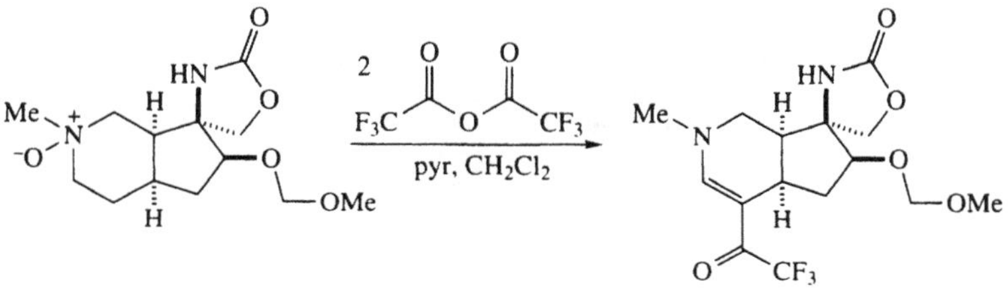

(x)   This is a hard one, but if you label all the atoms and concentrate on bonds made and bonds broken, you should be able to do it. One atom is already labeled to get you started.

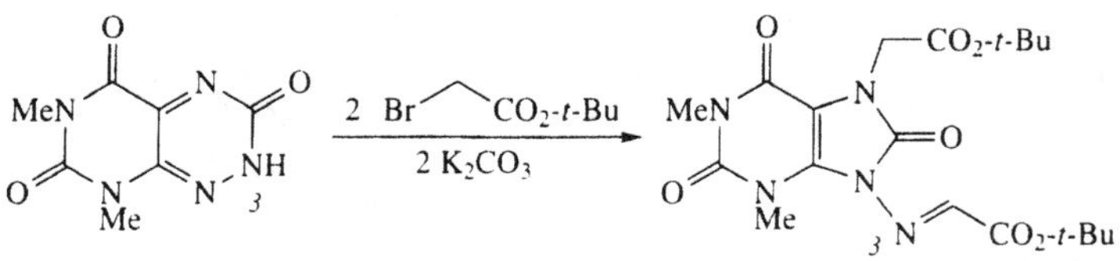

(y) Pip = piperidine, a saturated six-membered ring with one NH group.

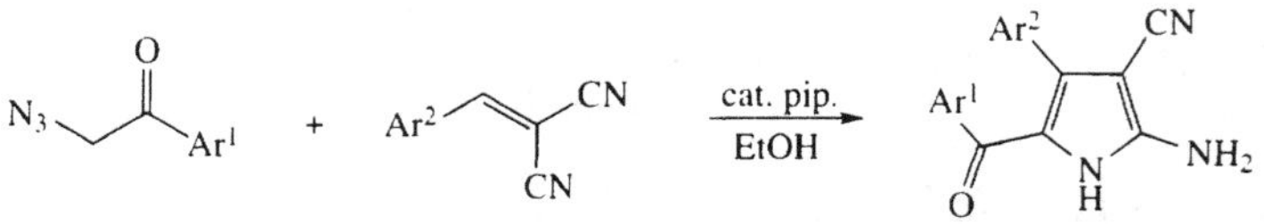

(z)

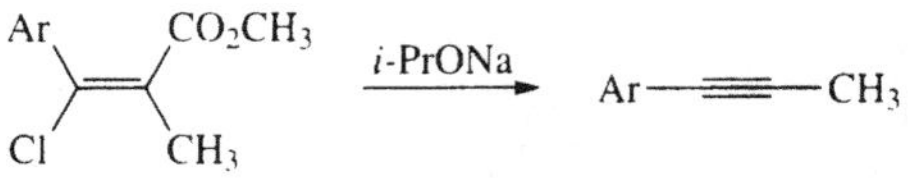

(aa)

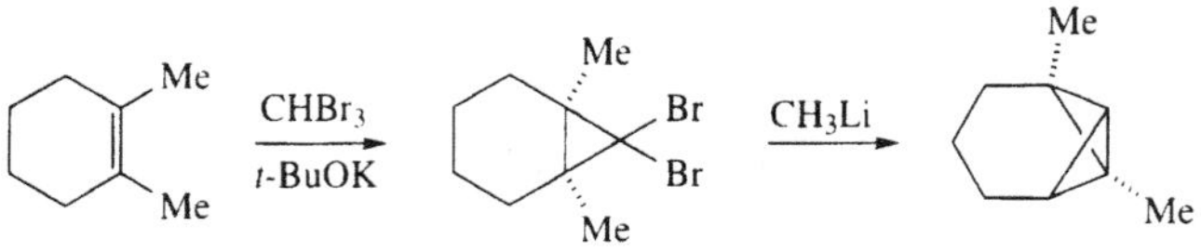

(bb)

(cc)

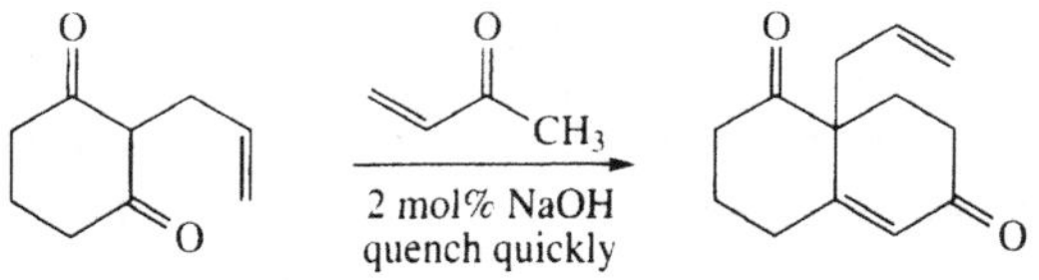

(dd) Under very similar conditions but using more NaOH, the Robinson annulation product is no longer the major one. *Numbering correctly is key
to solving this problem!*

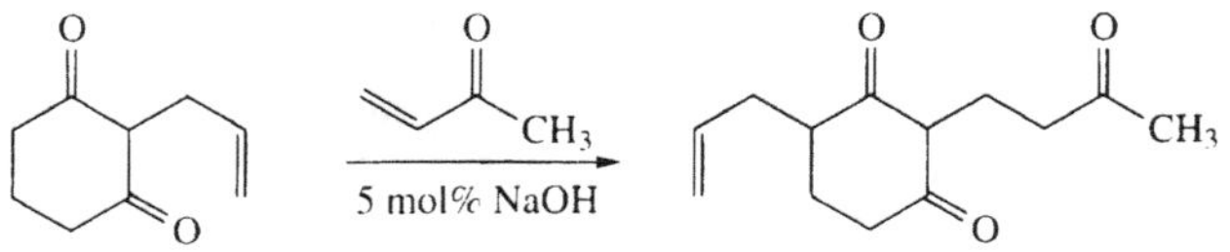

(ee)

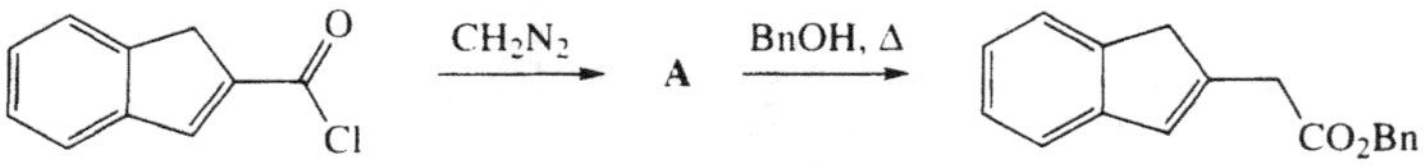

(ff)

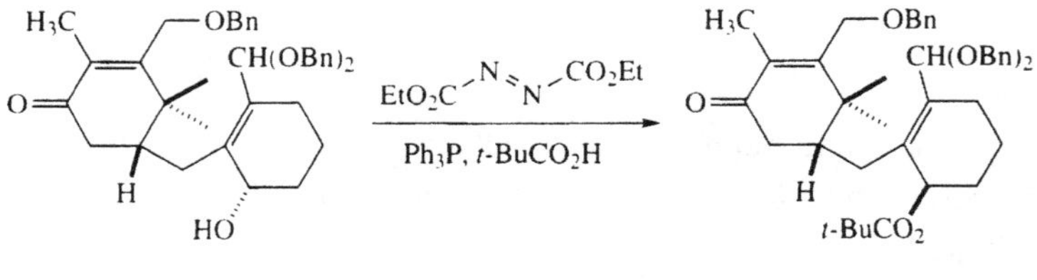

(gg) The first step is a Michael reaction. Use this information to number the atoms.

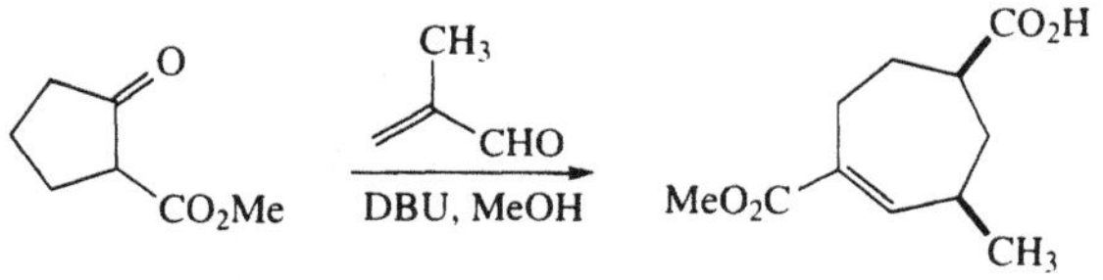

3. Indicate the product(s) of the following reactions and draw the mechanism. In some cases, no reaction occurs. Be sure to indicate the stereochemistry of the product where applicable.

(a) CH_3CH_2F $\xrightarrow{H_2O}$

(b) 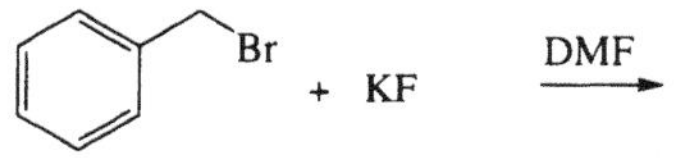 + KF $\xrightarrow{DMF}$

(c) 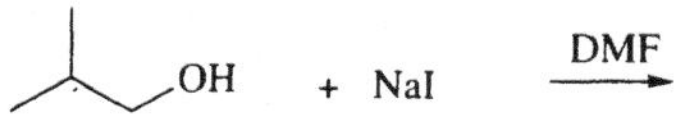 OH + NaI $\xrightarrow{DMF}$

(d) 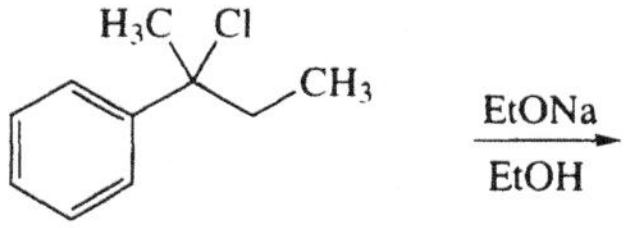 $\xrightarrow[EtOH]{EtONa}$

(e) 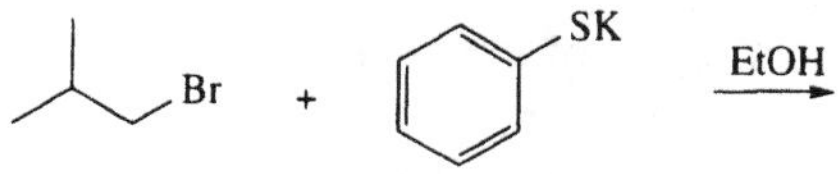 Br + SK $\xrightarrow{EtOH}$

(f) 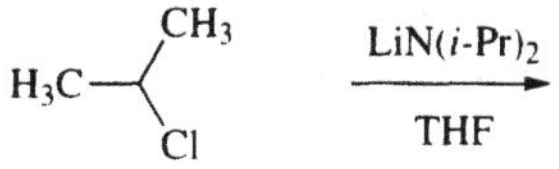 $\xrightarrow[THF]{LiN(i\text{-}Pr)_2}$

(g) 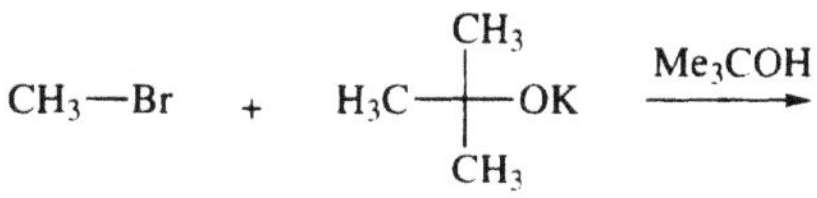 CH_3—Br + $\xrightarrow{Me_3COH}$

(h) 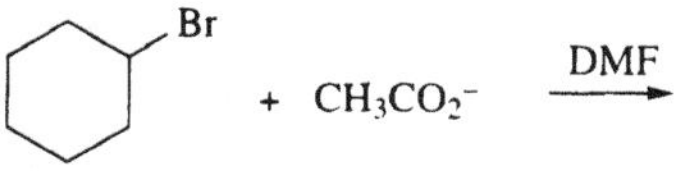 + $CH_3CO_2^-$ $\xrightarrow{DMF}$

(i) CH_3–S–CH_3 + CH_3I $\xrightarrow{ether}$

(j) 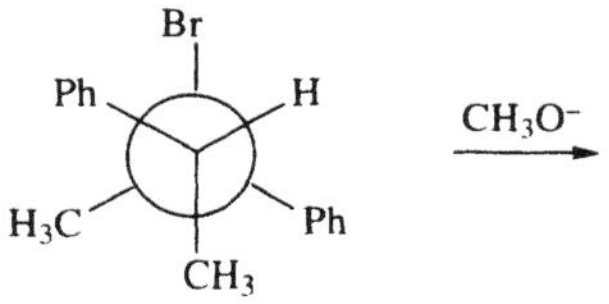 $\xrightarrow{CH_3O^-}$

(k)

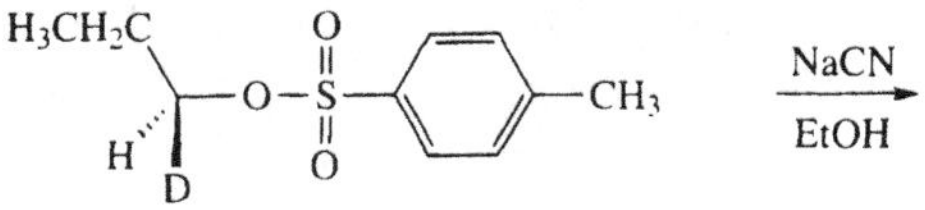

(l)

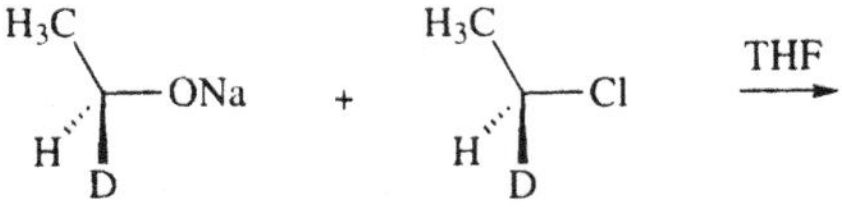

4. *Catalytic antibodies* were first prepared in the mid-1980s. In one example, a rabbit or mouse was immunized with phosphonate ester **A**, and pure anti-**A** antibodies were isolated from this animal's spleen cells. Some of the anti-**A** antibodies were found to catalyze the hydrolysis of the ester functionality of **B** at rates significantly faster than background.

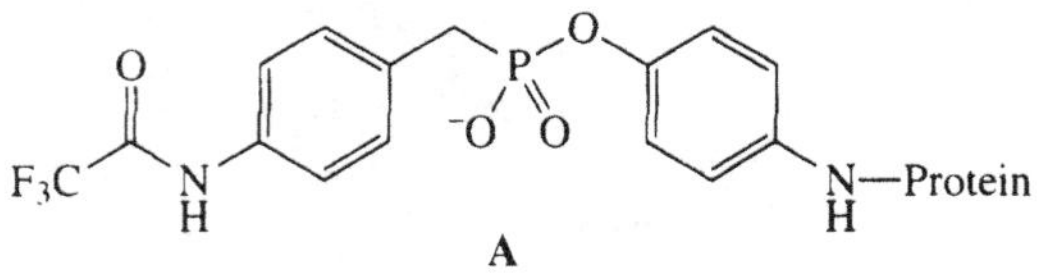

A

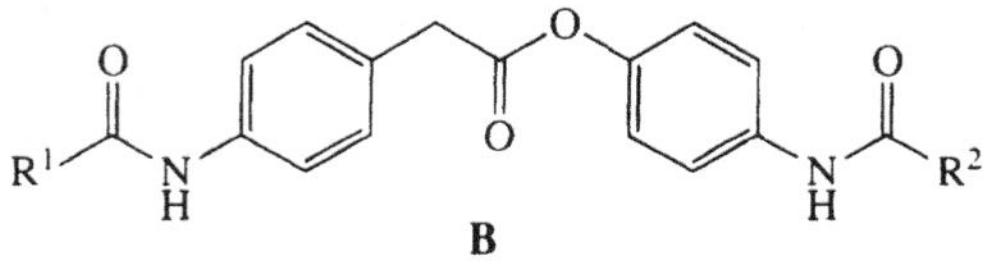

B

(a) Draw the mechanism for the hydrolysis of the ester functionality of **B** under basic conditions. What is the rate-determining step?

(b) Considering the parameters on which the rate of any chemical reaction depends, explain why antibodies to **A** catalyze (increase the rate of) the hydrolysis of **B**. Use a reaction coordinate diagram to illustrate your explanation.

3

Polar Reactions under Acidic Conditions

Most textbooks do not strongly differentiate polar mechanisms that occur under basic conditions from those that occur under acidic conditions, but the considerations that are brought to bear when drawing polar basic and polar acidic mechanisms are quite different. The typical reagents, the reactive intermediates, and the order of proton transfer steps all differ under acidic conditions from those seen under basic conditions. In this chapter the mechanisms that occur under acidic conditions will be discussed. A lot of time will also be spent discussing carbocations, which are central in these mechanisms.

3.1 Carbocations

A carbocation is an organic compound with a trivalent, six-electron C atom. The C atom is sp^2-hybridized, with an empty p orbital perpendicular to the plane of the three sp^2 orbitals. Carbocations are electron-deficient, and as such they are susceptible to attack by nucleophiles. Many (but not all) reactions of organic compounds that occur under acidic conditions involve carbocations or compounds for which important carbocationic resonance structures can be drawn.

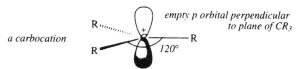

Carbocations were originally called "carbonium ions" by Americans and "carbenium ions" by Europeans. After a while the literature had become thoroughly confusing, so IUPAC stepped in and designated "carbonium ion" for pentavalent, positively charged C species such as CH_5^+, and "carbenium ion" for trivalent, positively charged C species such as CH_3^+. Unfortunately, this move only added to the confusion, as some Americans started to use the IUPAC convention, and others continued to use the older one. The term *carbocation* is now preferred for CR_3^+ because it is unambiguous and because it provides a nice analogy to *carbanion.*

* **Common error alert:** *When drawing reaction mechanisms involving carbocations, it is very important to obey Grossman's rule.* If you don't obey Grossman's rule, you are *guaranteed* to lose track of which atoms have three groups attached and which have four.

3.1.1   Carbocation Stability

Most carbocations are very unstable and can't be isolated except under very special circumstances. They exist only as transient, high-energy intermediates along
a reaction coordinate. Because they are high in energy, the Hammond postulate
(Chapter 1) suggests that carbocations should strongly resemble the transition states
leading to them. As a result, the rates of reactions that have carbocations as intermediates should be directly related to the stabilities of the carbocations.

* **Common error alert:** *If you draw a reaction mechanism that involves a carbocation that is extremely high in energy, the mechanism is likely to be unreasonable; conversely, a reaction mechanism that proceeds via a stable carbocation is likely to be reasonable.*

> *Every time* you draw a carbocationic intermediate, ask yourself: Is this a reasonably stable
> carbocation under these reaction conditions, or can a lower energy carbocation be drawn?

Most carbocations are high in energy, but some are higher in energy than others. There are four ways that carbocations can be stabilized: interaction of the
empty C(p) orbital with a nonbonding lone pair, interaction with a π bond, interaction with a σ bond (hyperconjugation), and by being part of an aromatic
system. Hybridization also affects the stability of carbocations.

The common stabilizing effects of lone pairs, π bonds, and σ bonds are all due to
the fact that two overlapping orbitals produce a bonding MO and an anti-bonding
MO. When a filled orbital such as a nonbonding orbital (n), a π bond, or a σ bond
overlaps with an empty C(p) orbital, a new bonding MO and a new antibonding MO
are created. The two electrons from the filled orbital go into the bonding MO, producing a net decrease in energy. The order of stabilization (n $>$ π $>$ σ) is due to the
fact that the interaction between the filled orbital and the empty C(p) orbital is strongest
when they are close in energy. Because the nonbonding C(p) orbital is higher in energy than either a bonding MO (σ or π) or a heteroatom-based nonbonding orbital,
the higher the energy of the filled orbital, the stronger the stabilization.

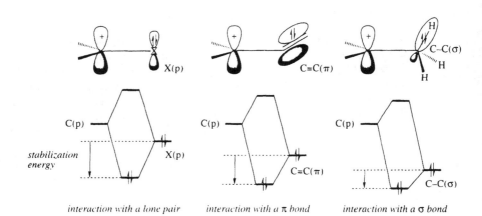

interaction with a lone pair   interaction with a π bond   interaction with a σ bond

• A nonbonding lone pair on a heteroatom stabilizes a carbocation more than any other interaction. A resonance structure can be drawn in which there is a π bond between the positively charged heteroatom and the adjacent cationic center. The heteroatom must be directly attached to the electron-deficient C for stabilization to occur; if not, destabilizing inductive effects take over. Even if the heteroatom is directly attached, if geometric constraints prevent overlap between the two orbitals, as in the bicyclic carbocation shown, then no stabilization occurs, and inductive effects take over.

resonance
stabilization
not possible

* **Common error alert:** *Lone-pair-bearing heteroatoms are usually electronegative, too, but when there is a competition between an inductive destabilizing effect and a resonance stabilization effect, resonance always wins!*

Lone-pair stabilization decreases as you go down the periodic table (e.g., O > S) and decreases as you go to the right (N > O > F), so N is the best resonance donor, with O a close second. Note that this pattern is not what you would expect from looking at trends in electronegativity. Heavier elements such as S and Cl are not nearly as good at stabilizing carbocations as are light elements, because their valence orbitals extend farther from the nucleus and do not overlap as well with the more compact C(p) orbital. F is a surprisingly good resonance donor, despite its electronegativity.

* **Common error alert:** *When a heteroatom stabilizes a carbocation by sharing its lone pair, it still has its octet and hence is* neither *electron-deficient* nor *electrophilic!* (Remember that when an electronegative atom has a formal positive charge and its full octet, the atoms adjacent to it are electrophilic.) Electronegative atoms like N, O, and F are capable of stabilizing carbocations by resonance exactly because they do not surrender their octet when they participate in resonance.

• Resonance with a C=C π bond is the next best way to stabilize a carbocation. (The C=O and C=N π bonds *destabilize* carbocations by a strong inductive effect, although resonance attenuates this effect by a small amount.) The more π bonds that can resonate, the better. Aromatic rings are good resonance stabilizers. Again, geometric constraints that prevent overlap also prevent stabilization.

• The third way to stabilize a carbocation is by overlap with an adjacent σ bond, or *hyperconjugation*, a fancy name for an ordinary phenomenon. The more σ bonds, the greater the stabilization. The empty C(p) orbital in Me_3C^+ has nine C–H bonds with which it can overlap (three overlap completely and six overlap partially), whereas the empty C(p) orbital in Me_2CH^+ has only six, and in $MeCH_2^+$ it has only three, leading to the order of stabilities of substituted alkyl carbocations: $3° > 2° > 1°$.

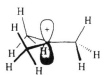
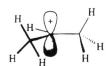

nine adjacent C–H σ bonds   *six adjacent C–H σ bonds*   *three adjacent C–H σ bonds*

* **Common error alert:** *If your mechanism has a 1° alkyl carbocation as an intermediate, it is almost certainly incorrect!*

Bonds from C to elements more electropositive than itself are even more stabilizing than C–H and C–C bonds. Thus, C–Si and C–Sn σ bonds are often used to stabilize carbocations. Again, these bonds must be *adjacent*, not attached, to the carbocationic center to have any stabilizing effect.

* **Common error alert:** *Sigma bonds* directly *attached to a carbocationic center do not stabilize the carbocation because they do not overlap with the empty C(p) orbital.* The bonds must be adjacent to the cationic center to have a stabilizing effect.

• Carbocations in which the empty C(p) orbital is part of an aromatic system (Chapter 1) are considerably stabilized over what one would expect just from resonance stabilization alone. The most important aromatic carbocation is the tropylium (cycloheptatrienylium) ion, which is so stable that one can buy its salts from commercial suppliers. Conversely, those carbocations in which the empty C(p) orbital is part of an antiaromatic system (e.g., cyclopentadienylium) are considerably destabilized.

tropylium
(aromatic)   *cyclopentadienyl cation*
(antiaromatic)

• Alkenyl (vinyl), aryl, and alkynyl carbocations are particularly unstable with respect to alkyl carbocations. Let's compare the isopropyl cation with the isopropenyl cation. In the latter, the central C has two σ bonds, one π bond, and one empty orbital, so it is sp-hybridized (linear). Both ions are stabilized by the $C(sp^3)$–H σ bonds of the CH_3 group on the right. In the isopropyl cation there is an additional interaction with $C(sp^3)$–H σ bonds on the left, whereas in the isopropenyl cation there is an additional interaction with $C(sp^2)$–H σ bonds on the left. Because $C(sp^2)$ orbitals are lower in energy than $C(sp^3)$ orbitals, the

$C(sp^2)-H$ σ bonds are lower in energy than the $C(sp^3)-H$ σ bonds, so the $C(sp^2)-H$ σ bonds are less stabilizing of the cationic center. In fact, 2° alkenyl carbocations ($R_2C=\overset{+}{C}R$) are only about as stable as 1° alkyl carbocations, and 1° alkenyl carbocations ($R_2C=\overset{+}{C}H$) are only about as stable as the CH_3^+ carbocation.

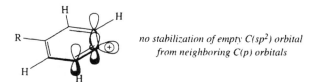

isopropyl cation   *isopropenyl cation*

Aryl and alkynyl carbocations are even less stable than 1° vinyl and methyl cations because they can't rehybridize to make an empty $C(p)$ orbital. In fact, both aryl and alkynyl carbocations can be generated only under very special circumstances and with the best of leaving groups (e.g., N_2).

* **Common error alert:** *If your mechanism has an alkenyl, alkynyl, or aryl carbocation as an intermediate, it is almost certainly incorrect.*

no stabilization of empty $C(sp^2)$ orbital from neighboring $C(p)$ orbitals

• Carbocations are *destabilized* by inductive electron-withdrawing elements and groups. Carbocations α to a carbonyl group ($R_2\overset{+}{C}C=O$), β-hydroxycarbocations ($HOCR_2\overset{+}{C}R_2$), and β-halocarbocations are all particularly unhappy species.

Carbocations occupy a unique place in the world of electron-deficient cations. Electron-deficient nitrogen and oxygen compounds (nitrenium and oxenium ions, respectively) are extremely unstable and are rarely or never seen. Electron-deficient boron and aluminum compounds are quite common, but these species are not cations. Silenium ions (R_3Si^+) are extremely unstable kinetically; still, it is useful to think of the R_3Si^+ group as a large, kinetically more stable H^+.

Obey Meier's rule: When in doubt, draw resonance structures! Often, carbocations are drawn in just one of their resonance forms, and sometimes not even in the stablest form. It is up to you to see the different resonance forms and to see how they impinge on the energy and reactivity of the carbocation.

3.1.2   Carbocation Generation; The Role of Protonation

There are three methods for generating carbocations: ionization of a $C-X^+$ σ bond, reaction of a lone pair on the heteroatom in a $C=X$ bond with a Lewis acid, and reaction of a $C=C$ π bond with a Lewis acid such as H^+ or a carbocation.

Carbocations can also be generated by one-electron oxidation of C=C π bonds. The carbocations generated by these methods are *radical cations*; they have an odd number of electrons. Reactions involving radical cations are discussed more thoroughly in Chapter 5.

• The R_3C-X^+ bond can ionize spontaneously to give R_3C^+ and :X. The likelihood of ionization depends on the stability of the carbocationic product R_3C^+, the leaving group X, and the solvent. Loss of H_2O from $Me_3C-\overset{..}{O}H_2$ occurs readily at room temperature, but ionization of $RCH_2-\overset{..}{O}H_2$ is very difficult because a very unstable carbocation would result. Note that X *usually* has a formal positive charge prior to ionization. Spontaneous departure of an anionic leaving group X^- is much less common, although it can happen in very polar (especially hydroxylic) solvents if the products are sufficiently stable. An anionic leaving group X^- is often converted into a neutral leaving group simply by protonating it; weak acids such as AcOH and pyrH$^+$ $^-$OTs (PPTS) are sometimes all that is required. Sometimes, strong Lewis acids such as BF_3, $AlCl_3$, $FeCl_3$, $ZnCl_2$, $SnCl_4$, or $TiCl_4$ are used to promote ionization. Silver(I) salts (AgNO$_3$, AgOTf, Ag$_2$O) can be used to promote the ionization of even relatively stable carbon–halogen bonds like C–Cl: the driving forces are establishment of the strong Ag–X bond and precipitation of AgX.

• A heteroatom in a C=X bond can use its lone pair to react with a Lewis acid to give a product for which a carbocationic resonance structure can be drawn. Protonation of a carbonyl compound belongs in this category. One of the lone pairs on O coordinates to H$^+$ to give a compound with two major resonance structures, one of which is carbocationic. The carbocationic resonance structure is not the *best* structure, but it tells the most about the *reactivity* of the ion. If the carbonyl C has heteroatoms directly attached (e.g., esters, carboxylic acids, and amides), more resonance structures can be drawn. Reactions of imines (Schiff bases) often begin by protonation of N.

* **Common error alert:** *The reaction of esters, amides, and carboxylic acids with electrophiles occurs on the carbonyl O.* Many more resonance structures can be drawn when the carbonyl O is protonated than when the noncarbonyl heteroatom is protonated.

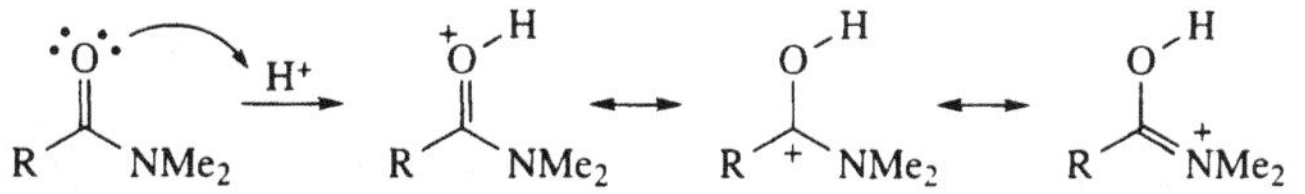

An apparent exception to the rule that the carbonyl O always reacts with Lewis acids is the reaction of acyl chlorides (RCOCl) with $AlCl_3$ to give acylium ions ($RC\equiv O^+$), which seems to proceed by coordination of $AlCl_3$ to Cl, not O. However, the generation of an acylium ion in this way requires more than one equivalent of $AlCl_3$, so it is thought that the first equivalent of $AlCl_3$ coordinates to the carbonyl O, and a second equivalent of $AlCl_3$ then coordinates to Cl.

• The π electrons in a C=C π bond can react with a Lewis acidic electrophile to give a carbocation. The simplest example is the reaction of an alkene with H^+. Note that one of the C atoms of the π bond forms the bond to H^+ using the electrons of the π bond, whereas the other C becomes electron-deficient and gains a formal positive charge. Other cationic electrophiles (carbocations, acylium ions, and Br^+) can react with C=C π bonds too.

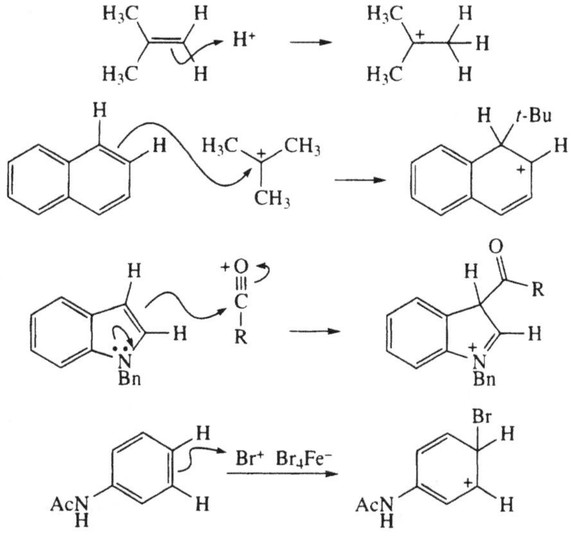

Both the first and second methods for generating a carbocation involve the reaction of a heteroatom with H^+ or another Lewis acid. The protonation of a heteroatom converts a π-bond electrophile into a more potent one, or it converts a mediocre leaving group into a better one. In Chapter 2 we saw that deprotonation of a substrate does not always occur at the most acidic site in polar basic mechanisms. Similarly, in polar acidic mechanisms, protonations do not always occur at the most basic site. To decide where to protonate a substrate, first determine which bonds need to be made and broken, then determine which of the atoms are nucleophilic and which are electrophilic. Then protonate a heteroatom either to convert a leaving group into a better one or to convert an electrophile into a more potent one.

3.1.3   Typical Reactions of Carbocations; Rearrangements

Regardless of how they are generated, carbocations always undergo one of three reactions: *addition* of a nucleophile, *fragmentation* to give a compound containing a π bond and a stable cation (often H^+, but not always) or *rearrangement*. These reactions compete with one another, and often slight differences in structure or reaction conditions tip the balance in favor of one reaction over another. For example, in the reactions of 3° alkyl carbocations, addition is favored when the nucleophile is a hydroxylic solvent (EtOH, H_2O, RCO_2H), while fragmentation is favored when R_3Si^+ can be cleaved or when the solvent is nonnucleophilic. It is difficult to predict which of these reactions will occur, and you will usually be asked merely to rationalize the formation of the observed product.

• In an *addition* reaction, a nucleophile donates a pair of electrons to the carbocation's empty orbital to form a new σ bond.

∗ **Common error alert:** *A lone-pair nucleophile bearing a proton (H_2O, ROH, RCO_2H, RNH_2, etc.), is always deprotonated immediately after it adds to the carbocation.* The two steps combined—addition, then deprotonation—are the microscopic reverse of the ionization of a carbon–heteroatom bond under acidic conditions. When the nucleophile is RCO_2H, the carbonyl O makes the new bond, not the O of the OH group.

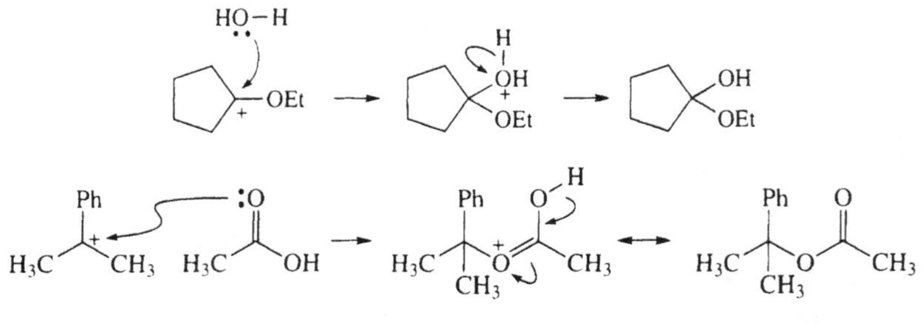

Pi-bond and σ-bond nucleophiles can also add to carbocations. In the former case, a new carbocation is formed, as discussed in Section 3.1.2.

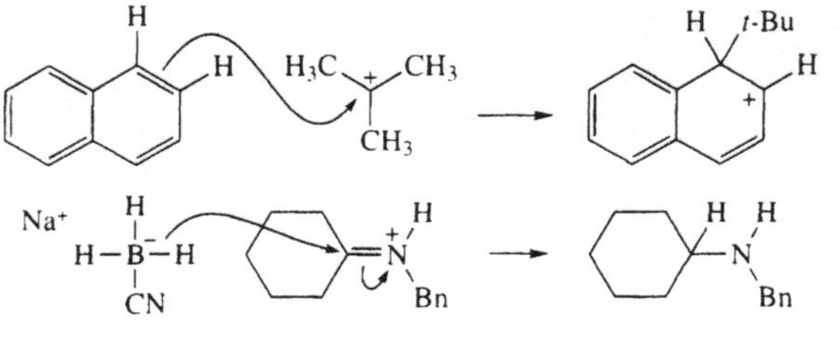

• Carbocations can undergo *fragmentation* to give a new electron-deficient cation and a species containing a π bond. In a fragmentation, the electrons in a bond adjacent to (but not attached to) the electron-deficient carbon atom move to form a π bond to the electron-deficient center ($^+C-X-Y \rightarrow C=X + Y^+$). The electron-deficient species (Y^+) that is generated can be H^+, a carbocation, or occasionally another species such as R_3Si^+ or R_3Sn^+. The intermediate atom X may be C or another atom like N or O. Fragmentation is the opposite of addition of a π-bond nucleophile to a Lewis acid electrophile. Note that the curved arrow showing the cleavage of the X–Y bond shows the movement of the electrons in the X–Y bond toward the cationic center. The group Y^+ is assumed to just float away upon loss of its bond to X.

* **Common error alert:** *Do not use the arrow to show the Y^+ fragment flying off into the wilderness!* If Y^+ floating away bothers you, show a weak base making a new bond to Y^+ at the same time as the X–Y bond is cleaved. Usually the base is assumed to be present and is not shown.

• Carbocations can also *rearrange*, much more so than carbanions. The longer the lifetime of the carbocation (i.e., the more acidic the reaction conditions), the greater the propensity to rearrange. Usually carbocations rearrange to give more stable carbocations, but rearrangements to less stable carbocations can occur at high temperatures. Also, sometimes a less stable carbocation in equilibrium with a more stable one is carried on to product selectively for kinetic reasons.

The major mechanisms for carbocationic rearrangements are *1,2-hydride shifts* and *1,2-alkyl shifts*. Both of these occur in the rearrangement of the cyclohexylium ion (a 2° carbocation) to the 1-methylcyclopentylium ion (a 3° carbocation). A 1,2-alkyl shift of the C3–C2 bond of cyclohexylium from C2 to C1 gives a cyclopentylmethylium ion (a 1° carbocation). This reaction is uphill in energy. However, a 1,2-hydride shift of the C1–H bond from C1 to C2 gives a much more stable product.

Example

Numbering: Only one CH_2 group, C8, is present in both starting material and product. Assume the C5−C6−C7−C8 ring in starting material and product doesn't change. C6 may still be attached to C3, C3 to C2 (or C4; they are equivalent, so it makes no difference), and that leaves C1 and C4. Assume C4 is still attached to C5 to give the final numbering of the product. Make: C1−C7, C2−C4, C3−OAc. Break: C2−C7, C3−C4, C1−OMs.

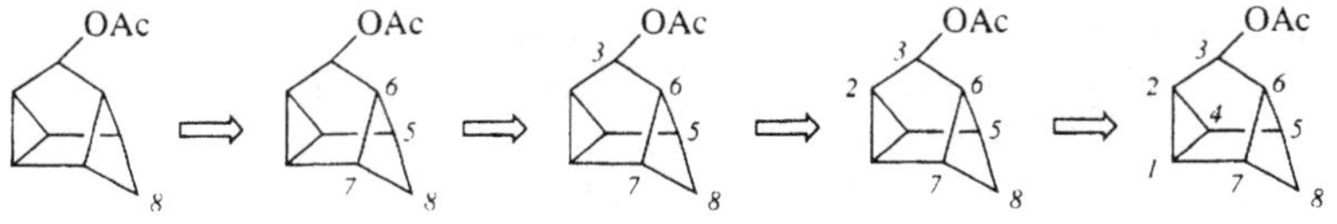

The cation is generated at C1. Then C7−C2 migrates to C7−C1, and C4−C3 then migrates to C4−C2. Addition of AcOH to the C3 carbocation and H$^+$ loss gives the product.

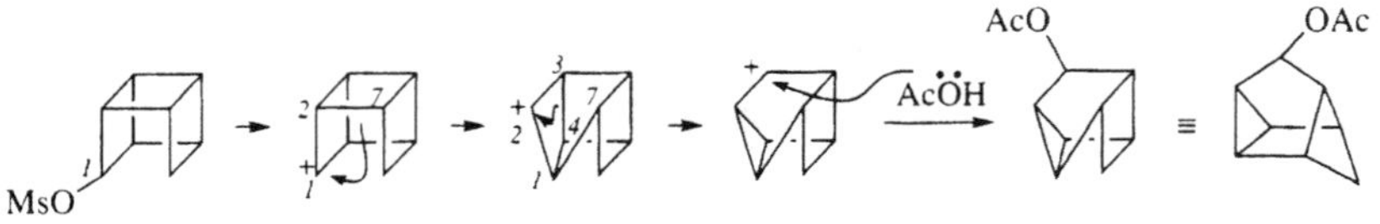

Note how the atoms are kept in the same relative positions in the drawings as the bonds are moved around. In rearrangement reactions especially, draw all intermediates exactly as the starting material was drawn. Only when you have made and broken all bonds in your list should you try to redraw the structure.

Transannular 1,5-hydride shifts are often seen across medium-sized (8-, 9-, and 10-membered) rings. The atoms directly across from each other in these rings are quite close together, so migration is facile. In general, though, most cationic rearrangements can be accounted for by a series of 1,2-shifts.

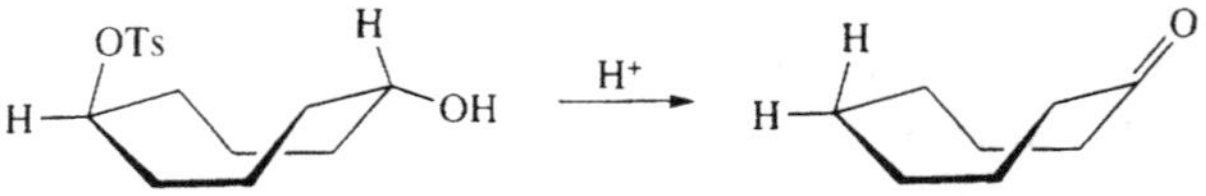

Often a 1,2-shift is *concerted* with loss of the leaving group. For example, consider the two diazonium ions shown. Loss of N_2 from either diastereomer gives a common carbocationic intermediate, and this carbocation can undergo a 1,2-alkyl shift to give the aldehyde or a 1,2-hydride shift to give the ketone. If both diastereomers rearranged via the carbocation, then each diastereomer should give the same ratio of aldehyde to ketone. In fact, one diastereomer gives only the aldehyde, and the other gives only the ketone. This result suggests that there is no carbocationic intermediate and that rearrangement is concerted with loss of the leaving group.

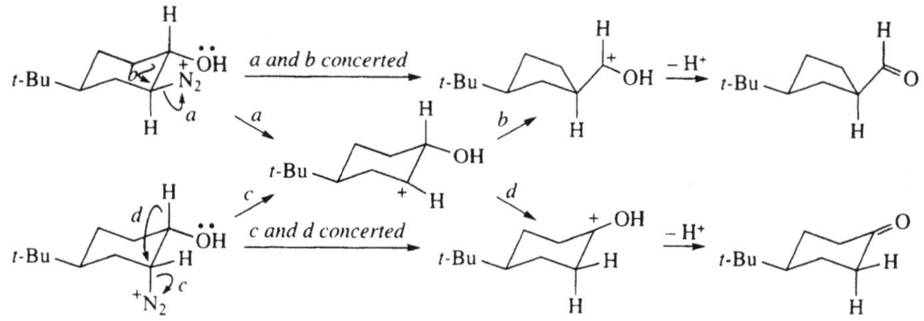

H
t-Bu — ⁺ÖH   *a and b concerted* → *t*-Bu — ⁺OH   — H⁺ → *t*-Bu — O
N₂
H   *a*
H   *a*
t-Bu — OH
b
H   OH

t-Bu — *c*   ⁺ H   *d*
c and d concerted
H
t-Bu — ÖH   H
⁺N₂   *c*
t-Bu — ⁺ H   *d*   ⁺ OH   — H⁺ → *t*-Bu — O   H
H

Why does the equatorial RN₂⁺ ion give only the aldehyde, and the axial RN₂⁺ ion only the ketone? When a 1,2-shift is concerted with loss of a leaving group, the migrating bond must be *antiperiplanar* to the C–X bond of the leaving group. (See Chapter 2 for a discussion of anti- and synperiplanar requirements in E2 elimination reactions.) In each diastereomeric diazonium ion in the example, the bond that is antiperiplanar to the C–N₂⁺ bond migrates selectively. (Draw a Newman projection or make a model if you have trouble seeing this.) The same phenomenon is observed in the *Beckmann rearrangement*, in which an oxime is converted into an amide after a 1,2-alkyl shift and addition of H₂O. The group that is trans to the OH group (i.e., antiperiplanar) shifts selectively.

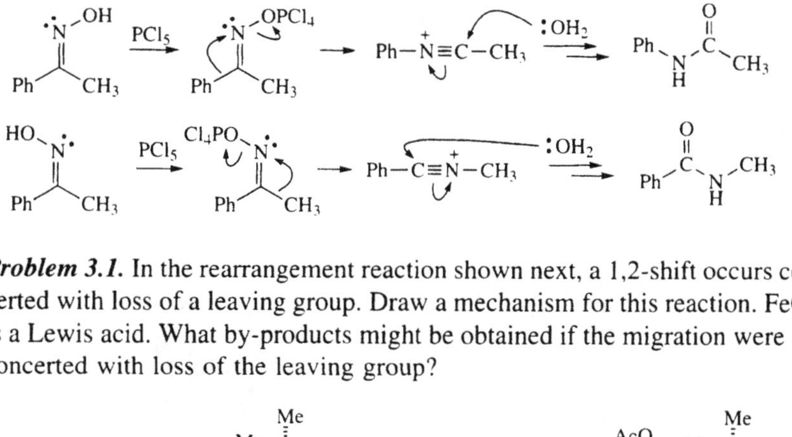

ÖH   OPCl₄
N — PCl₅ → N · → Ph–N≡C–CH₃   :OH₂ → Ph–N(H)–C(=O)–CH₃
Ph   CH₃   Ph   CH₃

HO   Cl₄PO
N· — PCl₅ → N· → Ph–C≡N–CH₃   :OH₂ → Ph–C(=O)–N(H)–CH₃
Ph   CH₃   Ph   CH₃

Problem 3.1. In the rearrangement reaction shown next, a 1,2-shift occurs concerted with loss of a leaving group. Draw a mechanism for this reaction. FeCl₃ is a Lewis acid. What by-products might be obtained if the migration were not concerted with loss of the leaving group?

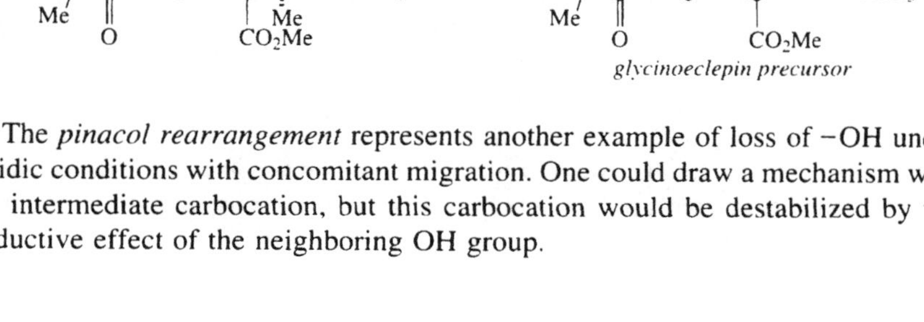

Me   OMe   Me   AcO   Me Me   Me
Me — O — ... OSiR₃   Ac₂O / FeCl₃ → Me — O — ... OSiR₃
Me   Me   CO₂Me   Me   CO₂Me
O   O
glycinoeclepin precursor

The *pinacol rearrangement* represents another example of loss of –OH under acidic conditions with concomitant migration. One could draw a mechanism with an intermediate carbocation, but this carbocation would be destabilized by the inductive effect of the neighboring OH group.

Example

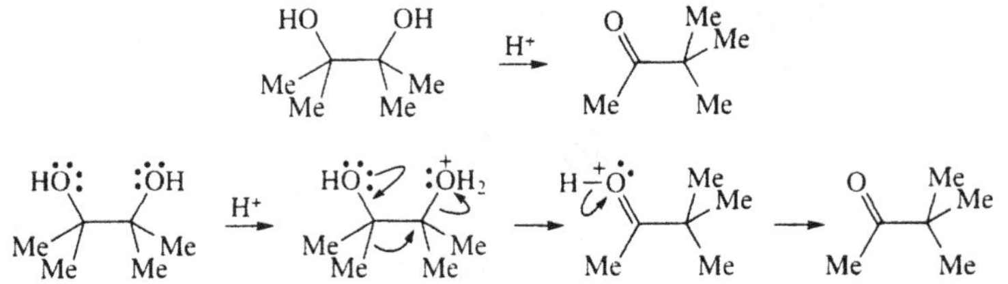

How does one decide whether loss of a leaving group is concerted with a 1,2-shift or whether a carbocation is formed before the 1,2-shift occurs? Under strongly acidic conditions, when carbocations are particularly long lived, non-concerted mechanisms are usually drawn. By contrast, a concerted mechanism is usually drawn when

• loss of the leaving group would give a particularly high energy carbocation; or

• the carbocation that one obtains *after* the 1,2-shift is particularly low in energy; or

• a group that is antiperiplanar to the leaving group shifts selectively, whereas other groups do not.

Phenyl and other unsaturated groups undergo 1,2-shifts also, but they use a two-step *addition–fragmentation* mechanism. A π bond from the phenyl group adds to the carbocation to give a three-membered ring intermediate called a *phenonium ion*. One of the bonds of the three-membered ring then fragments to give either the starting material or the product.

Example

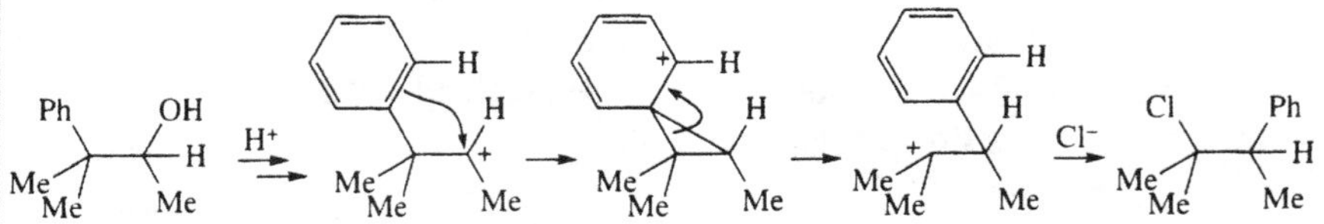

The concerted shift of a Ph group has a higher activation energy than the two-step shift because the C(sp^2) orbital on the migrating atom doesn't interact with the two other orbitals involved in the TS as strongly as an H(s) or C(sp^3) orbital would. The 1,2-shift of Ph groups is often faster than the 1,2-alkyl shift, even though it is nonconcerted and it involves a three-membered-ring intermediate. In the example, a lower energy carbocation would be obtained by a 1,2-methyl shift, but the phenyl group shifts more quickly.

It is sometimes difficult to discern that a mechanism involves a 1,2-alkyl shift. Look at your list of bonds broken and bonds made. If a group migrates from one atom to its next-door neighbor, you may have a 1,2-shift.

Problem 3.2. Draw a mechanism for the following rearrangement reaction. SnCl₄ is a Lewis acid. Treat the Me₃Si group as if it were a big proton.

Note how the group that migrates retains its stereochemistry! Retention of stereochemistry of the migrating group is a general feature of 1,2-shifts.

3.2 Substitution and β-Elimination Reactions at C(sp³)–X

3.2.1 Substitution by the S_N1 and S_N2 Mechanisms

Nucleophilic substitutions at C(sp³)–X under acidic conditions usually occur by the S_N1 mechanism. The S_N1 mechanism consists of two steps, not counting protonations and deprotonations. In the rate-determining step, the leaving group leaves to give a carbocation. In a subsequent fast step, the nucleophile *adds to* the carbocation (one of the three typical reactions of carbocations) to afford the product. The nucleophile may be anionic (as long as it is a weak base such as Br⁻ or I⁻) or neutral. In addition to these two steps, there is sometimes a protonation of the leaving group *before it leaves*, and there is sometimes a deprotonation of the nucleophilic atom *after it adds*. Protonation of the leaving group transforms it into a much better leaving group, especially when it is –OH or –OR. On the other hand, a neutral, protic nucleophile (e.g., H_2O, ROH, RCO_2H) always loses a proton to give a neutral species after it has added to the carbocation.

A carbocation is formed in the rate-limiting step of S_N1 substitution, so the rate of the reaction is directly related to the stability of the carbocation being formed. S_N1 substitution at C(sp³)–X occurs very readily at 3° centers, occasionally at 2° centers (especially when a cation-stabilizing group like a heteroatom or a π bond is directly attached), and at 1° alkyl centers only when a strongly cation-stabilizing group like O or N is attached. Substitutions in which the nucleophile is a π bond almost always proceed by the S_N1 mechanism, as π bonds are poor nucleophiles toward σ-bond electrophiles and rather good nucleophiles toward carbocations.

Example

It is not obvious at first sight whether the ring O atoms in the product come from the diol or the acetal. Draw the by-products! You will see that two equivalents of MeOH are obtained. This information suggests that the Me–O bonds of the acetal do not break and therefore that the ring O atoms in the product come from the diol. The first two steps of the reaction must be protonation of a MeO group and loss of MeOH to give a very stable carbocation.

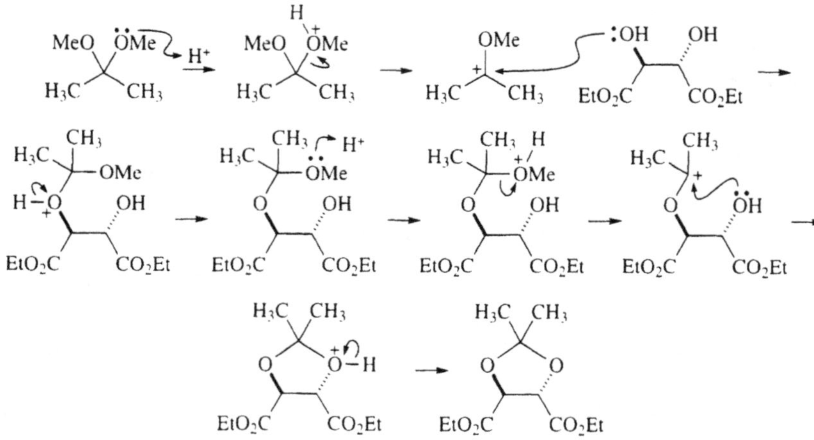

Problem 3.3. Draw mechanisms for each of the following substitution reactions. Remember that in each case, loss of the leaving group is preceded by coordination of the leaving group to a Lewis acid such as H^+.

(a)

(b)

Problem 3.4. Alkenyl and aryl halides do not undergo substitution reactions under acidic conditions. Why not?

Substitutions at an electrophilic C can occur by the S_N2 mechanism under acidic conditions only under special circumstances. The electrophilic C must be a 1° $C(sp^3)$–X electrophile (or CH_3–X) so it is unable to stabilize a carbocation, the nucleophile must be a *very* good one like Br^- or I^-, and the conditions must be very strongly acidic (e.g., refluxing 48% aq. HBr). One of the few examples of the S_N2 mechanism under acidic conditions is the conversion of 1° alcohols to halides with mineral acid. The strongly acidic conditions are required to convert the OH into a better leaving group that exists long enough to undergo the S_N2 reaction.

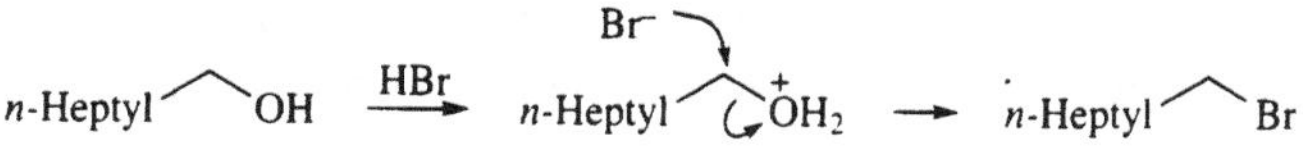

Problem 3.5. Draw a reasonable mechanism for the following substitution reaction. Which O is protonated first?

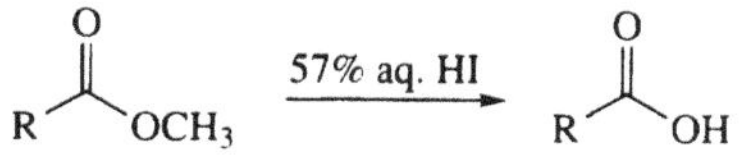

* **Common error alert:** *When reaction conditions are only weakly acidic, the S_N1
mechanism is almost always operative for substitutions at C.*

Example

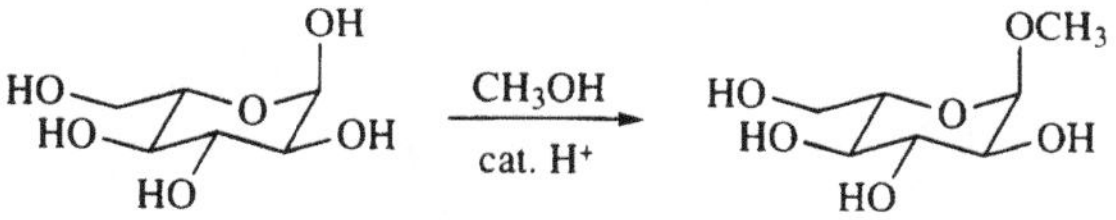

The mechanism one draws for this glycosylation reaction depends on whether
the O atom in the product that is attached to the CH_3 group is identified as
coming from CH_3OH or from the sugar. If it comes from the sugar, then substitution is occurring at the C of CH_3OH. There is no way a carbocation can
be formed at this C, so an S_N2 mechanism must be drawn.

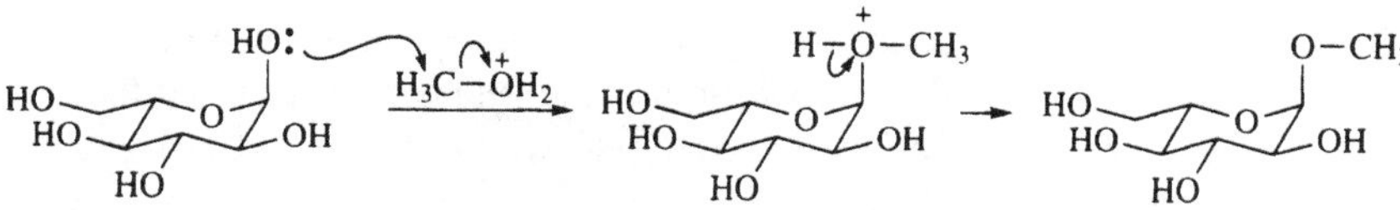

If it comes from CH_3OH, then substitution is occurring at the anomeric C of
the sugar. A very good carbocation (stabilized by O) can be formed at this C,
so an S_N1 mechanism is likely.

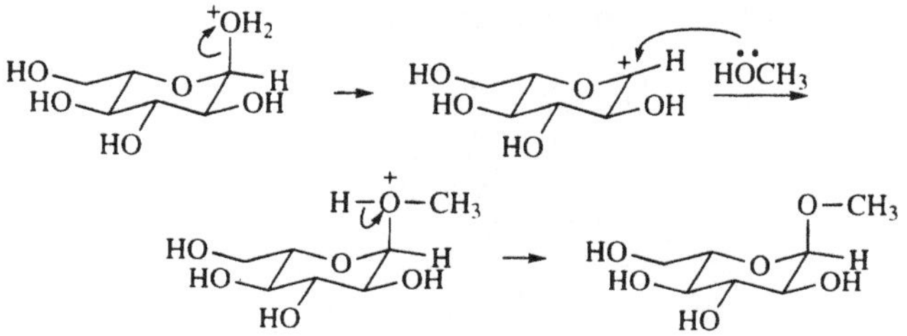

Which mechanism is better? S_N2 mechanisms require harsh acid conditions
and very good nucleophiles such as Br^- or I^-, neither of which is the case
here, so the S_N1 mechanism is much better than the S_N2 mechanism.

Substitutions at more electronegative elements like P and Br occur by the S_N2 mechanism under acidic conditions. These elements cannot form the stable carbocations required for an S_N1 mechanism.

Example

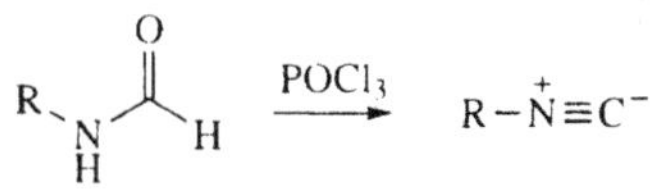

The elements of H_2O are removed from the amide with the help of $POCl_3$. The amide O is nucleophilic and P is electrophilic, so the first step is likely a simple S_N2 reaction at P with the amide O as nucleophile, with deprotonation of the nucleophile following. The C–O bond is cleaved next with assistance from the lone pair of N. Deprotonation of the nitrilium ion gives the isocyanide.

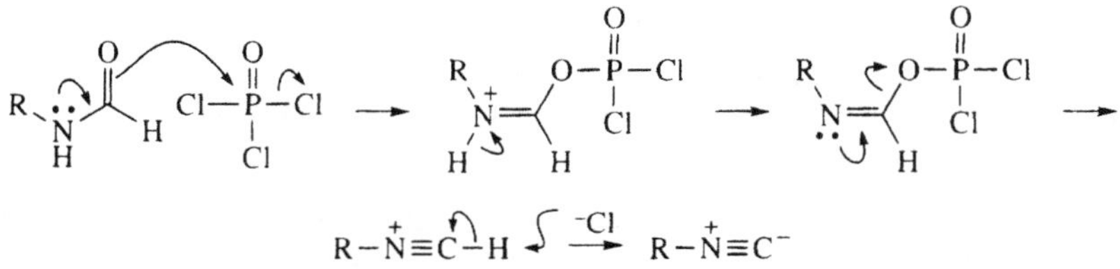

Substitution very rarely occurs by the $S_{RN}1$ mechanism under *acidic* conditions because radical anions are usually strongly basic. Substitution cannot occur by the S_N1 mechanism under *basic* conditions because the carbocationic intermediates are strongly Lewis acidic.

When S_N1 substitution occurs at a stereogenic center, loss of configurational purity is usually observed. The carbocationic intermediate is planar and there is no bias for approach of the nucleophile from one face or the other. Occasionally the loss of configurational purity is only partial for reasons that need not concern you here. However, sometimes an apparent S_N1 reaction proceeds stereospecifically. Such an outcome is usually the result of participation of neighboring π bonds or lone pairs, as in the example shown. You can think of these reactions as two subsequent S_N2 reactions. Even if a true carbocationic intermediate is formed in these reactions, it exists for less time than it takes to rotate about a C–C σ bond.

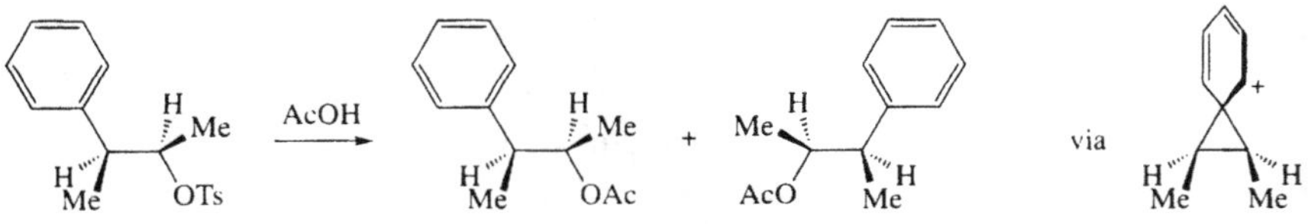

3.2.2   β-*Elimination by the E1 Mechanism*

Elimination can also occur at $C(sp^3)$–X centers under acidic conditions. The mechanism is called *E1*. The leaving group leaves, sometimes after protonation, to give a carbocation. The carbocation then *fragments* (one of the three reactions

of carbocations) to give a compound containing a π bond and a new cation, usually H⁺. The specific bond that fragments is always *adjacent* to the electron-
deficient C, *never attached to it*. The electrons in that bond migrate to form a π
bond. A weak base (either the solvent or the conjugate base of the strong acid
in the reaction mixture) is sometimes shown removing the H⁺. Other cationic
groups besides H⁺, most commonly R₃Si⁺, R₃Sn⁺, or a stabilized carbocation,
can be lost by fragmentation in this manner. In fact, given a choice between frag-
menting to lose H⁺ or R₃Si⁺, a carbocation will usually choose to lose the latter.

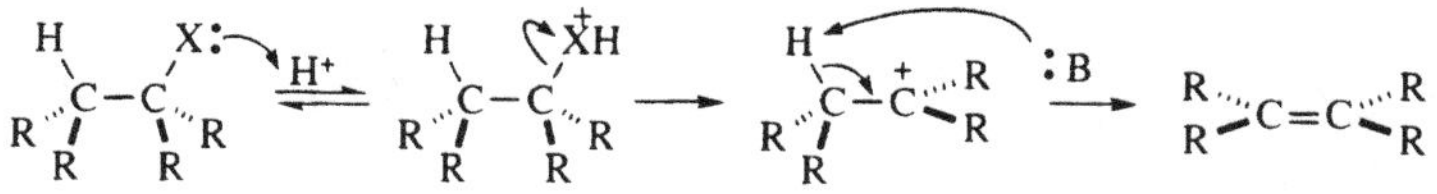

It is common practice not to show the weak base that removes H⁺, so that the carbocation seems to fragment spontaneously.

* **Common error alert:** *If you do not show the weak base that removes H⁺, be
sure that your curved arrow is drawn correctly.* The arrow shows how the *electrons in the C−H bond move to form a π bond. It should not show H⁺ flying
off into the wilderness!

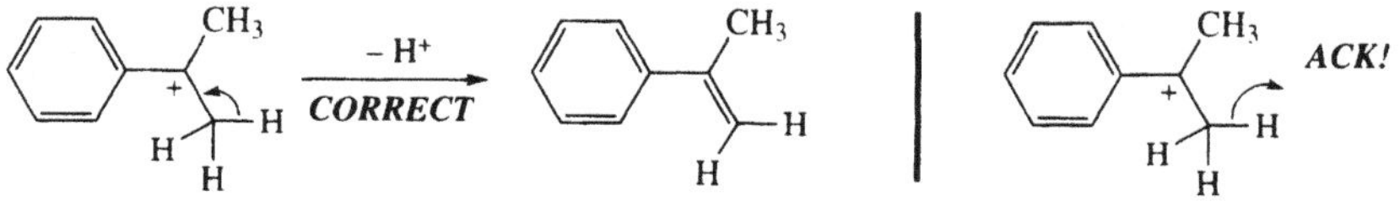

Elimination of small molecules like water and MeOH is often carried out under
acidic conditions. The reaction is often driven to completion by azeotropic distillation with benzene or petroleum ether.

Example

The elements of H₂O are eliminated. Under acidic conditions, eliminations occur by an E1 mechanism. First, the leaving group is protonated to convert it
into a better leaving group, and then it leaves to give a carbocation. Then fragmentation (one of the three typical reactions of carbocations) of a C−H bond
occurs to give a diene.

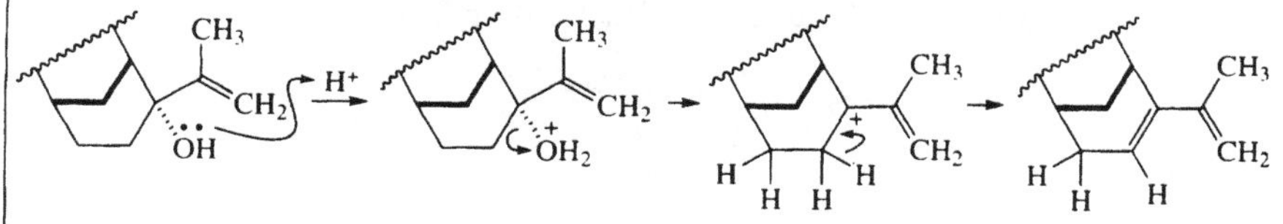

Unfortunately, this diene is not the desired product. Compare the diene with
the desired product. Which σ bonds differ between them? One C−H bond has

to be made, and another has to be broken. Because the conditions are acidic, first make a new C–H bond by protonating a C=C π bond. Then fragment the resulting carbocation to give the desired product.

Problem 3.6. Draw a reasonable mechanism for the following transformation.

$$\text{(cyclohexyl)}-CO_2Et \xrightarrow[\text{2) TsOH}]{\text{1) Me}_3\text{SiCH}_2\text{MgBr}} \text{product} -SiMe_3$$

3.2.3 Predicting Substitution vs. Elimination

A carbocationic intermediate is formed in both the S_N1 and the E1 mechanisms. After the carbocation is formed, addition of a nucleophile leads to an overall substitution reaction, whereas fragmentation leads to an overall elimination reaction, so there is a competition between the two modes of reaction. It is possible to predict whether substitution or elimination dominates, just as it is possible to predict whether S_N2 or E2 will predominate under basic conditions. Predicting which pathway dominates is easier under acidic conditions, though. *Addition is favored in hydroxylic (i.e., nucleophilic) solvents (RCO_2H, ROH, H_2O) and when the nucleophile is contained in the same molecule, whereas fragmentation is favored in aprotic solvents.* In other words, substitution occurs when the carbocation intermediate can be rapidly intercepted by a nucleophile, and elimination occurs when it cannot.

3.3 Electrophilic Addition to Nucleophilic C=C π Bonds

Alkenes that are substituted with alkyl groups or heteroatoms are good nucleophiles toward Brønsted acids (H–X) and heteroatom–heteroatom σ-bond electrophiles (Br_2, NBS, PhSeCl, RCO_3H). In an *electrophilic addition* reaction, the π bond acts as a nucleophile toward the electrophilic atom to give a carbocation, and then a nucleophile adds to the carbocation (one of the three reactions of carbocations) to give a product into which all of the atoms of the starting materials are incorporated.

Alkenes that are directly attached to lone-pair-bearing heteroatoms, especially N and O, are particularly reactive toward electrophiles. A resonance structure that shows a formal negative charge on the β-carbon can be drawn, and the intermediate carbocation is especially well stabilized. Conversely, alkenes substituted with electron-withdrawing groups are less reactive toward electrophiles.

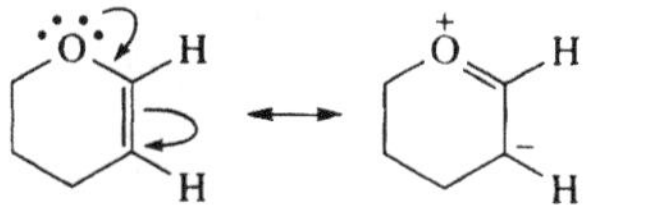

Mineral acids, carboxylic acids, alcohols, and H_2O all add across π bonds. Alcohols and H_2O are not sufficiently acidic to protonate the alkene in the first step, so a catalytic amount of a strong acid is used to promote the reaction.

Example

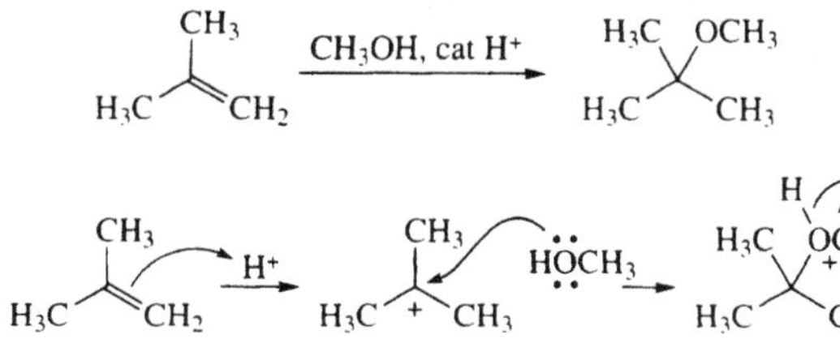

When the electrophilic atom is H^+ and the alkene is unsymmetrical, H^+ adds to the alkene so that the more stable carbocation (almost always the more substituted one) is formed. The nucleophile then adds to the carbocation to give the product. The observation that the nucleophile adds to the more substituted C of the double bond is known as *Markovnikov's rule*. Markovnikov's rule is simply an application of the Hammond postulate: the faster reaction is the one that leads to the intermediate lower in energy.

Problem 3.7. Dihydropyran (DHP) reacts with alcohols under acid catalysis to give tetrahydropyranyl (THP) ethers. The alcohols can be released again by treating the THP ether with MeOH and catalytic acid. Thus, the THP group acts as a *protecting group* for alcohols. Draw mechanisms for the formation and cleavage of the THP ether.

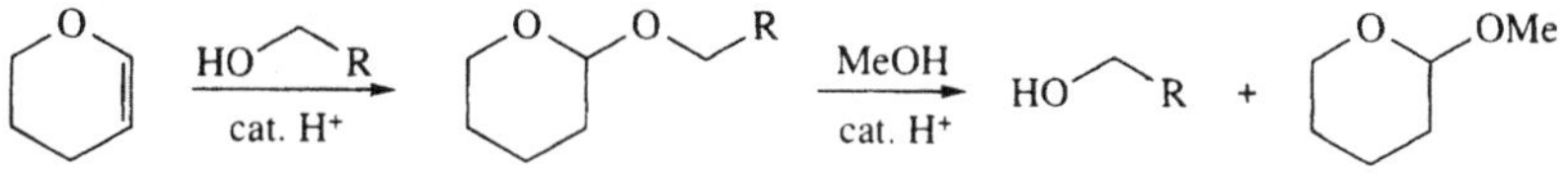

When a simple alkene reacts with an electrophilic, lone-pair-bearing heteroatom, a carbocation is not formed. Instead, both C atoms of the π bond form new bonds to the electrophilic atom to give a product containing a three-membered ring. The four electrons required for the two new σ bonds come from the π bond and one lone pair of the heteroatom. The reaction is stereospecifically syn (i.e., the two new bonds form on the same side of the double bond). For example, epoxides are formed in the reaction of alkenes with peracids (RCO_3H) such as mCPBA. A trans alkene gives a trans epoxide, and a cis alkene gives a cis epoxide.

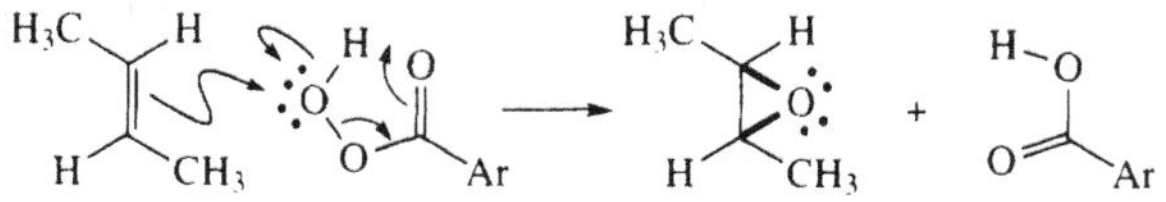

It is customary not to draw an arrow from the O lone pair back to one of the C atoms of the alkene.

Alkenes also react with sources of electrophilic halogen such as Br_2, NBS, and I_2 to give three-membered rings called *halonium* ions. Again, the four electrons required for the two new C–X bonds come from the alkene and a lone pair on the halogen. Halonium ions are in equilibrium with the open β-halocarbocations. For Br and I, the equilibrium favors the halonium ions greatly, but for Cl, the equilibrium favors the β-halocarbocation due to the greater electronegativity of Cl (less willing to share a pair of electrons) and the shorter C–Cl bonds. The β-halocarbocation is also favored when strongly electron-donating groups such as –OR or –NR$_2$ are present to stabilize the carbocation.

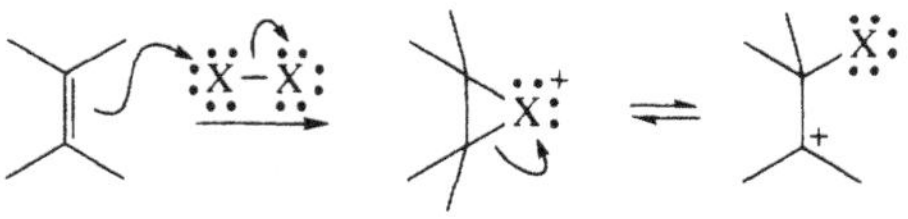

Again, it is customary not to draw an arrow from X back to one of the C atoms of the alkene.

Problem 3.8. Draw a mechanism for the following reaction. Will a halonium ion be an intermediate?

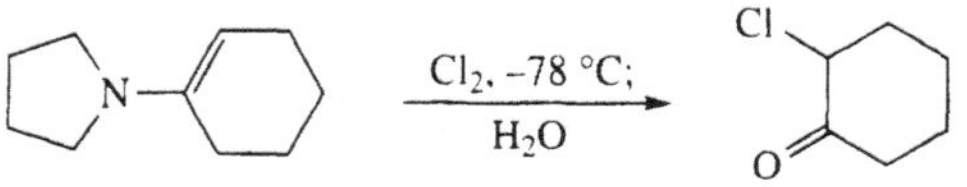

The halogen atom of a halonium ion has its octet and a formal positive charge, so the C atoms of the halonium ion are electrophilic and are susceptible to S_N2 nucleophilic attack. Because the S_N2 substitution reaction results in inversion at one of the C atoms, the overall reaction—addition of the electrophilic halogen and nucleophile across the π bond—usually proceeds with anti stereochemistry.

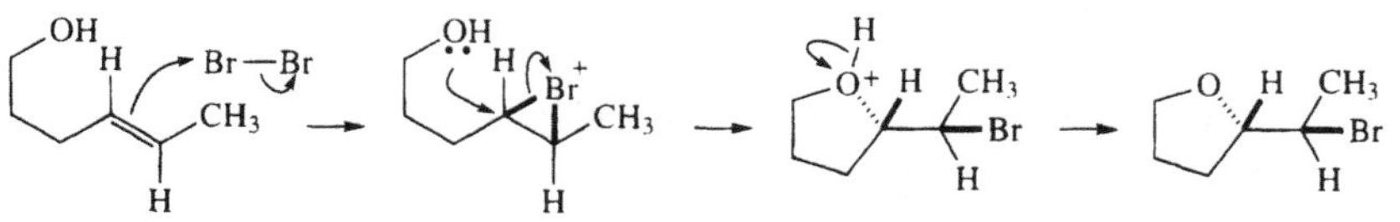

Sometimes anti addition isn't observed. Temporary participation of a neighboring group can give unusual stereochemical or regiochemical results.

Problem 3.9. Draw mechanisms that account for the formation of each product. The first step is the same in every mechanism. Be sure to think about stereochemistry!

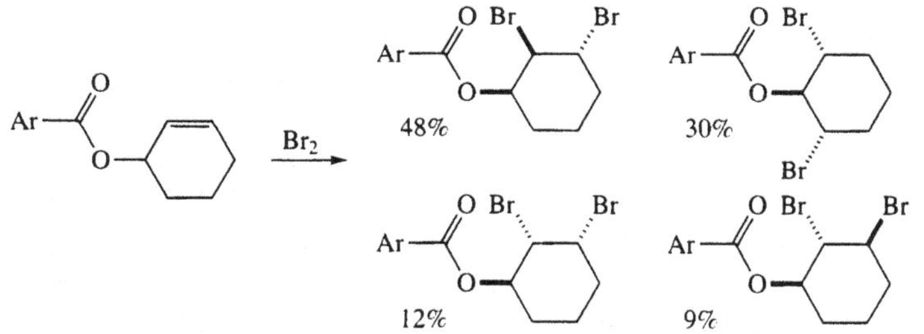

3.4 Substitution at Nucleophilic C=C π Bonds

3.4.1 Electrophilic Aromatic Substitution

The π bonds in aromatic compounds are also reactive toward electrophiles, although not nearly so much as alkenes. The aromatic ring attacks an electrophile to give an intermediate carbocation. The carbocation then undergoes fragmentative loss of H^+ (sometimes another cation) *from the same C to which the electrophile added* to re-form the aromatic system and give an overall *substitution* reaction. Thus, the predominant mechanism of substitution at aromatic rings under acidic conditions is electrophilic addition–elimination, sometimes referred to as S_EAr. The reaction of toluene and nitric acid is indicative.

Example

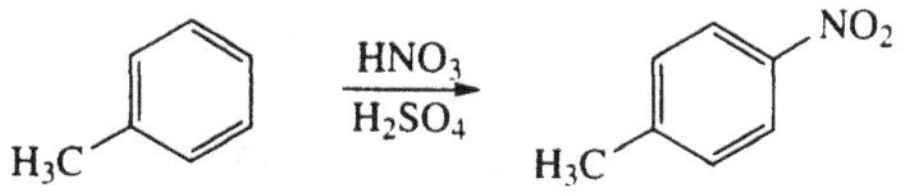

Toluene is a nucleophile, so HNO_3 must be electrophilic. The N=O bond in nitric acid is analogous to the C=O bond in carbonic acid ($HOCO_2H$). Just as $HOCO_2H$ decomposes to H_2O and electrophilic CO_2, so ONO_2H decomposes to H_2O and electrophilic $^+NO_2$.

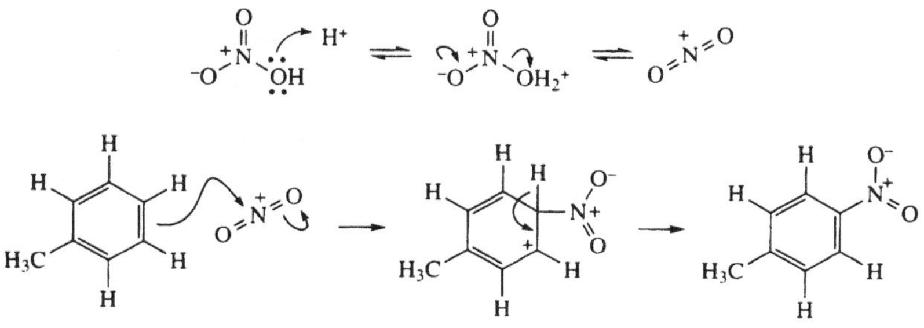

Arenes can be induced to react only with quite voracious electrophiles. For example, Br_2 reacts instantly with alkenes but is inert toward benzene. However, Br_2 reacts with benzene in the presence of the Lewis acid $FeBr_3$, which coordinates to Br_2 and makes Br^- a better leaving group. (A β-halocarbocation, not a bromonium ion, is an intermediate in the reaction of arenes with Br_2.) Aromatic rings are *sulfonated* (substituted with $-SO_3H$) by treatment with SO_3 and H_2SO_4, and *nitrated* (substituted with $-NO_2$) with HNO_3 and H_2SO_4.

New C-C bonds to arenes can be made by *Friedel–Crafts* reactions. Friedel–Crafts *alkylations* are traditionally executed with an alkyl chloride and catalytic $AlCl_3$ or an alkene and a strong Brønsted or Lewis acid; the key electrophilic species is a carbocation. Friedel–Crafts *acylations* are usually executed with an acyl chloride and an excess of $AlCl_3$; the key electrophilic species is an acylium ion ($RC\equiv O^+$). In the *Bischler–Napieralski reaction*, intramolecular attack on a nitrilium ion ($RC\equiv NR$) occurs.

Problem 3.10. Draw mechanisms for the following electrophilic aromatic substitution reactions.

(a)   A Friedel-Crafts
      alkylation

(b)   A Friedel-Crafts
      acylation

(c)   A Bischler-Napieralski
      reaction

The electronic nature of the substituents on an aromatic ring has a profound effect on the regiochemistry of electrophilic aromatic substitutions. Electrophiles preferentially attack ortho or para to electron-donating substituents such as $RO-$, R_2N-, $RCONH-$, and halo substituents, whereas they attack meta to electron-withdrawing groups such as carbonyl, $-CN$, $-NO_2$, and $-SO_3H$ groups. (Amino groups are protonated under the conditions of certain electrophilic aromatic substitution reactions, so they can be ortho–para or meta directors, depending on the reaction.) The directing ability of a group is directly related to its ability to stabilize or destabilize the intermediate carbocation in electrophilic aromatic substitutions.

Substitution para to an electron-donating group gives an intermediate with an excellent resonance structure...

... meta substitution does not!

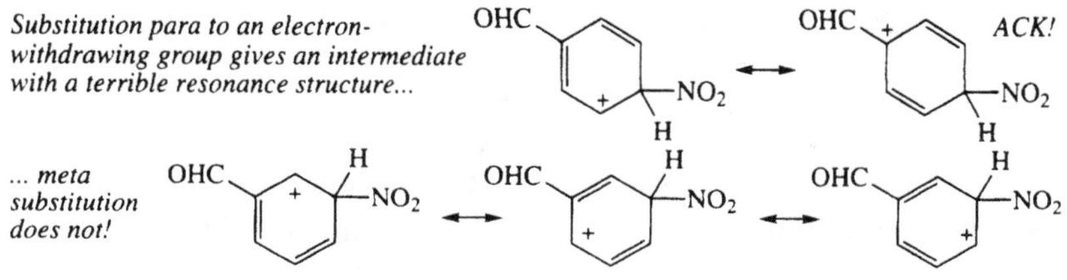

Substitution para to an electron-withdrawing group gives an intermediate with a terrible resonance structure...

ACK!

... meta substitution does not!

Charge-separated resonance structures for the starting materials also show that rings with electron-donating groups are most nucleophilic at the ortho and para positions, whereas rings with electron-withdrawing groups are least nucleophilic at these positions.

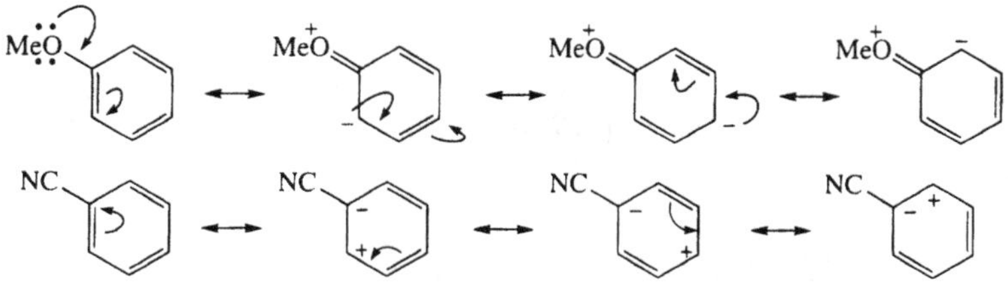

In most cases, the C of the arene that acts as a nucleophile has a H atom attached, so the fragmentation step of the substitution reaction gives H^+. Other fragments can be lost, though. For example, the Me_3Si group is an *ipso* director: despite its apparent steric bulk, it directs electrophiles to add to the C *to which it is attached*. The C–Si bond then fragments to reestablish aromaticity. Attack on the ipso C occurs selectively because it moves the C–Si bond out of the plane of the ring where it can stabilize the carbocation, whereas attack on any other C moves a C–H bond out of the plane of the ring, and a C–Si bond is more carbocation-stabilizing than a C–H bond. Alkyl groups that can form reasonably stable carbocations (*t*-Bu, *i*-Pr) can act as ipso directors, too.

Example

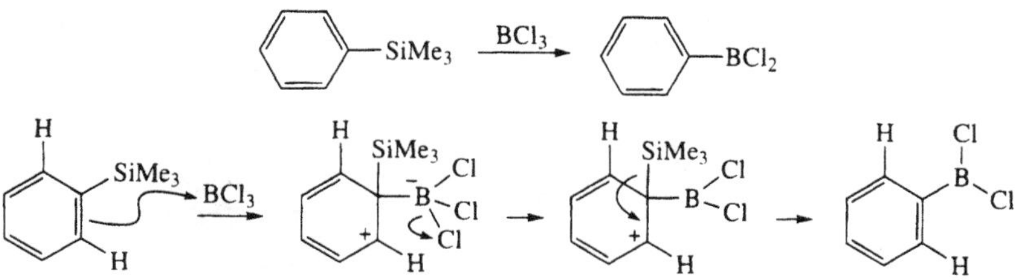

Sometimes, odd rearrangements in the course of aromatic substitution reactions can be explained by ipso substitution.

Example

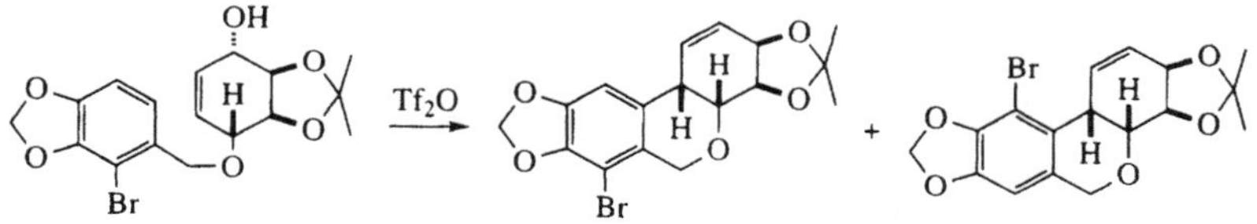

The first product results from a simple aromatic substitution reaction. The second product is much more interesting than the first, because it appears that the Br atom has migrated from the bottom to the top of the aromatic ring. A three-atom migration of a $C(sp^2)-Br$ bond is pretty unlikely, though, so perhaps something else is going on. Assuming that the C–Br bond does not break, careful numbering of the atoms reveals that the aryl C bound to the CH_2 group (C6) has undergone ipso substitution and that the aryl–CH_2 bond has undergone a 1,2-shift from C6 to C5.

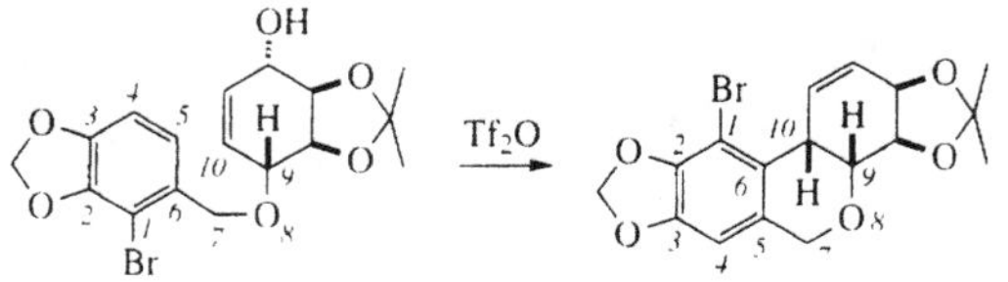

The OH group attacks triflic anhydride, displacing the fantastic leaving group TfO^- and converting the OH group into the same. Loss of TfOH from the substrate then gives an allyl cation.

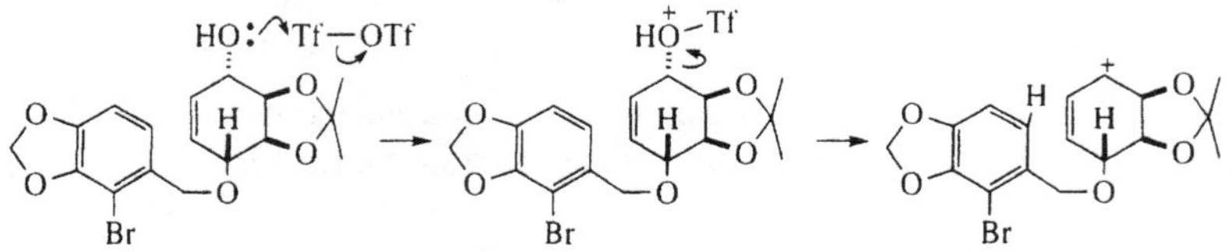

The first product is formed by attack of C5 on C10 of the allyl cation, followed by loss of H^+ from C5 to re-form the aromatic ring. The second product is formed by attack of C6 on C10 of the allyl cation. A 1,2-shift of C7 from C6 to C5 followed by loss of H^+ from C5 gives the product. In this particular reaction, instead of a 1,2-shift, a fragmentation (to give a stable $^+CH_2OR$ cation) followed by a second, conventional aromatic substitution is also reasonable.

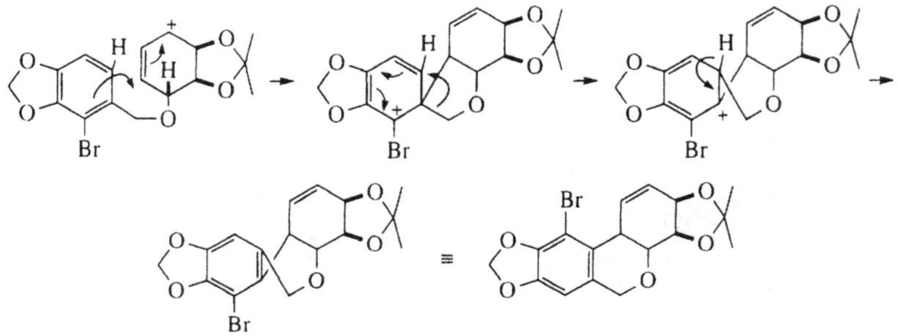

Problem 3.11. Draw mechanisms explaining the formation of both electrophilic aromatic substitution products.

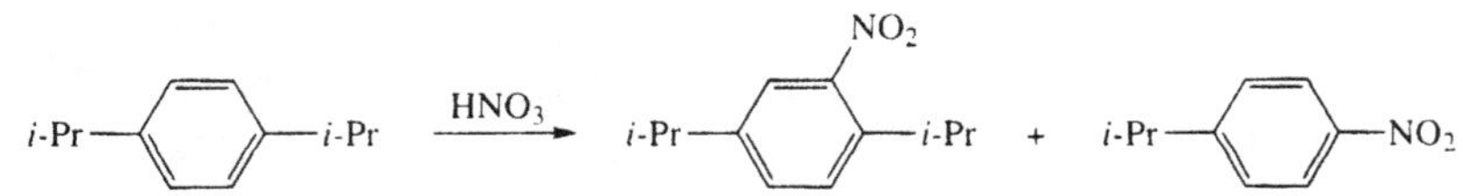

3.4.2   Aromatic Substitution of Anilines via Diazonium Salts

The NH_2 group of arylamines can be replaced by other groups by a special set of aromatic substitution reactions. When an arylamine is treated with sodium nitrite ($NaNO_2$) and acid, an aryldiazonium ion (ArN_2^+) is formed. The mechanism of the reaction begins with formation of the electrophile $^+N\equiv O$ from HNO_2. The aromatic amine attacks $^+N\equiv O$, and then dehydration occurs to give ArN_2^+.

Example

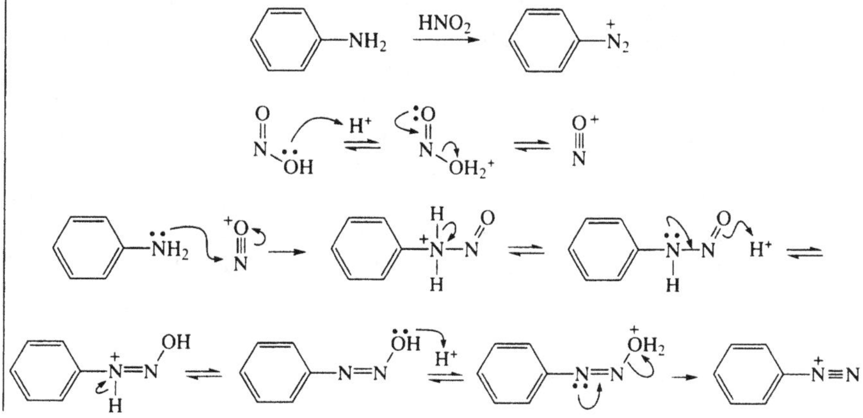

The aryldiazonium ion is usually prepared in situ and then immediately combined with a nucleophile to give a substitution product. The mechanisms of the substitution reactions vary greatly with the nucleophile, ranging all across the mechanistic spectrum of organic chemistry. For example, both hypophosphorous acid (H_3PO_2) and I^- react with diazonium ions by an $S_{RN}1$ mechanism (Chapter 2) to give the reduced arenes or aryl iodides, respectively.

Example

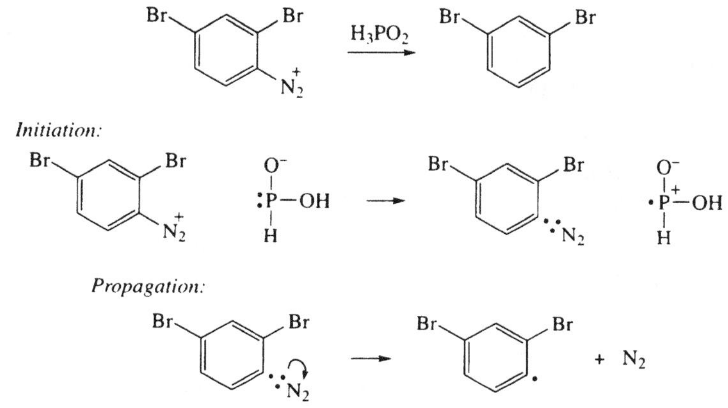

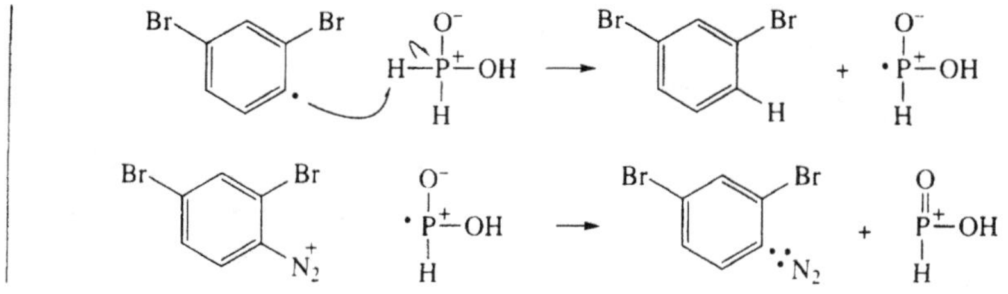

Copper(I) salts of anionic nucleophiles (CuBr, CuCl, CuCN) also react with diazonium ions to give the corresponding substitution products. These reactions probably proceed by a nonchain free-radical substitution mechanism similar to the one discussed earlier (Chapter 2). After coordination of the counterion of ArN_2^+ to CuCN, the Cu is sufficiently electron-rich to transfer an electron to ArN_2^+ to give $[ArN_2]$· and XCuCN. Loss of N_2 from $[ArN_2]$· gives Ar·, which then abstracts CN from XCuCN to give the observed product. A reasonable $S_{RN}1$ mechanism (Chapter 2) can also be drawn for this reaction, especially when both Cu(I) and Cu(II) salts are present in the reaction mixture, as can a metal-mediated mechanism involving oxidative addition and reductive elimination (Chapter 6).

Example

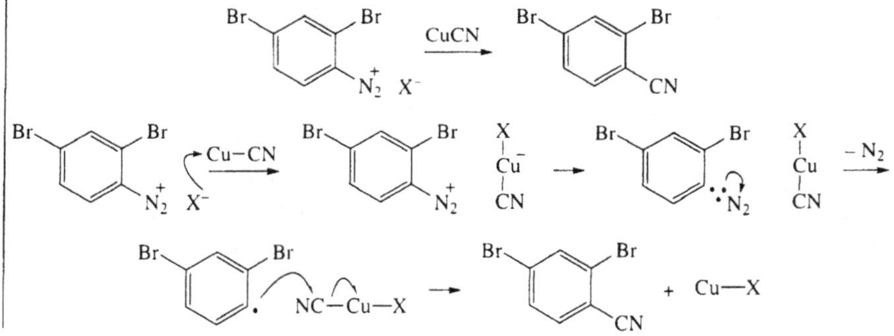

The nucleophiles H_2O and BF_4^- react with aryldiazonium ions to give phenols and aryl fluorides, respectively. Because neither H_2O nor BF_4^- are oxidizable enough to transfer an electron to the aryldiazonium ion, only S_N1 mechanisms are reasonable for these reactions.

* **Common error alert:** *These reactions are one of only a very few where it is reasonable to draw a very high-energy aryl cation.* The very large driving force for loss of N_2 accounts for the formation of this high-energy species.

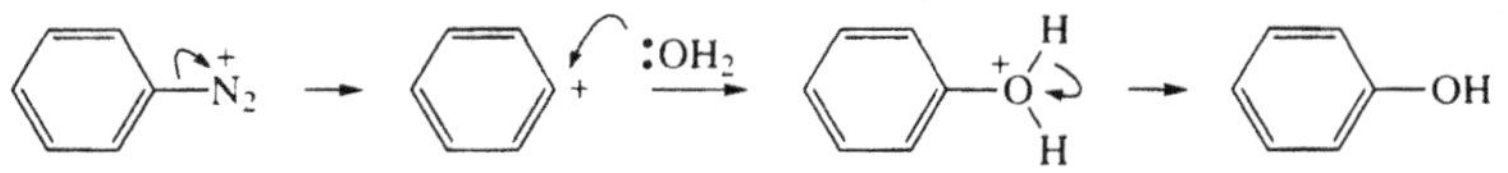

Finally, the terminal N of aryldiazonium ions is also electrophilic, and electron-rich aromatic compounds such as phenols and anilines undergo electrophilic

aromatic substitutions with ArN≡N⁺ as the electrophile to give azo compounds
(ArN=NAr').

Problem 3.12. Draw reasonable mechanisms for the following reactions:

(a)

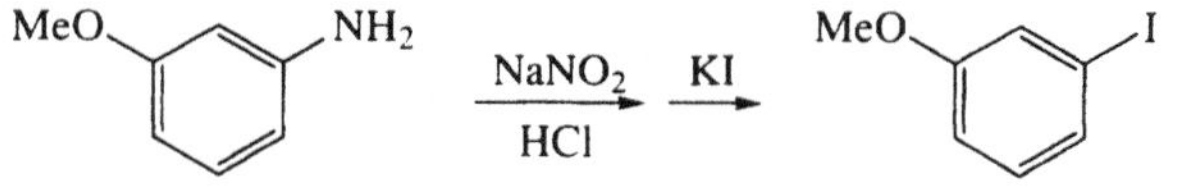

$$\text{MeO}\quad\quad\text{NH}_2 \xrightarrow[\text{HCl}]{\text{NaNO}_2\quad\text{KI}} \text{MeO}\quad\quad\text{I}$$

(b)

$$\text{MeO}\quad\quad-\text{NH}_2 \xrightarrow[\text{HCl}]{\text{NaNO}_2\quad\text{PhOH}} \text{MeO}\quad\quad-\text{N}\overset{\text{N}}{\underset{\text{N}}{\,}}-\quad-\text{OH}$$

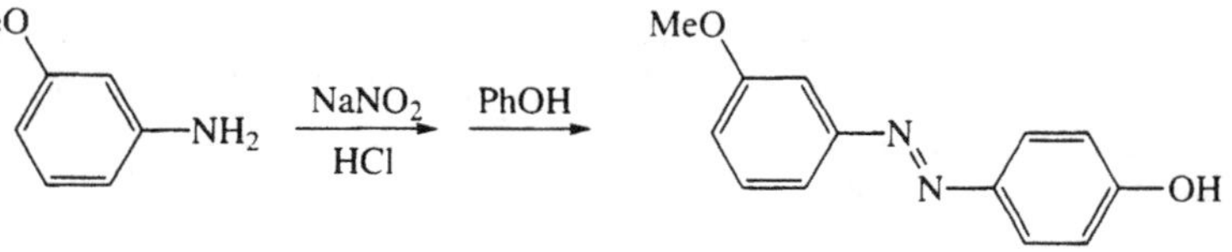

Aliphatic amines react with HNO_2 in the same way as arylamines to give diazonium salts, which react further by standard S_N1, S_N2, or rearrangement reactions.
α-Amino acids undergo substitution reactions with complete *retention* of configuration in the presence of HNO_2 (see Chapter 2). Likewise, acyl hydrazides react
with HNO_2 to give acyl azides, key intermediates in the Curtius degradation.

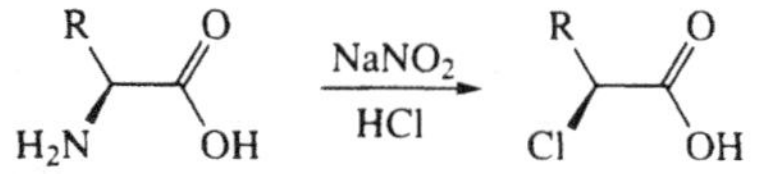

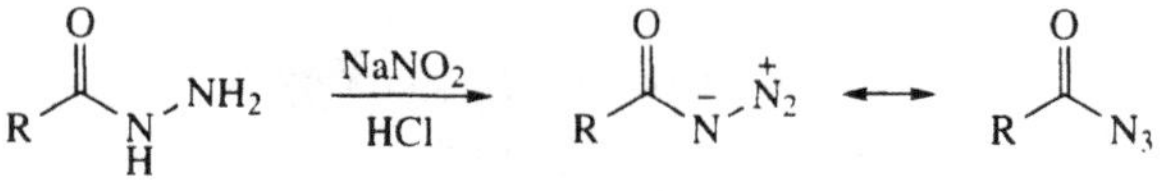

$$\underset{H_2N}{\overset{R}{\,}}\overset{O}{\underset{OH}{\,}} \xrightarrow[\text{HCl}]{\text{NaNO}_2} \underset{Cl}{\overset{R}{\,}}\overset{O}{\underset{OH}{\,}}$$

$$\underset{R}{\overset{O}{\,}}\underset{\underset{H}{N}}{\,}\text{NH}_2 \xrightarrow[\text{HCl}]{\text{NaNO}_2} \underset{R}{\overset{O}{\,}}\underset{N}{\overset{+}{\,}}\overset{-}{\,}\text{N}_2 \longleftrightarrow \underset{R}{\overset{O}{\,}}\text{N}_3$$

Problem 3.13. In the *Stiles reaction*, the highly reactive compound benzyne is
obtained by treatment of anthranilic acid (2-aminobenzoic acid) with $NaNO_2$
and HCl. Draw a mechanism for this reaction.

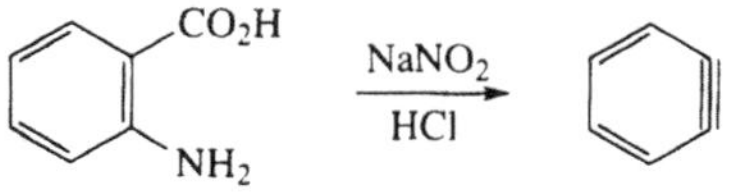

$$\underset{\text{NH}_2}{\overset{\text{CO}_2\text{H}}{\,}} \xrightarrow[\text{HCl}]{\text{NaNO}_2} \quad\bigcirc$$

3.4.3   *Electrophilic Aliphatic Substitution*

Alkenes undergo electrophilic substitution reactions by the same addition–
fragmentation mechanism as do arenes. Alkenylsilanes and -stannanes are especially good substrates for electrophilic substitution reactions because the carbocation intermediate is strongly stabilized by the C–Si or C–Sn bond. Fragmentation
of this bond then almost always occurs to give a desilylated or destannylated product. When the group that is lost is attached to the same C atom as the group that
is added, the overall reaction is called α-*substitution*. (Arene electrophilic substitution reactions are always α-substitution reactions because of the need to reestablish aromaticity after the carbocation intermediate is formed.)

Problem 3.14. Draw a mechanism for the following electrophilic substitution reaction.

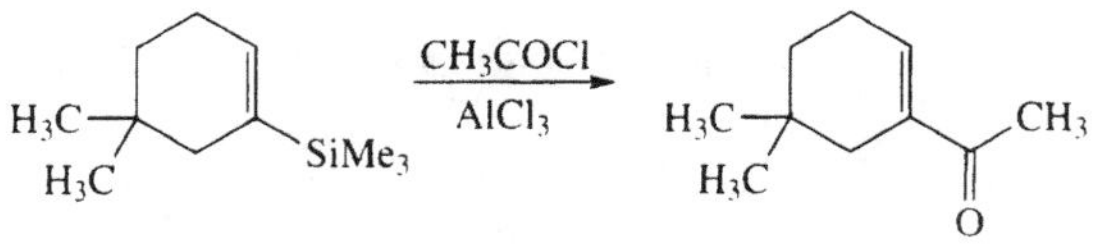

α-Substitution is not always observed when alkenes undergo electrophilic substitution. Sometimes the group that is lost to fragmentation is γ to the C atom where the new bond is formed. γ-Substitution is most commonly observed in the reactions of allylsilanes and -stannanes. The overall process is still a substitution with respect to the allylsilane, because a new σ bond is made at the expense of another σ bond, but the new bond is not made to the same atom as the old bond, and allylic transposition occurs.

Example:

The *Sakurai reaction.*

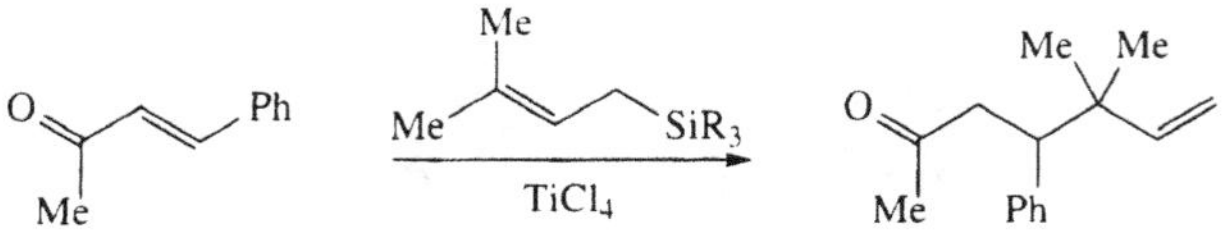

TiCl₄ is a Lewis acid; it coordinates to the carbonyl O, making the carbonyl and β-carbons into cations. The alkene then attacks the β-carbon in such a way as to put the new carbocation on the C adjacent to the C–Si bond. Fragmentation of the C–Si bond gives a Ti enolate. Aqueous workup protonates the enolate to give the observed product.

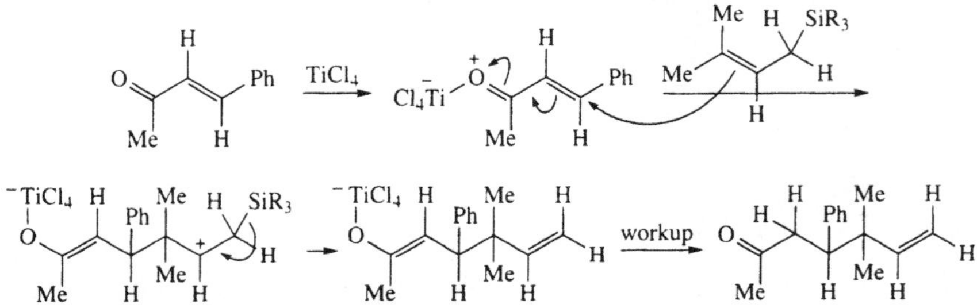

3.5   Nucleophilic Addition to and Substitution at Electrophilic π Bonds

3.5.1   Heteroatom Nucleophiles

Two major resonance structures can be drawn for a protonated carbonyl compound (more if the carbonyl C is σ-bound to heteroatoms). In the second-best resonance structure, the O is neutral, and C is both formally positively charged and electron-deficient. As a result, *carbonyl compounds undergo the same types of addition and substitution reactions under acidic conditions that they do under basic conditions!*

Carboxylic acids and carbonates and their derivatives undergo acid-catalyzed hydrolysis and alcoholysis reactions, in which the OH or OR group σ-bonded to the carbonyl C is replaced with another OH or OR group. The mechanism of the reaction is simple. Protonation of the carbonyl C occurs to give a (resonance-stabilized) carbocation. Addition of the nucleophile to the carbocation occurs to give a tetrahedral intermediate. The nucleophile is deprotonated, the leaving group is protonated, and the leaving group leaves to give a new carbocation. Finally, the unsubstituted O is deprotonated to regenerate a carbonyl compound. All of the steps are reversible, so the reaction can be driven in one direction or the other by executing the reaction in water or in alcohol.

Example

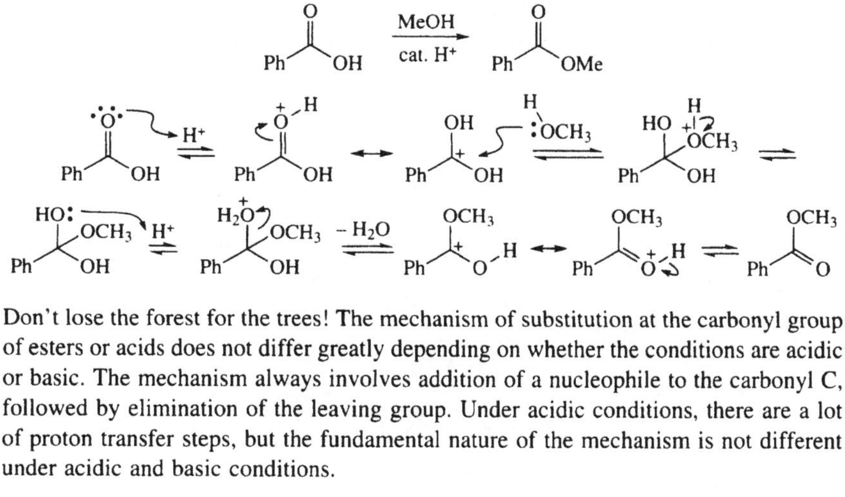

Don't lose the forest for the trees! The mechanism of substitution at the carbonyl group of esters or acids does not differ greatly depending on whether the conditions are acidic or basic. The mechanism always involves addition of a nucleophile to the carbonyl C, followed by elimination of the leaving group. Under acidic conditions, there are a lot of proton transfer steps, but the fundamental nature of the mechanism is not different under acidic and basic conditions.

Amides and nitriles are hydrolyzed to carboxylic acids by a similar mechanism, the latter proceeding through an amide. These hydrolyses are irreversible under acidic conditions, though, because the product amine is protonated under the reaction conditions and is made nonnucleophilic.

Problem 3.15. Draw mechanisms for the following carbonyl substitution reactions. In the second problem, think carefully about whence the O atoms in the product come.

(a)

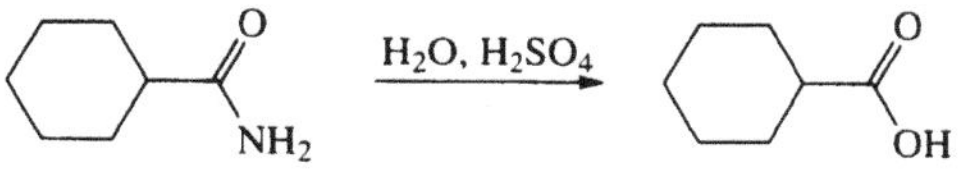

(b)

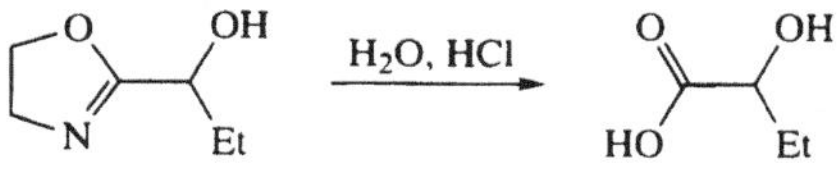

Ketones and aldehydes can be interconverted with acetals. (IUPAC now discourages the use of the word *ketal.*) Acetalizations are usually thermodynamically uphill, so the reaction is usually driven to completion by the removal of H_2O from the reaction mixture by azeotropic distillation or by addition of a dehydrating agent like 4 Å molecular sieves or triethyl orthoformate ($HC(OEt)_3$). Hydrolyses of acetals are usually executed simply by dissolving the acetal in water and a cosolvent such as THF and adding a catalytic amount of acid.

Example

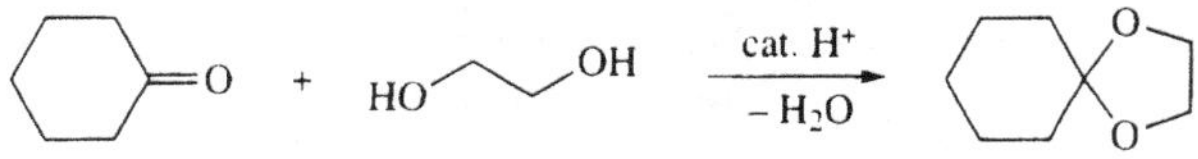

The mechanism of acetal formation consists of an addition followed by an S_N1 substitution. The first step is protonation of the carbonyl O to give a carbocation. An OH group in ethylene glycol adds to the carbocation and loses H^+ to give a tetrahedral intermediate called a *hemiacetal.* Protonation of the new OH group and loss of H_2O gives a new carbocation. Addition of the other OH group to the carbocation and deprotonation gives the acetal.

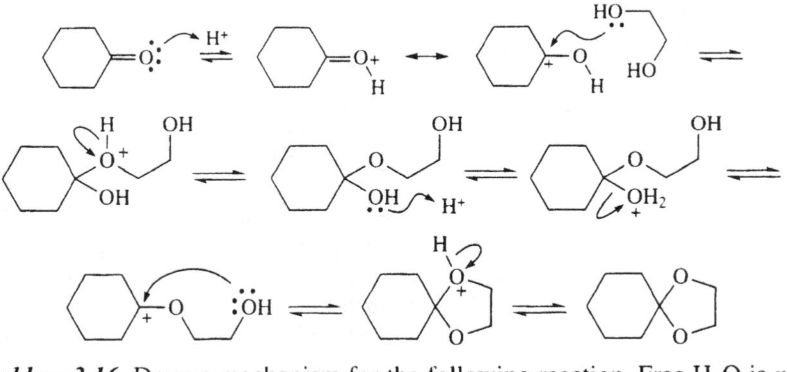

Problem 3.16. Draw a mechanism for the following reaction. Free H_2O is *not* an intermediate in the reaction. What is the by-product?

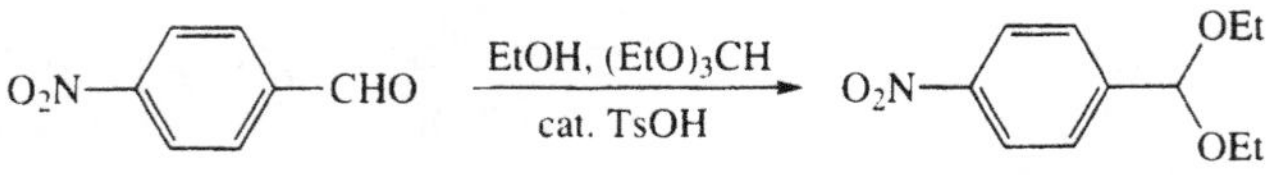

Ketones and aldehydes can also be converted to enol ethers if, after the loss of H_2O, the carbocation fragments with loss of H^+ to give the alkene. Because enol ethers are extremely reactive toward H^+, they are usually isolated only when the double bond is conjugated to an electron-withdrawing group, as in the conversion of β-diketones to vinylogous esters.

Example

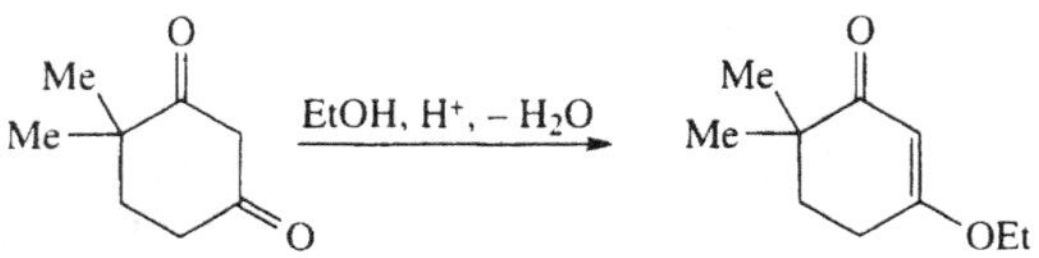

The diketone first tautomerizes to the keto–enol. Protonation of the ketone gives a very stable carbocation, to which EtOH adds. Deprotonation of the nucleophile, protonation of the OH leaving group, and loss of H_2O gives a new carbocation, which undergoes a fragmentation reaction with loss of H^+ to give the enol ether product.

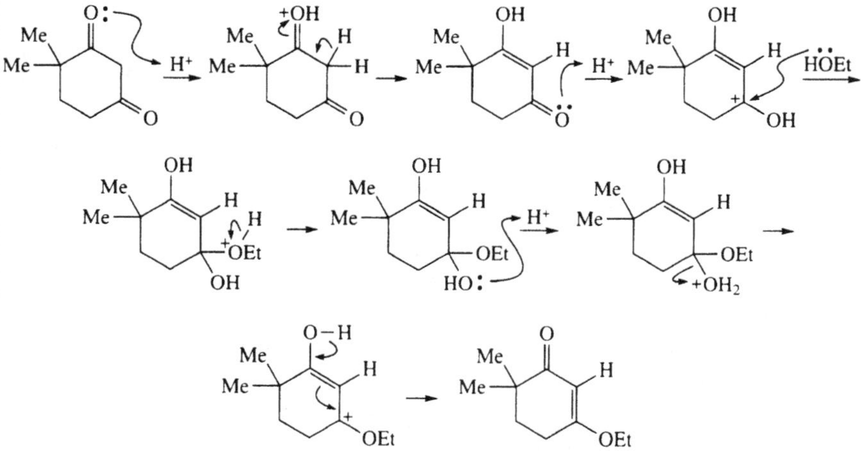

Why must the diketone tautomerize to the keto–enol before the addition of EtOH can occur? By the principle of microscopic reversibility, both enol ether formation and enol ether hydrolysis must proceed by the same sequence of steps (only in reverse of one another). Protonation of the enol ether, the first step in its hydrolysis, would certainly occur on the carbonyl O, not on C, so enol ether formation must proceed through the same intermediate.

Ketones/aldehydes, enols, and hemiacetals are all rapidly interconvertible under either acidic or basic conditions, but acetals and enol ethers are rapidly interconvertible with any of these species only under acidic conditions (Table 3.1). It is important for you to be able to recognize that any one of these species is

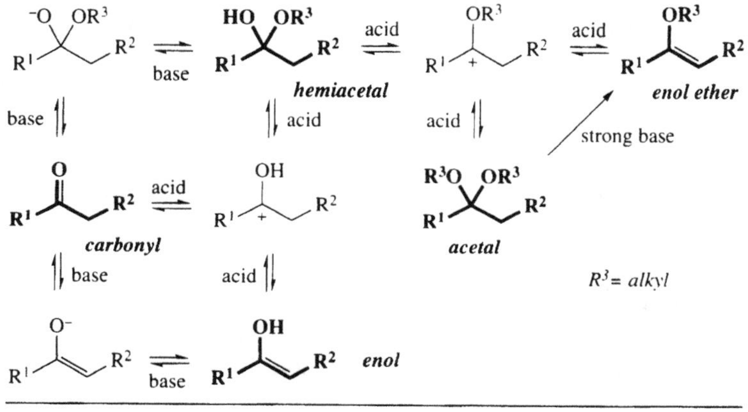

TABLE 3.1. Interconversions of ketones and their equivalents

functionally equivalent to any of the others under acidic conditions, and you need to be able to draw the mechanisms of their interconversion.

The *trimethylsilyl* enol ether (enol ether, R^3 = Me$_3$Si) can be converted directly into an enolate by treatment with CH$_3$Li or Bu$_4$N$^+$ F$^-$. Conversely, enolates can be converted directly into enol ethers when they are treated with certain particularly oxophilic electrophiles such as Me$_3$SiCl or even some acyl chlorides.

The reaction of a 1° or 2° amine with an aldehyde or ketone in the presence of sodium cyanoborohydride (NaBH$_3$CN) to give a 2° or 3° amine is called *reductive alkylation* (of the amine) or *reductive amination* (of the aldehyde or ketone) and is a valuable method for alkylating amines. The starting amine is mostly protonated and hence nonnucleophilic under mildly acidic conditions, but it is in equilibrium with a small amount of unprotonated amine that can act as a nucleophile toward the carbonyl compound. NaBH$_3$CN is an acid-stable source of H$^-$ that reacts only with iminium ions and not with carbonyl compounds or imines. Slightly acidic conditions are required to generate the iminium ion. The iminium ion can also be reduced by catalytic hydrogenation (Chapter 6).

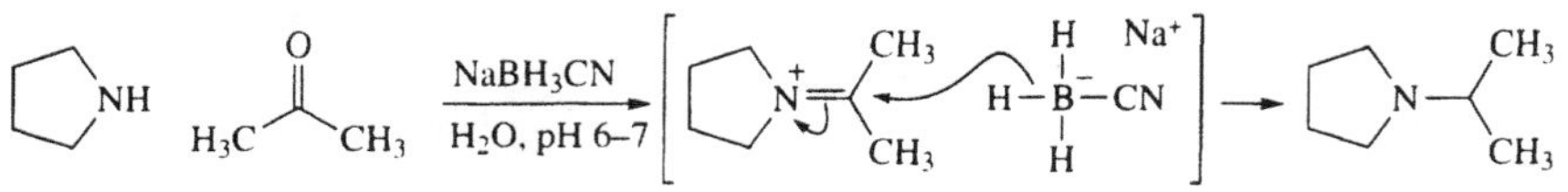

Problem 3.17. Draw the mechanism by which the iminium ion is formed in the above reaction.

3.5.2 Carbon Nucleophiles

The carbocation that is formed upon protonation of a carbonyl compound can lose H$^+$ from the α-carbon to give an enol. Enols are good nucleophiles. Thus, under acidic conditions, carbonyl compounds are *electrophilic at the carbonyl C and nucleophilic at the α-carbon and on oxygen*, just like they are under basic conditions. Resonance-stabilized carbonyl compounds such as amides and esters are much less prone to enolize under acidic conditions than less stable carbonyl compounds such as ketones, aldehydes, and acyl chlorides; in fact, esters and amides rarely undergo reactions at the α-carbon under acidic conditions.

Problem 3.18. Draw a mechanism for the following *Hell–Vollhard–Zelinsky* reaction.

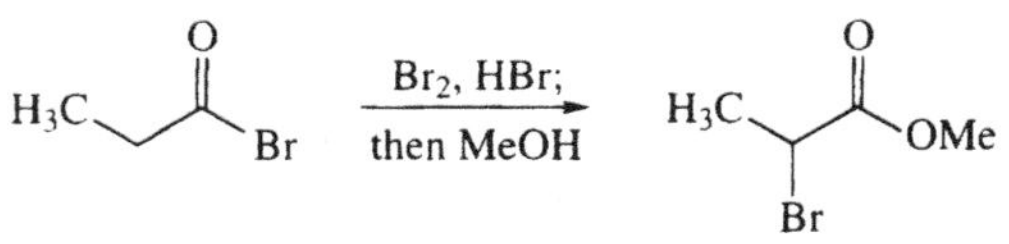

Enols are particularly reactive toward carbonyl compounds and toward electrophilic alkenes such as α,β-unsaturated carbonyl compounds. For example, the Robinson annulation proceeds under acidic conditions as well as under basic conditions, as shown in the next example. Under acidic conditions, the electrophilic carbonyl compound is protonated to make a carbocation before it is attacked by the enol.

Example

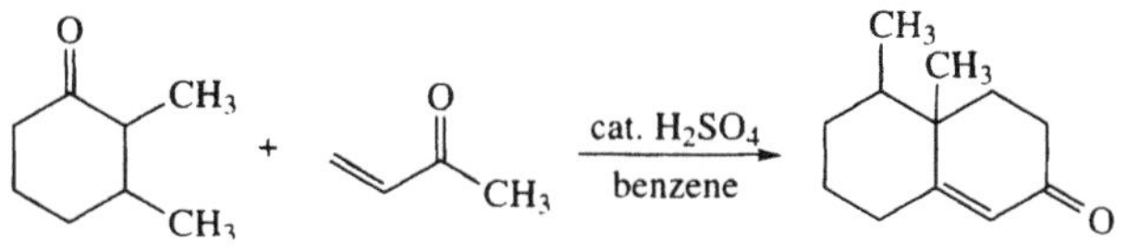

The Robinson annulation consists of a Michael addition, an aldol reaction, and a dehydration. In the Michael addition the nucleophilic ketone is converted into an enol by protonation and deprotonation. The enol then adds to the protonated Michael acceptor. Deprotonation of the positively charged O, protonation of C of the enol, and deprotonation of O then give the overall Michael addition product.

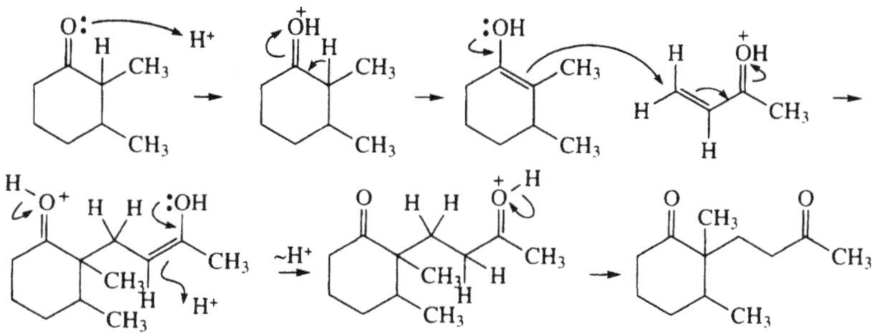

In the aldol reaction the nucleophilic ketone is first converted into an enol by protonation and deprotonation. The enol then adds to the protonated ketone in an aldol reaction. Deprotonation of the positively charged O gives the aldol.

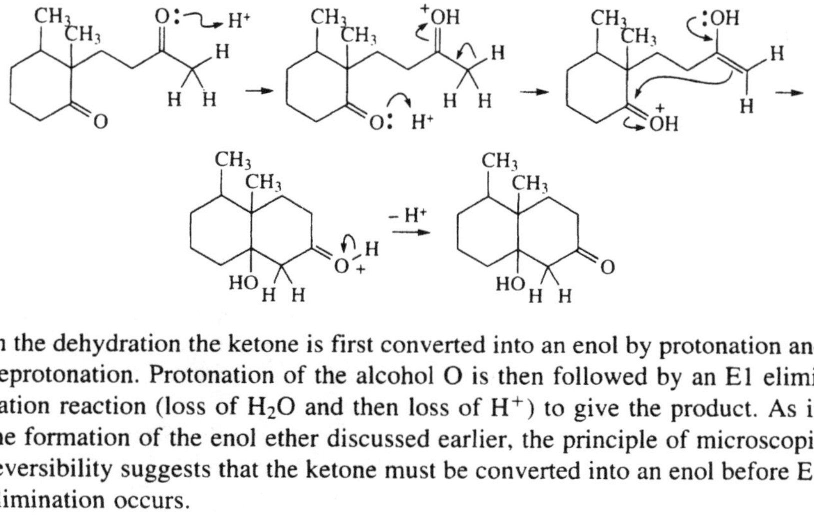

In the dehydration the ketone is first converted into an enol by protonation and deprotonation. Protonation of the alcohol O is then followed by an E1 elimination reaction (loss of H_2O and then loss of H^+) to give the product. As in the formation of the enol ether discussed earlier, the principle of microscopic reversibility suggests that the ketone must be converted into an enol before E1 elimination occurs.

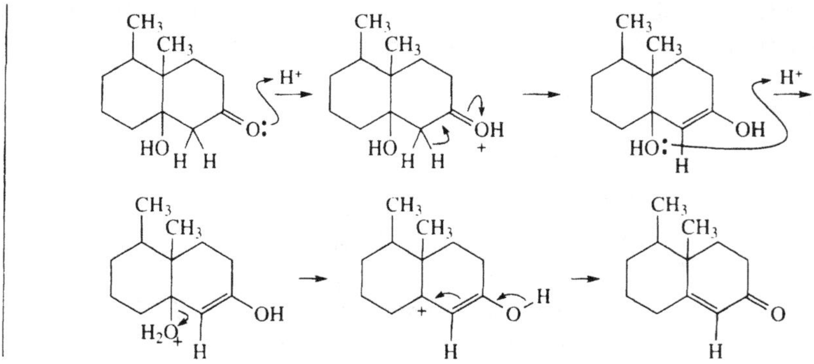

The *Mannich reaction* is another version of an aldol reaction that takes place under mildly acidic conditions. There are several variants, but all of them involve the reaction of a nucleophilic ketone with a primary or secondary ammonium ion ($R\overset{+}{N}H_3$ or $R_2\overset{+}{N}H_2$) and an aldehyde, usually formaldehyde (CH_2O). The Mannich reaction proceeds via an iminium ion ($R\overset{+}{N}H=CH_2$ or $R_2\overset{+}{N}=CH_2$), an electrophilic species that reacts with the enol form of the ketone to give a β-aminoketone. Sometimes E1 elimination of the amine (via the enol) follows to give an α,β-unsaturated ketone.

Example

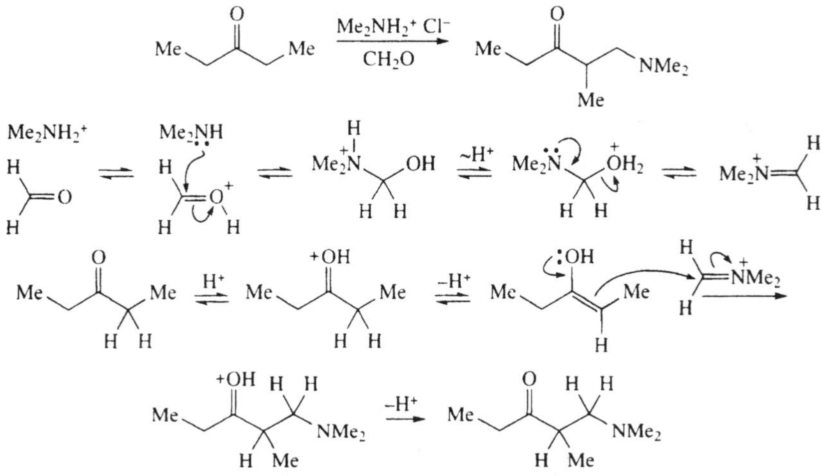

Two alternative mechanisms for the Mannich reaction can be written. In the first alternative mechanism, the ketone attacks the aldehyde directly in an aldol reaction, the aldol undergoes E1 elimination of H_2O (via the enol) to make an α,β-unsaturated ketone, and then the amine adds to the enone in a Michael fashion to give the product. Evidence against this mechanism: Ketones with only a single α-hydrogen undergo the Mannich reaction, but such ketones cannot be converted into α,β-unsaturated ketones.

In the second alternative mechanism, the ketone attacks the aldehyde directly in an aldol reaction, and then the β-carbon of the aldol undergoes S_N2 substitution, with the amine acting as the nucleophile. Evidence against this mechanism: S_N2 substitution reactions under acidic conditions require strongly acidic conditions, but the Mannich reaction takes place under only weakly acidic conditions.

Some carbonyl-based compounds (imines, carboxylic acids) are better electrophiles under acidic conditions than they are under basic conditions. Reactions using these compounds as electrophiles are usually executed under acidic conditions. On the other hand, enolates are always better nucleophiles than enols; when carbonyl compounds are required to react with electrophiles that are not particularly reactive, such as esters or alkyl bromides, basic conditions are usually used. Carbonyl compounds that are particularly low in energy (esters, amides) have such a small proportion of enol at equilibrium that they cannot act as nucleophiles at the α-carbon under acidic conditions. *Nevertheless, no matter whether acidic or basic conditions are used, carbonyl compounds are always nucleophilic at the α-carbon and electrophilic at the carbonyl carbon.*

Trimethylsilyl enol ethers, which are usually prepared by reaction of an enolate with Me₃SiCl, can be used as substitutes for enols in the Lewis-acid-catalyzed *Mukaiyama aldol* and *Michael reactions*. The mechanism of the Mukaiyama aldol or Michael reaction is very straightforward. After addition of the nucleophilic C=C π bond to the electrophile (activated by the Lewis acid), fragmentation of the O–SiMe₃ bond occurs; workup then provides the aldol or Michael adduct. The Me₃Si group, then, is treated as if it were nothing more than a big proton. The principal advantage of the Mukaiyama aldol and Michael reactions over the conventional versions is that the Mukaiyama versions allow esters and amides to act as nucleophiles under acidic conditions.

Example

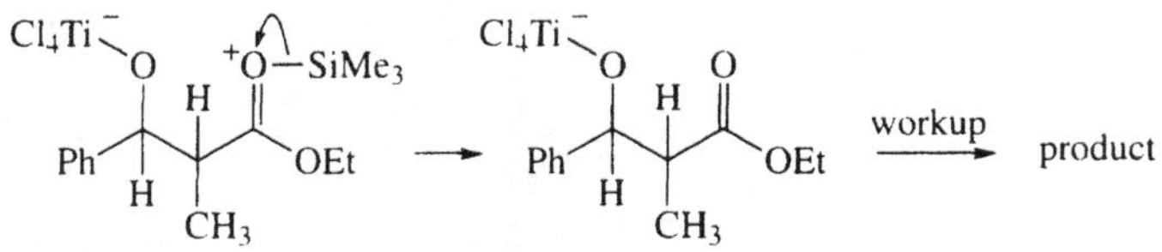

Retro-aldol and retro-Michael reactions occur under acidic conditions. The mechanisms are the microscopic reverse of the aldol and Michael reactions, as you would expect. One of the most widely used acid-catalyzed "retro-aldol" reactions is the decarboxylation of β-ketoacids, malonic acids, and the like. Protonation of a carbonyl group gives a carbocation that undergoes fragmentation to lose CO_2 and give the product. Decarboxylation does not proceed under basic conditions because the carboxylate anion is much lower in energy than the enolate product.

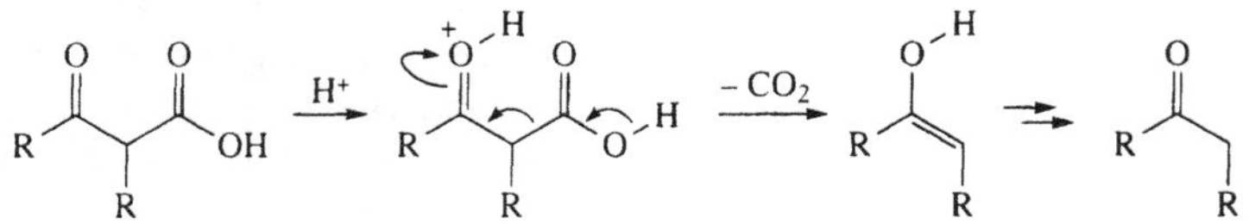

A cyclic TS for decarboxylation is often drawn. Such a mechanism can be described as a *retro-ene* reaction (see Chapter 4).

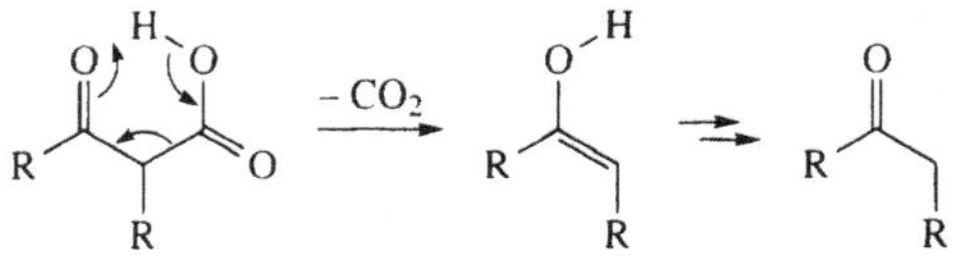

3.6   Summary

There are three methods for generating carbocations:

1. ionization of a $C-X^+$ σ bond (usually preceded by protonation of the leaving group);
2. reaction of a lone pair on the heteroatom in a $C=X$ π bond with a Lewis acid (typically H^+); and
3. reaction of a $C=C$ π bond with a Lewis acid (typically H^+ or a carbocation).

Carbocations undergo three typical reactions:

1. addition of a nucleophile (usually a lone-pair or π-bond nucleophile);
2. fragmentation to give a π bond and a new cation (often a carbocation or H^+); and
3. rearrangement (typically a 1,2-hydride or -alkyl shift).

The addition of a protic nucleophile to a carbocation is always followed by deprotonation of the nucleophile.

Substitution at C(sp³) usually occurs by an S$_N$1 mechanism under acidic conditions; the S$_N$2 mechanism is operative at C(sp³) only when the C(sp³) is 1°,
the nucleophile is very good, and the conditions are *very* strongly acidic.
(However, substitution at heteroatoms such as S and P usually occurs by an S$_N$2
mechanism under acidic conditions.)

Substitution at C(sp²) normally occurs by an addition–elimination mechanism
under acidic conditions. Electrophiles add to nucleophilic C=C π bonds, including most arenes, and nucleophiles add to C=N, C=O, and electrophilic C=C
π bonds. Arylamines can undergo substitution of the NH₂ group by a diazonium
ion mechanism.

Under both acidic and basic conditions, a carbonyl compound is electrophilic at the carbonyl carbon and nucleophilic at the α-carbon. Similarly,
under both acidic and basic conditions, an α,β-unsaturated carbonyl compound
is electrophilic at the carbonyl and β-carbons and nucleophilic at the α-
carbon after a nucleophile has already added to the β-carbon. Carbonyl mechanisms differ under acidic and basic conditions only in the protonation and
deprotonation steps: under acidic conditions, nucleophilic addition is always
preceded by protonation of the electrophile and followed by deprotonation of
the nucleophile.

PROBLEMS

1. Rank each set of compounds by the ease with which they ionize under acidic
   conditions.

(a)

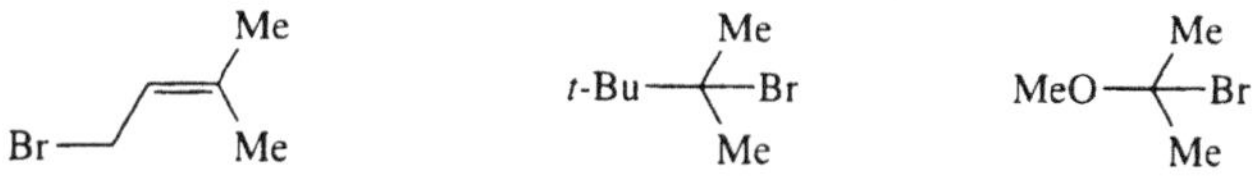

(b)

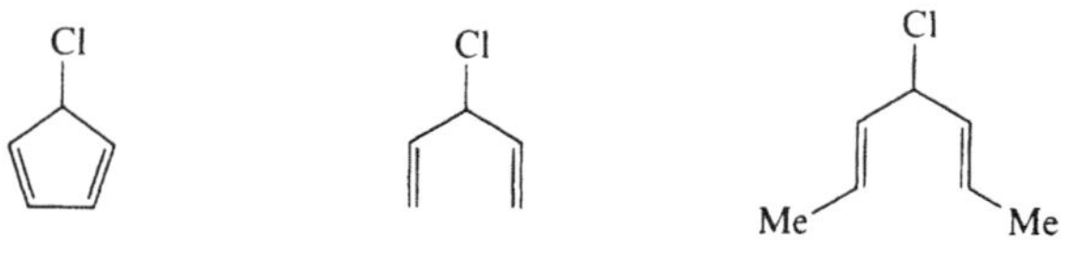

(c)

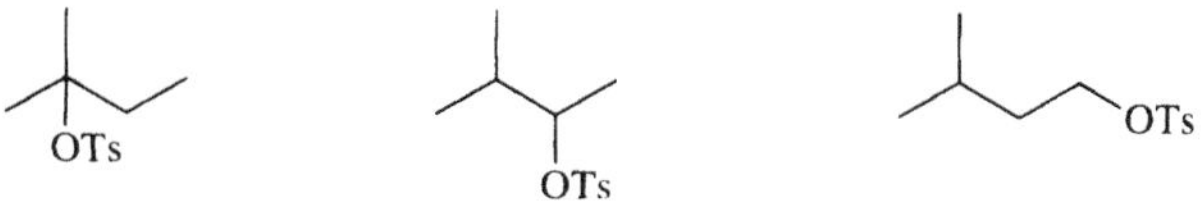

(d)

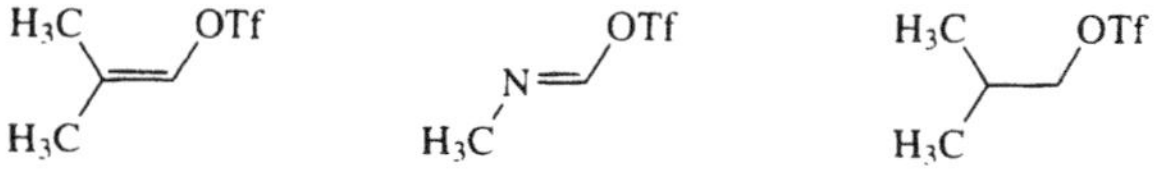

(e)

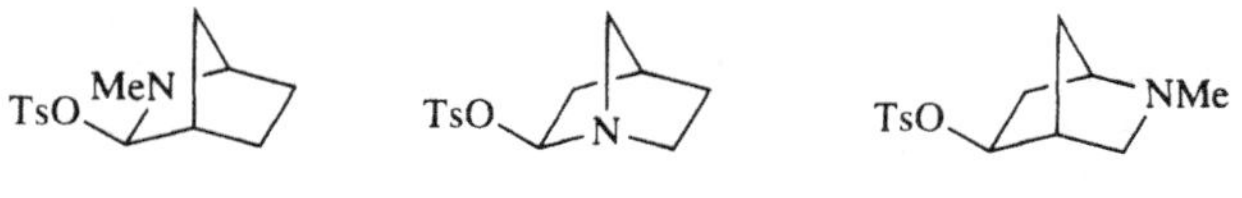

(f)

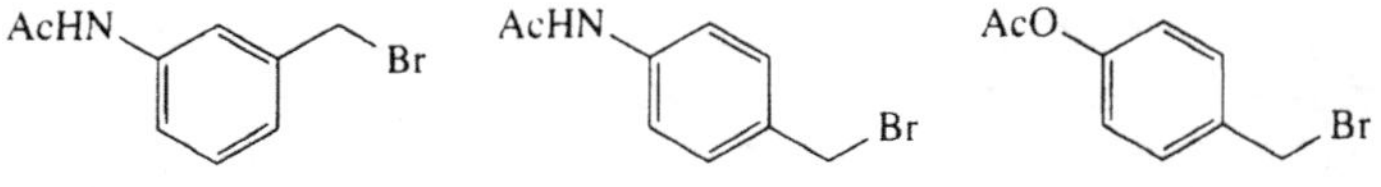

(g)

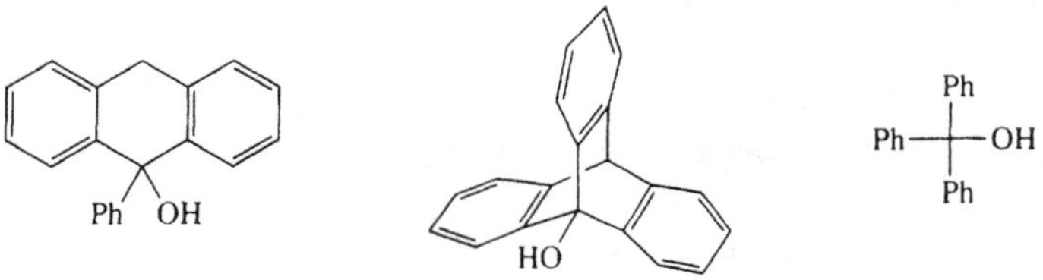

2. Indicate the major product of each of the following reactions, and draw a reasonable mechanism for its formation.

(a) 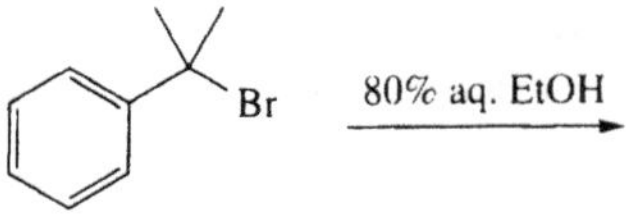 $\xrightarrow{\text{80\% aq. EtOH}}$

(b) 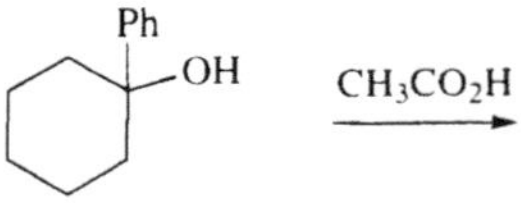 $\xrightarrow{\text{CH}_3\text{CO}_2\text{H}}$

(c) $\xrightarrow[\text{benzene}]{\text{cat. TsOH}}$

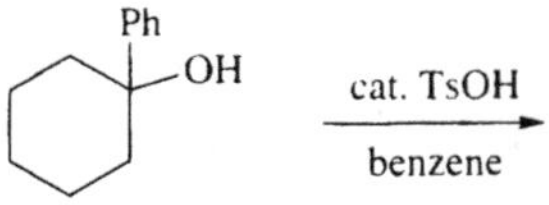

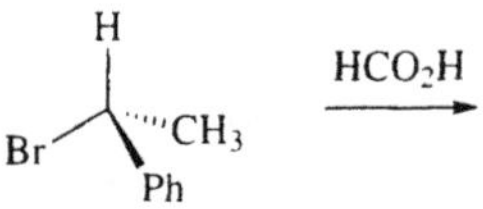

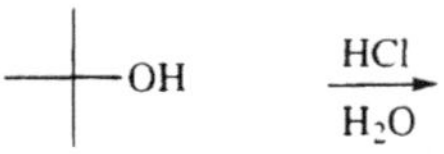

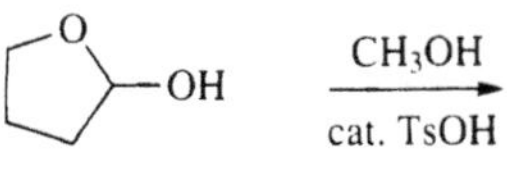

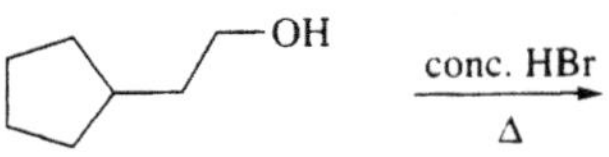

(d) $\xrightarrow{\text{HCO}_2\text{H}}$

(e) $\xrightarrow[\text{H}_2\text{O}]{\text{HCl}}$

(f) $\xrightarrow[\text{cat. TsOH}]{\text{CH}_3\text{OH}}$

(g) $\xrightarrow[\Delta]{\text{conc. HBr}}$

(h) $\xrightarrow{\text{dil. HBr}}$

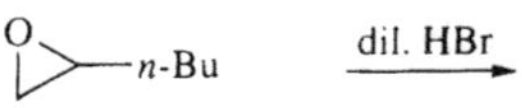

3. When β-caryophyllene (1) is treated with sulfuric acid in ether, a large assortment of products is formed. Initially, after 1 is consumed, compounds 2 through 4 are the major hydrocarbon products. As time goes on, compounds 2 through

4 disappear, and compounds **5** through **9** appear in the reaction mixture. At longer times, compounds **5** through **9** disappear, and compounds **10** through **12** become the final hydrocarbon products. A large number of alcohols were also formed. Draw mechanisms explaining the formation of each of these products. *Your mechanism should also account for the sequence of their formation.*

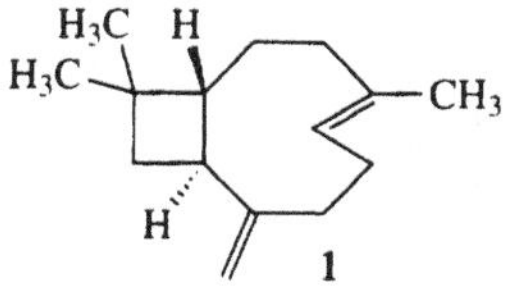
1

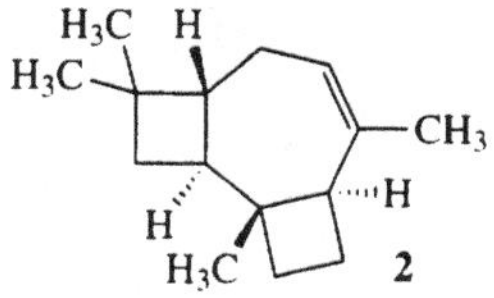
2

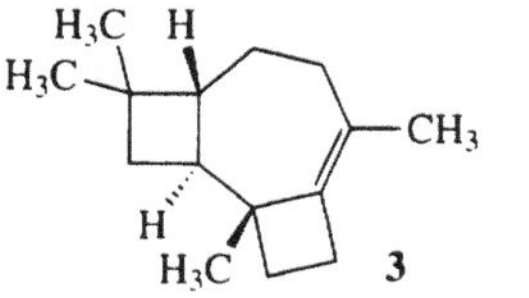
3

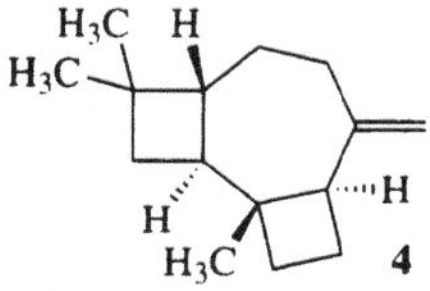
4

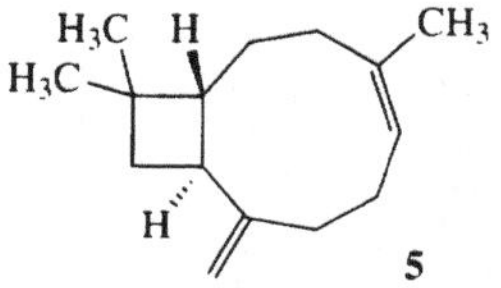
5

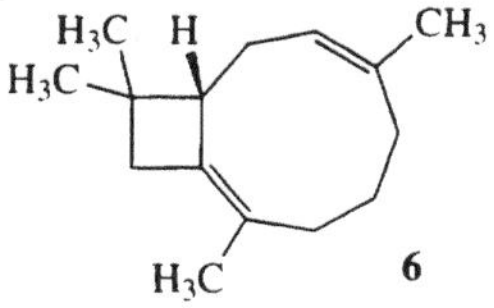
6

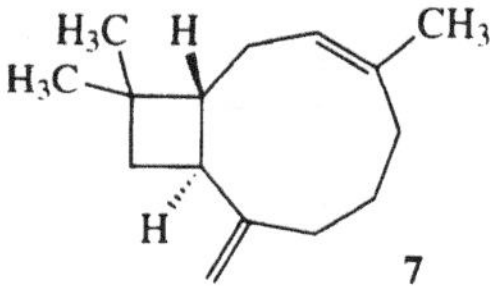
7

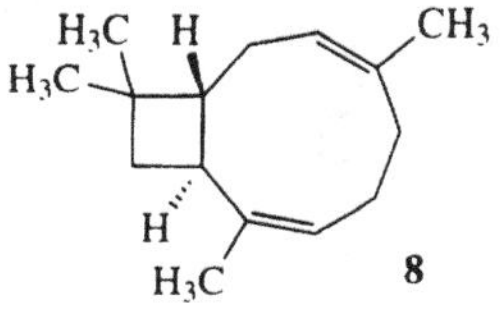
8

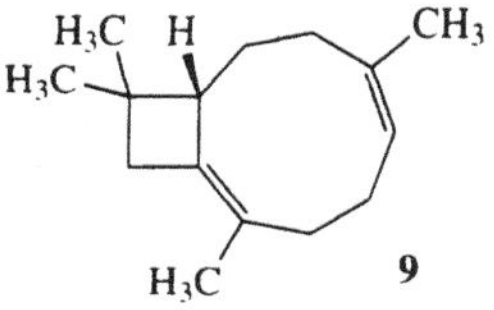
9

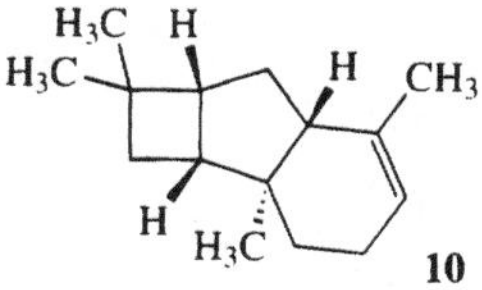
10

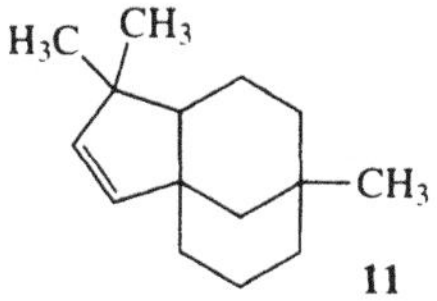
11

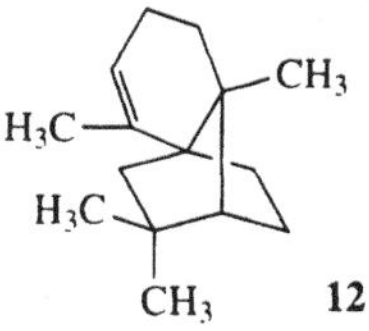
12

4. Draw mechanisms for the following reactions.

(a)

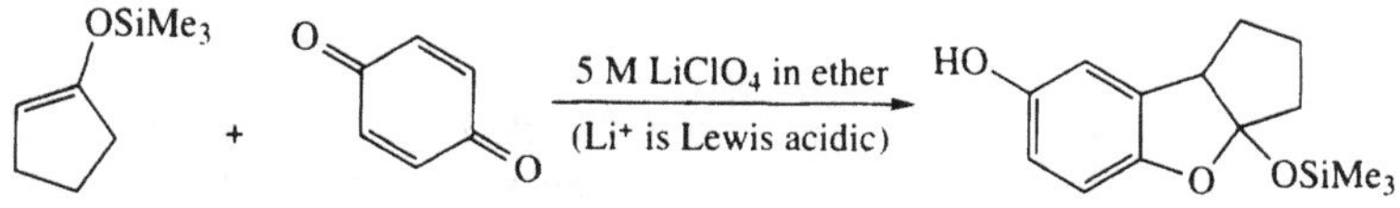

(b)

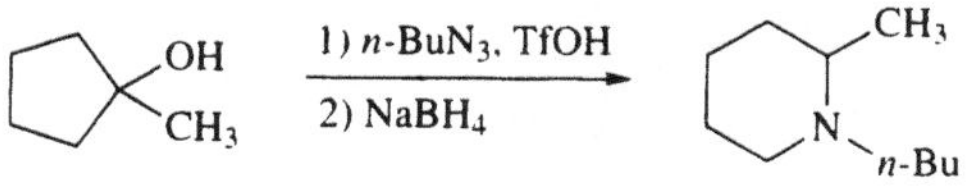

(c)

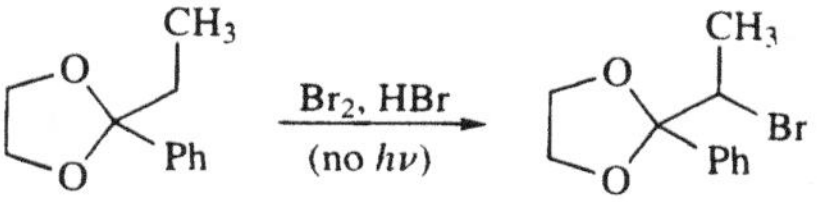

(d)

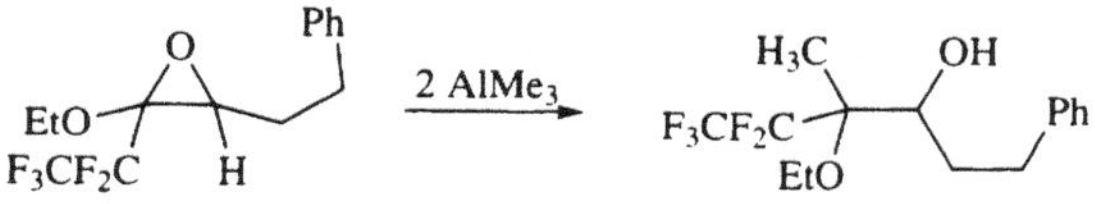

but

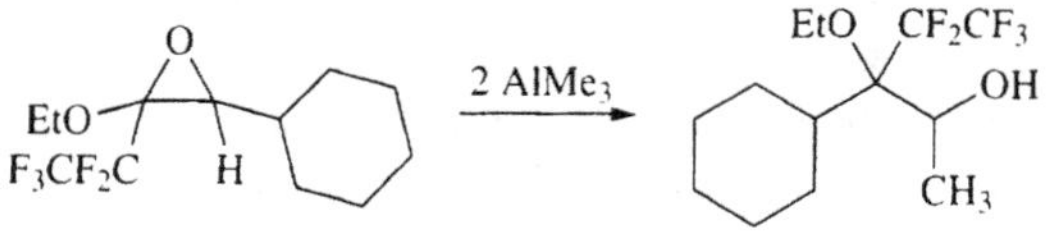

EtO—, F₃CF₂C, H (epoxide on cyclohexyl) → 2 AlMe₃ → EtO CF₂CF₃, cyclohexyl—C(OH)—CH₃

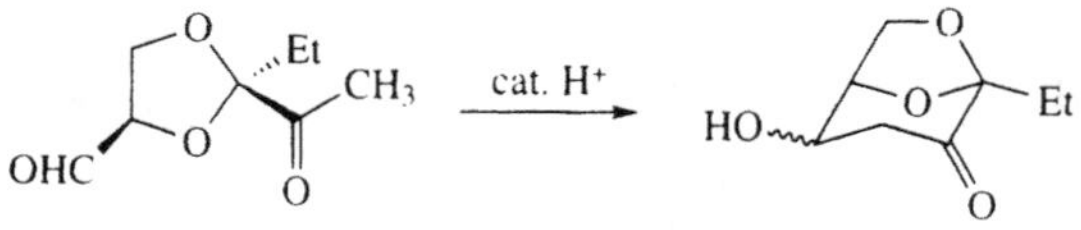

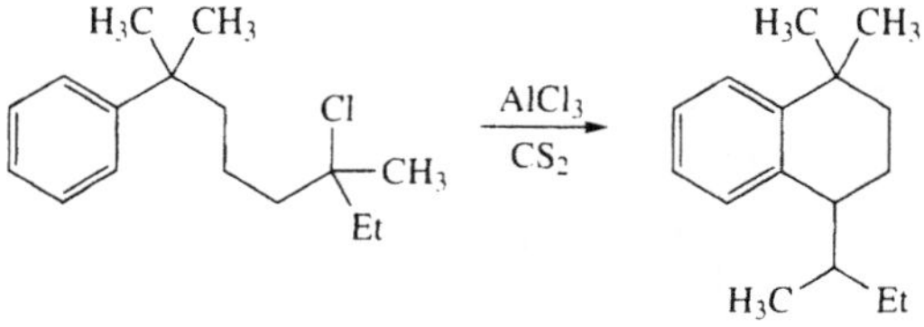

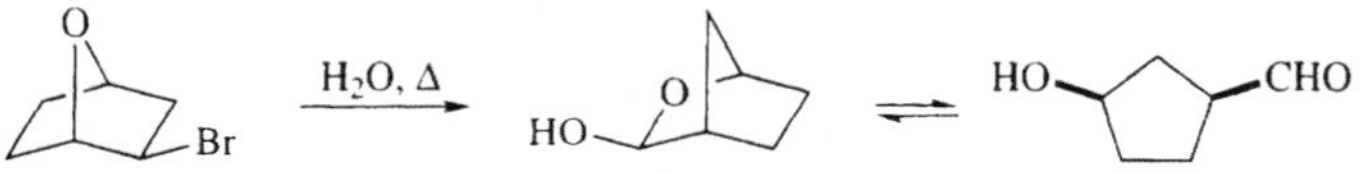

(e)   (dioxolane with Et, CH₃, OHC, O, C=O) → cat. H⁺ → (bicyclic, HO, O, O, Et, O)

(f)

H₃C CH₃ (phenyl with C(CH₃)₂ chain, Cl, —CH₃, Et) → AlCl₃ / CS₂ → (tetralin, H₃C CH₃, H₃C Et)

(g)

(bicyclic epoxide, Br) → H₂O, Δ → (bicyclic, HO, O) ⇌ HO—(cyclopentane)—CHO

(h)

(pyrrolidine, CO₂H, CH₃, N, Ph) → 1) NBS 2) NaOCH₃ → (pyrrolidine, CO₂CH₃, CH₃, N, epoxide O, Ph)

NBS = (succinimide with Br, O, N, O)

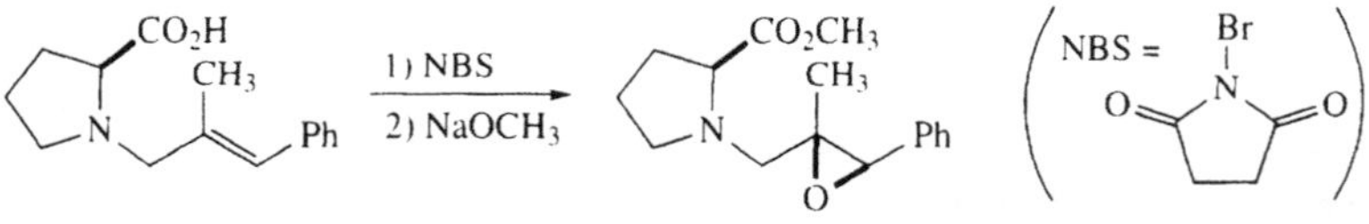

(i)

(ascorbic acid structure: OH, H, OH, O, OH, HO, O=) + (O, CHO, H₃C, enal) → cat. H⁺ → (product: O=, O, H, OH, O, O, OH, OH, H₃C—)

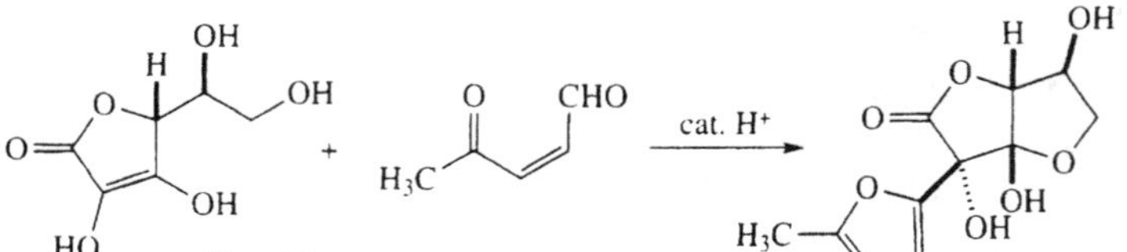

ascorbic acid

(j)   The product, one example of an azo dye, was used for a long time as the coloring agent in margarine until it was discovered that it was carcinogenic.

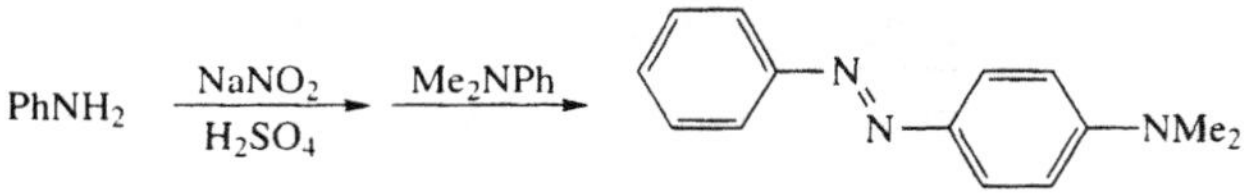

PhNH₂ → NaNO₂ / H₂SO₄ → Me₂NPh → (Ph—N=N—C₆H₄—NMe₂)

(k)   When p-nitroaniline and salicylic acid are used as starting materials in the reaction in (j), Alizarin Yellow R, an azo compound used to dye wool, is obtained. What is the structure of Alizarin Yellow R?

(l)   The following reaction helps explain why piperidine is a federally controlled substance.

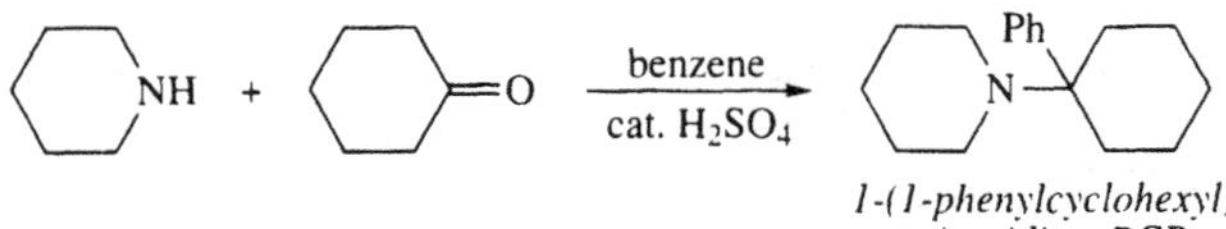

1-(1-phenylcyclohexyl)-piperidine, PCP

(m)

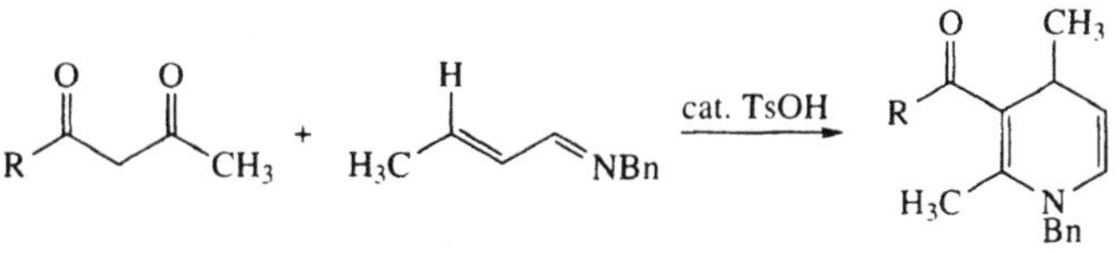

(n)

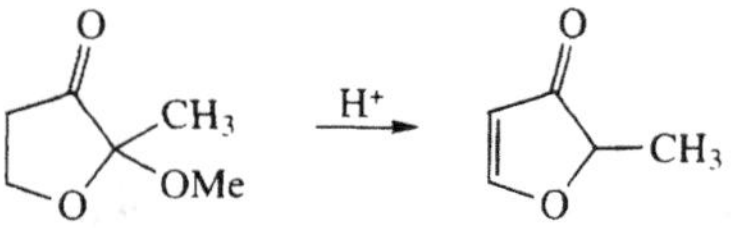

(o)

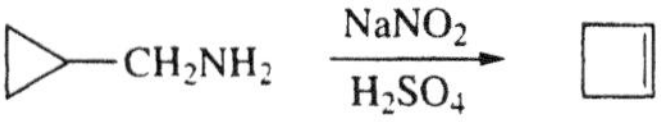

(p) The following reaction is the key step in the biosynthesis of cholesterol. Compounds that inhibit the enzyme that catalyzes the reaction have been studied intensively recently because of their potential as anticholesterol drugs.

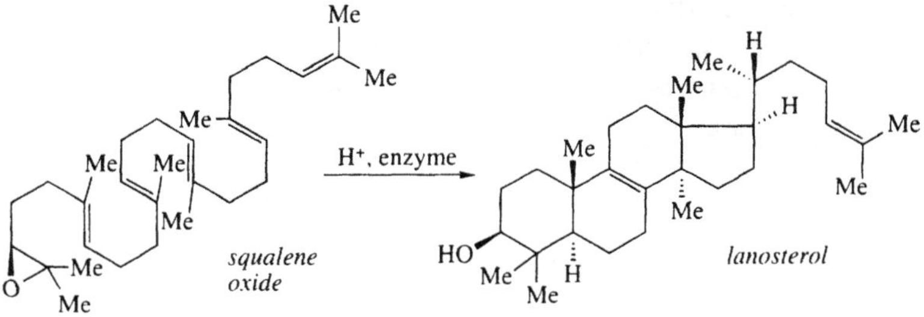

squalene oxide

lanosterol

(q) The Wallach rearrangement.

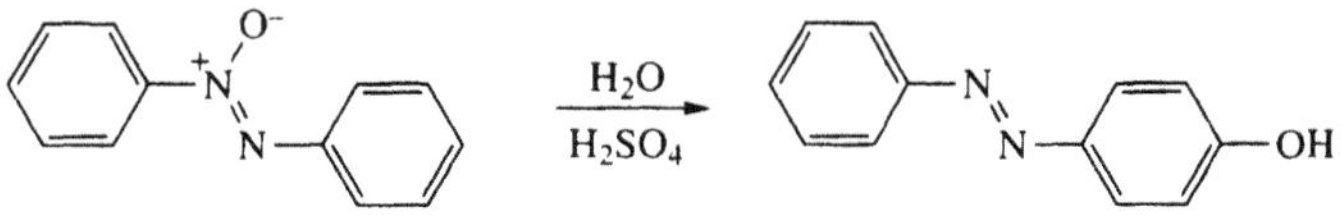

The following pieces of information should help you: (1) When ^{18}O-labeled water is used as solvent, the product is labeled. (2) When one N in the starting material is selectively labeled with ^{15}N, the label is found to be randomly distributed in the product. (3) The key intermediate is dicationic.

(r) Think of the Me_3Si^+ group as a big proton.

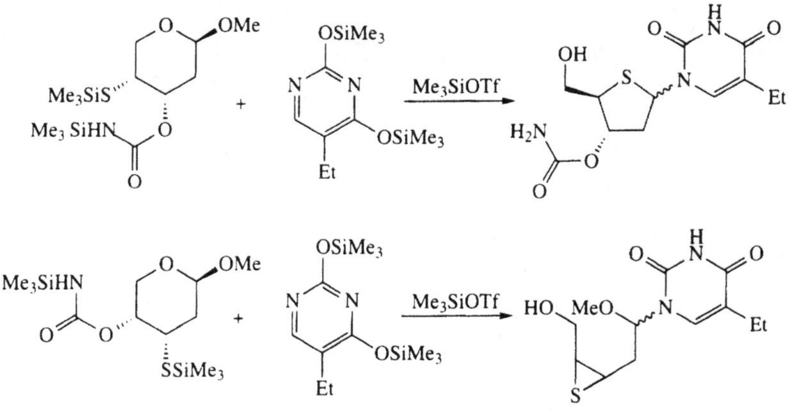

(s) First draw a mechanism for formation of the five-membered ring. Then draw a mechanism by which the five-membered ring is converted into the six-membered ring simply by heating (no acid present).

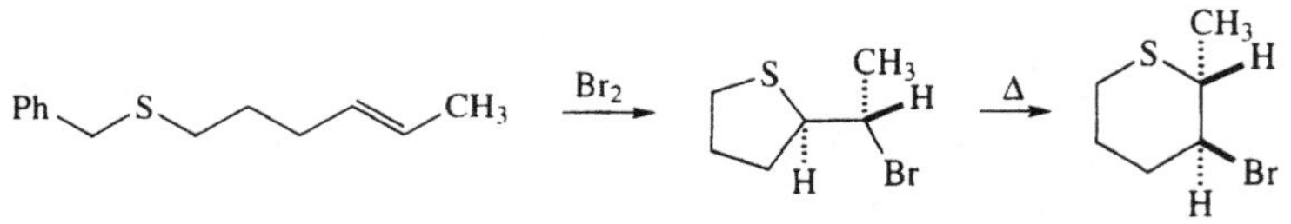

(t) The following two pieces of information should help you draw the mechanism. (1) The rate-determining step involves cleavage of the C–Cl bond. (2) The rate decreases *dramatically* as the number of C atoms between N and Cl increases.

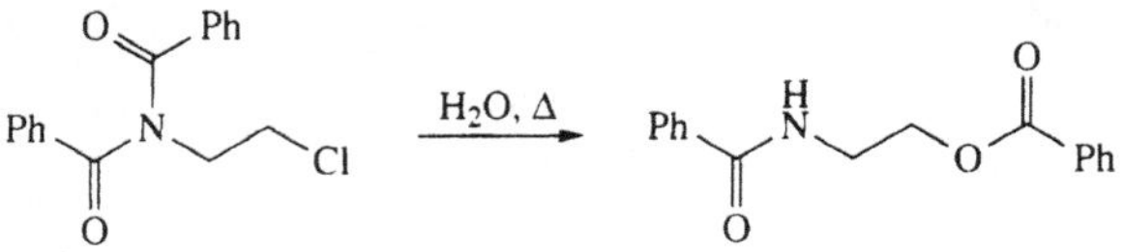

(u)

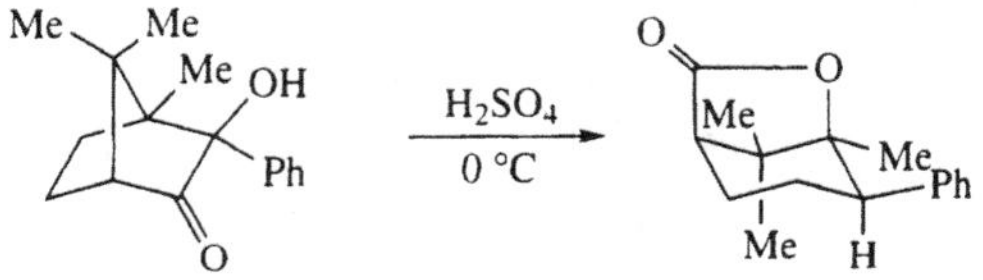

(v)

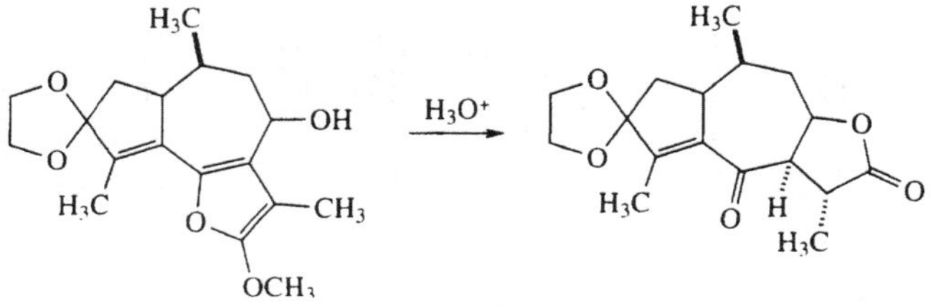

(w)

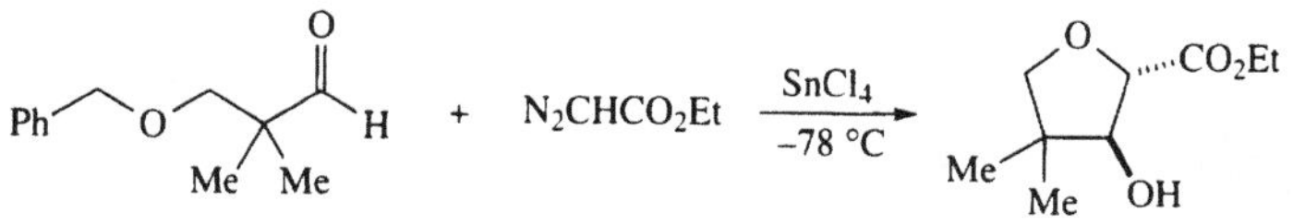

(x) SnCl$_2$ is a Lewis acid.

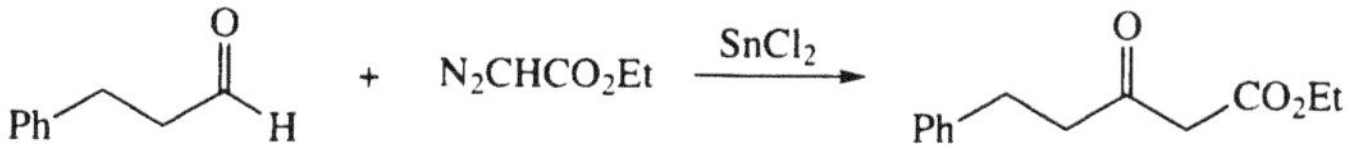

(y) *Note the stereochemistry of the product!* What does this tell you about the mechanism?

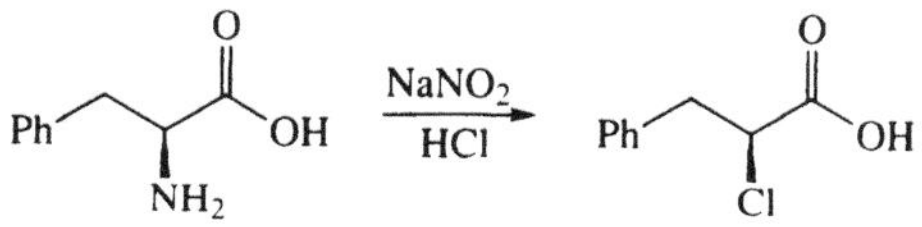

(z)

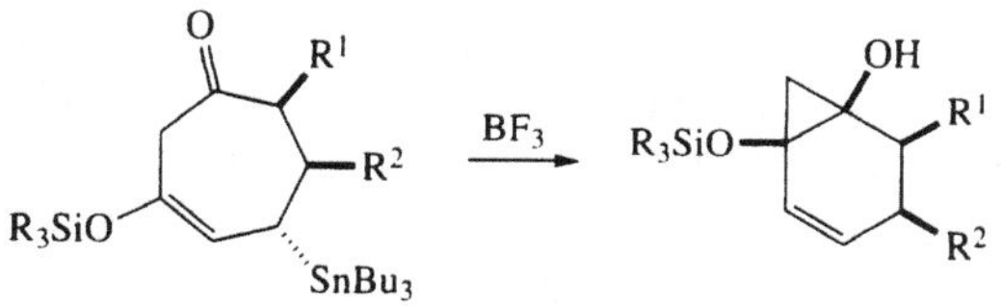

(aa)

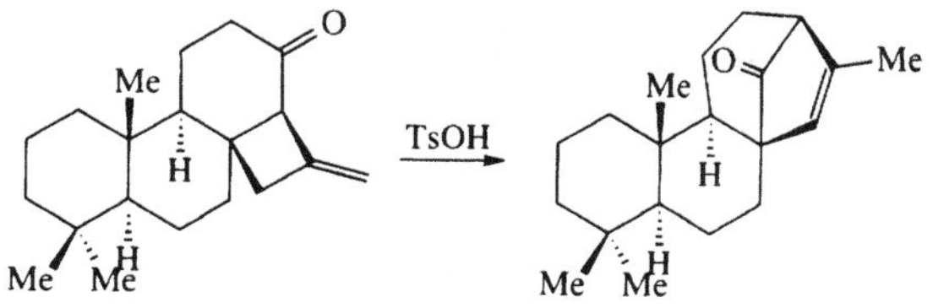

(bb)

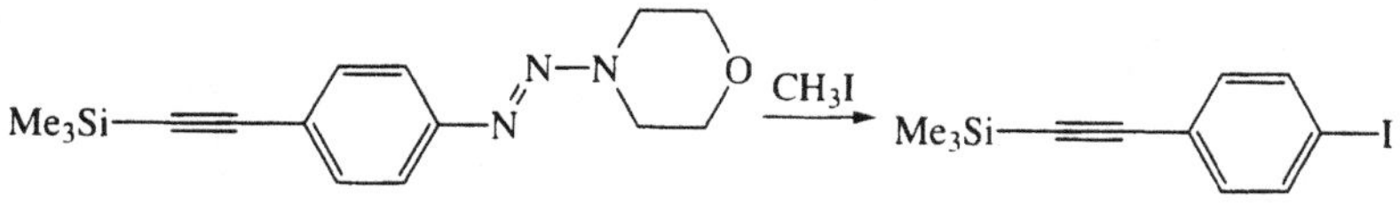

(cc)

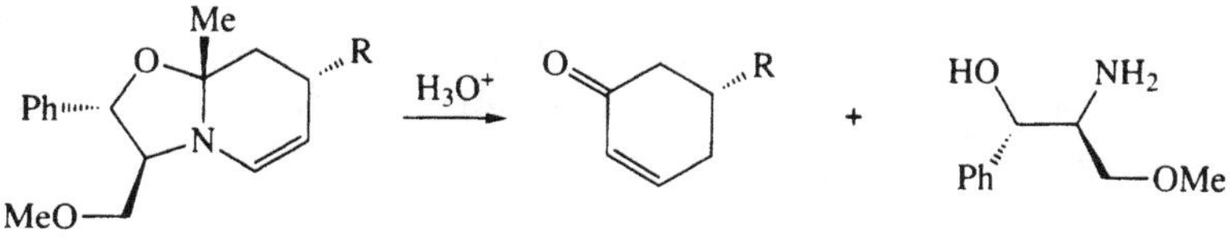

(dd)

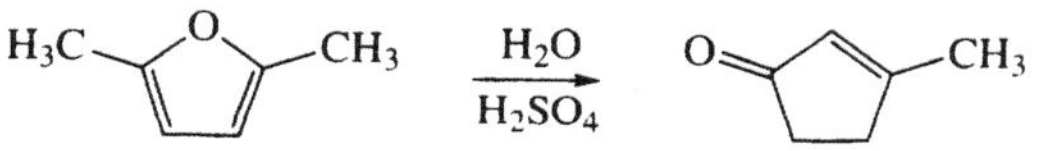

4

Pericyclic Reactions

4.1   Introduction

A pericyclic reaction is a reaction in which bonds are formed or broken at
the termini of one or more conjugated π systems. The electrons move around
in a circle, all bonds are made and broken simultaneously, and no intermediates intervene. The requirement of concertedness distinguishes pericyclic
reactions from most polar or free-radical reactions, although for many pericyclic reactions reasonable alternative stepwise mechanisms can also be
drawn.

For each class of pericyclic reactions two or more of the following characteristics will be discussed: the *typical reactions*, *regioselectivity*, *stereoselectivity*,
and *stereospecificity*. The discussions of typical reactions and stereospecificity
will help you recognize when pericyclic reactions are occurring in a particular
chemical reaction. The discussions of regioselectivity, stereoselectivity, and stereospecificity will allow you to *predict* the structures and stereochemistry of the
products obtained from pericyclic reactions.

4.1.1   Classes of Pericyclic Reactions

Before discussing specific reactions, it is important to learn how to describe pericyclic reactions. There are four major classes of pericyclic reactions: *electro-*
cyclic reactions (ring openings or ring closings), *cycloadditions*, *sigmatropic re-*
arrangements, and *ene reactions*.

In an *electrocyclic ring closing*, a σ bond forms between the termini of a
conjugated π system. An *electrocyclic ring opening* is the reverse reaction,
in which a C–C σ bond breaks to give a conjugated π system in which the
termini used to be bound. Empty or filled p orbitals can participate in the π
system. Electrocyclic reactions are further subclassified as 2 π electrocyclic,
4 π electrocyclic, etc., according to how many electrons are involved in
the reaction. As in all pericyclic reactions, the electrons move around in

a circle. Note how the π bonds shift upon formation or cleavage of the key σ bond.

4 π electrocyclic ring opening

6 π electrocyclic ring opening/ closing

2 π electrocyclic ring closing

You may have trouble seeing that the last example of an electrocyclic reaction is a two-electron and not a four-electron reaction. The new σ bond is formed between the termini of a three-atom π system. That three-atom system contains two electrons. The lone pairs of O are not part of the three-atom π system, so they are not included in the electron count of the electrocyclic reaction.

When σ bonds are formed between the ends of two π systems to give a cyclic product, the reaction is called a *cycloaddition*. The reverse reaction is called a *retro-cycloaddition*. Cycloadditions are further classified as [m + n] according to the number of atoms in each component.* Again, it is important to note not only the number of atoms but also the number of electrons involved in the process. You are already familiar with the six-electron [4 + 2] cycloaddition, the *Diels–Alder reaction*. Four-electron [2 + 2] cycloadditions are less common, for reasons that will be discussed, but ketenes undergo them readily. The [3 + 2] cycloadditions (or 1,3-dipolar cycloadditions) are a very important class of six-electron cycloadditions that are used to make a wide variety of five-membered heterocycles. Other cycloadditions, including [8 + 2], [4 + 3], and [6 + 4] cycloadditions, are also known.

---

*Two conventions for naming cycloadditions are used in the literature. The older convention (and the one used in this textbook) is that m and n denote the number of *atoms* in each component. Woodward and Hoffmann altered the convention to make m and n denote the number of *electrons* in each component. The number of electrons and the number of atoms are the same for reactions involving neutral species such as the Diels–Alder reaction, but they are not the same for reactions involving charged or dipolar species. For example, the 1,3-dipolar cycloaddition is a [3 + 2] cycloaddition according to the older convention and a [4 + 2] cycloaddition according to the newer one. Always be careful to note which convention is being used.

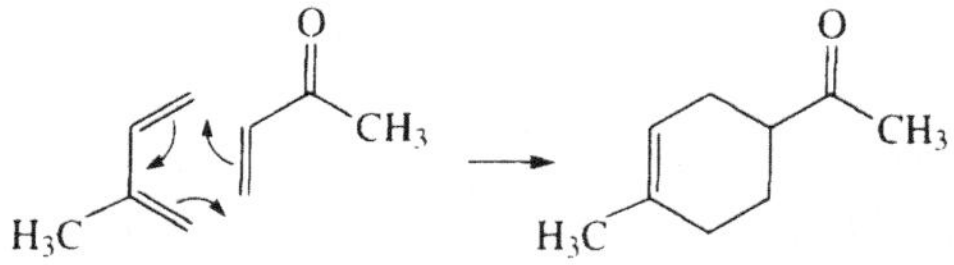

[4 + 2] cycloaddition
(Diels-Alder)

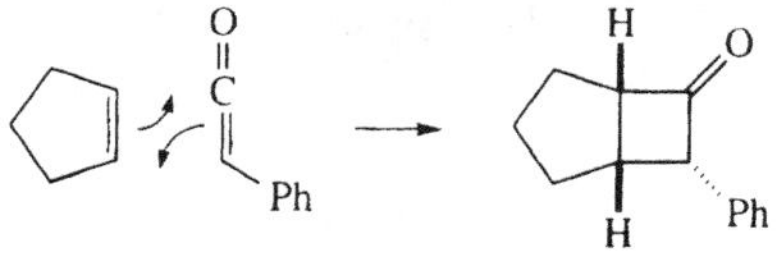

[2 + 2] cycloaddition

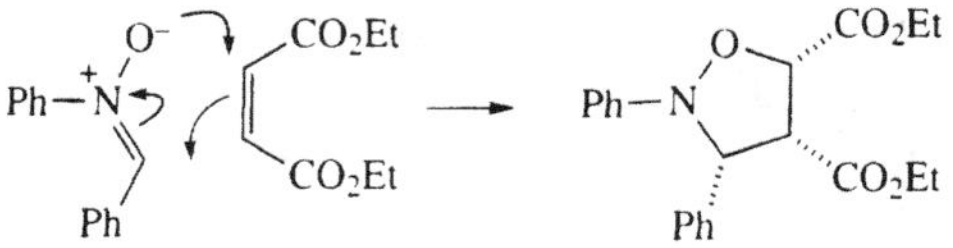

[3 + 2] cycloaddition
(1,3-dipolar)

Cheletropic reactions (e.g., [2 + 1] cycloadditions and [4 + 1] retro-cycload-ditions) are a special class of cycloadditions in which one of the components is a single atom. The one-atom component must have one full and one empty or-bital; it may be a carbene ($\pm CR_2$), SO_2 ($O=\ddot{S}^+-\bar{O} \longleftrightarrow \bar{O}-\ddot{S}^{2+}-\bar{O}$), or $C\equiv O$ ($\bar{C}\equiv \overset{+}{O}$ $\longleftrightarrow \pm C=O$). The [4 + 1] cycloadditions usually proceed in the retro direction with the expulsion of a stable small molecule such as CO or SO_2.

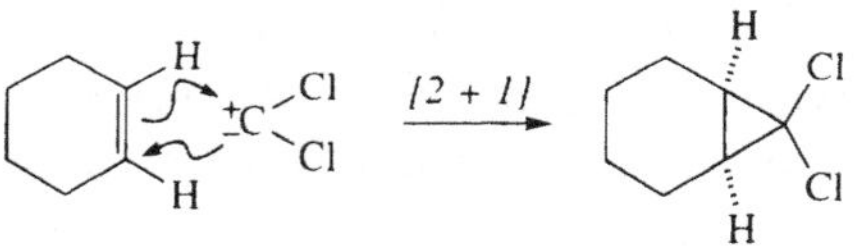

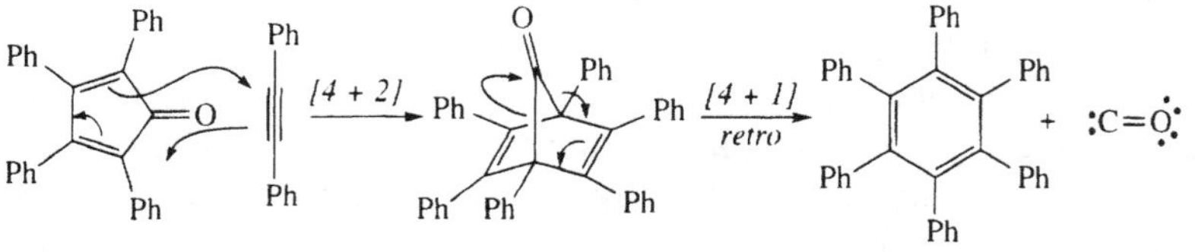

Sigmatropic rearrangements involve the cleavage of a σ bond connecting the end of one fragment with the end of another, and concerted formation of another σ bond at the other ends of the fragments. The σ bond seems to migrate, hence the name of the reaction. Sigmatropic rearrangements are subclassified as [n,m] sigmatropic rearrangements, where n and m are the number of atoms (not elec-trons) in each of the fragments. The very common [3,3] sigmatropic rearrange-ment is called the *Cope rearrangement*. When one of the atoms in the π system is O, the [3,3] sigmatropic rearrangement is called the *Claisen rearrangement* (not be confused with the Claisen condensation). (The initial product, a cyclo-hexadienone, isomerizes to a phenol, thus driving the reaction to the right.) The cationic [1,2] sigmatropic rearrangement (i.e., the 1,2-alkyl or -hydride shift) has already been encountered (Chapter 3). It is important to note the number of elec-trons involved in sigmatropic rearrangements, too. The [3,3] and [1,5] rearrangements are six-electron reactions, the cationic [1,2] rearrangement is a two-electron reaction, and the [1,3] rearrangement is a four-electron reaction.

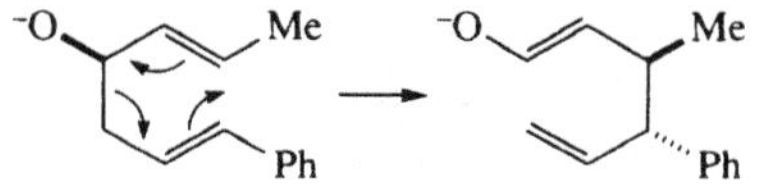

[3,3] sigmatropic (Cope) rearrangement

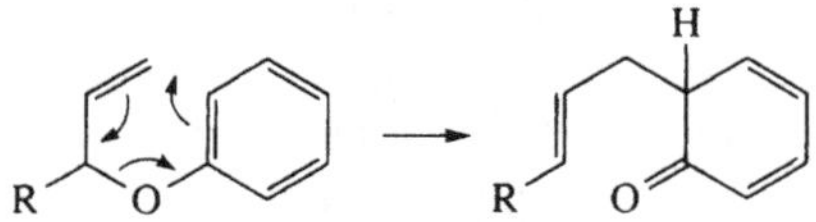

[3,3] sigmatropic (Claisen) rearrangement

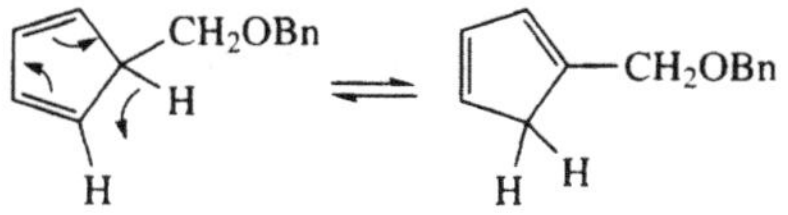

[1,5] sigmatropic rearrangement

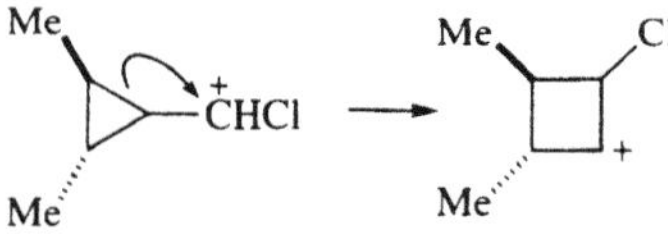

cationic [1,2] sigmatropic rearrangement (hydride or alkyl shift)

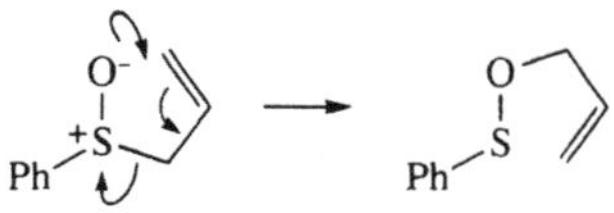

[2,3] sigmatropic rearrangement

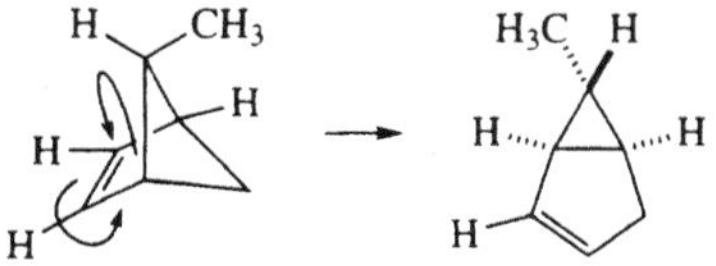

[1,3] sigmatropic rearrangement

Students seem to have more trouble recognizing and naming sigmatropic rearrangements than any other reaction. There is no net change in the number of σ or π bonds in a sigmatropic rearrangement. To name a sigmatropic rearrangement, draw a dotted line between the atoms where a bond forms, and draw short squiggles through the center of the bond that breaks and the dotted line. Put a heavy dot at each atom involved in σ bond formation or cleavage. Then count the number of atoms around the π system from one dot to another inclusive, being careful not to cross a squiggle.

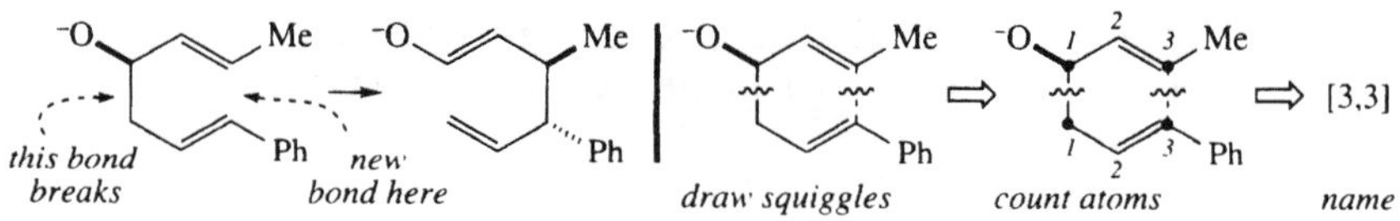

this bond breaks

new bond here

draw squiggles

count atoms

name

⇒ [3,3]

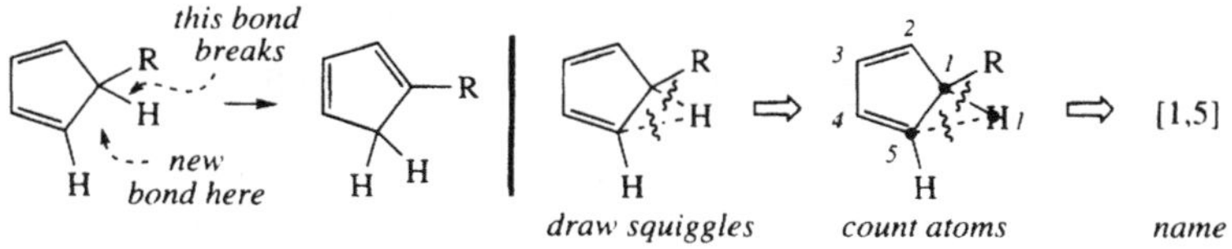

this bond breaks

new bond here

draw squiggles

count atoms

name

⇒ [1,5]

The *ene reaction* is always a six-electron reaction. It shares some characteristics with the [4 + 2] cycloaddition and the [1,5] sigmatropic rearrangement.
Like the [4 + 2] cycloaddition, a four-atom component reacts with a two-atom
component, but in the ene reaction, the four-atom component consists of a π
bond and an allylic σ bond (usually to a H atom), not two π bonds. Like a [1,5]
sigmatropic rearrangement, a σ bond migrates, but the ene reaction also produces one σ bond at the expense of one π bond. When the reaction occurs between a C=C π bond and a C=C π bond with an allylic hydrogen, it is called
the *Alder ene reaction*. The reaction may or may not be intramolecular. *Retro-
ene reactions* are commonly seen, too, as in the thermal elimination of acetic
acid from alkyl acetates. The *hetero-ene reaction* looks like a cross between a
[3 + 2] cycloaddition and a [1,4] sigmatropic rearrangement. *Retro-hetero-ene
reactions* are used in some elimination procedures that are very important synthetically.

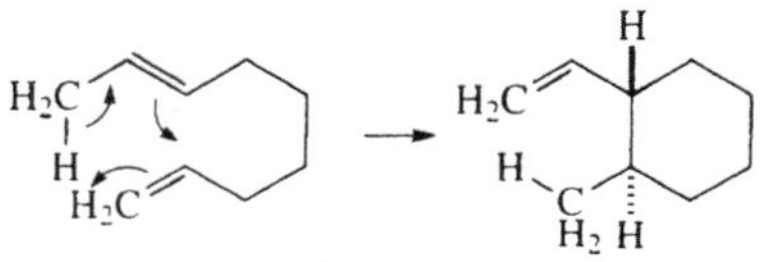

Alder ene reaction

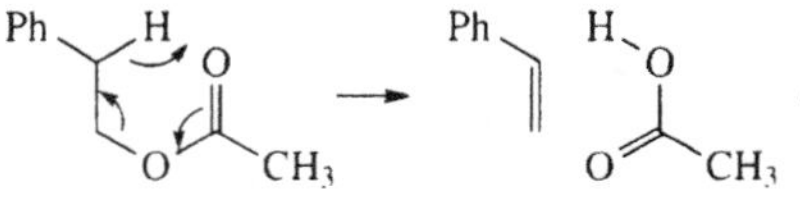

retro-ene reaction

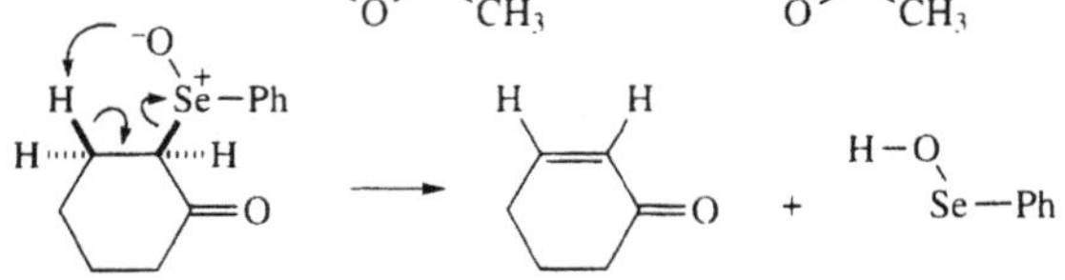

*retro-hetero-ene
reaction*

The key characteristics of the four major types of pericyclic reactions are sum-
marized in Table 4.1.

* **Common error alert:** *The pericyclic reactions are superficially similar to one
another, and students sometimes have trouble distinguishing the different types
from one another.* Archetypical six-electron examples of each class of pericyclic
reaction are shown in the following figure. (Note the degeneracy of the archetypical sigmatropic rearrangement!) Each archetypical reaction has three arrows moving in a circle, but each results in a different change in the patterns of bonding.

electrocyclic reaction

cycloaddition

sigmatropic rearrangement

ene reaction

TABLE 4.1. Bond changes in pericyclic reactions

Reaction class	Changes in bond types
Electrocyclic reactions	one π bond $\rightleftarrows$ one σ bond
Cycloadditions	two π bonds $\rightleftarrows$ two σ bonds
Sigmatropic rearrangements	one σ bond $\longrightarrow$ new σ bond
Ene reactions	one π bond $\rightleftarrows$ one σ bond *and* one σ bond migrates

When drawing the change in bonding patterns in pericyclic reactions, it doesn't matter whether you draw the electrons moving clockwise or counterclockwise, because pericyclic reactions are not characterized by the movement of electron density from electron-rich to electron-poor or high-energy to low-energy sites. The curved arrows are merely meant to show the change in bonding patterns as the reaction proceeds from starting material to product. However, when one component in the reaction has an atom with a formal negative charge, as in the [3 + 2] cycloaddition, the [2,3] sigmatropic rearrangement, or the retro-hetero-ene reaction, it is important to begin your electron-pushing there.

* **Common error alert:** *Students often draw fictional one-step pericyclic mechanisms for reactions that require more than one step.* A good guard against this sort of error is to be sure that you can name any pericyclic step in your mechanism. For example, for the following reaction, you may be tempted to draw a four-electron, one-step mechanism involving the C–C σ bond of the first starting material and the π bond of the second. Such a mechanism, however, cannot be classified in the scheme that has just been discussed, and therefore it is likely to be incorrect.

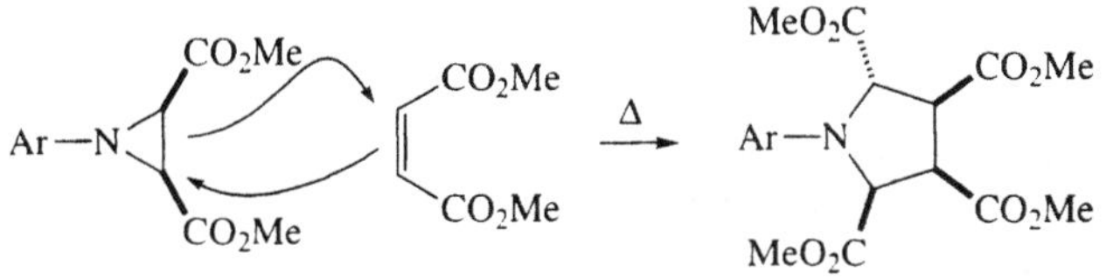

an unnamable and hence fictitious pericyclic reaction

Stereospecificity, the property that the stereochemistry of the starting materials determines the stereochemistry of the product, is one of the hallmarks of pericyclic reactions. It is possible to draw two-step nonconcerted, polar or free-radical mechanisms for many pericyclic reactions, but these two-step mechanisms fail to account for the stereospecificity of the reactions. For example, a two-step polar mechanism can be drawn for the Diels–Alder reaction between 2-methoxybutadiene (a nucleophile) and ethyl *cis*-crotonate (an electrophile). This mechanism proceeds through a dipolar intermediate in which one new σ bond has formed. In this intermediate, there is free rotation about the two C atoms of the dienophile, so the cis stereochemical relationship between the Me and CO_2Et groups is expected to be lost in the product. In fact, though, the product is exclusively cis. This finding does not completely rule out a polar mechanism—it is possible that the intermediate exists but that ring closure occurs more quickly than rotation about the σ bond—but it does limit the lifetime of the dipolar intermediate to such an extent that one can say *practically* that it does not exist.

loss of stereochemical integrity likely

Many stereochemical results like these, and the consistent observation of nonthermodynamic products that would not be expected from two-step mechanisms, have accumulated to support the operation of the pericyclic mechanism in most cases. Recent advances in experimental and theoretical techniques have allowed chemists to probe the mechanisms of pericyclic reactions at time scales far shorter than the time required to rotate about a C–C σ bond, and these experiments have supported concerted mechanisms for some pericyclic reactions and nonconcerted mechanisms for others. In practical terms, though, if no experimental or theoretical evidence rules it out, Occam's razor compels one to propose a concerted mechanism in favor of a nonconcerted one.

Pericyclic reactions can proceed under acidic or basic conditions. For example, the oxy-Cope rearrangement is greatly accelerated under basic conditions, and the Diels–Alder reaction is greatly accelerated by Lewis acids. Often a series of polar reactions is used to synthesize an unstable intermediate, which then undergoes a pericyclic reaction to reveal the product. In other words, a good command of polar mechanisms (Chapters 2 and 3) is essential to understanding how to draw pericyclic mechanisms.

Like all reactions, pericyclic reactions are reversible in principle (even though they may be irreversible in practice). The forward and reverse reactions *always go through the same transition state.* As an analogy, if you wanted to travel from Lexington, Ky., to Richmond, Va., you would choose the path that went through the lowest gap in the Appalachian mountains. If you wanted to go from Richmond back to Lexington, you would choose the same route, only in reverse. The path you chose would not depend on which direction you were traveling. Reactions obey the same principle.

4.1.2 Polyene MOs

You may think that there's not much to say about the "no-mechanism" pericyclic reactions, but there is. First, how they proceed stereochemically and even whether they proceed at all depends on whether the reaction is conducted *thermally* or *photochemically.* For example, many [2 + 2] cycloadditions proceed only photochemically, whereas all [4 + 2] cycloadditions proceed thermally. Second, all pericyclic reactions proceed stereospecifically, but the stereochemistry of the products sometimes depends on the reaction conditions. For example, 2,4,6-octatriene gives *cis*-5,6-dimethylcyclohexadiene upon heating and *trans*-5,6-dimethylcyclohexadiene upon photolysis. *These phenomena can be explained by examining the MOs of the reactants.* The rules governing whether pericyclic reactions proceed and the stereochemical courses when they do proceed are known as the *Woodward–Hoffmann* rules.

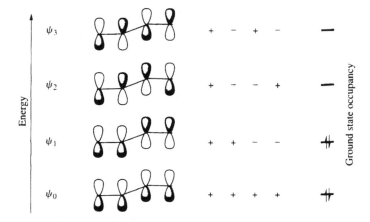

The Woodward–Hoffmann rules can be rationalized by examining the properties of the *frontier MOs* of the reactants, i.e., the HOMOs (highest occupied
molecular orbitals) and LUMOs (lowest unoccupied molecular orbitals) of the
reactants. In order to understand pericyclic reactions, then, you need to be able
to construct the MOs of a polyene system from the constituent p orbitals.

Consider 1,3-butadiene. Four p orbitals interact, each with its neighbors, to
produce four MOs. These four orbitals are called ψ_0, ψ_1, ψ_2, and ψ_3. In the lowest energy MO, ψ_0, every p orbital overlaps constructively with its neighbor to
give an MO in which there are no *nodes*, or changes in sign between orbitals. In
the highest energy MO, ψ_3, every p orbital overlaps destructively with its neighbor to give an MO in which there are three nodes. In the intermediate
energy MOs, ψ_1 and ψ_2, there are one and two nodes, respectively, and the MOs
are constructed accordingly. Under thermal conditions, the four electrons derived
from the four AOs occupy ψ_0 and ψ_1; ψ_1 is the HOMO, and ψ_2 is the LUMO.

The signs of the AOs do not tell the whole story. In each MO, each AO has a coefficient associated with it that reflects the magnitude of its contribution to the MO. The
coefficient can range from 0 to 1. The squares of the coefficients of one atom's AO in
each MO add up to 1. The coefficients of the MOs can provide important information
about reactivity.

Note that the signs of the terminal orbitals alternate from like, to opposite, to
like, to opposite. A physical chemist would say that the symmetries of the MOs
alternate between *symmetric* and *antisymmetric*. This property is universal among
the MOs of polyenes. *You do not need to construct the complete MOs of any
polyene* to determine the signs of the termini of the HOMO and LUMO, on which
the properties of pericyclic reactions largely depend.

In 1,3,5-hexatriene, six AOs produce six MOs, the signs of whose termini alternate from like, to opposite, to like, etc. The number of nodes increases with

each MO. In the ground state of 1,3,5-hexatriene, ψ_2 is the HOMO and ψ_3 is the LUMO. Note that in ψ_3 and ψ_4, the p orbitals of the second and fifth C atoms of the chain have a coefficient of zero, indicating that the p orbitals from those C atoms do not contribute at all to ψ_3 and ψ_4. Two of the nodes in both ψ_3 and ψ_4 pass through the centers of these two C atoms.

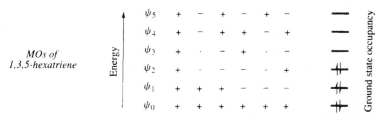

The symmetries of the MOs of conjugated π systems with odd numbers of atoms also alternate. The allyl system has three MOs: ψ_0 (symmetric), ψ_1 (antisymmetric), and ψ_2 (symmetric). In the allyl cation, ψ_0 is the HOMO and ψ_1 is the LUMO, whereas in the allyl anion, ψ_1 is the HOMO and ψ_2 is the LUMO. The pentadienyl system has five MOs. In the pentadienyl cation, ψ_1 (antisymmetric) is the HOMO and ψ_2 (symmetric) is the LUMO, whereas the pentadienyl anion, ψ_2 is the HOMO and ψ_3 (antisymmetric) is the LUMO.

4.2   Electrocyclic Reactions

4.2.1   Typical Reactions

In an electrocyclic reaction, a new σ bond forms (or breaks) between the termini of an acyclic conjugated π system to give a cyclic compound with one fewer (or more) π bond. Like all pericyclic reactions, electrocyclic reactions are reversible in principle. The ring-closed compound is usually lower in energy, because it has a σ bond in place of a π bond, but not always.

Cyclobutenes are in electrocyclic equilibrium with 1,3-butadienes. 1,3-Butadienes are lower in energy, because cyclobutenes are very strained, and it is therefore possible to convert a cyclobutene to a butadiene *thermally*. The reverse reaction, the conversion of a 1,3-butadiene to a cyclobutene, does not usually proceed under thermal conditions, as it involves an uphill climb in energy. However, the more conjugated 1,3-butadienes absorb light at longer wavelengths than cyclobutenes, so it is possible to convert a 1,3-butadiene to a cyclobutene *photochemically* by choosing a wavelength at which the butadiene absorbs and the cyclobutene does not.

higher in energy

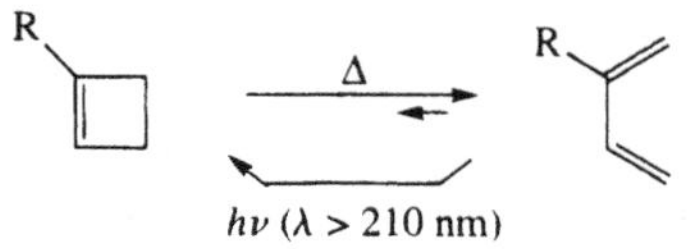

absorbs hv at longer wavelengths

hv (λ > 210 nm)

The equilibrium between the cyclobutene and the 1,3-butadiene is shifted in favor of the cyclobutene in benzocyclobutenes. The electrocyclic ring opening of a benzocyclobutene, an aromatic compound, leads to an *o-xylylene*, a nonaromatic and reactive compound. *o*-Xylylenes are useful intermediates in Diels–Alder reactions.

1,3-Cyclohexadienes are in electrocyclic equilibrium with 1,3,5-hexatrienes. Neither compound is strained, and the cyclohexadiene has one more σ bond than the hexatriene, so the cyclohexadiene is lower in energy. The hexatriene is more conjugated than the cyclohexadiene, so it is more reactive photochemically. Under both thermal and photochemical conditions, then, the cyclohexadiene is favored over the hexatriene.

higher in energy, absorbs at longer wavelengths

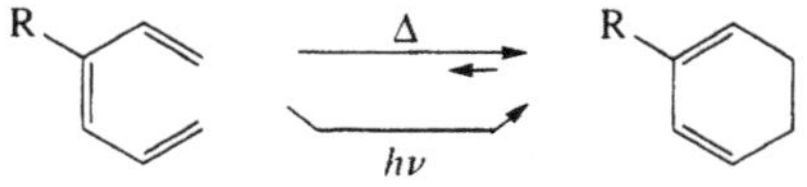

The equilibrium between the 1,3-cyclohexadiene and the 1,3,5-hexatriene is less lopsided when ring closure creates a strained ring. For example, the equilibrium constant for the interconversion of 1,3,5-cycloheptatriene and norcaradiene (bicyclo[4.1.0]hepta-2,4-diene) is close to 1, and the interconversion occurs rapidly at room temperature. The ring *opening* is facile because the strain of the three-membered ring is relieved, and the ring *closing* is facile because the ends of the π system are held close together and because a C–C σ bond is gained at the expense of a π bond.

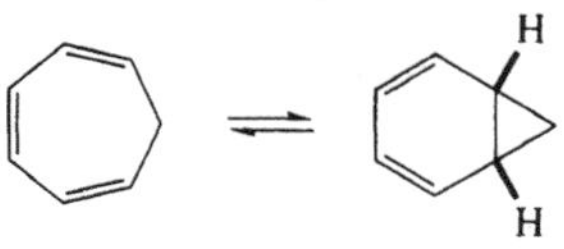

Allyl and pentadienyl cations participate in electrocyclic ring openings and closings, too. (The corresponding anions undergo similar reactions, but they are less important synthetically.) Less highly conjugated cations are produced upon ring closure of these species, but the decrease in cation stabilization is compensated by the gain of a C–C σ bond. In the case of the allyl system, the increase in strain upon ring closing must also be considered.

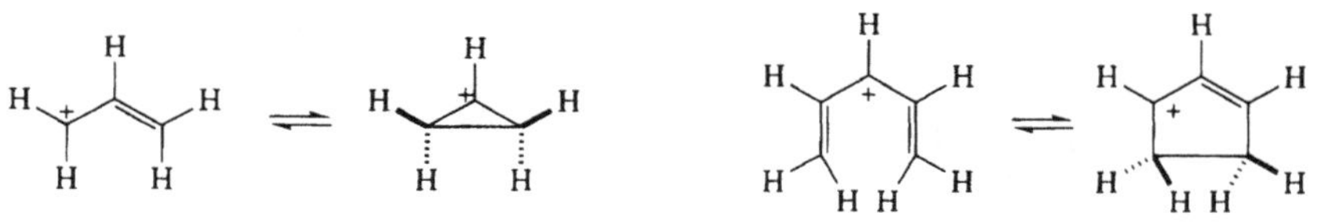

In the *Nazarov cyclization*, a divinyl ketone undergoes electrocyclic ring closing upon treatment with a Lewis acid such as $SnCl_4$ to give a cyclopentenone.
The allylic cation obtained upon electrocyclic ring closure can undergo fragmentation of either of two different σ bonds to give the product. The lower energy product is usually obtained, but the tendency of the $C-SiMe_3$ bond to fragment preferentially to the $C-H$ bond can be used to put the double bond in the
thermodynamically less favorable position.

Example

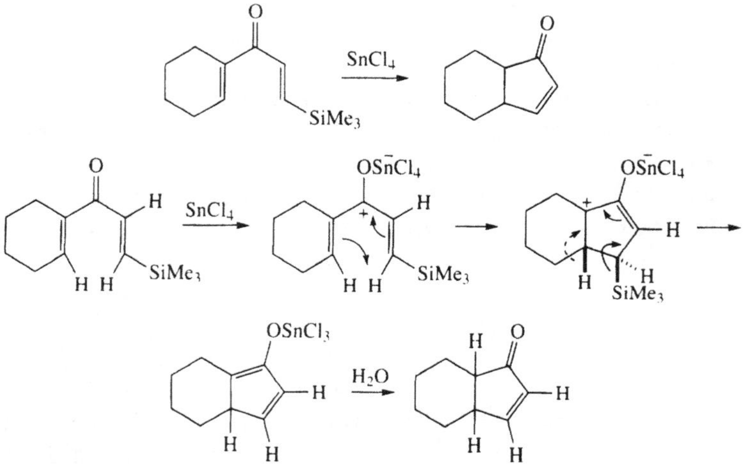

Problem 4.1. Draw a reasonable mechanism for the following preparation of
pentamethylcyclopentadiene, a useful ligand for transition metals.

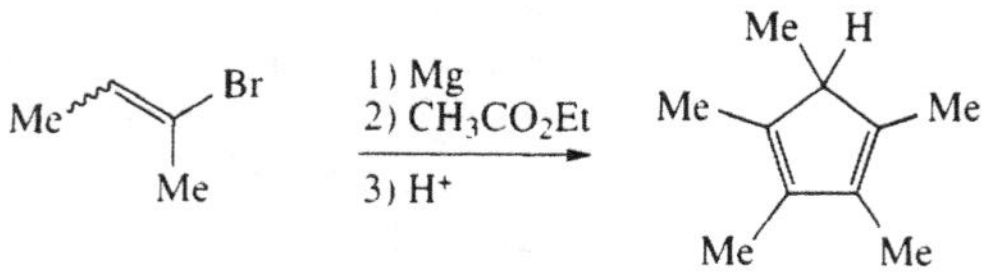

The electrocyclic equilibrium between cyclopropyl cations and allyl cations normally favors the allyl cation because the cyclopropyl cation is both more strained
and less conjugated than the starting allylic cation. For example, the cyclopropyl
halide shown undergoes concerted loss of Br^- and electrocyclic ring opening to
the allyl cation upon distillation. (In principle, a ring opening that is not concerted
with loss of the leaving group could be drawn, but the cyclopropyl cation is very
high in energy.) Addition of Br^- to the allylic cation then gives the product.

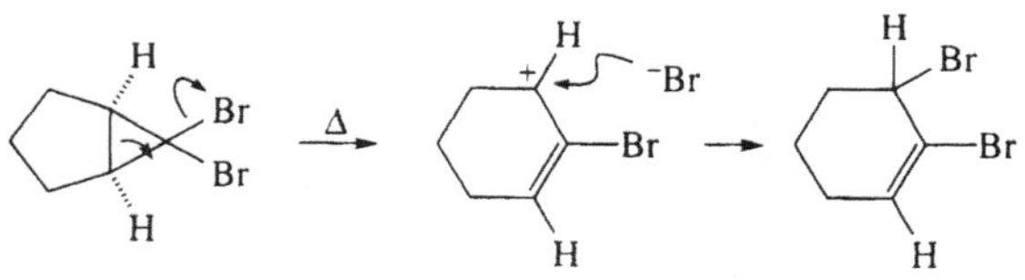

The carbene derived from halogen–metal exchange between RLi and a dibromocyclopropane can also undergo a two-electron electrocyclic ring opening to give an allene.

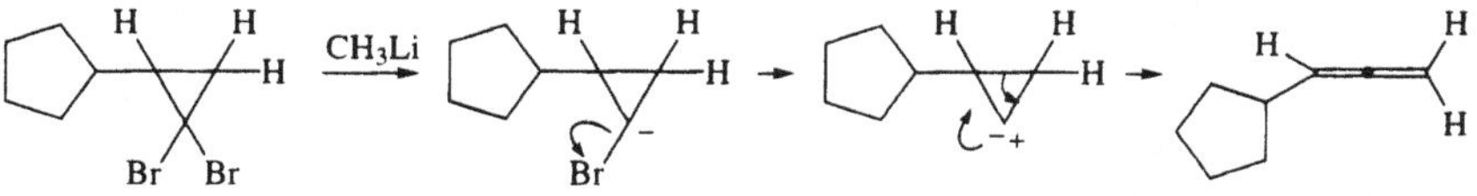

Cyclopropanones are in electrocyclic equilibrium with *oxyallyl cations*. The cyclopropanone is generally lower in energy, but the oxyallyl cation is not so much higher in energy that it is kinetically inaccessible. Oxyallyl cations can undergo cycloadditions, as will be discussed later.

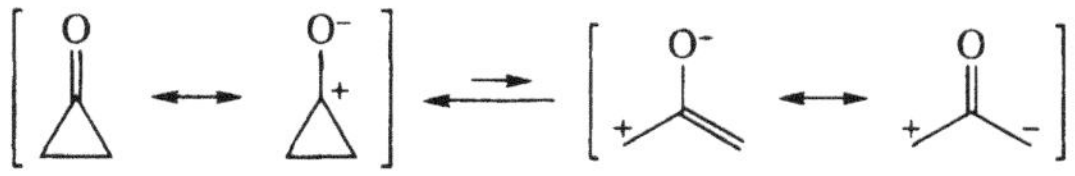

α-Haloketones rearrange to carboxylic acids in the *Favorskii rearrangement*. When the ketone is cyclic, ring contraction results.

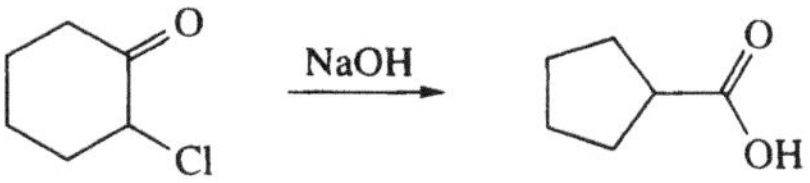

Two reasonable mechanisms can be drawn for the Favorskii rearrangement. In the first mechanism, the ketone acts as an *acid*. Deprotonation of the α carbon gives an enolate. The enolate undergoes concerted expulsion of Cl^- and two-electron electrocyclic ring closing to give a cyclopropane intermediate. Addition of HO^- to the carbonyl group of this very strained ketone is followed by strain-relieving elimination of an alkyl group to give a carbanion, which is probably quenched by solvent even as it forms.

Mechanism 1:

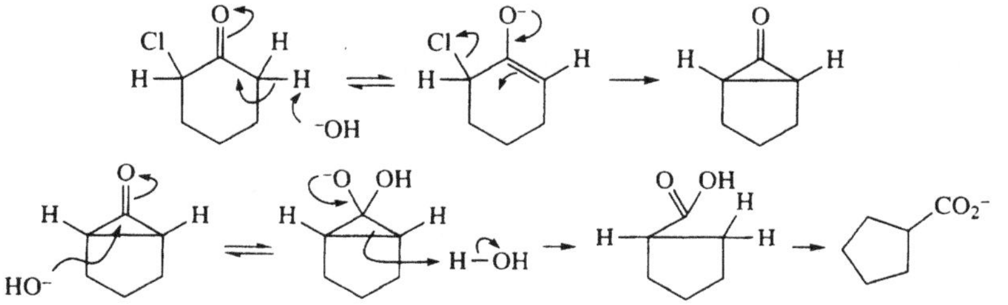

In the second mechanism, which was discussed in Chapter 2, the ketone acts as an *electrophile*. Addition of HO^- to the carbonyl group and then migration of C to its electrophilic neighbor with expulsion of Cl^- provides the product.

Mechanism 2:

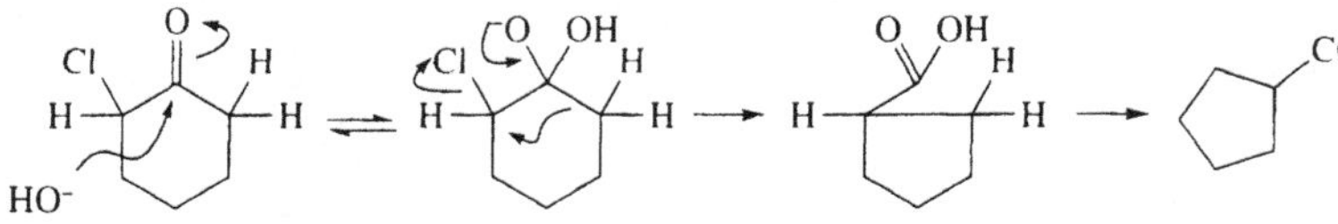

Mechanism 2 (the semibenzilic acid mechanism) looks better, but labeling studies show that the two C atoms α to the ketone become equivalent in the course of the reaction, which is consistent only with mechanism 1 (the electrocyclic mechanism). The rearrangement of α-chloro-α-phenylacetone to methyl hydrocinnamate is also consistent only with the electrocyclic mechanism; if the semibenzilic mechanism were operative, then methyl 2-phenylpropionate would be the product.

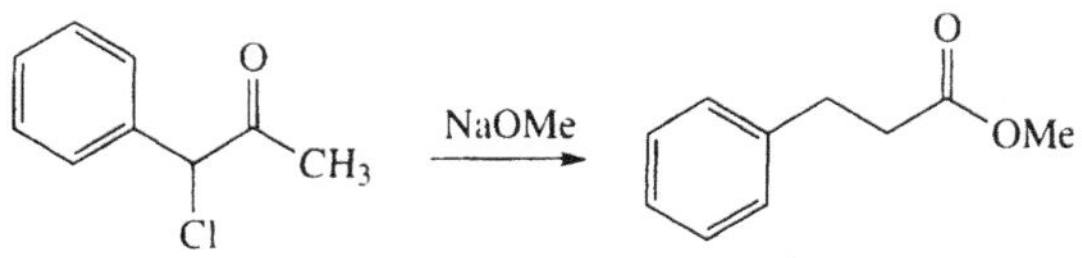

However, the rearrangement of α-chloroacetophenone to diphenylacetic acid *must* proceed by a semibenzilic acid mechanism, because a cyclopropane can't be formed.

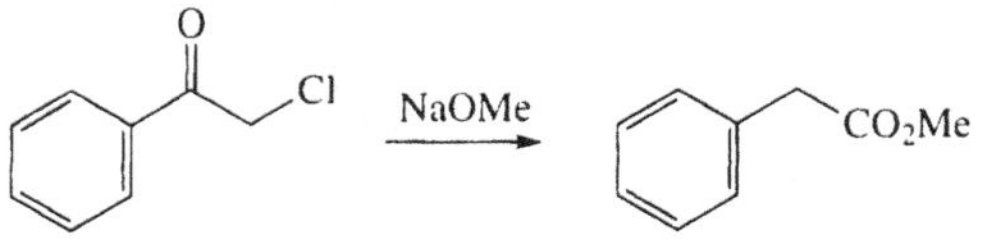

To summarize, when H atoms are present on the α-carbon opposite the leaving group, the electrocyclic mechanism usually operates; when they are not, the semibenzilic mechanism operates. Why does the electrocyclic mechanism proceed more quickly than the "more reasonable" semibenzilic mechanism for enolizable α-haloketones? Deprotonation and electrocyclic ring closing are both very rapid reactions—the latter even when a strained ring is formed—and they must simply be faster than HO^- addition and migration, despite what our "chemical intuition" tells us.

Problem 4.2. Draw a reasonable mechanism for the following reaction. A Favorskii rearrangement is involved.

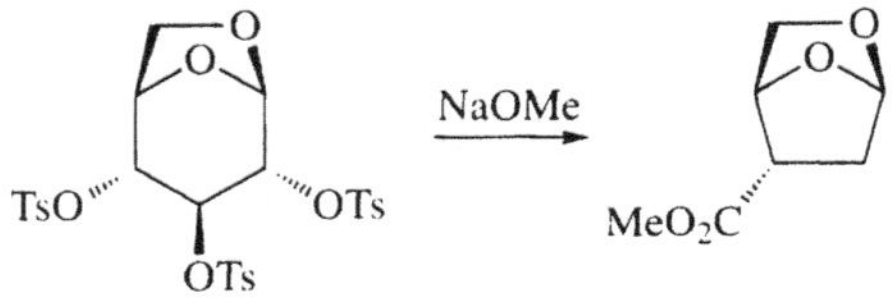

The electrocyclic ring closing of the allyl "cation" in the Favorskii rearrangement proceeds only because the cyclopropyl cation in the product is quenched by O⁻. Charge neutralization also provides the driving force for the electrocyclic ring closing of vinylcarbene to give cyclopropene. The ring closing can be drawn as either a two-electron or a four-electron process, depending on whether the empty orbital or the full orbital of the carbene is conjugated to the π bond.

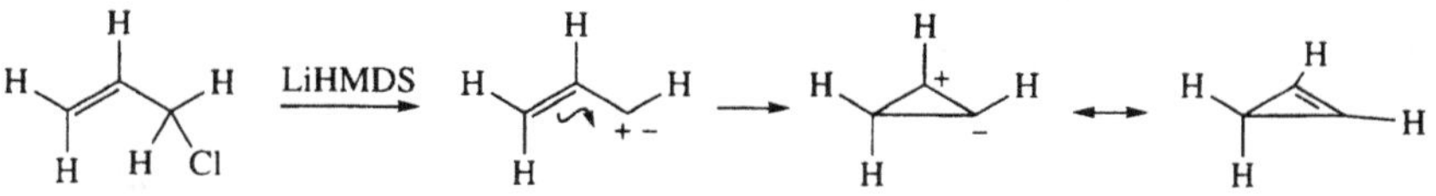

The key to identifying an electrocyclic ring *closing* is to look for the formation of a new bond at the ends of a conjugated π system. The key to identifying an electrocyclic ring *opening* is to look for the cleavage of a σ bond joining the allylic positions of a conjugated π system. Three-membered rings are closed from or opened to allylic systems by electrocyclic reactions.

Example

The following rearrangement reaction was proposed to be the biosynthetic pathway to the natural product endiandric acid A.

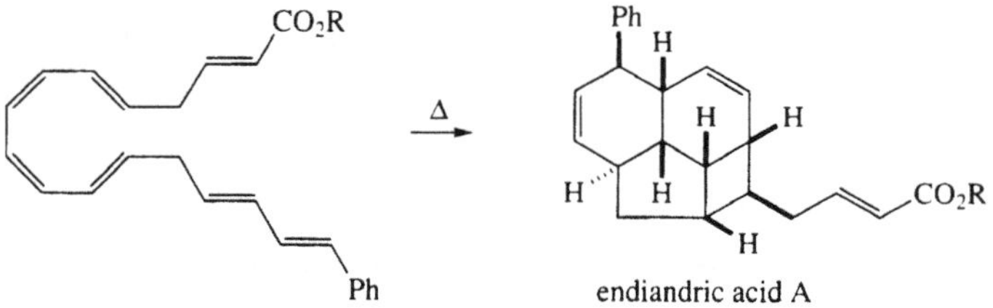

endiandric acid A

If you number the atoms, you will see that bonds need to be made at C5−C12, C6−C11, C9−C17, and C10−C14.

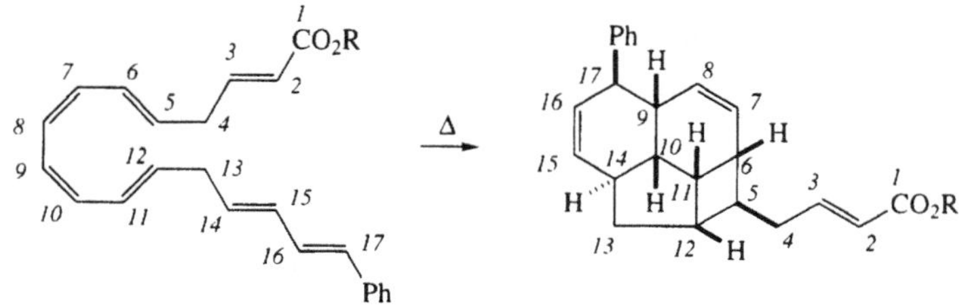

Sometimes it helps to draw dashed lines where you need to make bonds.

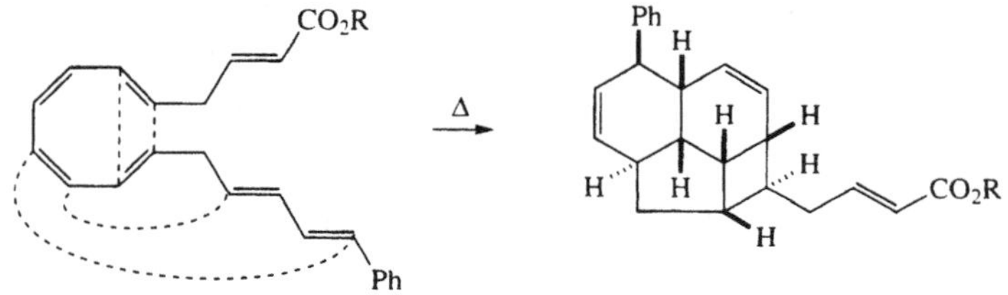

The starting material contains a 1,3,5,7-tetraene group (C5 to C12), and one of the bonds you need to make, C5–C12, is between the termini of that system. An eight-electron electrocyclic ring closing forms that bond. The electrocyclic reaction also creates a new 1,3,5-hexaene group (C6 to C11), and another bond you need to make, C6–C11, is at the termini of that system. A six-electron electrocyclic ring closing forms that bond.

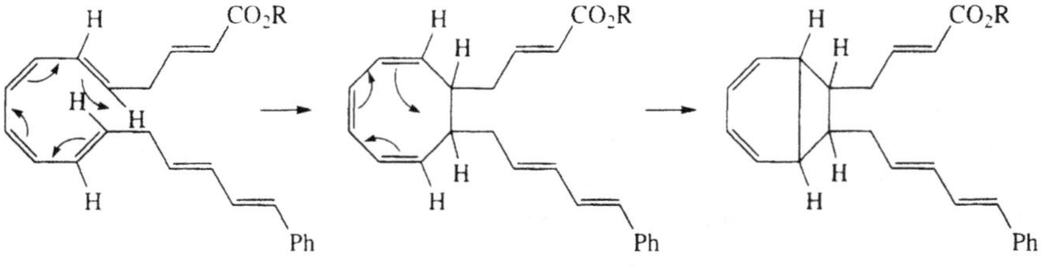

The last two bonds are formed by a Diels–Alder reaction ([4 + 2] cycloaddition) between the C9=C10 π bond and the diene from C14 to C17.

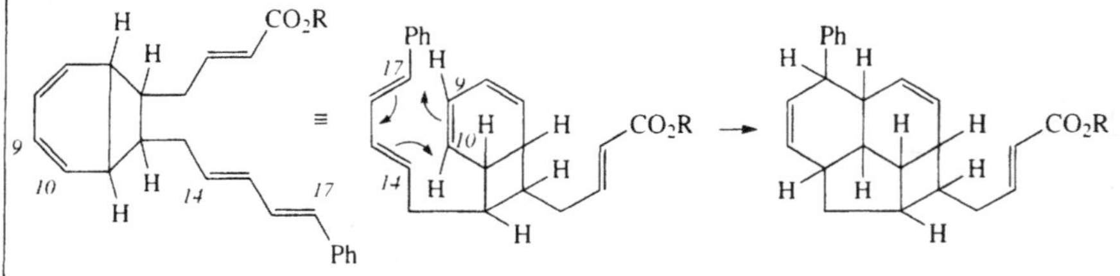

You may have been tempted to form the C5–C12 and C6–C11 bonds simultaneously by a [2 + 2] cycloaddition. You will soon see, though, that [2 + 2] cycloadditions of simple C=C π bonds are forbidden to occur under thermal conditions by the Woodward–Hoffmann rules.

Problem 4.3. When phorone is treated with base, isophorone is obtained by an electrocyclic ring closing. Draw a reasonable mechanism.

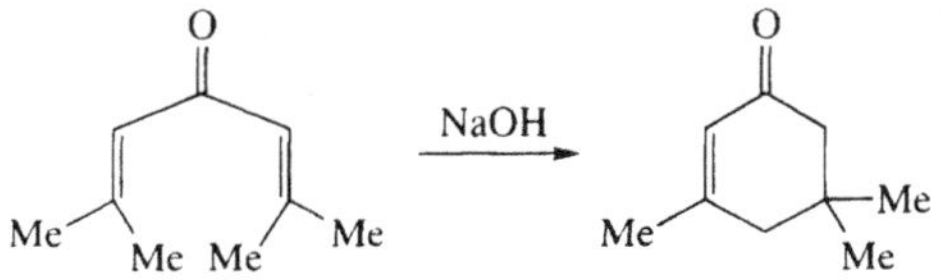

Problem 4.4. 4-Vinylcyclobutenones are very unstable compounds, isomerizing rapidly to phenols. Draw a reasonable mechanism involving electrocyclic reactions.

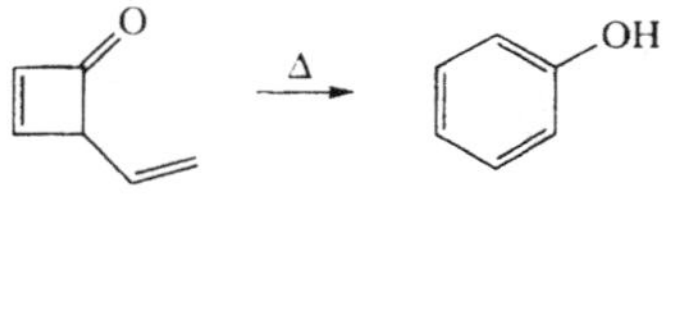

4.2.2 Stereospecificity

Of all the pericyclic reactions, the stereochemical course of electrocyclic reactions is easiest to understand because electrocyclic reactions are unimolecular and involve only one array of orbitals. First, consider the electrocyclic ring closing of a butadiene, a four-electron reaction. The butadiene has four substituents at the termini. When the butadiene is in its reactive s-cis conformation, two of the groups can be described as *in* groups (because they are on the concave face of the arc defined by the four atoms of the butadiene group), and two as *out* groups (because they are on the convex face of the same arc). In the electrocyclic reaction, the termini of the butadiene must rotate in order for the p orbitals at the termini to overlap and form a bond. Two stereochemical results are possible. *The stereochemical result of the reaction depends on how the two termini rotate.* If the termini rotate in the same direction (*conrotatory*), then the two out groups become trans to each other, the two in groups become trans to each other, and any out group becomes cis to any in group.

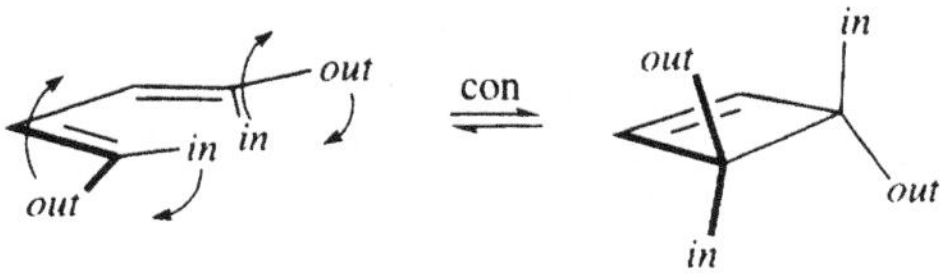

On the other hand, if they rotate in opposite directions (*disrotatory*), then the two out groups become cis to each other, the two in groups become cis to each other, and any out group becomes trans to any in group.

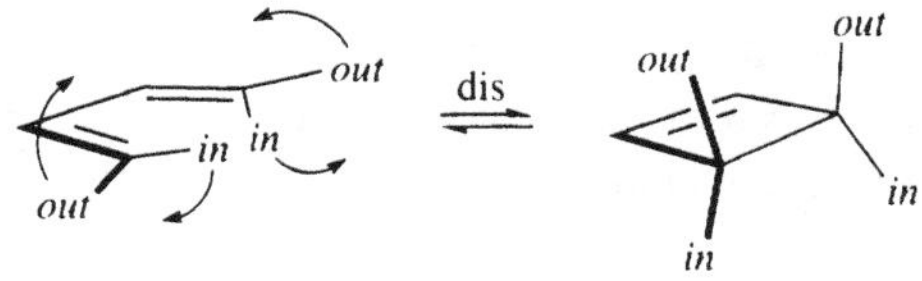

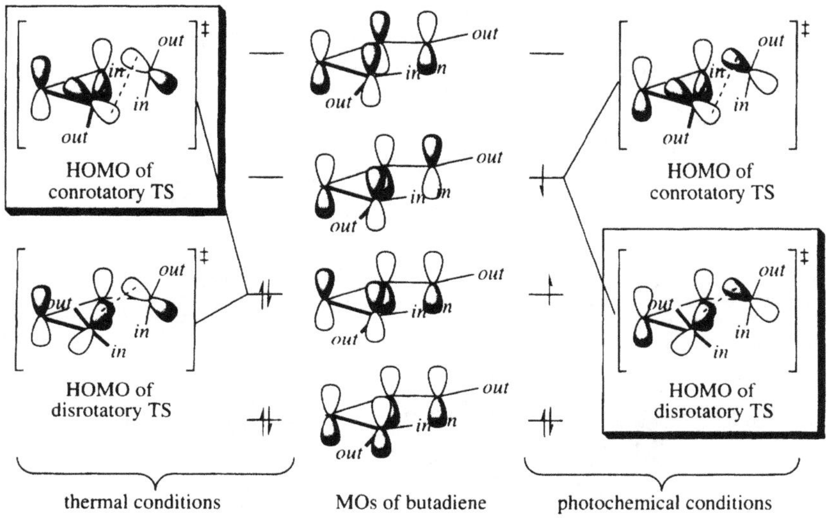

HOMO of conrotatory TS		HOMO of conrotatory TS
HOMO of disrotatory TS		HOMO of disrotatory TS
thermal conditions	MOs of butadiene	photochemical conditions

It happens that under thermal conditions the electrocyclic ring closing of butadienes always proceeds by the conrotatory pathway. There are several ways to rationalize this observation; we use the frontier MO method here. Under thermal conditions, the HOMO of 1,3-butadiene is ψ_1, and the LUMO is ψ_2. The electrocyclic ring closing proceeds so that there is a constructive interaction where bond-making takes place as the HOMO goes through the TS. Because the HOMO ψ_1 has a bonding interaction between the termini of the π system in the conrotatory TS and an antibonding interaction between the termini of the π system in the disrotatory TS, the reaction proceeds in a conrotatory fashion.

Under photochemical conditions, the electrocyclic ring closing of butadienes always proceeds by the disrotatory pathway, which is the opposite of the result under thermal conditions. The FMOs make the stereochemical result easy to understand. Under photochemical conditions, an electron is promoted from the HOMO ψ_1 to the LUMO ψ_2, so ψ_2 becomes the HOMO. Molecular orbital ψ_2 has an antibonding interaction between the termini of the π system in the conrotatory TS but a bonding interaction between the termini of the π system in the disrotatory TS, so the reaction proceeds in a disrotatory fashion.

It is easier to see the FMO interactions in electrocyclic ring closing reactions than in ring opening reactions. However, the TSs for ring closing and ring opening are the same, so the stereochemistry of cyclobutene ring opening is conrotatory under thermal conditions and disrotatory under photochemical conditions—the same as the stereochemistry of butadiene ring closing.

In summary, for butadienes and cyclobutenes: four-electron, thermal, conrotatory; four-electron, photochemical, disrotatory. The easiest way to visualize the stereochemical result is to make a fist, use your thumbs to designate substituents at the termini of the π system, and rotate your fists to determine the stereochemical result upon disrotatory or conrotatory ring closure or opening.

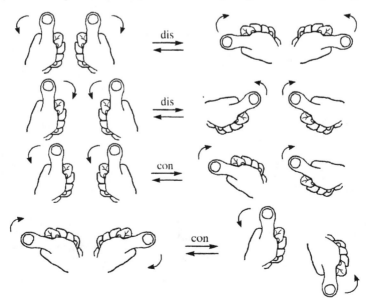

The stereochemistry of electrocyclic ring closings of 1,3,5-hexatrienes is opposite that observed in 1,3-butadienes: under thermal conditions, they proceed by the disrotatory pathway, whereas under photochemical conditions, they proceed by the conrotatory pathway. The disrotatory TS derived from the thermal HOMO (ψ_2, symmetric) of 1,3,5-hexatriene and the conrotatory TS derived from the photochemical HOMO of 1,3,5-hexatriene (ψ_3, antisymmetric) have bonding interactions at the termini. Again, the TS for six-electron ring opening is the same as the TS for six-electron ring closing, so ring openings of 1,3-cyclohexadienes are also disrotatory under thermal conditions and conrotatory under photochemical conditions.

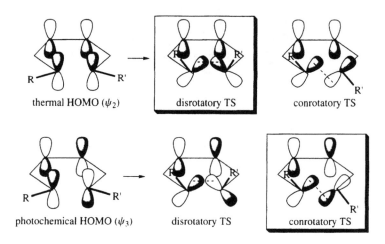

You may begin to discern a pattern. The Woodward–Hoffmann rules for electrocyclic reactions are as follows (Table 4.2):

TABLE 4.2. Woodward—Hoffmann Rules for
Electrocyclic Reactions

Number of electron pairs	Δ	$h\nu$
Odd	disrotatory	conrotatory
Even	conrotatory	disrotatory

Electrocyclic reactions involving an odd number of electron pairs proceed through a *disrotatory* TS under thermal conditions and a *conrotatory* TS under photochemical conditions. Electrocyclic reactions involving an even number of electron pairs proceed through a *conrotatory* TS under thermal conditions and a *disrotatory* TS under photochemical conditions. In practice, you need only remember "even-thermal-con" (or any other combination) to generate the entire table.

Problem 4.5. Cycloheptatriene undergoes rapid electrocyclic ring closure under thermal conditions. Determine the structure and stereochemistry of the product.

Problem 4.6. Confirm that the proposed mechanism of formation of endiandric acid A is consistent with the observed stereochemistry of the product.

Problem 4.7. Draw the HOMO for the pentadienyl cation $H_2C=CH-CH=CH-CH_2^+$, and determine whether it should undergo disrotatory or conrotatory ring closing under thermal conditions. Then do the same for the pentadienyl anion $H_2C=CH-CH=CH-CH_2^-$. Are the stereochemical courses of these reactions consistent with the Woodward–Hoffmann rules?

The Woodward–Hoffmann rules for electrocyclic reactions can also be formulated using the terms *suprafacial* and *antarafacial* (Table 4.3). A π system is said to react *suprafacially* in a pericyclic reaction when the bonds being made to the two termini of the π system are made to the same face of the π system. It reacts *antarafacially* when the bonds are made to opposite faces of the π system. In electrocyclic reactions, disrotatory reactions are suprafacial, and conrotatory reactions are antarafacial.

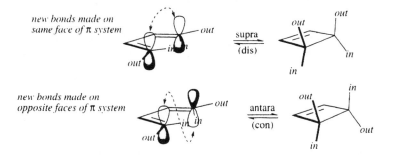

TABLE 4.3. Woodward—Hoffmann Rules for
Electrocyclic Reactions

Number of electron pairs	Δ	hν
Odd	suprafacial	antarafacial
Even	antarafacial	suprafacial

The value of the terms *suprafacial* and *antarafacial* is that, unlike *disrotatory* and *conrotatory*, they can also be used to describe the way that π systems react in cycloadditions and sigmatropic rearrangements. Most importantly, when a π system reacts suprafacially, its out groups become cis in the product; when it reacts antarafacially, they become trans. Note that in disrotatory electrocyclic reactions, the out groups become cis, and in conrotatory electrocyclic reactions, the out groups become trans.

The property that the stereochemical result of an electrocyclic reaction is absolutely predictable is called *stereospecificity*. A stereospecific reaction will give you one stereochemical result when a cis starting material is used, and the opposite result when a trans starting material is used. Other examples of stereospecific reactions include S_N2 substitutions, catalytic hydrogenation of alkynes or alkenes, and dihydroxylation and bromination of alkenes.

The Woodward–Hoffmann rules can be used to predict the stereochemical result of any electrocyclic ring opening or closing. There are a few conditions to the
Woodward–Hoffmann rules of which you must be aware. First, the Woodward–
Hoffmann rules apply only to concerted, pericyclic reactions. If an apparent electrocyclic reaction actually proceeds by a nonconcerted mechanism, the rules do
not apply. Second, a reaction can be forced to proceed through the higher energy
TS if the lower energy one is raised prohibitively high in energy by geometric
constraints. In these cases, the starting material is usually more stable than one
would expect. For example, Dewar benzene is surprisingly stable, because for it
to decompose to benzene, it must go through the "disallowed" disrotatory TS;
the "allowed" conrotatory ring opening gives a trans double bond in the six-membered ring, which is prohibitively high in energy.

The biosynthesis of vitamin D_2 also illustrates the Woodward–Hoffmann rules.
Mammals prepare precalciferol from ergosterol by a six-electron electrocyclic
ring opening. The ring opening *must* proceed in conrotatory fashion (otherwise
a trans double bond in a six-membered ring would be obtained), so it requires
light. When sunlight never shines on one's skin, vitamin D_2 synthesis is inhib-
ited, and rickets is contracted.

ergosterol   Z-tachysterol (precalciferol)

Of course, too much sunlight is bad for you, too, due to another photochemical pericyclic reaction that will be discussed later.

Two-electron electrocyclic ring openings are also stereospecific. 1-Chloro-1,3-
di-*t*-butylacetone, upon treatment with base, can undergo electrocyclic ring closing to give di-*t*-butylcyclopropanone. The intermediate dipole can exist in three
different diastereomeric forms. The lowest energy one has the *t*-Bu groups pointing outward. This out,out diastereomer undergoes disrotatory electrocyclic ring
closing under the thermal conditions to give only the *higher energy* cis product,
which is in fact the product that is observed first. In time, more trans product is
observed as the cis product undergoes stereospecific electrocyclic ring opening
to the lower energy out,out dipole, which can undergo unfavorable equilibra-
tion to the higher energy out,in dipole, which can close to the lower energy, trans
product!

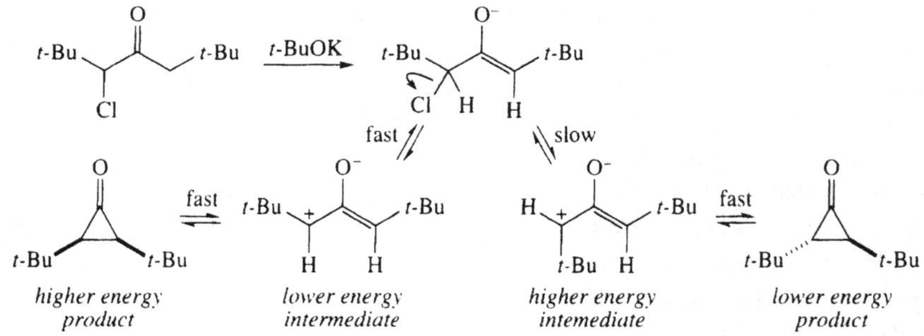

t-Bu, *t*-Bu   *t*-BuOK   *t*-Bu, *t*-Bu

Cl

Cl H H

fast   slow

t-Bu, *t*-Bu   fast

O⁻   O⁻   O

t-Bu   *t*-Bu   H   *t*-Bu   *t*-Bu

t-Bu   *t*-Bu   H   H   *t*-Bu H   *t*-Bu   *t*-Bu

higher energy *lower energy* *higher energy* *lower energy*
product *intermediate* *intermediate* *product*

4.2.3 Stereoselectivity

Consider the electrocyclic ring opening of *trans*-3,4-dimethylcyclobutene. This
compound opens in a conrotatory fashion under thermal conditions. Two products might be obtained from conrotatory ring openings allowed by the Woodward-
Hoffman rules, but in fact only the trans,trans product is obtained because there
are severe steric interactions in the TS leading to the cis,cis product.

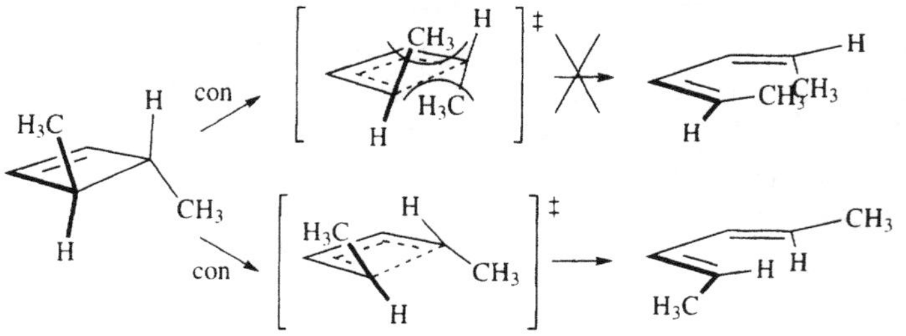

H_3C H con   CH₃   H   H
    H₃C   CH₃H₃
    H

CH₃ con   H₃C H   CH₃   H₃C   CH₃
    CH₃   H H
    H   H₃C

The phenomenon that one "allowed" TS is preferred over the other is called
torquoselectivity, a special kind of stereoselectivity. The ring opening of *trans*-
3,4-dimethylcyclobutene is *stereospecific* because the cis,trans isomer is never
obtained, but it is *stereoselective* because the trans,trans isomer is selectively obtained over the cis,cis isomer.

Torquoselectivity is observed in a few other contexts. The ring opening of *cis*-
bicyclo[4.2.0]octa-2,4-diene can occur in two disrotatory modes in principle. One
of those modes, however, gives a product that has two trans double bonds in an
eight-membered ring. This mode is disfavored with respect to the other disrotatory mode, which gives an all-cis product.

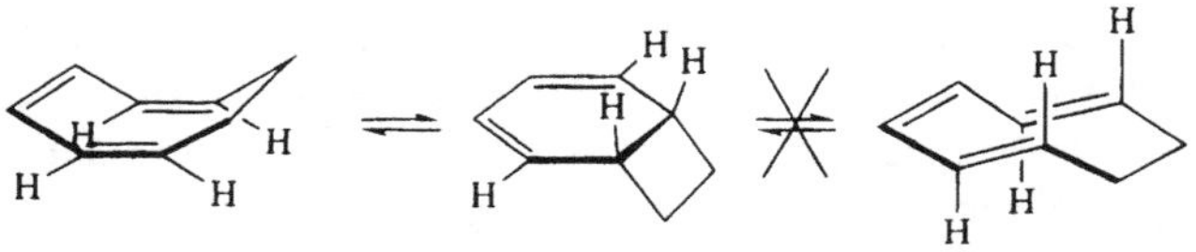

H   H   H
H   H   H   H
H   H   H   H
H   H   H   H
        H   H
        H

Halocyclopropanes undergo electrocyclic ring openings. Consider the two
diastereomers of 1-bromo-*cis*-2,3-dimethylcyclopropane. If the Br⁻ left first

to give a cyclopropyl cation, and then the ring opening occurred, then the *trans*-bromo and the *cis*-bromo compounds would give the same intermediate cyclopropyl cation; therefore, the allyl cation in which the two Me groups were "out" would be obtained from *either* diastereomer. Moreover, the *cis*-bromo compound would be expected to give the product more quickly because of steric encouragement of departure of the leaving group. In fact, though, the two isomers give isomeric allyl cations, and the *trans*-bromo compound reacts much more rapidly than the *cis*-bromo compound. This result suggests that the two compounds do not proceed through a common intermediate produced by loss of Br⁻; therefore, the loss of the leaving group and the ring opening must be concerted. Why does the *trans*-bromo compound react more quickly? The orbitals that constitute the breaking σ bond prefer to turn in the direction so that their large lobes overlap with the back lobe of the C–Br bond in the transition state (i.e., backside displacement of Br⁻). In the *cis*-bromo compound, when the orbitals of the breaking σ bond turn in this way, the two Me groups bump into each other in the TS, so the reaction is much slower.

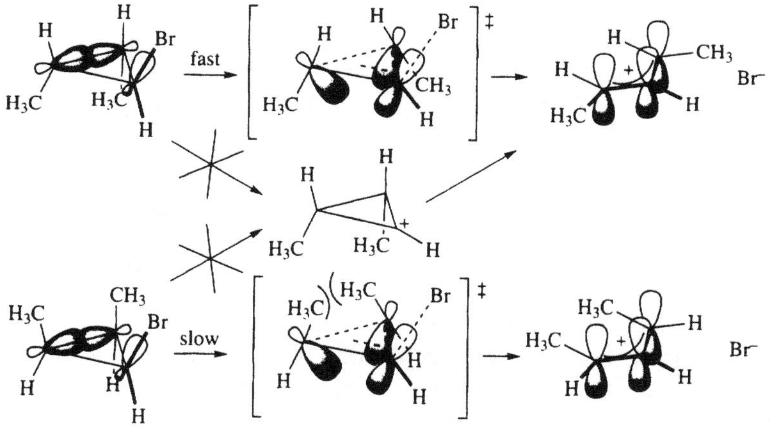

There are many cases in which the torquoselectivity is not nearly so easy to explain or predict. For example, which diastereomer is predominantly obtained from the electrocyclic ring opening of *cis*-3-chloro-4-methylcyclobutene? The answer is not obvious. Steric effects could easily be offset by electronic effects. Calculations of TS energies can sometimes give reasonably good predictions.

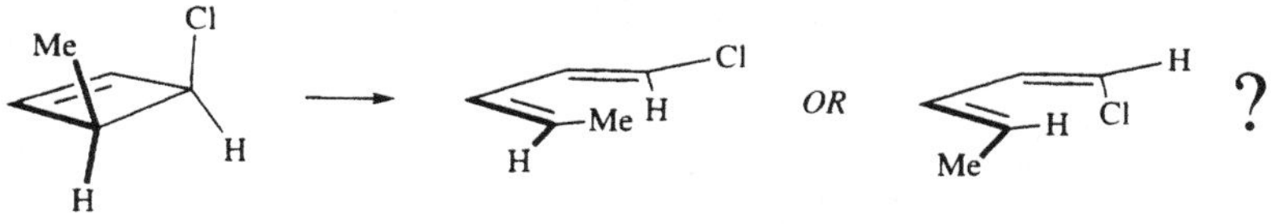

4.3   Cycloadditions

4.3.1   Typical Reactions

4.3.1.1   The Diels–Alder Reaction

The best-known cycloaddition is the *Diels–Alder reaction*, a six-electron, [4 + 2] cycloaddition. A 1,3-diene reacts with the π bond of a *dienophile* to give a six-membered ring with one π bond. Two new σ bonds are formed at the expense of two π bonds. The classic Diels–Alder reaction gives a carbocyclic ring, but hetero-Diels–Alder reactions, in which the new ring is heterocyclic, are widely used also.

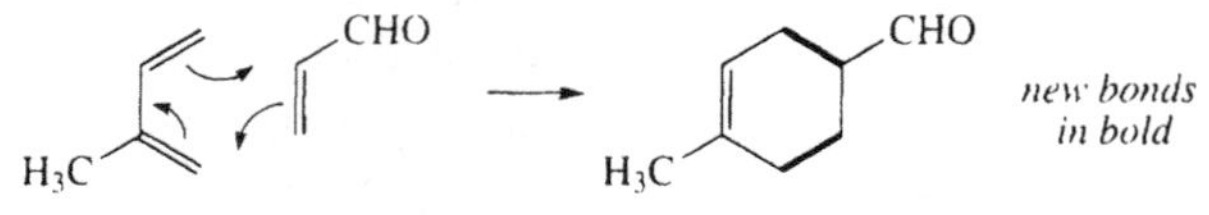

The product of the Diels–Alder reaction is a cyclohexene that has two new bonds in a 1,3-relationship. Look for the presence of such a ring in the product to determine that a Diels–Alder reaction has taken place.

Example

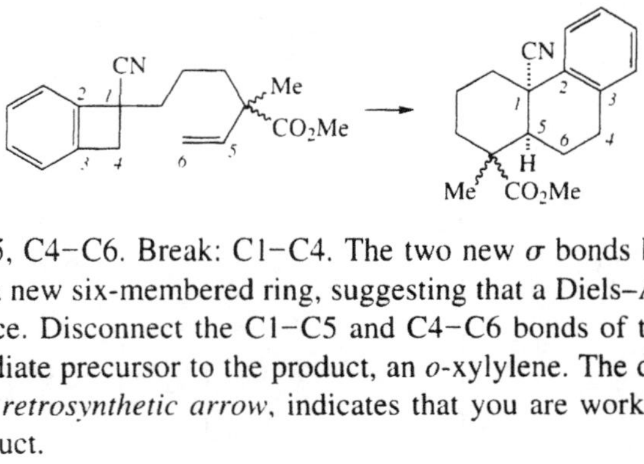

Make: C1–C5, C4–C6. Break: C1–C4. The two new σ bonds have a 1,3-relationship in a new six-membered ring, suggesting that a Diels–Alder reaction has taken place. Disconnect the C1–C5 and C4–C6 bonds of the product to see the immediate precursor to the product, an *o*-xylylene. The double-bodied arrow ($\Rightarrow$), a *retrosynthetic arrow*, indicates that you are working backward from the product.

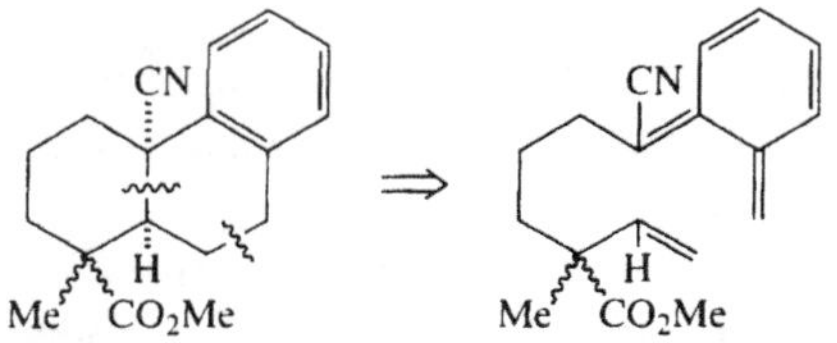

To get from the starting material to the *o*-xylylene, the C1–C4 bond needs to be broken. It can be broken by an electrocyclic ring opening of the cyclobutene. The entire mechanism in the forward direction, then, is as follows:

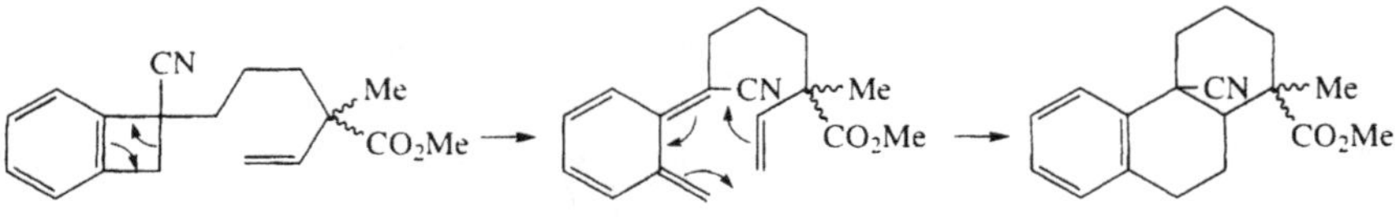

The Diels–Alder reaction requires that the two double bonds of the diene be coplanar and pointing in the same direction (i.e., in the s-cis conformation). Rotation about the σ bond between the internal C atoms of the diene interconverts the s-cis and s-trans conformations. In the s-trans conformation, there are fewer steric interactions between the in groups at the termini of the diene, so dienes normally reside predominantly in this lower energy (by about 4.0 kcal/mol) conformation.

* **Common error alert:** *The cis or trans* configurations *of the double bonds of a 1,3-diene substituted at the terminal C atoms should not be confused with the* s-cis *and* s-trans *conformations* of the diene.

The barrier to converting the s-trans conformation to the s-cis conformation contributes to the overall activation barrier for Diels–Alder reactions. Structural factors that increase the proportion of diene in its s-cis conformation increase the rate of the Diels–Alder reaction, and factors that increase the proportion of diene in its s-trans conformation decrease the rate of the reaction. Cyclopentadiene is one of the best dienes for the Diels–Alder reaction partly because it cannot rotate out of its s-cis conformation. In fact, cyclopentadiene undergoes [4 + 2] cycloaddition to itself so readily that it lasts only a few hours at 0 °C. *o*-Xylylenes are especially good dienes both because of their enforced s-cis geometry and because a nonaromatic starting material is transformed into an aromatic product. By contrast, dienes in which one of the double bonds is cis are poor substrates for Diels–Alder reactions because steric interactions between the in substituents in the s-cis conformation are particularly severe, and dienes whose s-trans conformation is enforced do not ever undergo the Diels–Alder reaction.

Similarly, acyclic 2,3-diazadienes (azines) have never been observed to undergo Diels–Alder reactions because the lone pairs on N repel one another, making the s-cis conformation about 16 kcal/mol higher in energy than the s-trans conformation. Cyclic azines, though, undergo Diels–Alder reactions with facility.

Because two σ bonds are produced at the expense of two π bonds, the Diels–Alder reaction is normally exothermic in the forward direction, but the *retro-Diels–Alder* reaction is facile when one of the products is N_2, CO_2, or an aromatic ring.

Example

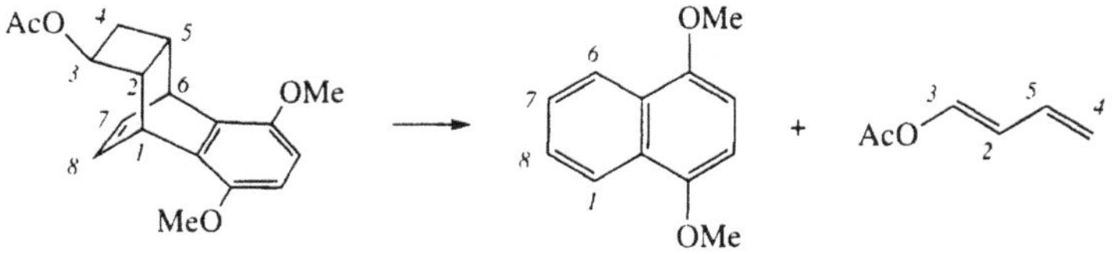

Make: none. Break: C1–C2, C3–C4, C5–C6. The C1–C2 and C5–C6 bonds are in a 1,3-relationship in a six-membered ring, and there is a C7=C8 π bond in a 1,3-relationship to each of them in that ring. Therefore, the C1–C2 and C5–C6 bonds can be broken by a retro-Diels–Alder reaction to give the naphthalene and a cyclobutene. Cleavage of the C3–C4 bond then occurs by an electrocyclic reaction.

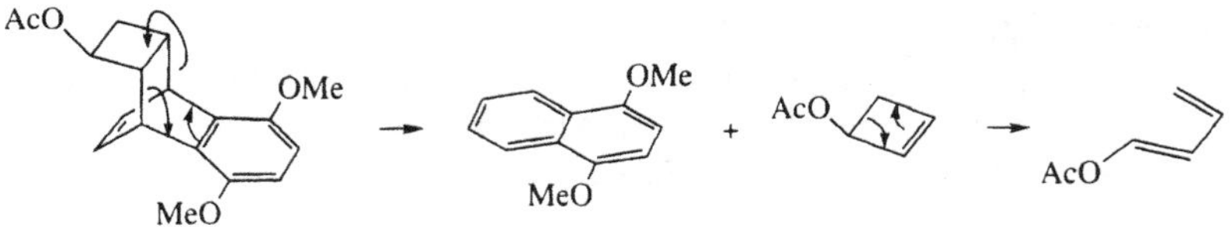

Most Diels–Alder reactions occur with what is called *normal electron-demand*, in which an *electron-rich* (nucleophilic) *diene* reacts with an *electron-poor* (electrophilic) *dienophile*. The dienophile may be substituted with carbonyl, CN, sulfonyl, NO_2, or any other electron-withdrawing group. Dienophiles substituted with two electron-withdrawing groups (diethyl fumarate, maleic anhydride, benzoquinone) are particularly good substrates for the Diels–Alder reaction. Dienophiles that are not electron-poor undergo Diels–Alder reactions with electron-rich dienophiles under rather drastic conditions, if at all, although the rate of such a reaction can be increased by making it intramolecular. However, compounds with very strained double bonds (benzyne, norbornadiene) make good dienophiles even when they are not substituted with electron-withdrawing groups. Alkenes make better dienophiles than alkynes, all other things being equal.

Good Diels-Alder dienophiles

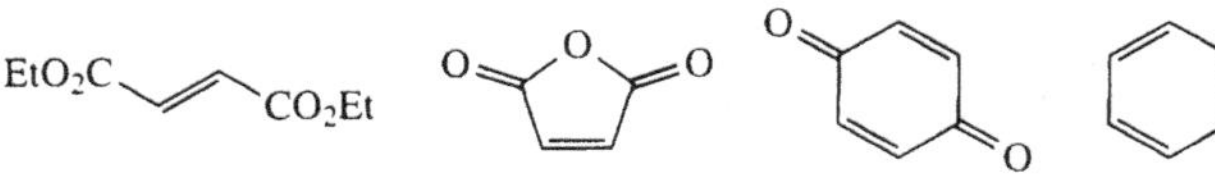

Dienes substituted with RO and R_2N groups (e.g., Danishefsky's diene, 1-methoxy-3-trimethylsilyloxy-1,3-butadiene) are particularly good substrates for Diels–Alder reactions, but alkyl-substituted dienes and even butadiene itself are common substrates. Benzene rings are very poor dienes in Diels–Alder reactions, because they lose aromaticity upon cycloaddition, but less aromatic compounds such as anthracene undergo Diels–Alder reactions more readily. Again, both cyclopentadiene and *o*-xylylenes are especially good dienes in Diels–Alder reactions.

Good Diels-Alder dienes

Frontier MO theory can be used to understand the dependence of the rate of the Diels–Alder reaction on the electronic nature of the substrates. As in any reaction, the rate of the Diels–Alder reaction is determined by the energy of its TS. In the TS of most Diels–Alder reactions, the *HOMO of the diene* interacts with the *LUMO of the dienophile*.

The energy of the Diels–Alder TS is directly related to the strength of the interaction between the MOs, which is in turn related to the difference in energy between the two MOs. The smaller the difference in energy between the two MOs, the stronger their interaction, the lower the energy of the TS, and the faster the reaction. The HOMO$_{diene}$, being a bonding orbital, is lower in energy than the LUMO$_{dienophile}$. Substitution of the diene with electron-donating groups raises the energy of the HOMO$_{diene}$, bringing it closer to the energy of the LUMO$_{dienophile}$, and substitution of the dienophile with electron-withdrawing groups lowers the energy of the LUMO$_{dienophile}$, bringing it closer to the energy of the HOMO$_{diene}$. Either of these substitutions brings the two orbitals closer in energy, increasing the rate of the reaction.

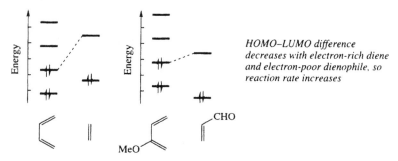

HOMO–LUMO difference decreases with electron-rich diene and electron-poor dienophile, so reaction rate increases

We mentioned earlier that alkynes are poorer dienophiles than the correspondingly substituted alkenes. The C≡C bond in an alkyne is shorter than the C=C bond in the corresponding alkene, so the C(p)–C(p) overlap is better, and thus the alkyne's HOMO is lower in energy and its LUMO higher in energy than the alkene's. The LUMO of the alkyne interacts more poorly with the HOMO$_{diene}$

than the lower energy LUMO of the alkene, so the Diels–Alder reaction of the alkyne is slower.

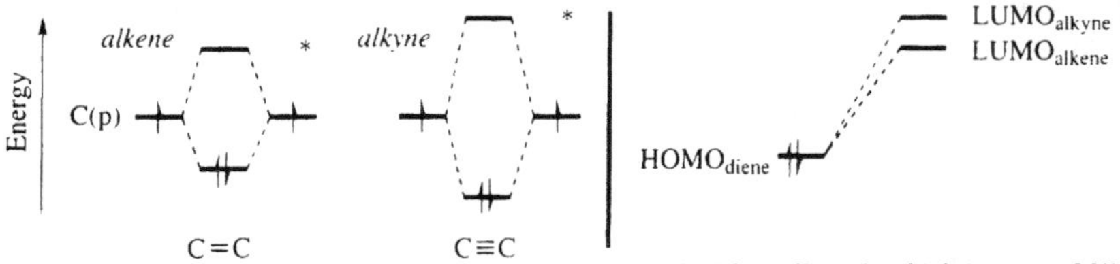

C–C distance is shorter in alkynes than in alkenes, so C(p)–C(p) interaction is stronger ...

... therefore alkyne has higher-energy LUMO, interacts more weakly with HOMO of diene, and reacts more slowly in Diels–Alder reaction

The energy of the $LUMO_{dienophile}$ can also be lowered by the use of a Lewis acid catalyst. The Lewis acid coordinates to the O of a carbonyl group substituent on the dienophile, lowering the energy of the $LUMO_{dienophile}$ and leading to an increased reaction rate. The Lewis acid also increases the regioselectivity and the stereoselectivity of the reaction. Use of a chiral Lewis acid can lead to an enantioselective Diels–Alder reaction. The development of catalytic asymmetric Diels–Alder reactions is an active area of research.

* **Common error alert:** *It is easy to assume that a reaction mechanism is polar acidic when a Lewis acid is present, but Lewis acids are often used to promote cycloadditions, too.*

Note that the dependence of the rate of a Diels–Alder reaction on the electronic match between $HOMO_{diene}$ and $LUMO_{dienophile}$ is a kinetic effect, not a thermodynamic one. The reaction of an electron-rich diene with an electron-rich dienophile is just as thermodynamically favorable as its reaction with an electron-poor dienophile, but it does not proceed at a perceptible rate.

Very electron poor dienes can undergo Diels–Alder reactions with electron-rich dienophiles in the *inverse electron-demand* Diels–Alder reaction. The dominant interaction in the TS of inverse electron-demand Diels–Alder reactions is between the $LUMO_{diene}$ and the $HOMO_{dienophile}$.

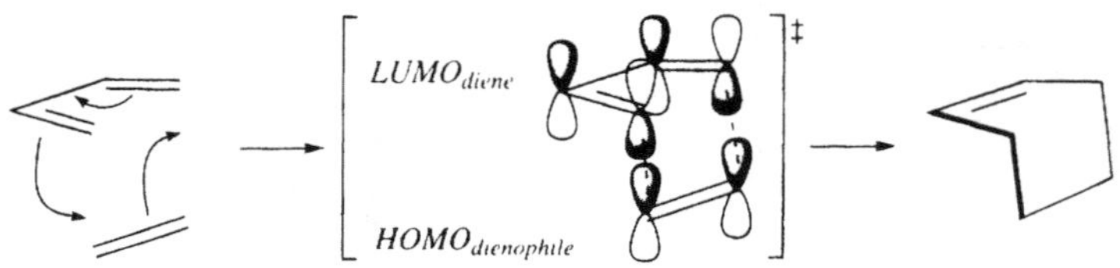

For example, the reaction of hexachlorocyclopentadiene with norbornadiene proceeds readily by an inverse electron-demand process to give Aldrin, an insecticide that has now been banned because of its environmental persistence.

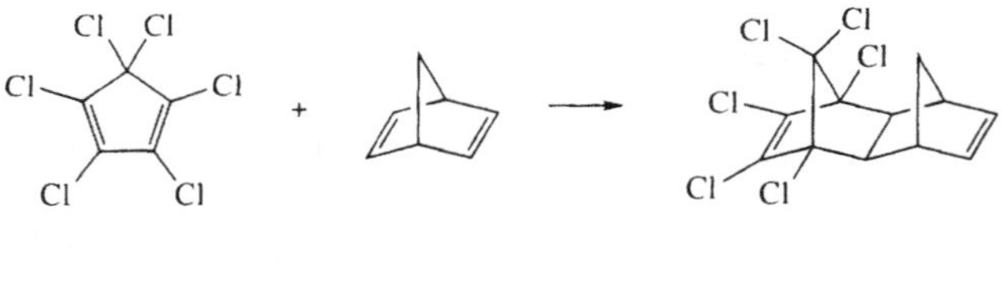

Dienes containing heteroatoms such as N and O undergo inverse electron-demand Diels–Alder reactions with electron-rich dienophiles such as enol ethers and enamines. The relatively low energy of the heteroatom p orbitals dramatically lowers the energy of both the $HOMO_{diene}$ and $LUMO_{diene}$.

Example

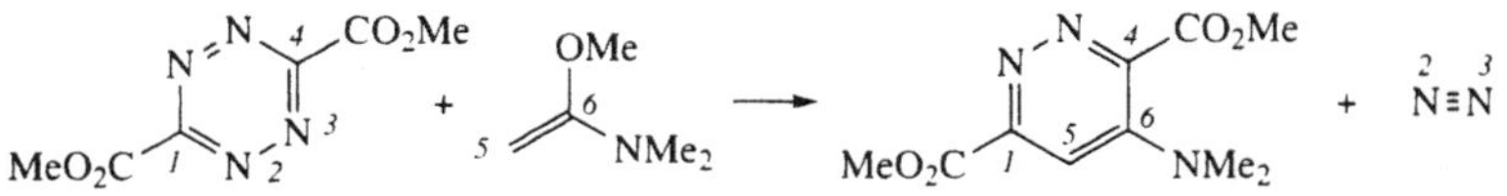

Make: C1–C5, C4–C6. Break: C1–N2, C4–N3, C6–OMe. The two new σ bonds have a 1,3-relationship in the six-membered ring, suggesting a Diels–Alder reaction as the first step. The Diels–Alder adduct can undergo a retro-Diels–Alder reaction to cleave the C–N bonds and give N_2 as a by-product.

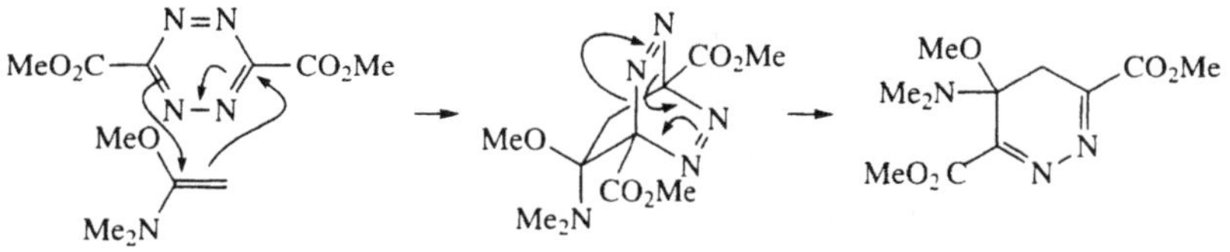

An elimination reaction (probably E1) then gives the observed product.

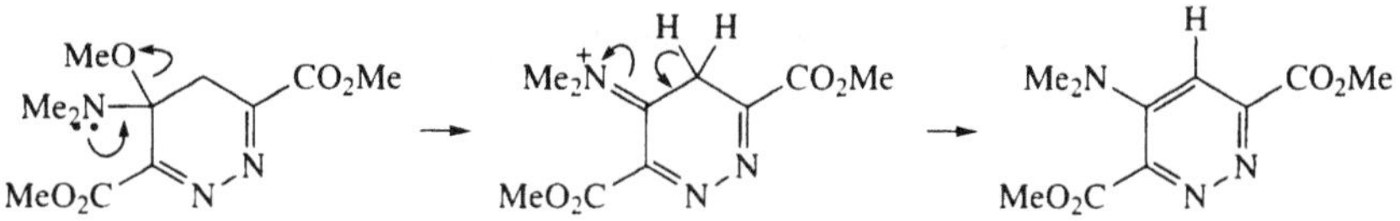

Lewis acid catalysis can lower the energy of $LUMO_{diene}$ in heterodienes even further, as in the cycloaddition of enol ethers to α,β-unsaturated carbonyl compounds. This reaction provides an important method for the synthesis of glucals, dehydrated carbohydrates that are important building blocks in the synthesis of polysaccharides.

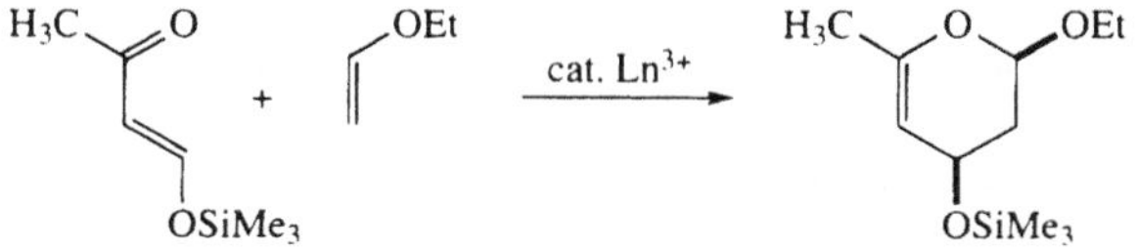

Heteroatomic dienophiles such as aldehydes and imines also participate in Diels–Alder reactions. Heteroatomic dienophiles have low-energy MOs, so they undergo normal electron-demand Diels–Alder reactions with electron-rich dienes. Singlet O_2 ($^1O_2$, O=O) also undergoes normal electron-demand Diels–Alder reactions. Atmospheric O_2 is a triplet, best described as a 1,2-diradical ($\cdot$O–O$\cdot$), whereas $^1O_2$ has a normal O=O π bond and no unpaired electrons. Triplet O_2 is converted to the higher energy singlet form by $h\nu$ in the presence of a sensitizer such as Rose Bengal.

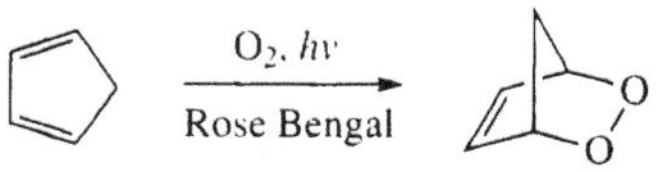

4.3.1.2   Other Cycloadditions

1,3-Dipoles react with alkenes and alkynes (*dipolarophiles*) in 1,3-dipolar cycloadditions (a.k.a. [3 + 2] cycloadditions) to give five-membered heterocycles. Many agrochemicals and pharmaceuticals contain five-membered heterocycles, and the dipolar cycloaddition is an important synthetic route to these compounds.

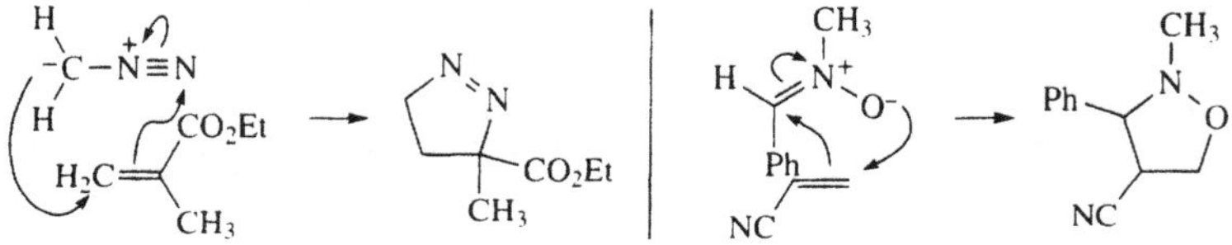

The three-atom component of the cycloaddition, the 1,3-dipole, is a compound for which a relatively stable resonance structure can be drawn in which one terminus has a formal positive charge (and is electron-deficient) and the other terminus has a formal negative charge. All of the common 1,3-dipoles have a heteroatom (N or O) in the central position in order to stabilize the electron-deficient terminus.

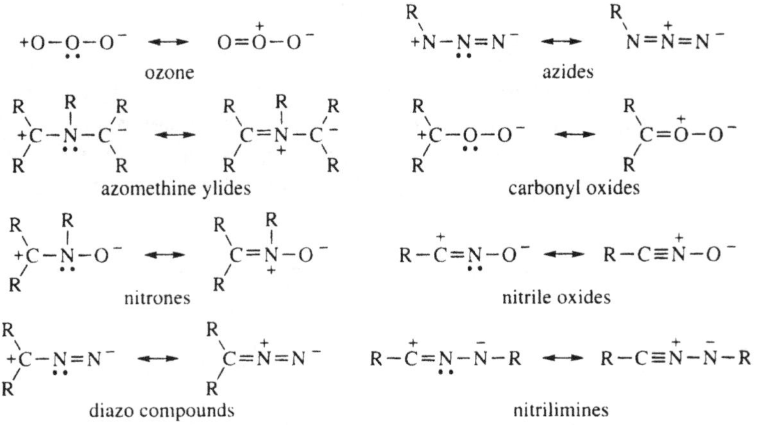

The five-membered heterocyclic product is the key to identifying a 1,3-dipolar cycloaddition. Many 1,3-dipoles are not stable, so they are generated by a series of polar reactions and then react in situ without being isolated.

Example

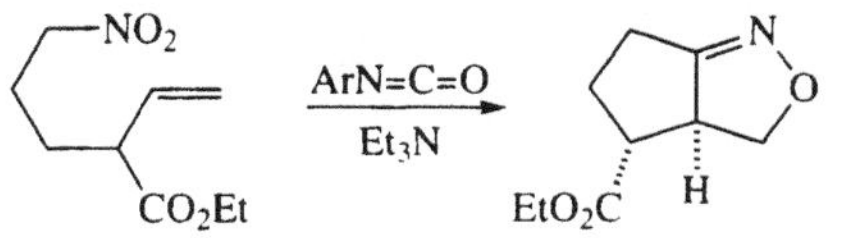

The reaction produces a five-membered heterocycle, suggesting a 1,3-dipolar cycloaddition. What penultimate intermediate would undergo a 1,3-dipolar cycloaddition to give the observed product? The two-atom component of the dipolar cycloaddition is the C=C π bond, so the three-atom component must be C–N–O. The C is likely to be the (+) terminus and the O the (−) terminus of the dipole. The 1,3-dipole in this reaction is a *nitrile oxide*, an unstable functional group that must be generated in situ.

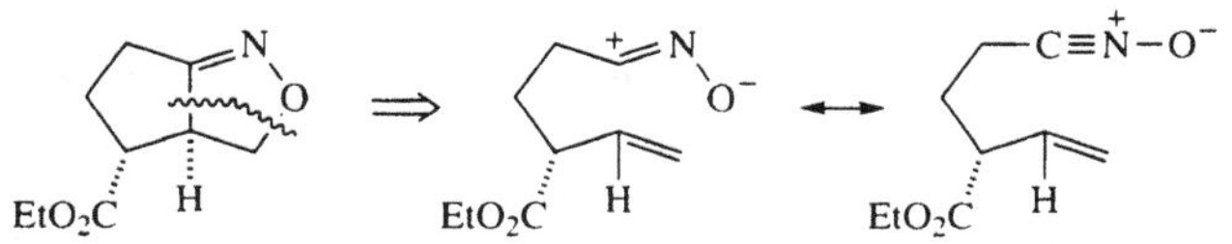

How is the nitrile oxide formed? The elements of water must be eliminated from the nitro compound, and an N–O bond must be cleaved. The O of the NO$_2$ group is not a leaving group, so the role of ArNCO must be to convert it into one. Nitro compounds are quite acidic (pK_a = 9), so deprotonation by Et$_3$N is the first step. Attack of O$^-$ on the electrophilic C of the isocyanate, protonation of N, and then E2 elimination gives the nitrile oxide, which undergoes the [3 + 2] cycloaddition to give the product.

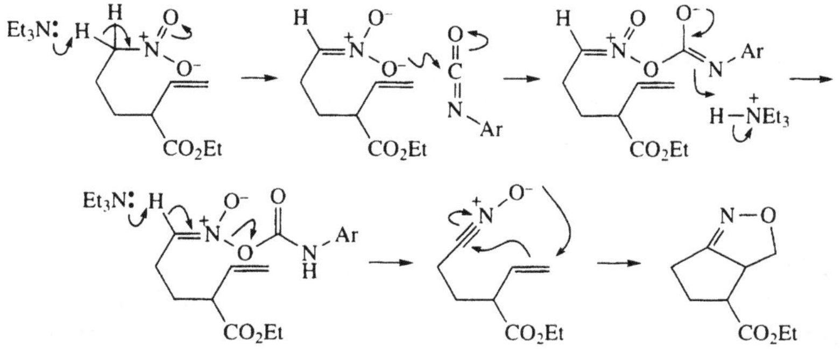

Note how this problem is solved by working backwards one step from the product. This technique is very useful for solving pericyclic mechanism problems.

Problem 4.8. Draw a reasonable mechanism for the following reaction that involves a 1,3-dipolar cycloaddition. *Hint:* Work backward from the product before working forward.

The selectivity of 1,3-dipoles for electron-rich or electron-poor dipolarophiles is complex. Very electron poor dipoles such as ozone react most quickly with electron-rich and slowly with electron-poor dipolarophiles. Other dipoles, such as azides, react quickly with very electron poor dipolarophiles, slowly with dipolarophiles of intermediate electronic character, and quickly with very electron rich dipolarophiles. This "U-shaped reactivity" is due to a crossover in the nature of the reaction from LUMO$_{dipole}$/HOMO$_{dipolarophile}$-controlled to HOMO$_{dipole}$/LUMO$_{dipolarophile}$-controlled as the dipolarophile becomes more electron-poor. At intermediate electronic character, neither the HOMO$_{dipole}$/LUMO$_{dipolarophile}$ nor the LUMO$_{dipole}$/HOMO$_{dipolarophile}$ interaction is particularly strong, and the reaction proceeds slowly.

The reaction of ozone with alkenes is one of the most useful 1,3-dipolar cycloadditions. Ozone undergoes [3 + 2] cycloaddition to the alkene to give a 1,2,3-trioxolane, which immediately decomposes by a [3 + 2] retro-cycloaddition to give a carbonyl oxide and an aldehyde. When the ozonolysis is carried out in the presence of an alcohol, the alcohol adds to the carbonyl oxide to give a hydroperoxide acetal. In the absence of alcohol, though, the carbonyl oxide undergoes another [3 + 2] cycloaddition with the aldehyde to give a 1,2,4-trioxolane.

1,2,4-Trioxolanes are isolable, but they can explode when heated gently. Neither 1,2,4-trioxolanes nor hydroperoxide acetals are usually isolated, though. They are decomposed in situ in one of three ways: by gentle reduction (Me$_2$S, Ph$_3$P, H$_2$ over Pd, or Zn/HCl) to give two aldehydes, by a stronger reduction (NaBH$_4$ or LiAlH$_4$) to give two alcohols, or oxidatively (H$_2$O$_2$ and acid) to give two carboxylic acids. (Obviously, if the alkene is tri- or tetrasubstituted, ketones and not aldehydes are obtained.) Today the Me$_2$S reduction is the most widely used method, as it gives the most valuable products (aldehydes) in high yields, and the by-product (DMSO) is easily removed.

Problem 4.9. Draw a mechanism for the reaction of Me₂S with a 1,2,4-triox-olane to give two aldehydes.

The [2 + 2] cycloadditions are widely used reactions. There are basically three situations in which [2 + 2] cycloadditions are seen: when the reaction is promoted by light, when one of the components is a ketene ($R_2C=C=O$) or another *cumu-lene* (e.g., $RN=C=O$), or when one of the components has a π bond between C and a second-row or heavier element (e.g., $Ph_3P=CH_2$ or $Cp_2Ti=CH_2$).

• The [2 + 2] photocycloaddition of two alkenes is widely used to form cy-clobutanes. The reaction proceeds in the forward direction because the product cannot absorb light of the wavelengths that the starting material can absorb. In the *Paterno–Büchi* reaction, one of the two-atom components is a ketone or an aldehyde instead of an alkene. Why these cycloadditions require light to proceed will be discussed later.

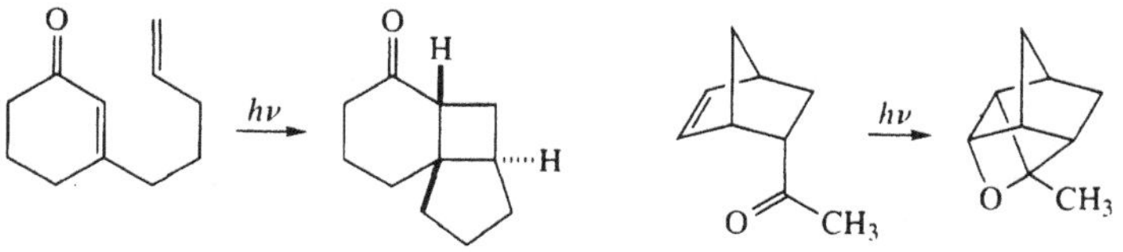

The light-induced [2 + 2] cycloaddition can occur in vivo. Two adjacent thymi-dine residues in DNA can undergo a [2 + 2] cycloaddition to give a *thymine dimer*. DNA repair enzymes excise the dimer and usually repair it correctly, but occasionally they make a mistake, and a mutation occurs. The mutation can lead to skin cancer. Sunlight is essential for your health (for the ergosterol-to-precalciferol electrocyclic ring opening), but not too much sunlight!

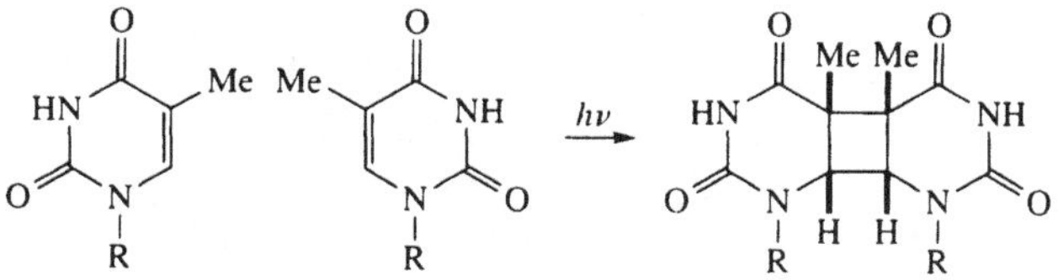

Problem 4.10. The thymine dimer shown in the preceding example is produced by a [2 + 2] cycloaddition of two C=C bonds. A different, perhaps more muta-genic thymine dimer is produced by a different [2 + 2] cycloaddition. Draw a detailed mechanism for the formation of the second type of thymine dimer.

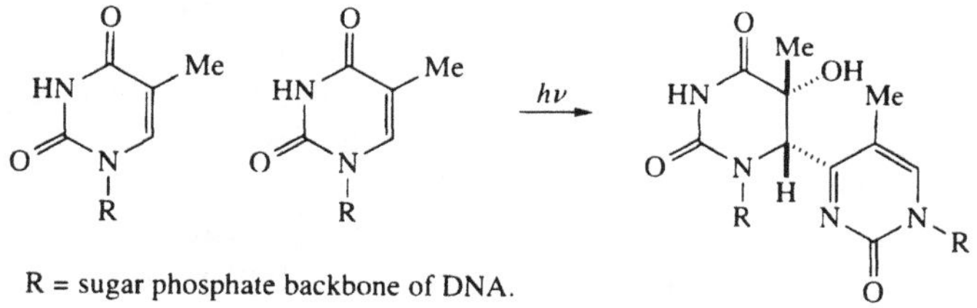

R = sugar phosphate backbone of DNA.

• The ketene–alkene cycloaddition gives cyclobutanones in a *thermal* reaction. Most ketenes are not kinetically stable, so they are usually generated in situ, either by E2 elimination of HCl from an acyl chloride or by a Wolff rearrangement of an α-diazoketone (see Chapter 2).

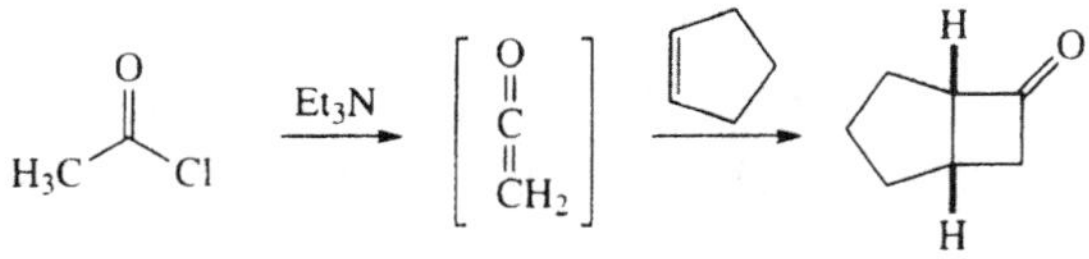

Other cumulenes such as isocyanates RN=C=O can also undergo thermal [2 + 2] cycloadditions. The [2 + 2] cycloaddition of an isocyanate and an alkene is a useful route to β-lactams, the key functional group in the penicillin and cephalosporin antibiotics, as is the [2 + 2] cycloaddition of a ketene and an imine.

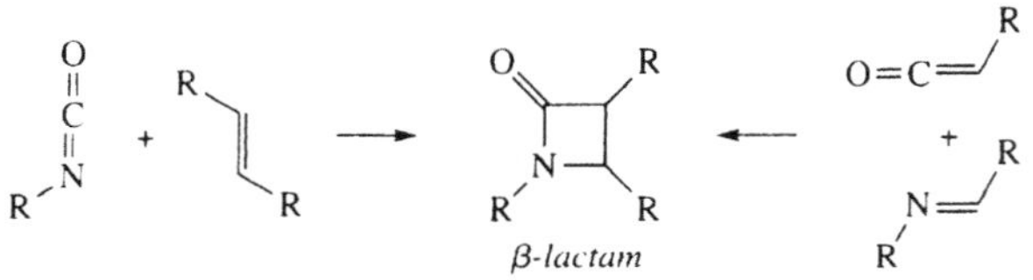

Ketenes dimerize by a [2 + 2] cycloaddition in the absence of another substrate. The electron-rich C=C π bond combines with the electron-poor C=O π bond to give a β-lactone.

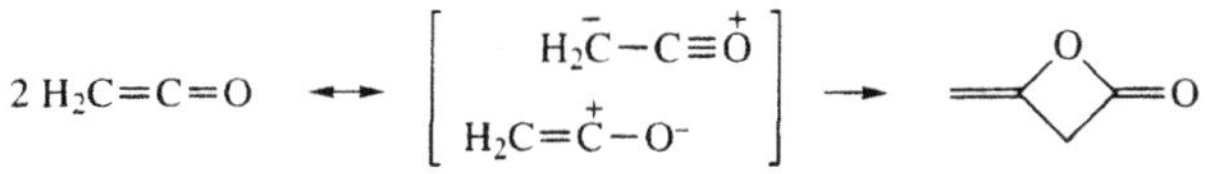

• The most important example of the third type of [2 + 2] cycloaddition is the Wittig reaction ($Ph_3P=CH_2 + R_2C=O \rightarrow Ph_3P=O + R_2C=CH_2$). The phosphorane adds to the ketone to give a phosphaoxetane, either by a concerted, [2 + 2] cycloaddition or by a two-step, polar process involving a *betaine* (pronounced BAY-tah-een) intermediate. The phosphaoxetane then undergoes [2 + 2] retro-cycloaddition to give $Ph_3P=O$ and $R_2C=CH_2$.

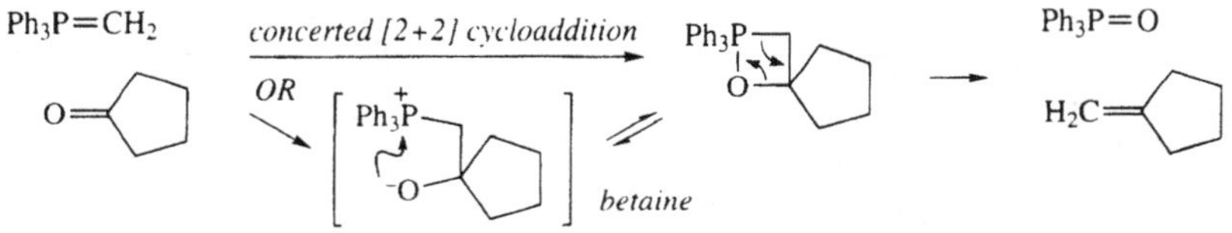

There are other kinds of cycloadditions, too. The [4 + 1] cycloaddition, a cheletropic reaction, usually goes in the retro direction for entropic reasons. 3-Sulfolene (butadiene sulfone, 2,5-dihydrothiophene 1,1-dioxide) undergoes a [4 + 1] retro-cycloaddition to generate SO_2 and 1,3-butadiene, which can undergo a Diels–Alder reaction with a dienophile. It is much more convenient to use 3-sulfolene instead of 1,3-butadiene itself, as the latter compound is a gas that is prone to polymerization.

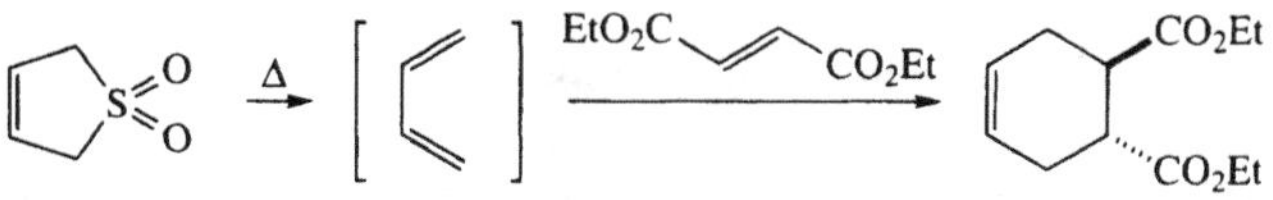

Cyclopentadienones are very prone to do Diels–Alder reactions because of their enforced s-cis conformation and their antiaromaticity. The Diels–Alder reaction of a substituted cyclopentadienone and an alkyne is followed immediately by a [4 + 1] retro-cycloaddition to generate CO and an aromatic compound.

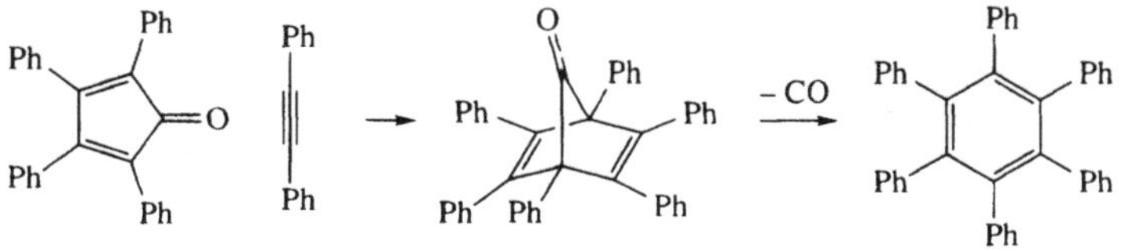

Problem 4.11. A cyclopentadienone is the starting point for a cascade of cycloadditions and retro-cycloadditions in the following reaction. Draw a reasonable mechanism. *Hint*: Number the atoms and draw the by-products!

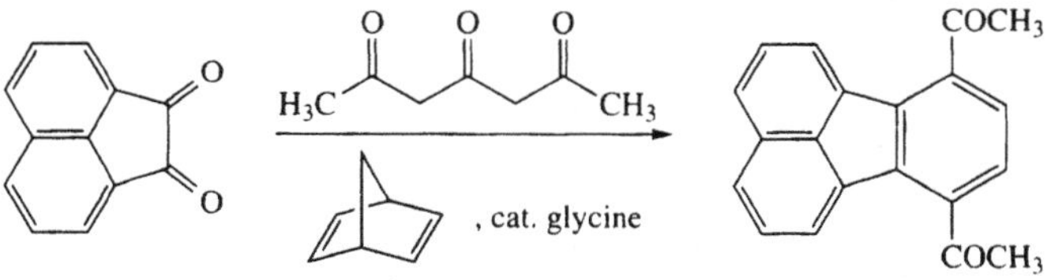

The [2 + 1] cycloaddition of carbenes and alkenes to give cyclopropanes was discussed in Chapter 2. Other cycloadditions are less common, although [4 + 3], [4 + 4], [6 + 4], [8 + 2] and many other cycloadditions are certainly known. The [4 + 3] cycloaddition in particular involves an allyl cation as the three-atom component and an electron-rich diene as the four-atom component.

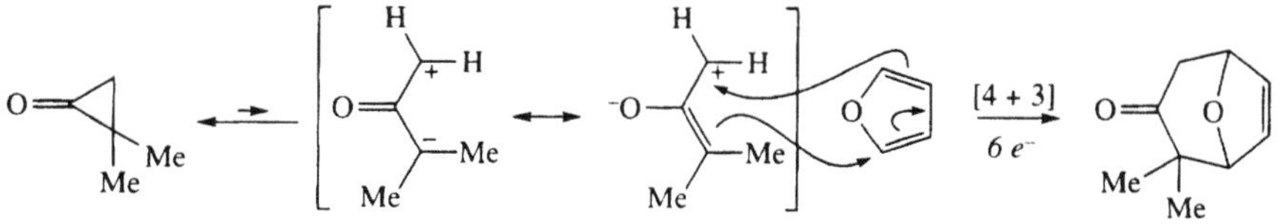

The following are some key ways of identifying cycloaddition reactions:

• All cycloadditions (except cheletropic reactions) form two new σ bonds to the termini of two π systems.

• If you see a new six-membered ring containing two new bonds with a 1,3-relationship and a π bond, think Diels–Alder! A six-membered ring fused to a benzene ring is often made by the [4 + 2] cycloaddition of an *o*-xylylene and a dienophile or by a [4 + 2] cycloaddition of benzyne and a 1,3-diene.

• If you see a five-membered *heterocycle*, especially one containing N, think 1,3-dipolar cycloaddition! One of the ring heteroatoms is the central atom of the 1,3-dipole.

• If you see a four-membered ring, think [2 + 2] cycloaddition, especially if the ring is a cyclobutanone (ketene) or light is required (photochemically allowed). Ketenes and other cumulenes undergo [2 + 2] cycloadditions with special facility. An oxetane (four-membered ring with one O) is often obtained from the [2 + 2] photocycloaddition of a carbonyl compound and an alkene.

Example

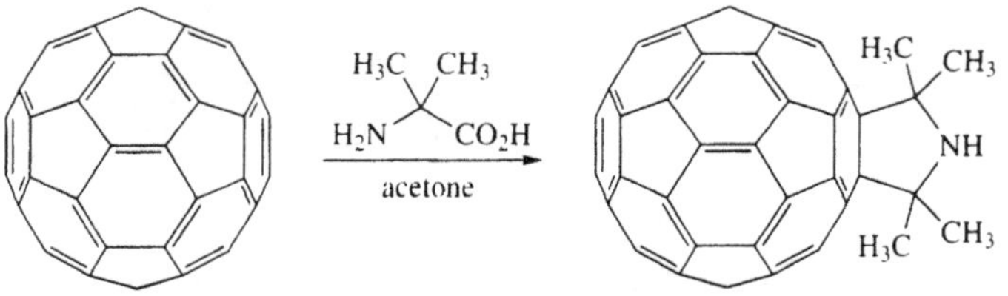

A five-membered heterocycle is formed, suggesting a 1,3-dipolar cycloaddition. A 1,3-dipole almost always has a heteroatom in the 2-position, and the heterocycle here has only one heteroatom, so the 1,3-dipole must be the azomethine ylide.

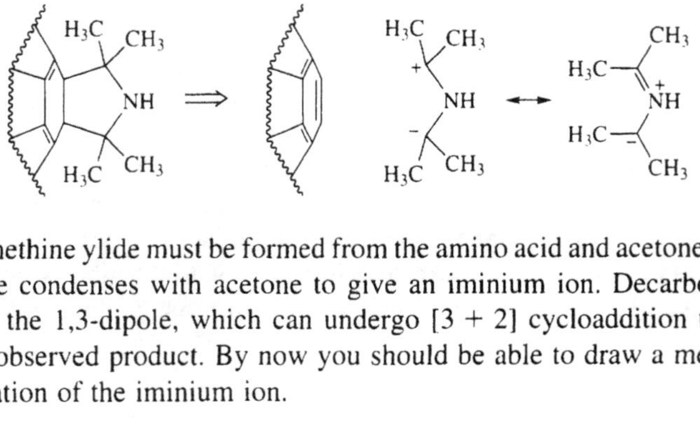

The azomethine ylide must be formed from the amino acid and acetone. Briefly, the amine condenses with acetone to give an iminium ion. Decarboxylation produces the 1,3-dipole, which can undergo [3 + 2] cycloaddition to C_{60} to give the observed product. By now you should be able to draw a mechanism for formation of the iminium ion.

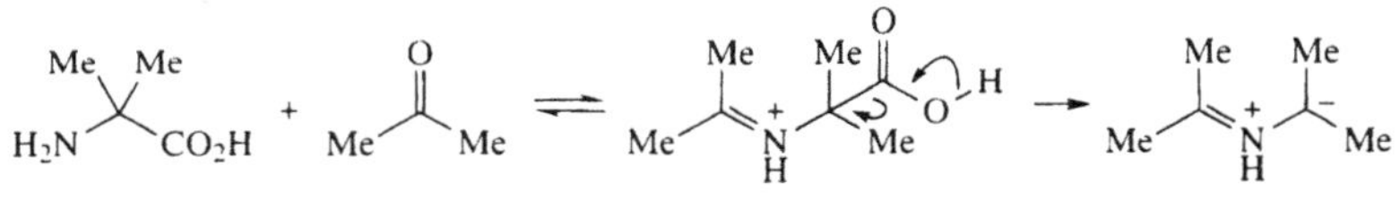

Problem 4.12. The following reaction proceeds by a series of cycloadditions and retro-cycloadditions. Draw a reasonable mechanism.

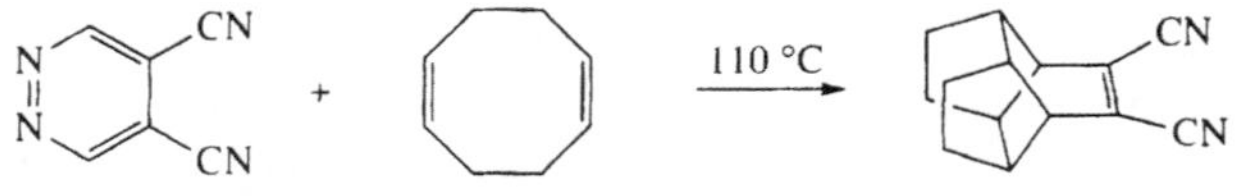

4.3.2   Regioselectivity

In the Diels–Alder reaction of *trans*-1-methoxybutadiene with ethyl acrylate, either the 1,2- or the 1,3-disubstituted product can be obtained in principle. The 1,3-disubstituted product is thermodynamically more stable (less steric hindrance), but the kinetic product is the 1,2-disubstituted compound. The easiest way to explain this phenomenon is to note that C4 of the diene is nucleophilic, and C2 of the dienophile (i.e., the β-carbon) is electrophilic. Combination of $C4_{diene}$ with $C2_{dienophile}$ gives the observed product.

The same explanation can be used to rationalize the outcome of the reaction of isoprene (2-methyl-1,3-butadiene) with acrolein (2-propenal). In this case, though, because the diene is 2-substituted, $C1_{diene}$ is nucleophilic and combines with $C2_{dienophile}$ to give the observed product.

In fact, because most Diels–Alder reactions proceed by the reaction of nucleophilic dienes with electrophilic dienophiles, the following rule can be formulated as follows: Diels–Alder reactions proceed to put the most electron donating substituent on the diene and the most electron withdrawing substituent on the dienophile either "ortho" or "para" to one another. This "ortho–para rule" misuses the terms ortho and para, which really apply only to benzene rings, but it allows one to remember the regioselectivity of these reactions fairly easily.

> The regioselectivity of an *intramolecular* cycloaddition depends more on geometric constraints within the substrate than electronic preferences.

Sometimes, the ortho–para rule doesn't work well, especially when two substituents on the diene have competing directing abilities. For example, a PhS substituent at $C1_{diene}$ is more strongly directing than a MeO substituent at $C2_{diene}$. By resonance arguments, the opposite should be the case, as MeO is a better resonance donor than PhS.

The only way the result can be explained is by looking at the coefficients of the orbitals of the HOMO of the diene and the LUMO of the dienophile. When the p orbitals of the diene combine in a bonding and antibonding way to make the HOMO and the LUMO, they don't contribute an equal weight to both MOs (unless the dienophile is symmetrical). (The situation is analogous to the one where $C(sp^3)$ and $O(sp^3)$ combine to form a σ bond, the $O(sp^3)$ orbital contributes more to the σ orbital, and the $C(sp^3)$ orbital contributes more to the $\sigma*$.) *The Diels–Alder reaction proceeds so that the diene terminus with the largest coefficient in the diene HOMO interacts with the dienophile terminus with the largest coefficient in the dienophile LUMO.* It happens that the influence of substituents on C1 and C4 of the diene on the orbital coefficients is greater than that of substituents on C2 or C3. With simple dienes the prediction from orbital coefficient arguments is usually the same as from the ortho–para rule—that the most electron withdrawing substituent on the diene and the most electron donating substituent on the dienophile are "ortho" or "para" in the product—but the orbital coefficient rule explains many cases that can't be explained simply by looking at resonance structures.

The regioselectivity of inverse electron-demand Diels–Alder reactions, 1,3-dipolar cycloadditions, and other cycloadditions can similarly be explained by resonance and orbital coefficient arguments. Determining which end of a 1,3-dipole is "nucleophilic" and which end is "electrophilic" can be dicey, though. In nitrones ($R_2C=NR-O$), the O is the nucleophilic end, so it reacts with the electrophilic end of a dipolarophile.

The problem with orbital coefficient arguments is that they require a calculation, whereas resonance arguments can be made using pen and paper. On the other hand, the increased power of personal computers in recent years has made it possible for almost anyone to do a simple orbital coefficient calculation.

4.3.3   Stereospecificity

Some cycloadditions proceed thermally, whereas others require $h\nu$. The dependence of certain cycloadditions on the presence of light can be explained by examining interactions between the MOs of the two reacting components. Frontier MO theory suggests that the rate of cycloadditions is determined by the strength of the interaction of the HOMO of one component with the LUMO of the other. In normal electron-demand Diels–Alder reactions, $HOMO_{diene}$ (ψ_1) interacts with $LUMO_{dienophile}$ (ψ_1). There is positive overlap between the orbitals where the two σ bonds form when both components of the reaction react from the same face of the π system (*suprafacially*).

thermal, normal electron demand

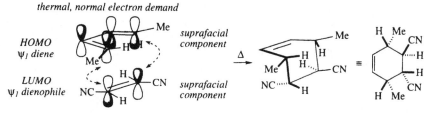

shows positive overlap where new bonds can form

Under inverse electron demand, $LUMO_{diene}$ (ψ_2) interacts with $HOMO_{dienophile}$ (ψ_0). Again, there is positive overlap between the orbitals at both termini of the two π systems when both components of the reaction react suprafacially.

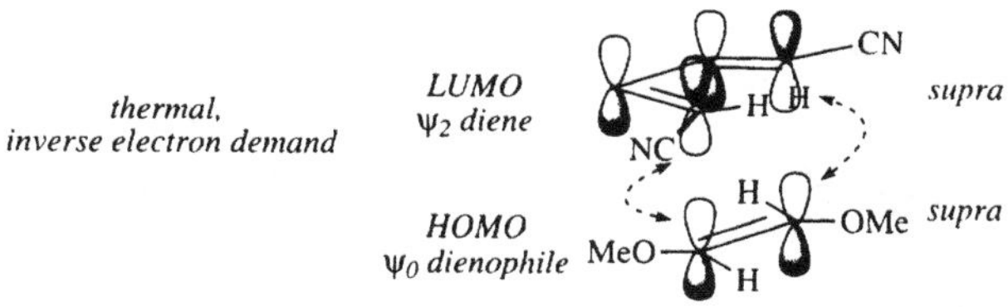

thermal,
inverse electron demand

$LUMO$
ψ_2 *diene*

$HOMO$
ψ_0 *dienophile*

supra

supra

By contrast, under photochemical conditions and normal electron demand, $HOMO_{diene}$ changes from ψ_1 to ψ_2. In this case, positive overlap at both termini of the π systems can occur only if one of the π components reacts *antarafacially*. This situation is very difficult to achieve geometrically, and hence six-electron cycloadditions do not proceed photochemically.

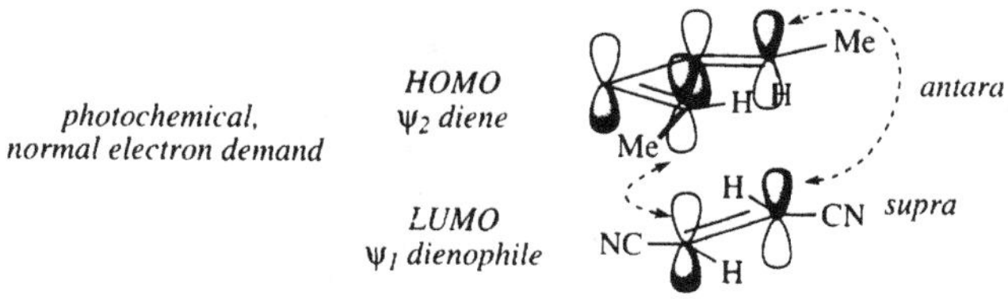

photochemical,
normal electron demand

$HOMO$
ψ_2 *diene*

$LUMO$
ψ_1 *dienophile*

antara

supra

The stereochemical relationships among substituents in a suprafacial component of a cycloaddition are preserved in the cycloadduct. Groups that are cis (or trans) to one another in the dienophile become cis (or trans) to one another in the product. The two out groups in the diene become cis to one another in the product, as do the two in groups. Because one diastereomeric starting material gives one diastereomeric product, cycloadditions are said to be *stereospecific*.

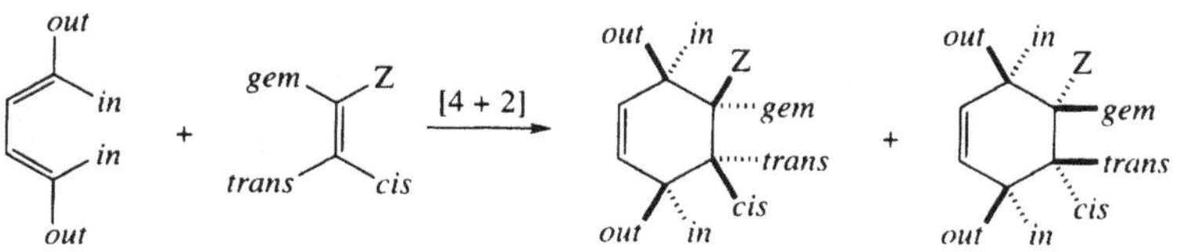

Note that two stereoisomeric products that are consistent with the Woodward–Hoffmann rules can be obtained. The Woodward–Hoffmann rules allow you to predict the stereochemical relationship between substituents derived from the *same* component. They do not allow you to predict the relationship between substituents derived from *different* components. Guidelines for predicting the latter kind of relationship will be discussed shortly.

An analogous picture can be drawn for [3 + 2] cycloadditions. These six-electron reactions can be either $HOMO_{dipole}/LUMO_{dipolarophile}$-controlled or $HOMO_{dipolarophile}/LUMO_{dipole}$-controlled. In either case, the maximum number of bonding interactions between the termini occurs when both components are suprafacial.

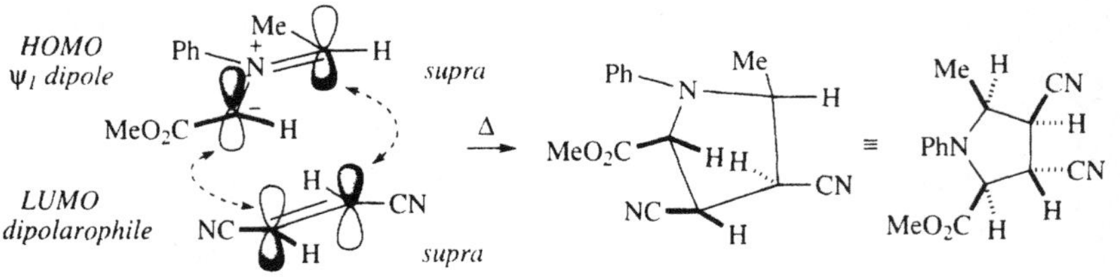

As in the Diels–Alder reaction, the stereochemistry of both starting materials is preserved in the products. The two out groups on the dipolarophile become cis in the product, as do the two in groups. The stereochemical relationships among the substituents of the dipolarophile are also preserved in the product.

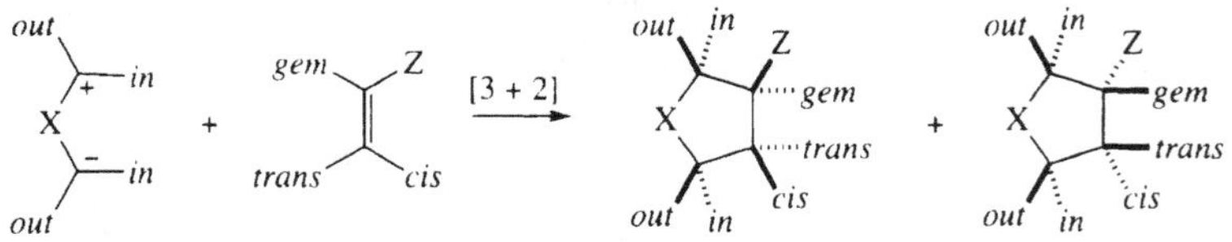

Which groups on the termini are in or out is less clear for 1,3-dipoles than it is for 1,3-dienes. The three atoms of the dipole form an arc. The terminal substituents pointing to the concave side of the arc are the in groups, and the substituents pointing to the convex side are the out groups.

The suprafaciality of the 1,3-dipole component is important only when both ends of the dipole are C(sp^2) atoms (i.e., in azomethine ylides and carbonyl ylides). If either end of the dipole is N, O, or C(sp), the termini have no stereochemical relationship in either the dipole or the cycloadduct.

Example

Predict the stereochemistry of the product of the following reaction.

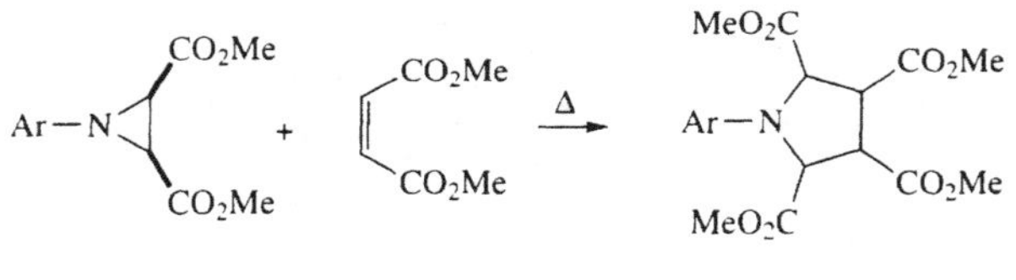

The product is clearly derived from a [3 + 2] cycloaddition. Disconnection of the appropriate bonds leads back to an azomethine ylide intermediate, which can be formed by electrocyclic ring opening of the aziridine.

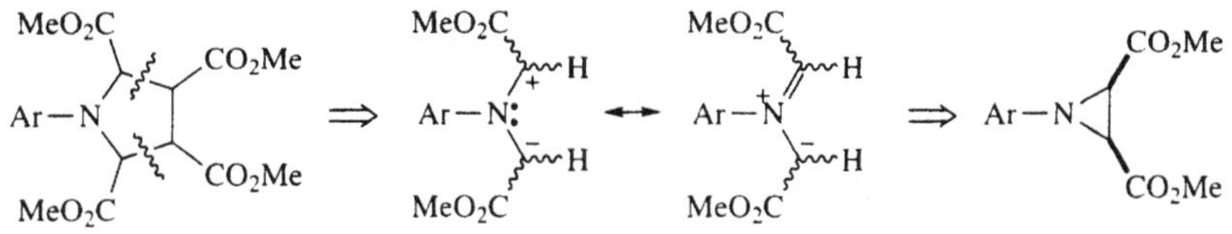

As for the stereochemistry, the electrocyclic ring opening of the aziridine is a four-electron process and is therefore conrotatory under thermal conditions. One CO_2Me group becomes an in group and one an out group. The suprafacial nature of the [3 + 2] cycloaddition means that these two groups become trans in the product. The two ester groups on the dipolarophile, of course, retain their stereochemistry. The orientation of the esters in the product is "down, up, up, up" (or vice versa).

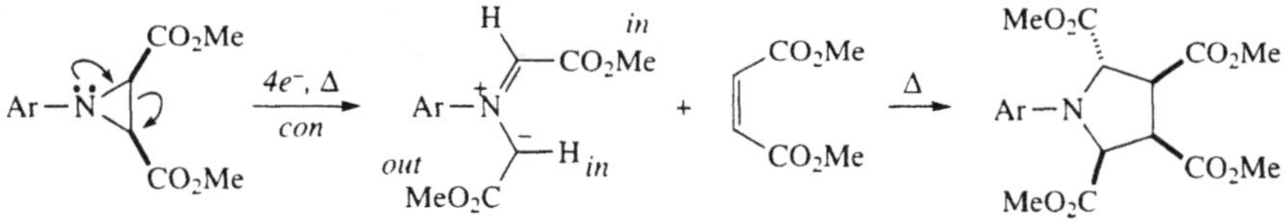

Under thermal conditions, the TS of the [2 + 2] cycloaddition is made up of ψ_1 (the LUMO) of one component and ψ_0 (the HOMO) of the other. Positive overlap between the orbitals at both termini of the π systems can be obtained only if one of the components reacts antarafacially. This orientation is very difficult to achieve geometrically, and hence [2 + 2] cycloadditions do not normally proceed under thermal conditions. However, under photochemical conditions, one of the components has an electron promoted from ψ_0 to ψ_1. Now the HOMO–LUMO interaction is between ψ_1 of the photoexcited component and ψ_1 of the unexcited component, and thus both components can be suprafacial in the TS. The [2 + 2] cycloaddition of most alkenes and carbonyl compounds do in fact proceed only upon irradiation with light.

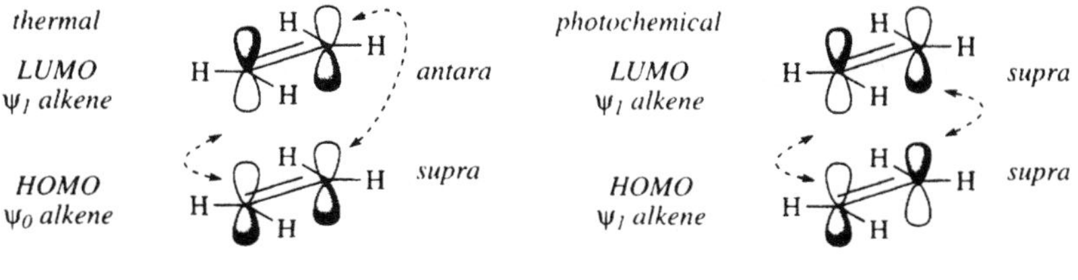

Three classes of [2 + 2] cycloadditions do proceed under thermal conditions. Ketenes ($R_2C=C=O$) undergo concerted cycloadditions to alkenes under thermal conditions because the ketene can react *antarafacially* with an alkene that reacts *suprafacially*. The two termini of the $C=C$ π bond of the ketene react from opposite faces of the π bond, creating positive overlap between the orbitals at both termini of the two π systems. The antarafacial nature of the ketene does not have any stereochemical consequences, as there is no cis–trans relationship in the ketene to preserve in the product. The alkene component of the [2 + 2] cycloaddition with ketenes, however, reacts suprafacially, and its stereochemistry is preserved in the product.

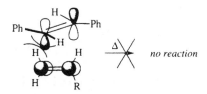

Why is it that ketenes are able to react antarafacially, and alkenes are not? After all, the π orbital of the ketene that reacts antarafacially is also present in an alkene. In ketenes, one of the C atoms has only the sterically undemanding O atom attached to it. Common alkenes have sterically demanding substituents on both ends of the antarafacial component. The substituents at one end of the antarafacial alkene jut directly into the path of the other alkene in the TS, sterically inhibiting the reaction.

The second kind of thermally allowed [2 + 2] cycloaddition occurs when one of the atoms involved is a second-row or heavier element, as in the Wittig reaction. Whether the first step of the Wittig reaction actually proceeds in a concerted fashion is a matter of debate, but the point here is that a concerted mechanism is a reasonable possibility. Moreover, there is no controversy over whether the second step of the mechanism is a concerted [2 + 2] retro-cycloaddition.

Why aren't the [2 + 2] cycloaddition and retro-cycloaddition of the Wittig reaction disallowed? The Woodward–Hoffmann rules state that when the symmetries of the MOs of the reactants are mismatched in ways that we have discussed, the TS for the reaction is raised very high in energy, and the reaction is therefore "disallowed." In principle, even a disallowed reaction can proceed at sufficiently high temperatures; the reason this approach usually doesn't work is that other reactions usually take place at temperatures far below those required for the disallowed process. (However, even a symmetry-disallowed reaction involving only C=C π bonds can proceed at reasonable temperatures if it is particularly favorable, as in the electrocyclic ring opening of Dewar benzene to benzene.) Now, the Wittig reaction proceeds at ordinary temperatures because when π bonds to elements heavier than C, N, or O are involved in pericyclic reactions, factors other than mismatches in MO symmetries become much more important to the energy of the TS. In other words, the concerted mechanism for the Wittig reaction *is* disallowed by the Woodward–Hoffmann symmetry rules, but it is not *as* disallowed as the [2 + 2] cycloadditions of C=C or C=O π bonds. Other factors, including the poor overlap of the orbitals comprising the C=P π bond, the very high energy of that bond, and the very low energy of the P–O σ bond, lower the energy of the TS of the Wittig reaction enough that it proceeds at ordinary temperatures.

Metal alkylidenes (M=CR$_2$) undergo thermally allowed [2 + 2] cycloadditions with alkenes during *olefin metathesis* (see Chapter 6) and other reactions for similar reasons.

In the third class of thermally allowed [2 + 2] cycloadditions, a very electron rich alkene is allowed to react with a very electron poor diene. These reactions almost certainly proceed stepwise through a zwitterionic or diradical intermediate, so they are not pericyclic reactions, and no violation of the Woodward–Hoffmann rules occurs. [2 + 2] Cycloadditions that proceed by a stepwise mechanism are nonstereospecific.

To summarize, most [2 + 2] cycloadditions are light-promoted. The only concerted thermal [2 + 2] cycloadditions involve a ketene or other cumulene, or a compound in which a heavy atom such as P or a metal is doubly bound to another element.

* **Common error alert:** *If neither component of a thermal [2 + 2] cycloaddition is a member of these classes, the cycloaddition must proceed by a stepwise mechanism.*

Example

Provide two reasons why the following mechanism for the formation of endiandric acid A is unreasonable.

endiandric acid A

Reason 1: [2 + 2] Cycloadditions do not occur under thermal conditions unless one of the components is a ketene or has a heavy atom. No light is specified in this reaction, and ambient light is not sufficient to promote a [2 + 2] cycloaddition. Reason 2: Both π bonds that participate in the purported [2 + 2] cycloaddition have trans H substituents. In the product, though, the two bottom H atoms of the cyclobutane are cis. Therefore, one of the π bonds would have to be reacting antarafacially, which is geometrically impossible.

The Woodward–Hoffmann rules for cycloadditions (Table 4.4) are as follows. Both components of a cycloaddition involving an odd number of electron pairs are suprafacial under thermal conditions; under photochemical conditions, one component must be antarafacial. Both components of a cycloaddition involving an even number of electron pairs are suprafacial under photochemical conditions; under thermal conditions, one component must be antarafacial.

TABLE 4.4. Woodward–Hoffmann Rules for Cycloadditions

Number of electron pairs	Δ	$h\nu$
Odd	suprafacial–suprafacial	suprafacial–antarafacial
Even	suprafacial–antarafacial	suprafacial–suprafacial

Photochemical suprafacial-antarafacial reactions are very rare because the geometrically simpler thermal reaction is likely to occur instead.

Problem 4.13. What do the Woodward–Hoffmann rules suggest about the facial reactivity of the components of the following thermal [6 + 4] cycloaddition and thermal [4 + 3] cationic cycloaddition?

The application of the Woodward–Hoffmann rules to cheletropic reactions is not straightforward. In the [2 + 1] cycloaddition of singlet carbenes to alkenes, the stereochemistry of the alkene is preserved in the product, so the alkene must react suprafacially. The Woodward–Hoffmann rules suggest that the carbene component of this thermal, four-electron reaction must react antarafacially. However, what this means for a species lacking a π system is difficult to interpret.

[2 + 1] cycloaddition suprafacial with respect to alkene

The [4 + 1] retro-cycloaddition proceeds suprafacially with respect to the four-atom unit, with cis substituents on the terminal carbons of the four-atom unit becoming out in the product. The one-atom component must react suprafacially, but again, what this means for a species lacking a π system is difficult to interpret.

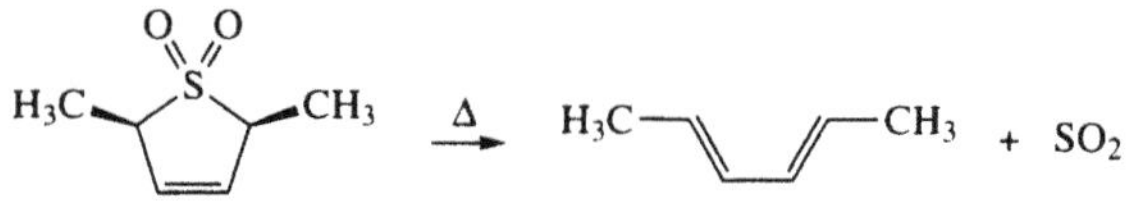

Not all cheletropic reactions proceed suprafacially with respect to the larger component. In the following [6 + 1] retro-cycloaddition, the cis Me groups become out and in in the product. Results such as these are difficult to rationalize or predict a priori.

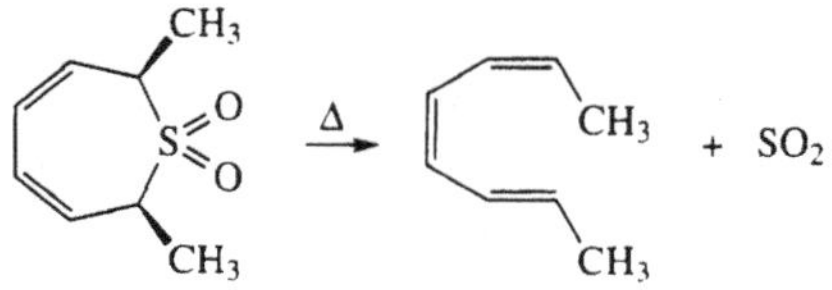

4.3.4   Stereoselectivity

Consider the Diels–Alder reaction between 1-methoxybutadiene and ethyl acrylate. The major product has the MeO and CO_2Et groups on adjacent C atoms. A reaction that is stereospecific with respect to each component could give either the cis or the trans product. The TS leading to the product in which the substituents are trans is clearly less sterically encumbered than the other TS, and so one would predict that the trans product is predominantly obtained. However, the major product is the one in which the groups are cis.

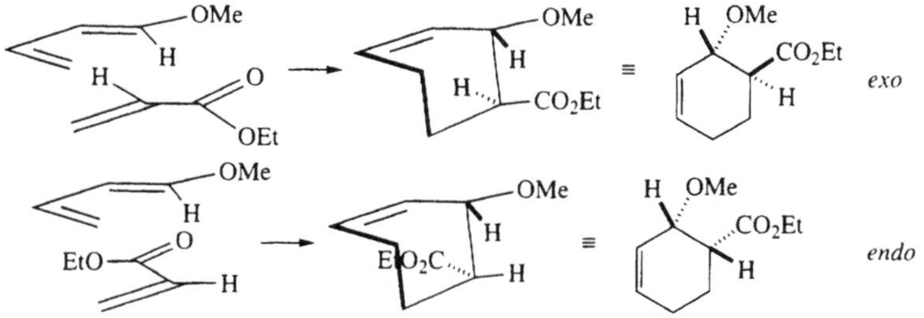

Diels–Alder reactions generally proceed selectively via the TS in which the most powerful electron-withdrawing group on the dienophile is *endo*, i.e., sitting underneath the diene, as opposed to pointing away from it. This phenomenon is known as the *endo rule*.

The Diels–Alder reactions of dienophiles with cyclopentadiene to give norbornenes were among the earliest studied. The fact that the endo norbornene (the one in which the electron-withdrawing group derived from the dienophile was cis to the two-carbon bridge) was always obtained was called the endo rule. The rule was later extended to Diels–Alder reactions not involving cyclopentadiene, and the terms exo and endo then changed their meaning.

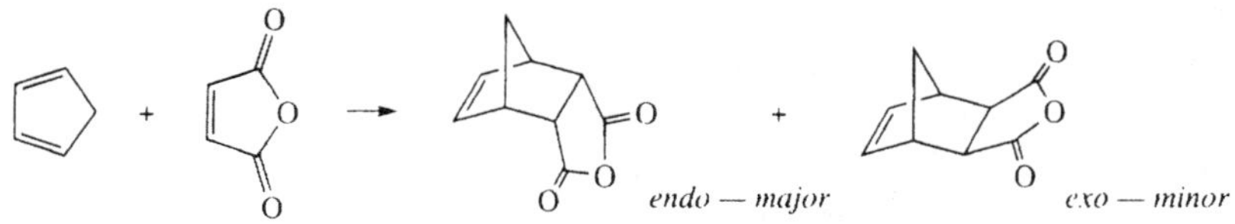

The *out-endo-cis rule* is a device for drawing the products of Diels–Alder reactions with stereochemistry consistent with the endo rule. The first word refers
to the orientation of a group on a terminus of the diene. The second word refers
to a substituent on the dienophile (usually the most electron withdrawing one).
The third word gives the stereochemical relationship of these two groups in the
product. Thus, the out group on the diene and the endo group on the dienophile
are cis in the product. Either of the first two words can be toggled to its opposite, as long as the third word is toggled, too: out-exo-trans, in-endo-trans, and
in-exo-cis. To apply the out-endo-cis rule, you must determine which group on
the dienophile is the endo group. Application of the out-endo-cis rule then gives
the stereochemistry of the major product.

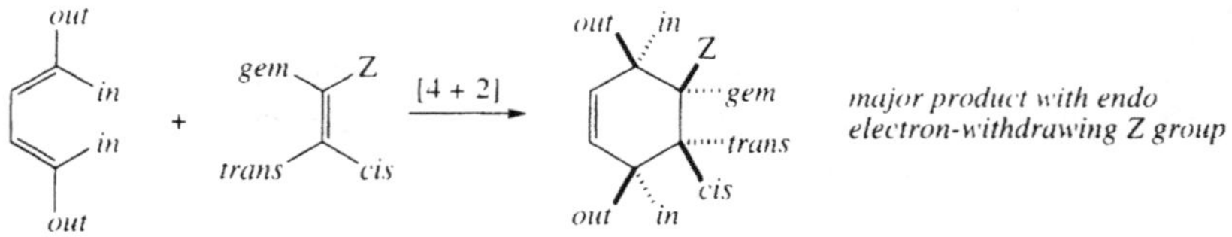

Example

Predict the stereochemistry of the major product of the following Diels–Alder
reaction.

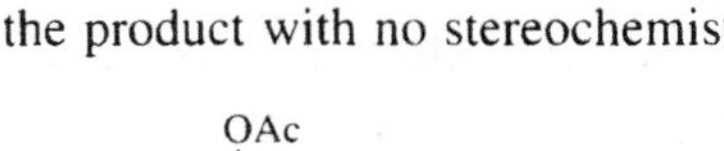

First redraw the starting material and product so that they are in the proper
conformation (s-cis for the diene) and mutual orientation. Orient the two so
that the strongest electron donating group on the diene is in a 1,2- or 1,4-
relationship with the strongest electron withdrawing group on the dienophile.
Then draw the product with no stereochemistry indicated.

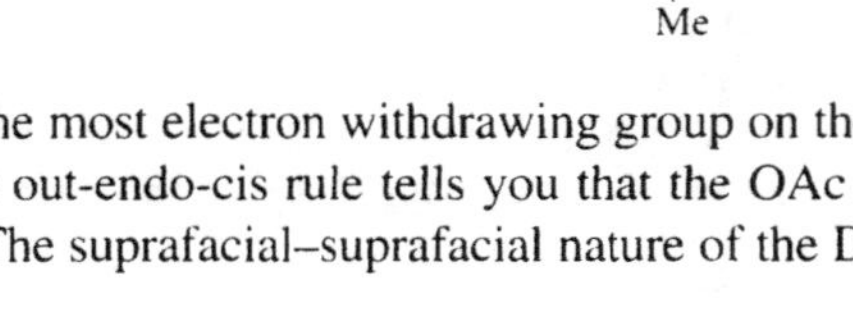

Stereochemistry: The most electron withdrawing group on the dienophile prefers
to be endo, and the out-endo-cis rule tells you that the OAc and CN groups are
cis in the product. The suprafacial–suprafacial nature of the Diels–Alder reaction

tells you that the two out groups on the diene become cis, and that the stereo-chemistry of the dienophile is preserved in the product, so the OAc group (out) and the Me group opposite it (in) are trans in the product, and the CN group and the Me group next to it are also trans in the product. Draw the OAc up (or down, it doesn't matter), and the rest of the stereochemistry follows.

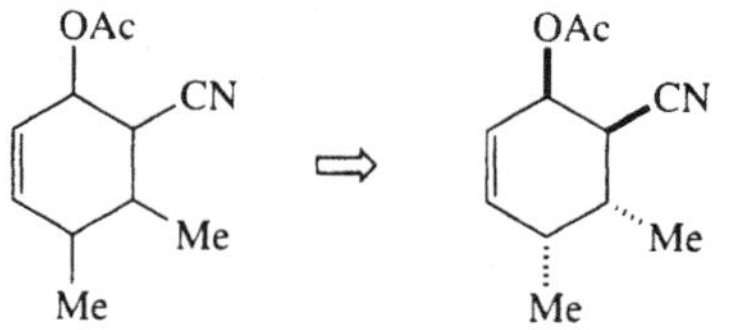

* **Common error alert:** *Be certain to preserve the stereochemistry about all three π bonds!* When a diene is given in its s-trans conformation, as in the preceding example, students often isomerize the π bonds inadvertently when they try to re-draw it in its s-cis conformation. A correct application of the out-endo-cis rule then leads to an incorrect answer. This common error can be prevented by obey-ing Grossman's rule.

The endo rule applies equally to inverse electron-demand Diels–Alder reac-tions. In these reactions, the most electron donating group on the dienophile is preferentially endo. The out-endo-cis rule applies, too.

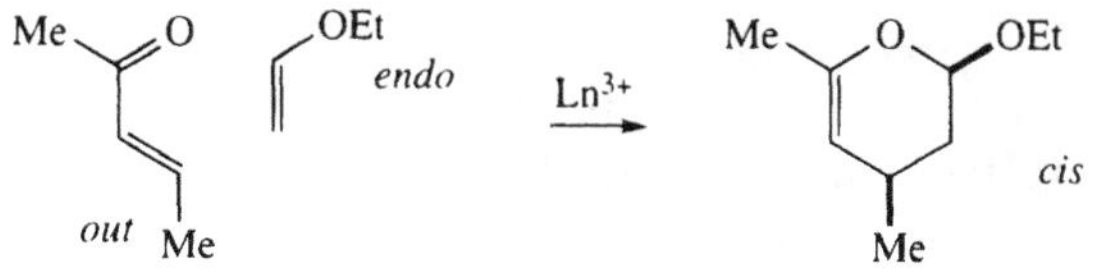

The endo/exo ratio can vary a lot depending on reaction conditions and the substrates. The ratio increases in favor of the endo product when Lewis acids are used to accelerate the Diels–Alder reaction. Increased steric interactions can turn the tide in favor of the exo product. In intramolecular Diels–Alder reactions, the ratio depends more on conformational preferences than anything else, and either exo or endo products may be obtained predominantly, depending on the reaction conditions and the particular substrate.

Why is the more sterically encumbered TS lower in energy? The most widely accepted explanation cites *secondary orbital interactions*. In the more crowded approach, the orbitals of the carbonyl group of the dienophile can interact with the orbital on C2 of the diene. The secondary orbital interactions in the endo TS are energetically favorable, so the kinetic product is the more crowded, less ther-modynamically stable endo product.

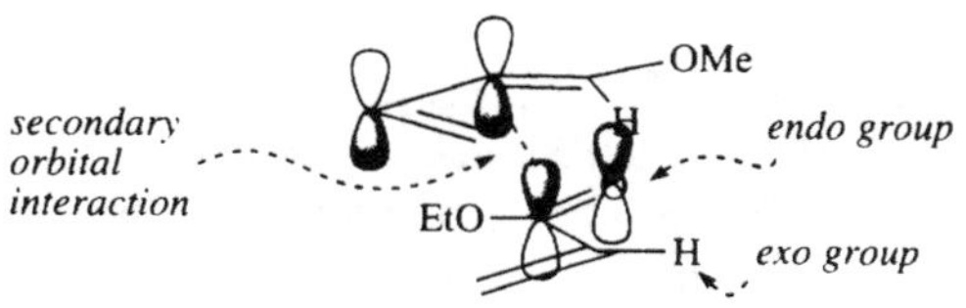

The preference for the endo TS can also be rationalized by invoking effects other than secondary orbital interactions. For example, the dipoles associated with the in C–H bond of the diene and the electron-withdrawing group of the dienophile interact most favorably when the electron-withdrawing group is endo. Lewis acids increase the endo selectivity by polarizing the electron-withdrawing group and thus increasing the magnitude of the dipole. The endo selectivities of Diels–Alder reactions of certain substrates can also be explained by steric and solvent effects.

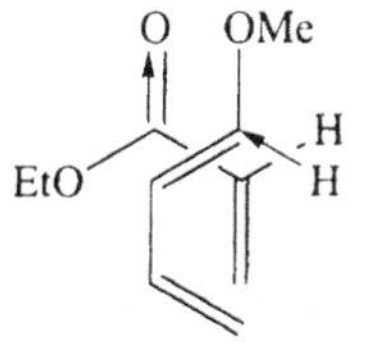

endo TS: dipoles aligned favorably 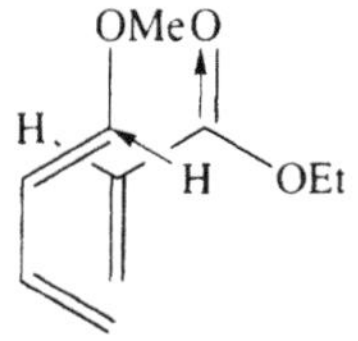   *exo TS: dipoles aligned unfavorably*

Endo selectivity is a kinetic phenomenon. If an equilibrium is established between the higher energy endo and lower energy exo products, then predominantly exo products will be seen. The cycloaddition of maleic anhydride and furan proceeds very rapidly, even at very low temperatures, to give only the exo product. The unusually low energy of furan (an aromatic compound) allows the retro-Diels–Alder reaction of the endo product to proceed at a reasonable rate. Even though the rate of formation of the endo product is faster than the rate of formation of the exo product, establishment of an equilibrium between starting materials and products leads to a thermodynamic ratio that favors the exo product.

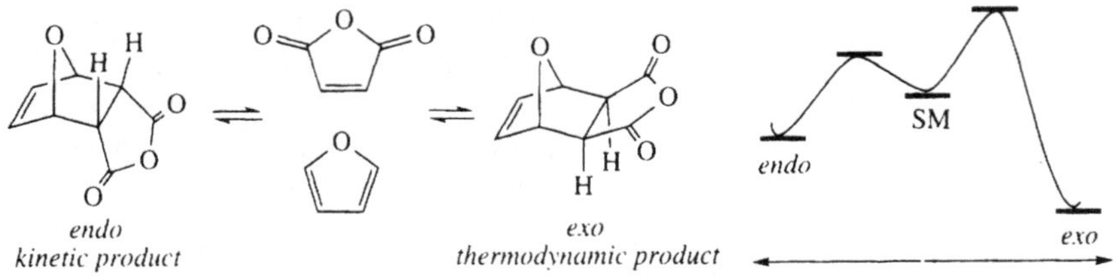

endo
kinetic product

exo
thermodynamic product

1,3-Dipolar cycloadditions give predominantly endo products, too. Again, the out-endo-cis rule applies.

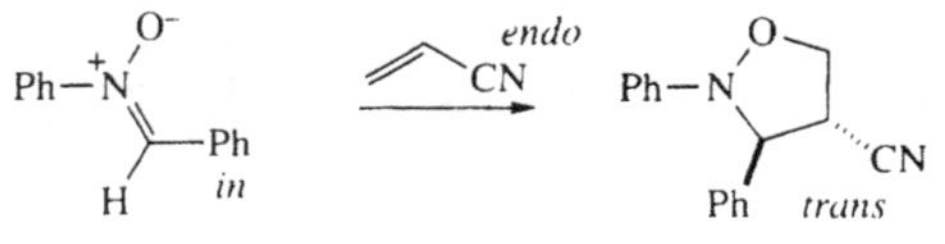

[2 + 2] Photocycloadditions also proceed to give predominantly endo products. For example, the light-induced dimerization of thymidine occurs with endo selectivity. The carbonyl group of one ring positions itself over the other ring, and the more sterically crowded stereoisomer is obtained.

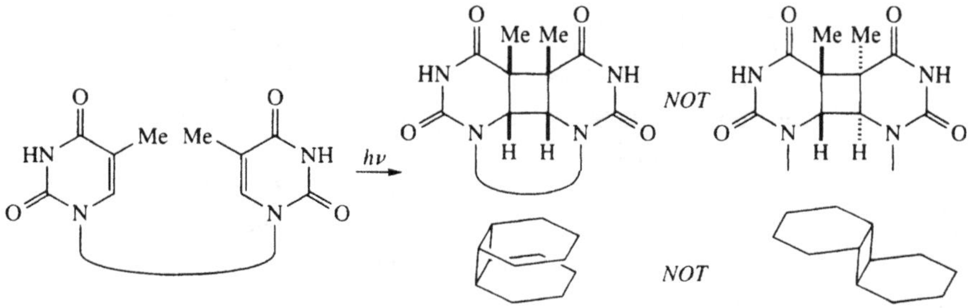

In the [2 + 2] cycloaddition of monosubstituted ketenes to cycloalkenes, two products can also be obtained. The thermodynamic product, where the R group is on the convex face of the bicyclic system, is *not* obtained predominantly. In fact, the reaction is "masochistic": the larger the R group, the greater the proportion of the thermodynamically less stable product.

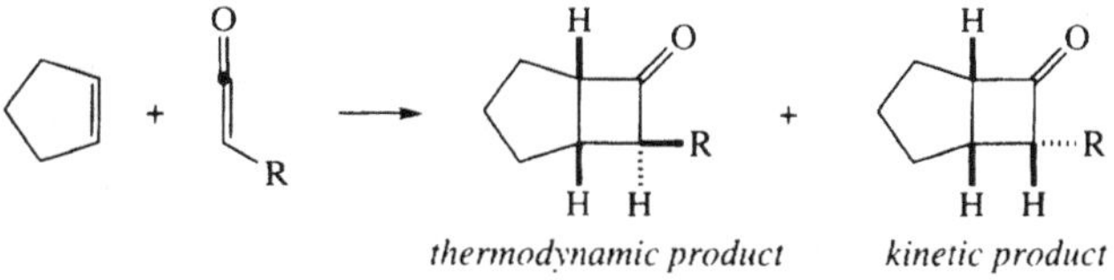

thermodynamic product   *kinetic product*

Because the ketene must react antarafacially, the alkene approaches the ketene with the two π bonds nearly perpendicular to each other. The smaller portion of the ketene, the C=O group, sits under the ring of the alkene, whereas the larger portion, the CHR group, resides farther away. The H and R groups may be oriented with respect to the alkene in two ways. In the lower energy orientation, the H substituent on the ketene points up toward the alkene, while the larger R group points down. The antarafacial reactivity of the ketene causes this lower energy orientation to lead to the higher energy product, in which the R group is on the *concave* face of the bicyclic system. The larger the R group, the lower the relative energy of the TS that has the R group pointing away from the cycloalkene, and the greater the proportion of higher energy product.

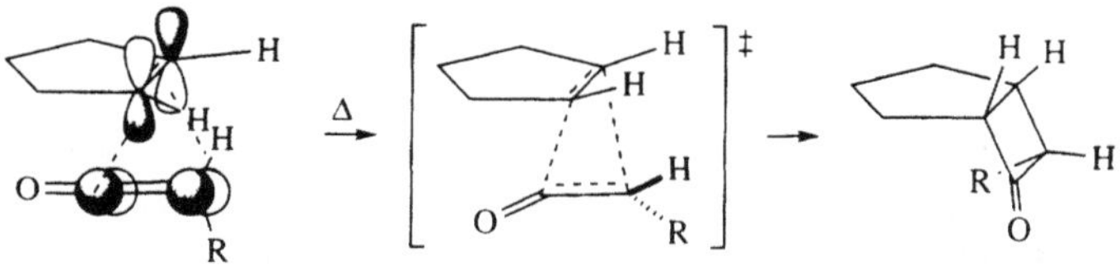

4.4 Sigmatropic Rearrangements

4.4.1 Typical Reactions

A sigmatropic rearrangement produces a new σ bond at the expense of a σ bond, so this reaction is the most inherently reversible of all pericyclic reactions. The position of the equilibrium depends on the relative thermodynamic and kinetic stabilities of the starting material and products. Most synthetically useful sigmatropic rearrangements are two- or six-electron processes.

The [3,3] sigmatropic rearrangements (Cope and Claisen rearrangements) are the most widely used sigmatropic rearrangements and are probably the most widely used pericyclic reactions after the Diels–Alder reaction. In the Cope rearrangement, a 1,5-diene isomerizes to another 1,5-diene. In the Claisen rearrangement, an allyl vinyl ether (a 1,5-diene in which one C atom is replaced with O) isomerizes to a γ,δ-unsaturated carbonyl compound (another 1,5-diene in which one C atom is replaced with O). Both the Cope and Claisen rearrangements normally require temperatures of 150 °C or greater to proceed, although certain types of substitution can lower the activation barrier.

Cope
rearrangement   *Claisen*
                                                        rearrangement

The Cope rearrangement of the simplest 1,5-diene, 1,5-hexadiene, is degenerate: the starting material is identical with the product, and the equilibrium constant for the rearrangement is 1. Substituents may shift the equilibrium to one side or the other. For example, the equilibrium between 3,4-dimethyl-1,5-hexadiene and 2,6-octadiene lies on the side of the more substituted π bonds.

The position of the equilibrium of the Cope rearrangement is pushed even further toward one side when cleavage of the σ bond relieves ring strain. The relief of ring strain also lowers the activation barrier for the rearrangement. For these reasons, *cis*-1,2-divinylcyclopropane is stable only at very low temperatures.

The position of the Cope equilibrium can also be altered by removing the product 1,5-diene from the reaction mixture. In the *oxy-Cope rearrangement*, a 3-hydroxy-1,5-diene undergoes the Cope rearrangement to give an enol, which isomerizes quickly to a δ,ε-unsaturated carbonyl compound. The latter compound is a 1,6-diene, not a 1,5-diene, so it is incapable of undergoing the Cope rearrangement in the retro direction. The reverse isomerization of the δ,ε-unsaturated carbonyl compound to the enol does not occur quickly enough for the retro-Cope rearrangement to proceed.

Oxy-Cope rearrangements proceed at especially low temperatures when the alcohol is deprotonated. The *anionic oxy-Cope* rearrangement is accelerated compared with the neutral reaction because the negative charge is more delocalized in the TS than in the starting material. The driving force for the anionic oxy-Cope rearrangement is no longer removal of the product diene from the equilibrium but simply delocalization of the negative charge. The starting alcohol is usually deprotonated by KH because the O–K bond is much weaker than the O–Na or O–Li bond; the crown ether 18-crown-6 is often added to isolate the K^+ counterion from the alkoxide even further.

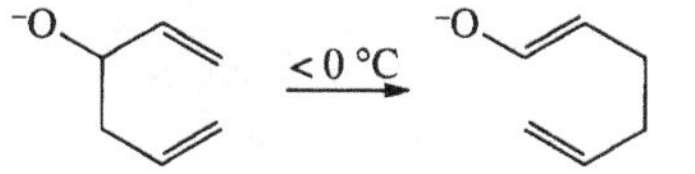

negative charge more delocalized
in TS and product than in SM

The Claisen rearrangement is driven in the forward direction by formation of a carbonyl π bond. The Claisen rearrangement originally referred only to the isomerization of an *O*-allylphenol to a 2-allylphenol. One might expect that in this particular case, the equilibrium would lie on the side of the aromatic compound, not the carbonyl. However, the carbonyl quickly tautomerizes (by a nonconcerted mechanism!) to the aromatic 2-allylphenol, which can't undergo the reaction in the reverse direction.

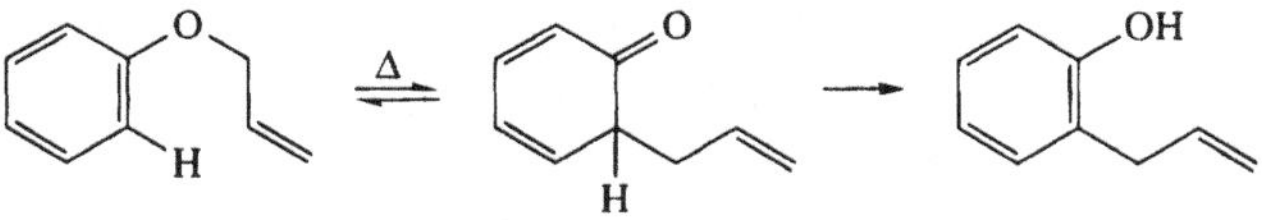

The key to identifying Cope and Claisen rearrangements is the 1,5-diene in the starting material or in the product. A γ,δ-unsaturated carbonyl compound (a 1,5-heterodiene) can be made by a Claisen rearrangement, and a δ,ϵ-unsaturated carbonyl compound can be made by an oxy-Cope rearrangement.

Example

The orthoester Claisen rearrangement.

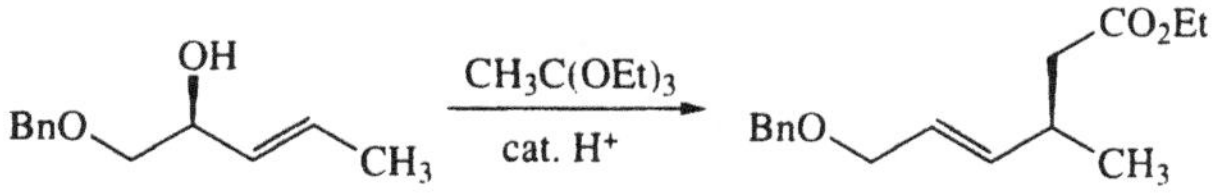

Label the atoms and draw the by-products.

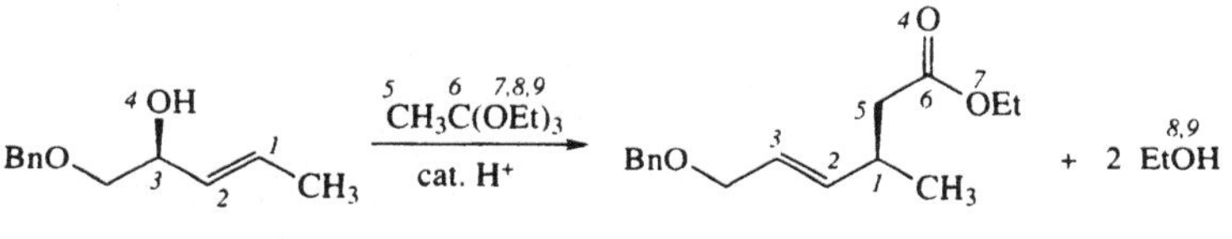

Make: C1–C5, O4–C6. Break: C3–O4, C6–O8, C6–O9. The product has a 1,5-disposition of two π bonds, C2=C3 and O4=C6. Working retrosynthetically, the C1–C5 bond can be made and the C3–O4 bond broken by a [3,3] sigmatropic (Claisen) rearrangement of a 1,5-diene precursor.

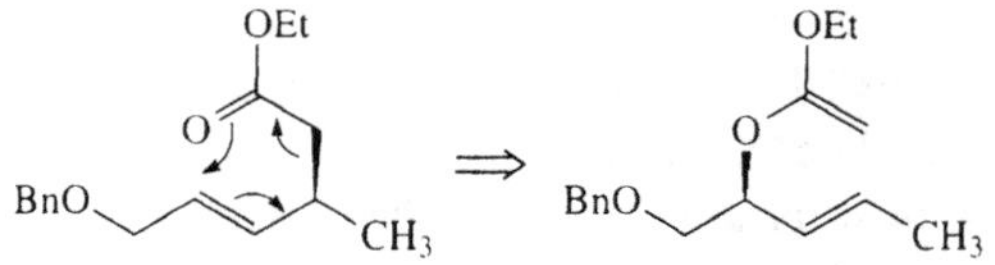

The immediate precursor to the product can be made from the alcohol and the orthoester by S_N1 substitution followed by E1 elimination.

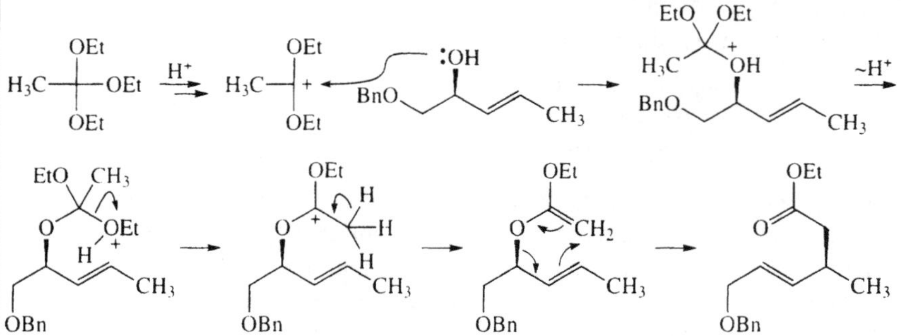

The [1,5] sigmatropic rearrangement of H atoms around a cyclopentadienyl group is an extremely facile process because the ends of the pentadienyl system are held closely to one another. The different isomers of a substituted cyclopentadiene are in rapid equilibrium (on the laboratory time scale) above 0 °C. A synthesis of prostaglandins was developed in the late 1960s in which the first two steps were (1) alkylation of cyclopentadiene with benzyloxymethyl bromide to give a 5-substituted cyclopentadiene, and (2) a Diels–Alder reaction. A dienophile that would react with the cyclopentadiene below −20 °C was sought, because at higher temperature the cyclopentadiene isomerized into its more stable 1- and 2-substituted isomers.

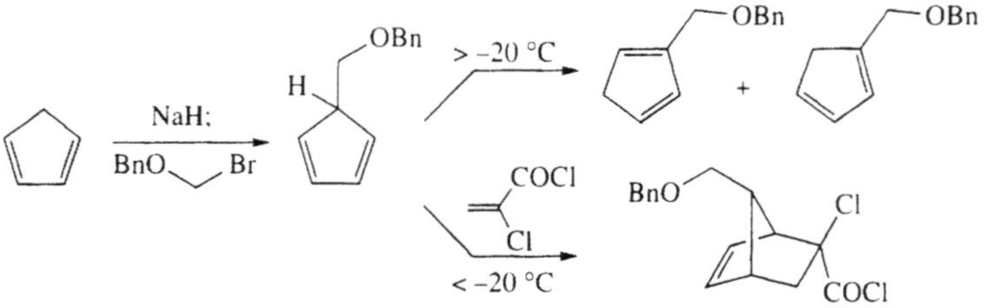

The [1,5] *alkyl* shift is also seen in cyclopentadienes, but much higher temperatures (usually >200 °C) are required. The C(sp^3) orbital is much more directional than the H(s) orbital, so the simultaneous overlap with two orbitals that is required in the TS is not as favorable as it is with H.

Problem 4.14. The following reaction involves [1,5] sigmatropic rearrangements. Draw a reasonable mechanism. Remember to obey Grossman's rule!

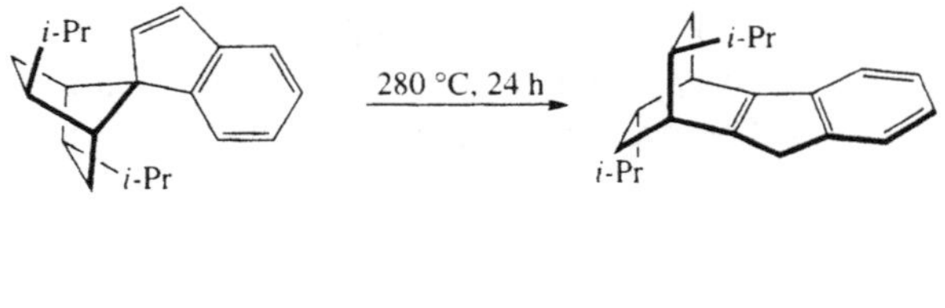

The [2,3] sigmatropic rearrangement involves the rearrangement of a 1,2-dipole to a neutral compound. An allyl group migrates from the positive end to the negative end of the dipole to neutralize the charges.

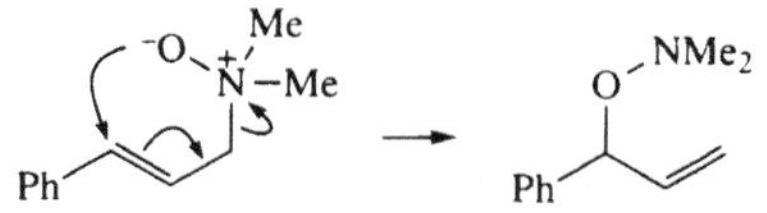

Amine oxides, sulfoxides, and selenoxides all undergo the [2,3] sigmatropic rearrangement. The equilibrium between allyl sulfoxides and allyl sulfenates lies on the side of the sulfoxide, but the equilibrium can be pushed toward the sulfenate by reduction of its O–S bond.

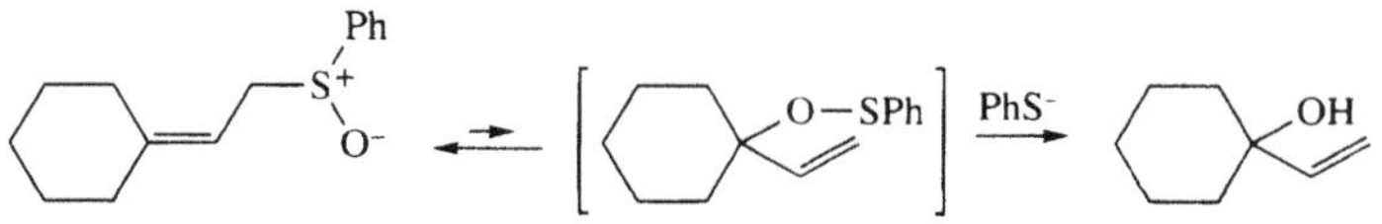

Sulfonium ylides (allylŜR–C̄R$_2$) and ammonium ylides (allylN̂R$_2$–C̄R$_2$) are also substrates for the [2,3] sigmatropic rearrangement. Both ylides are usually generated in situ by deprotonation of the sulfonium or ammonium salt.

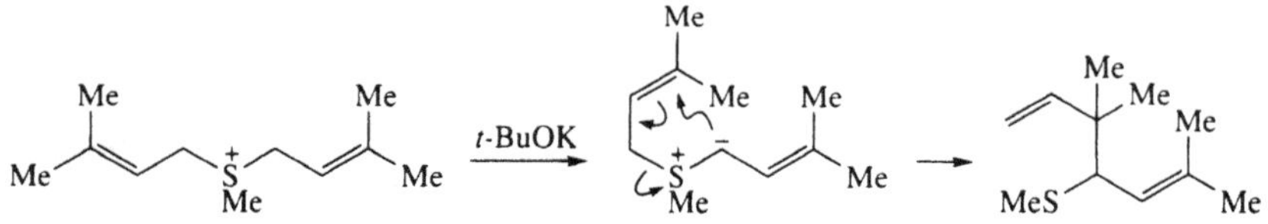

Problem 4.15. Draw a mechanism for the following [2,3] sigmatropic rearrangement. Why does this particular reaction proceed under such mildly basic conditions?

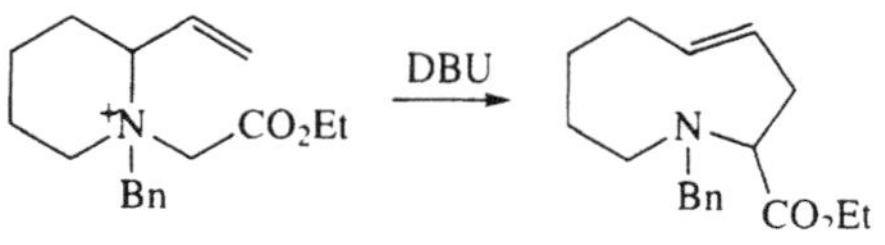

The [2,3] sigmatropic rearrangement of alloxycarbanions is known as the *Wittig rearrangement* (not to be confused with the Wittig reaction). The requisite carbanions can be prepared by transmetallation of a stannane (tin compound). Stannanes, unlike organolithium compounds, are stable, isolable, chromatographable compounds, but they are easily converted to the organolithium compound with BuLi.

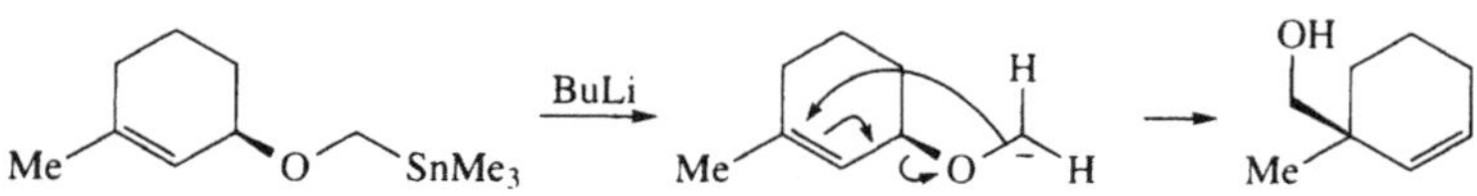

Allyl propargyl ethers are also good substrates for the Wittig rearrangement. Deprotonation of the propargyl position is relatively facile.

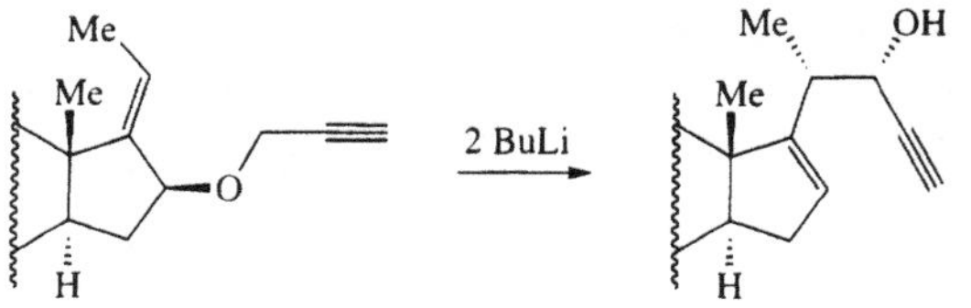

The key to identifying a [2,3] sigmatropic rearrangement is that an allylic group migrates from a heteroatom to an adjacent atom (which may be C or another heteroatom).

Problem 4.16. Upon mixing an *N*-chloroaniline and an α-mercaptoester, a nucleophilic substitution occurs, followed by a [2,3] sigmatropic rearrangement. An ortho-alkylated aniline is ultimately obtained. Draw the mechanism for this reaction.

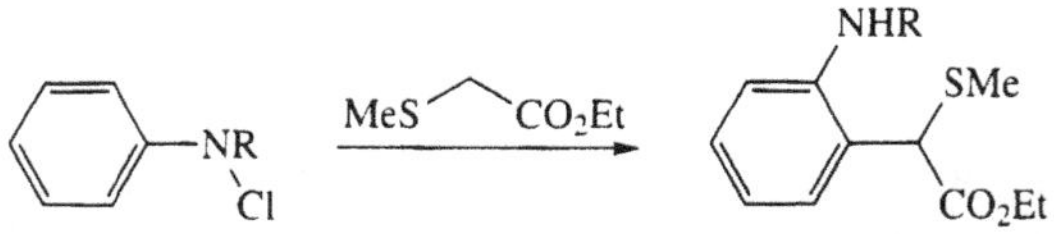

Four-electron, [1,3] sigmatropic rearrangements are very rare. Occasionally an alkyl group can undergo a concerted 1,3-shift, but H atoms *never* undergo concerted [1,3] sigmatropic rearrangements.

* **Common error alert:** *The tautomerization of an enol to a carbonyl compound appears to be a [1,3] sigmatropic rearrangement at first glance, but the reaction is catalyzed by base or acid and* never *proceeds by a concerted pericyclic mechanism.* These observations will be rationalized in the next section.

The [1,7] sigmatropic rearrangement, an eight-electron process, is also quite rare. A 1,3,5-hexatriene can undergo a six-electron electrocyclic ring closure more rapidly and with greater release of energy than it can a [1,7] sigmatropic rearrangement. A very important [1,7] sigmatropic rearrangement, though, occurs in the human body. Precalciferol (provitamin D_2) (which you may remember is made by a conrotatory electrocyclic ring opening of ergosterol) is converted to ergocalciferol (vitamin D_2) by a [1,7] sigmatropic H shift.

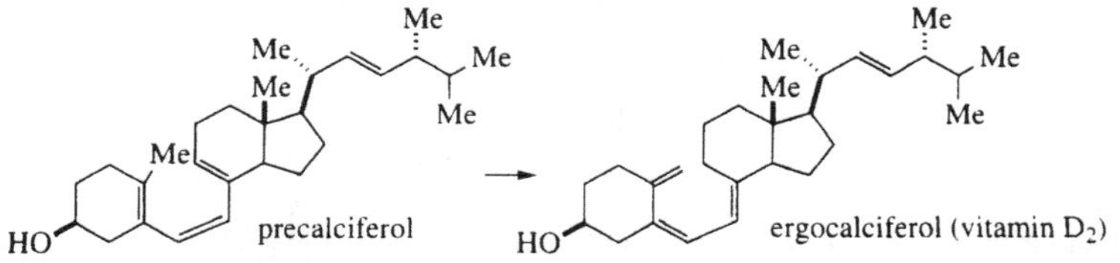

All the sigmatropic rearrangements discussed so far occur under thermal conditions. Photochemical sigmatropic rearrangements are extremely rare.

4.4.2   Stereospecificity

In a sigmatropic rearrangement, bonds are made and broken at the ends of two conjugated systems. The Woodward–Hoffmann rules for sigmatropic rearrangements must take both components into account.

The most common sigmatropic rearrangements are [1,2] cationic, [1,5], and [3,3] rearrangements. These rearrangements proceed readily under thermal conditions. In the cationic [1,2] hydride shift, a thermal reaction, the MOs of two components must be examined. The one-atom component, the H atom, has one orbital, the 1s orbital. The two-atom component has two orbitals, ψ_0 and ψ_1. Two electrons must be distributed among these three orbitals. It is clear that the H(1s) orbital is lower in energy than the antibonding orbital ψ_1, but it is not clear whether H(1s) or ψ_0 is lower in energy. If H(1s) is lower in energy, the two electrons will both go in H(1s); if ψ_0 is lower in energy, they will both go in ψ_0; and if the two MOs are equal in energy, one electron may go in each. Nevertheless, no matter how the two electrons are distributed between the two components, *the dominant HOMO–LUMO interaction in the TS is between the 1s orbital of the one-atom component and ψ_0 of the two-atom component*. The H atom is always classified as a suprafacial component, as the 1s orbital is monophasic, so positive overlap in the TS between orbitals where bond-making and bond-breaking take place is produced when the two-atom component is suprafacial.

[1,2] cationic supra–supra

Likewise, in the cationic [1,2] alkyl shift, both components must be suprafacial for there to be positive overlap in the TS between orbitals where bond-making and bond-breaking take place. The migrating group retains its configuration because of the requirement for suprafaciality.

[1,2] cationic supra–supra

The dominant FMO interaction in the TS of the [1,5] sigmatropic rearrangement, a six-electron reaction, is between the H(1s) orbital and ψ_2 of the five-atom component, no matter how the six electrons are divided between the two components. (An interaction between two half-filled orbitals is shown). Again, positive overlap in the TS between orbitals where bond-making and bond-breaking take place is produced when the five-atom component reacts suprafacially.

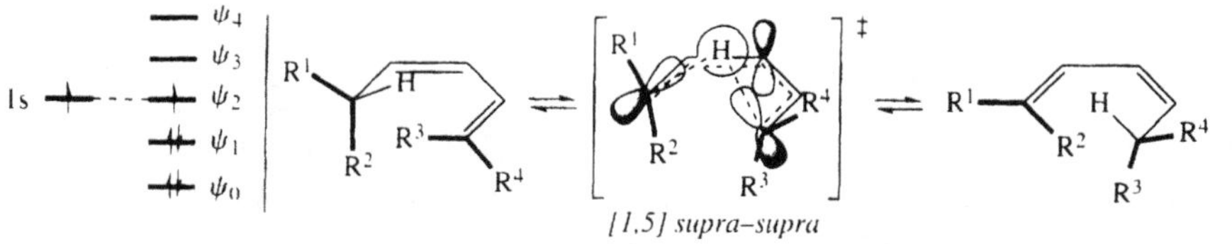

$[1,5]$ supra–supra

The [3,3] sigmatropic rearrangement is a six-electron reaction. No matter how the six electrons are distributed between the two three-atom components, the dominant FMO interaction in the TS is between ψ_1 of one component and ψ_1 of the other component. The reaction proceeds suprafacially with respect to both components.

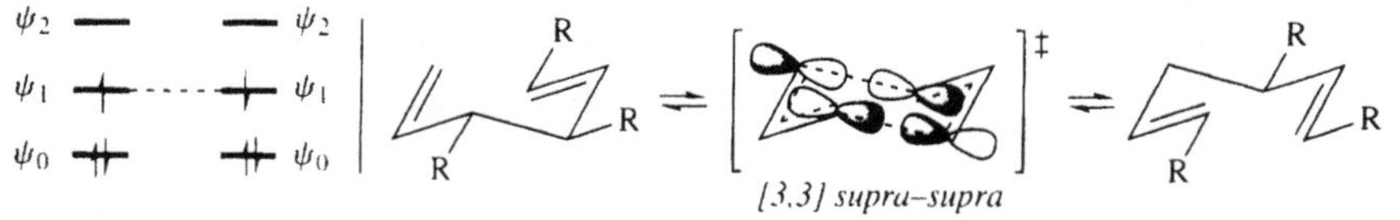

$[3,3]$ supra–supra

Likewise, the dominant FMO interaction in the TS of the [2,3] sigmatropic rearrangement is between ψ_1 of both components. The reaction proceeds suprafacially with respect to both components.

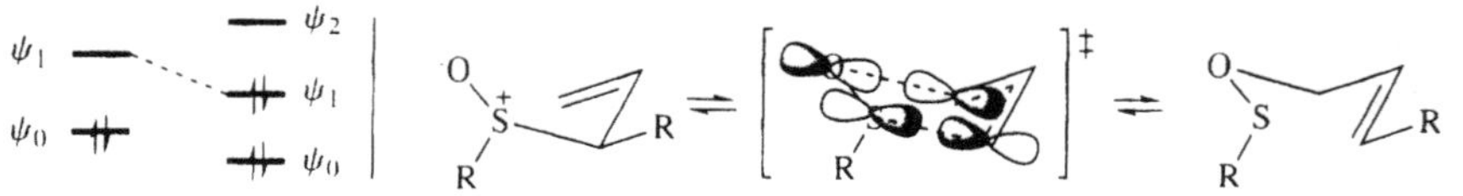

$[2,3]$ supra–supra

By contrast, in the [1,2] anionic H shift, the dominant FMO interaction is between the 1s orbital of the one-atom component and ψ_1 of the two-atom component. For there to be positive overlap between orbitals where bond-making and bond-breaking take place, the two-atom component must react antarafacially. Thus, for this shift to be a thermally allowed process, the H atom must have partial bonds to the top and bottom faces of the C_2 unit simultaneously. Because this arrangement is geometrically impossible, [1,2] anionic H shifts are thermally disallowed reactions.

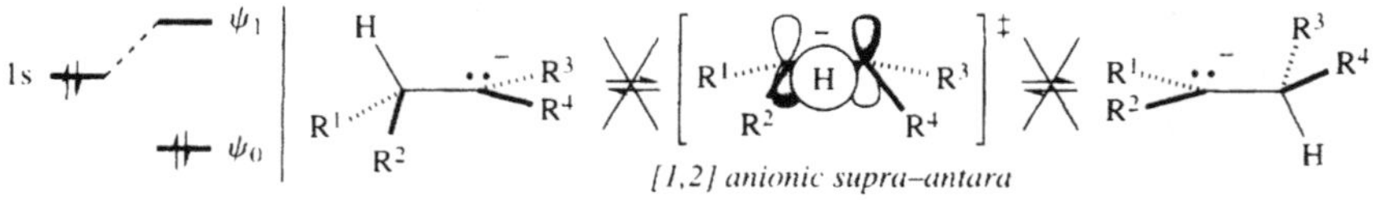

$[1,2]$ anionic supra–antara

A thermally allowed anionic [1,2] shift can occur if the one-atom component reacts antarafacially. In the case of the alkyl shift, the C(p) orbital of the migrating R group can react antarafacially, with the configuration inverting upon migration. In fact, the geometric requirements for anionic [1,2] alkyl shifts are so stringent that the reactions are extremely rare.

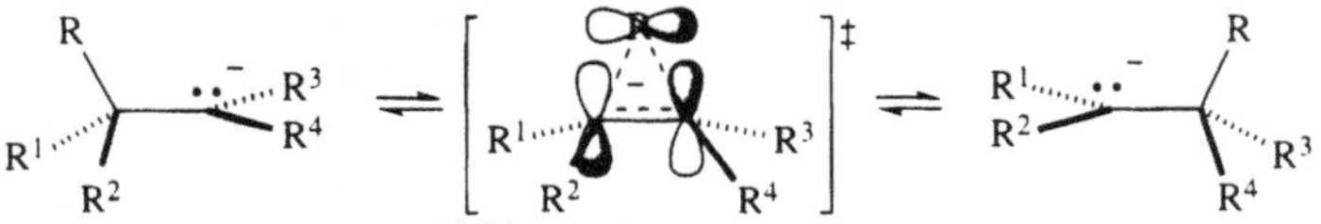

[1,2] anionic supra-antara

Thermal [1,3] H shifts such as the concerted rearrangement of enols to carbonyl compounds are disallowed. The allylic C–C–O unit itself can only react suprafacially, as it is geometrically impossible for the H(1s) orbital to bond simultaneously to a top lobe on one terminus and a bottom lobe at the other terminus, and the H atom itself must also react suprafacially, as the H(1s) orbital has only one lobe. The Woodward–Hoffmann rules, though, say that one of the two components of this four-electron rearrangement must react antarafacially for it to be allowed. Therefore this rearrangement reaction always proceeds through a nonconcerted mechanism and requires acidic or basic catalysis.

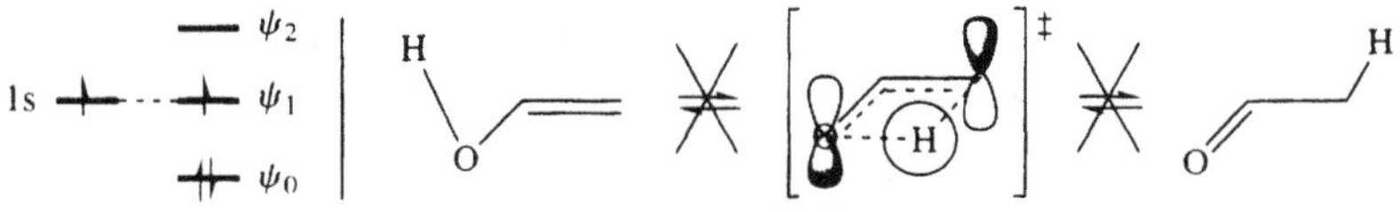

[1,3] antara–supra

* **Common error alert:** *A mechanism involving a concerted [1,3] sigmatropic rearrangement of H is almost always incorrect.*

On the other hand, a thermal [1,3] alkyl shift is allowed because the alkyl group can adopt an antarafacial orientation in the TS. The two lobes of the C(p) orbital of the migrating atom bond to opposite ends of the allylic system in the TS, and the configuration of the migrating group inverts.

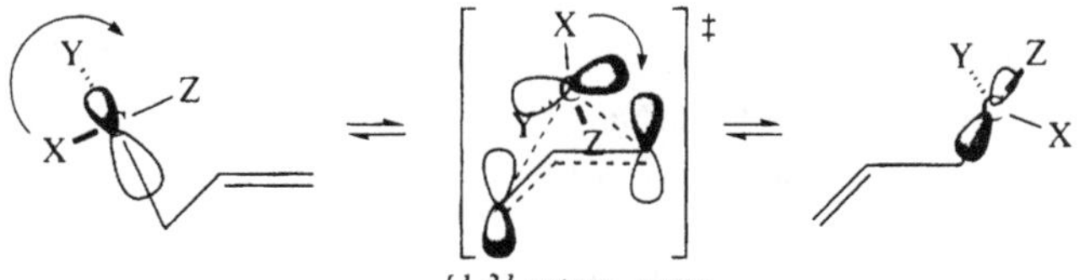

[1,3] antara–supra

Because the [1,3] sigmatropic rearrangement of alkyl groups has such stringent geometric requirements, it is quite rare. Only a handful of examples of [1,3] sigmatropic rearrangements are known. In one such example, the configuration of the migrating C inverts, demonstrating its antarafacial reactivity in the rearrangement.

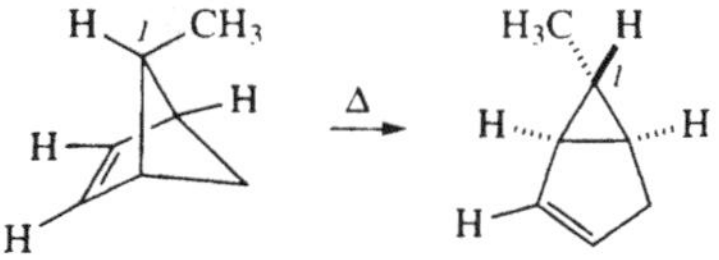

* **Common error alert:** *A mechanism involving a [1,3] sigmatropic rearrangement of an alkyl group, though not impossible, should be viewed with suspicion.*

A thermal [1,3] sigmatropic rearrangement is allowed only if one component is antarafacial, but a photochemical [1,3] sigmatropic rearrangement is expected to be allowed when it proceeds suprafacially with respect to both components. The stereochemical requirement changes because under photochemical conditions, the HOMO of the three-atom component is ψ_2 (symmetric), not ψ_1 (antisymmetric).

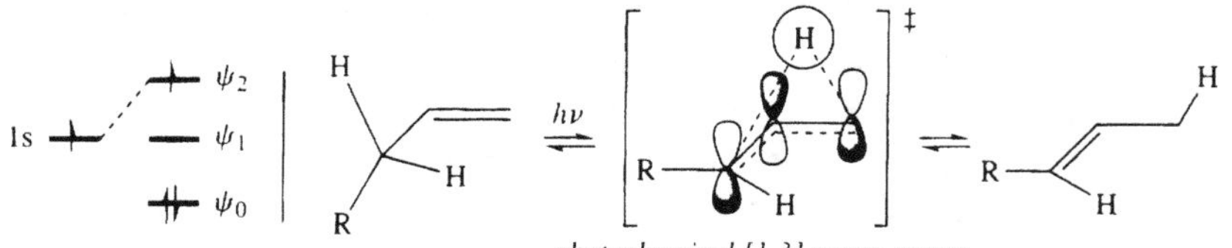

photochemical [1,3] supra–supra

Photochemical sigmatropic rearrangements are rare, but in one example, a photochemical [1,3] sigmatropic rearrangement proceeds suprafacially with respect to both components, resulting in retention of configuration about the migrating one-atom component, a stereogenic alkyl group. The reaction fails to proceed at all under thermal conditions.

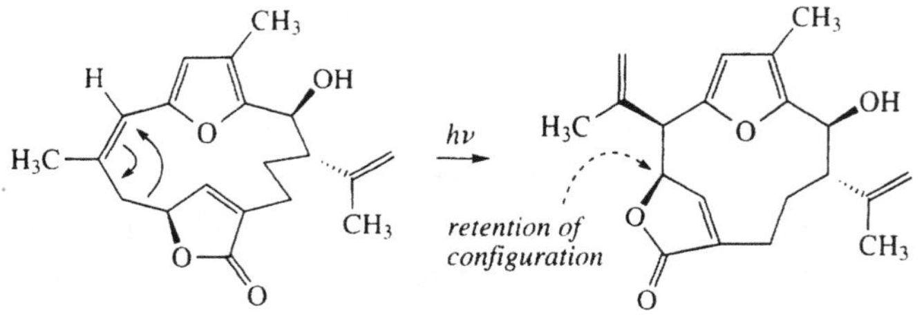

retention of configuration

A thermal, concerted [1,7] H shift is sometimes observed in acyclic systems because the 1,3,5-triene system is floppy enough to allow the H to migrate from the top face to the bottom face (making the triene the antarafacial component). The precalciferol–calciferol rearrangement is thermally allowed for this reason. (However, because precalciferol is formed in its excited state, it is quite possible that it undergoes a *photochemical* [1,7] sigmatropic rearrangement that is suprafacial with respect to both components.) In cyclic compounds like cycloheptatriene, though, geometric constraints prevent the H from migrating from the top to the bottom face of the seven-carbon π system, and the shift does not occur.

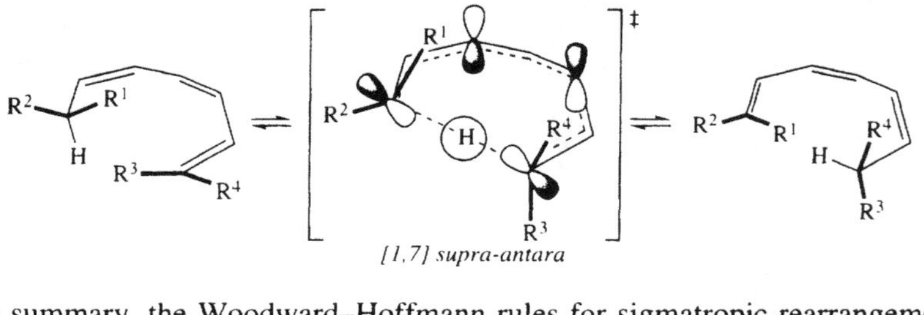

[1,7] supra–antara

In summary, the Woodward–Hoffmann rules for sigmatropic rearrangements (Table 4.5) are as follows. Both components of a sigmatropic rearrangement involving an odd number of electron pairs are suprafacial under thermal conditions; under photochemical conditions, one component must be antarafacial. Both components of a sigmatropic rearrangement involving an even number of electron pairs are suprafacial under photochemical conditions; under thermal conditions, one component must be antarafacial.

TABLE 4.5. Woodward–Hoffmann Rules for Sigmatropic Rearrangements

Number of electron pairs	Δ	hν
Odd	suprafacial–suprafacial	suprafacial–antarafacial
Even	suprafacial–antarafacial	suprafacial–suprafacial

The stereochemical consequences of facial selectivity are manifested most clearly in [3,3] sigmatropic rearrangements. Consider the orthoester Claisen rearrangement, in which an allylic alcohol is converted into a γ,δ-unsaturated ester. The intermediate ketene acetal undergoes a [3,3] sigmatropic rearrangement that is suprafacial with respect to both three-atom components, so the new bond between C1 and C6 and the old bond between C3 and C4 must both be to the same face of the C4–C5–C6 component. In practice, this means that because the C3–C4 bond in the starting material comes out of the plane of the paper, the C1–C6 bond in the product also comes out of the plane of the paper. The same argument holds for [2,3] sigmatropic rearrangements.

Problem 4.17. A small amount of 4-allylphenol is often obtained from the Claisen rearrangement of allyl phenyl ether. Draw a concerted mechanism for this reaction, name the mechanism, and determine whether the suprafacial–suprafacial rearrangement is thermally allowed or disallowed by the Woodward–Hoffmann rules. If it is not allowed, draw a multistep mechanism for the reaction.

Problem 4.18. Examples of the *Stevens rearrangement* and the nonallylic Wittig rearrangement are shown. What do the Woodward–Hoffmann rules say about the nature of these reactions? Offer two *explanations* (not necessarily mechanisms) of why these reactions proceed. (*Hint*: Think of the conditions on the applicability of the Woodward–Hoffmann rules.)

4.4.3   *Stereoselectivity*

Cope rearrangements are required to be suprafacial with respect to both components, but multiple stereochemical results are still possible. Cope rearrangements have a six-membered ring in the TS. The ring can be in one of four conformations (two chair and two boat). These different conformers can lead to different stereoisomeric products. Consider the following possible conformations of the disubstituted compound. Cope rearrangement via one chair conformation leads to a product in which the Ph-substituted π bond is cis and the CH_3-substituted π bond is trans. Rearrangement via the other chair conformation leads to a product in which the CH_3-substituted π bond is cis and the Ph-substituted π bond is trans. Rearrangement of the same compound via one boat conformation leads to a product in which both π bonds are trans. Rearrangement via the other boat conformation leads to a product in which both π bonds are cis. In other words, all four stereochemistries about the C=C bonds can be obtained from a single diastereomeric starting material. We must emphasize that all four possible stereochemistries are consistent with a suprafacial arrangement of both components in the TS.

chair TS 1

chair TS 2

boat TS 1

boat TS 2

Of course, one of the four possible TSs is usually quite a bit lower in energy than the others, and this leads to *stereoselectivity* in the Cope rearrangement. The chair TSs are preferred to the boat TSs, just as they are in cyclohexanes, and chair TS 2, in which the large Ph group is pseudoequatorial and the smaller Me is pseudoaxial, is lower in energy than chair TS 1, in which Ph is pseudoaxial and Me is pseudoequatorial. However, certain substitution patterns can cause one of the boat TSs to be lower in energy. For example, if the Me and Ph groups were replaced with *t*-Bu groups, which are much too large ever to be pseudoaxial, then boat TS 1 would be preferred.

Most Cope rearrangements, especially those of simple 1,5-dienes, proceed through a chair TS, and the stereochemical consequences are easily analyzed. For example, the following (*R*,*E*) allylic alcohol undergoes an orthoester Claisen re-arrangement to give mostly the (*E*,*R*) product; the alternative chair TS, which leads to the (*Z*,*R*) product, has a pseudoaxial BnOCH$_2$ group and so is higher in energy.

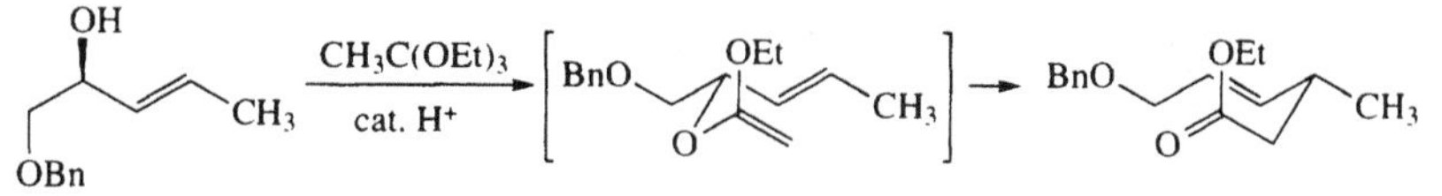

Example

Draw the product of the following oxy-Cope rearrangement (an aldehyde) with the correct stereochemistry.

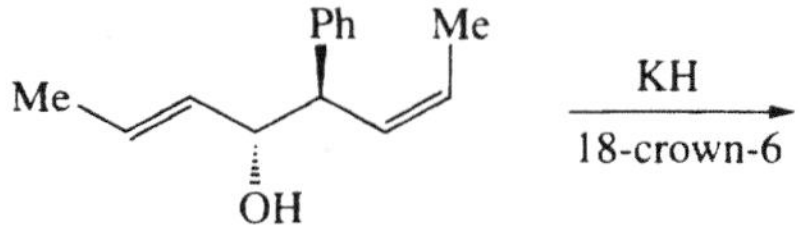

The first step is to draw the starting material in the proper conformation for the rearrangement, *without altering its stereochemistry*. The starting material is drawn in the figure such that the C2−C3 and C4−C5 bonds are in the s-trans conformation (antiperiplanar). Redraw the compound so that these two bonds are s-cis (eclipsed). After the rotation, the Ph group, formerly up, will now be down. Then rotate about the C2−C3 and C4−C5 bonds so that C1 and C6 are near one another, being careful to preserve the double bond geometry. Finally, rotate the entire structure so that the C3−C4 bond is vertical. You now have a *top view* of your starting material.

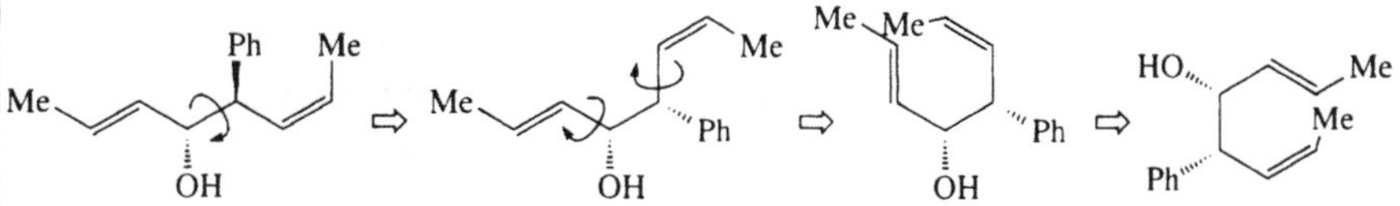

Now you need to draw the compound in its chair conformation. Draw a chair with only five sides, and then draw the two double bonds at the ends of the chain.

The easiest way to draw a five-sided chair is as follows. Draw a shallow V, draw an upside-down V of the same size below and to the left of the first V, and connect the ends of the two Vs on one side.

Although the termini of the 1,5-diene are sp^2-hybridized, their two substituents occupy pseudoaxial and pseudoequatorial positions. Draw the substituents of the double bonds in your chair, again being careful to preserve the double bond geometry. Note that the upper double bond in your *top view* drawing is now in the back of the chair, and the lower double bond is now in the front. Finally, draw the four bonds (two axial and two equatorial) to the C(sp³) centers of your chair, and then fill in the substituents, being careful that the "down" substituents point down and the "up" substituents point up. You have now drawn your compound in a correct chair conformation.

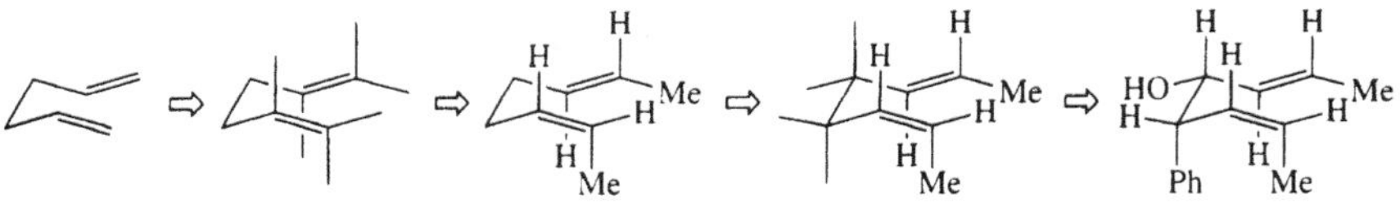

OOPS! The chair conformation you have drawn has the largest C(sp³) substituent, the Ph group, in an axial orientation! This means that you have not drawn the lowest energy chair conformation. The easiest solution to this problem is to switch the configurations of both C(sp³) centers so that you have the wrong enantiomer (but still the right diastereomer), draw the rearrangement reaction, and then switch the configurations of both C(sp³) centers in the product back to the correct enantiomeric series to give you the product.

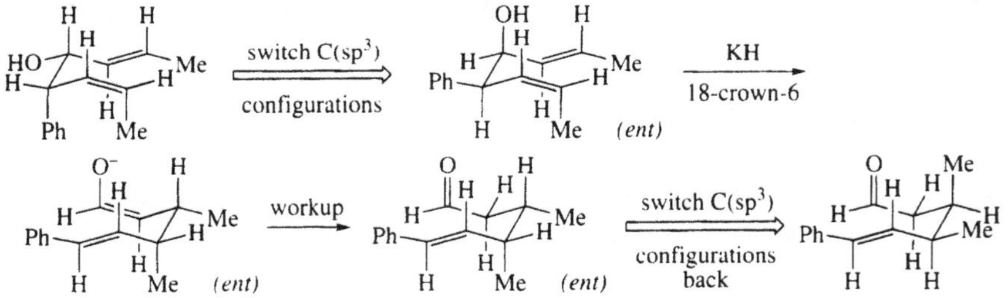

A more rigorous way of dealing with the problem is to flip the chair. When you flip the chair, all the axial substituents at C(sp³) centers become equatorial, and vice versa. Be careful to preserve the stereochemistry about the double bonds: the pseudoaxial and pseudoequatorial substituents of the double bonds retain their orientation in the flipped chair!

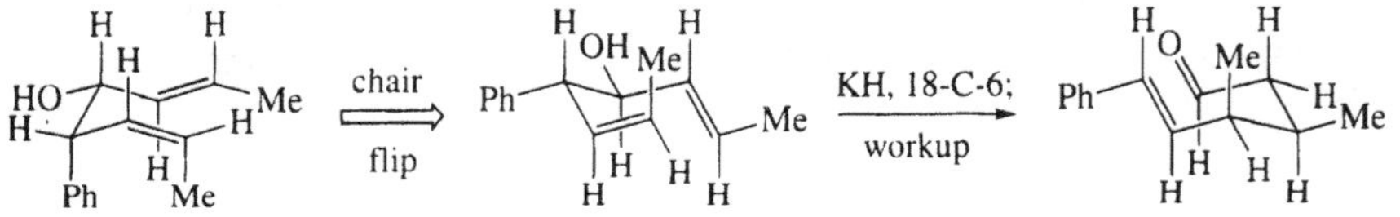

Of course, if your original chair has the largest C(sp³) substituent in the correct, equatorial position, you don't need to worry about switching configurations or flipping the chair.

Problem 4.19. A mechanism involving a [3,3] sigmatropic rearrangement (an aza-Cope rearrangement) followed by a Mannich reaction can be drawn for the following reaction. Draw the mechanism. Then predict the stereochemistry of the product obtained when the aza-Cope rearrangement proceeds through a chair conformation.

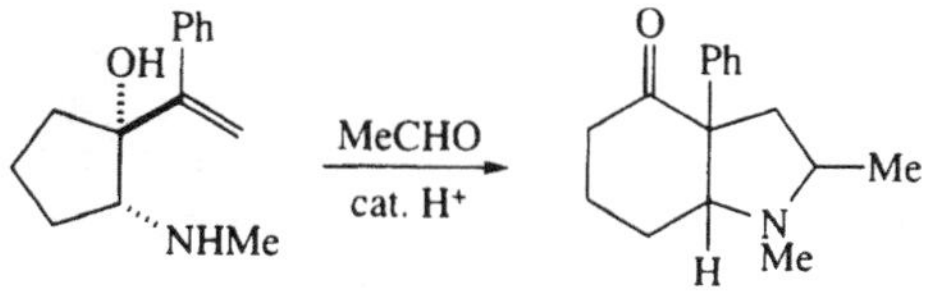

Not every Cope or Claisen rearrangement can proceed through a chair TS. Sometimes the chair TS is raised prohibitively high in energy compared with the boat TS. For example, *cis*-1,2-divinylcyclopropane undergoes a Cope rearrangement to give 1,4-cycloheptadiene. If the starting material underwent rearrangement through the chair TS, one of the π bonds in the seven-membered ring in the product would be trans, so the reaction proceeds through the boat TS.

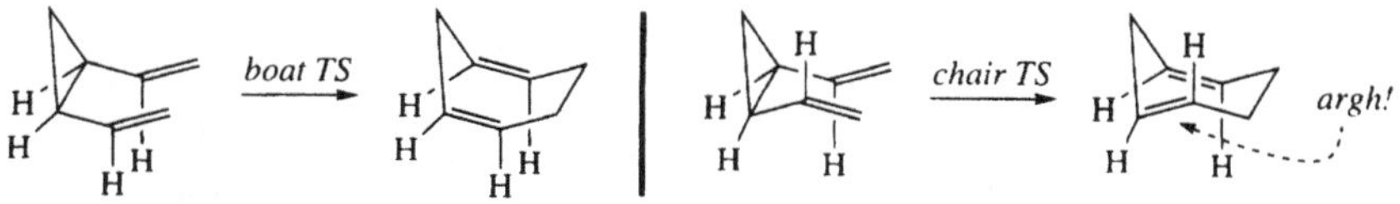

For any 1,5-diene other than a simple, acyclic one, molecular models are usually necessary to determine whether the Cope or Claisen rearrangement will proceed through a chair or a boat TS.

Cope and Claisen rearrangements are not the only sigmatropic rearrangements that show stereoselectivity. For example, consider the [1,5] sigmatropic rearrangement of the following diene. It has the (S) configuration about the stereocenter and the (E) configuration about the double bond distal to the stereocenter. Because the reaction is suprafacial with respect to the five-atom component, the H can move from the top face at one terminus to the top face at the other, or it can move from the bottom face at one terminus to the bottom face at the other. In the first case, the (E,S) product is obtained; in the second case, the (Z,R) product is obtained. The (E,S) product would probably be obtained selectively, as it has the large Ph group in the thermodynamically preferred position. Neither the (E,R) product nor the (Z,S) product is obtained because of the stereospecific nature of the rearrangement.

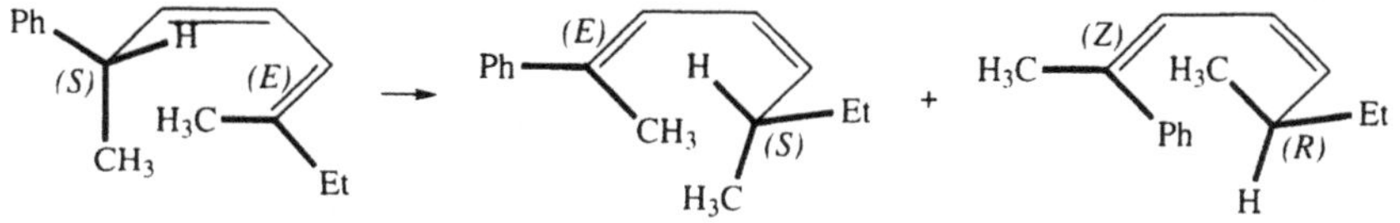

A similar argument can be made for the [1,7] sigmatropic rearrangement. The (S,E) starting material rearranges antarafacially with respect to the seven-atom component. The H can move from the bottom face at one terminus to the top face at the other or it can move from the top face at one terminus to the bottom face at the other. Either the (E,R) or the (Z,S) product, but neither the (E,S) nor the (Z,R) product, is obtained.

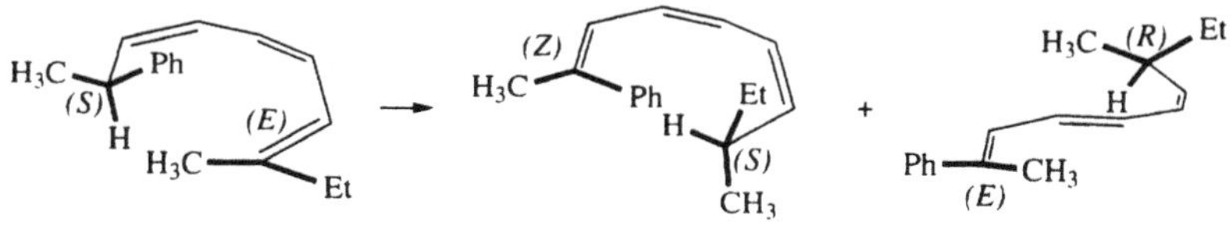

4.5   Ene Reactions

The ene reaction shares some characteristics with both the Diels–Alder reaction and the [1,5] sigmatropic rearrangement. The ene reaction is always a six-electron reaction. Like the Diels–Alder reaction, it has one four-electron component, the *ene*, and one two-electron component, the *enophile*. The two-electron component is a π bond. The four-electron component consists of a π bond and an allylic σ bond. The atom at the terminus of the σ bond is usually H; the other five atoms involved in the ene reaction may be C or heteroatoms. Because the ene reaction involves six electrons, it is suprafacial with respect to all components.

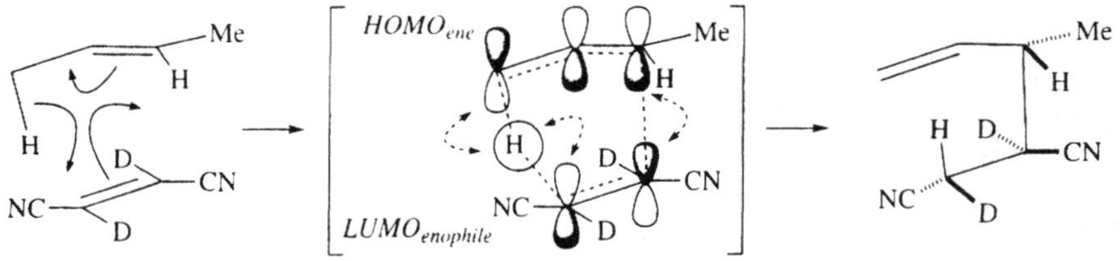

The suprafacial reactivity of the enophile means that the two new bonds to the enophile form to the same face. When the enophile is an alkyne, the two new σ bonds in the product are cis to one another.

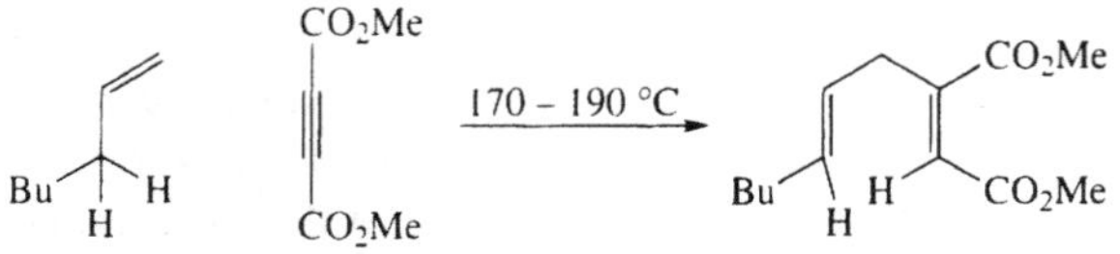

Ene reactions involving five C atoms and one H usually require very high temperatures (>200 °C) to proceed. The reaction occurs at much lower temperature if the H atom is replaced with a metal atom in the *metalla-ene* reaction. The metal may be Mg, Pd, or another metal.

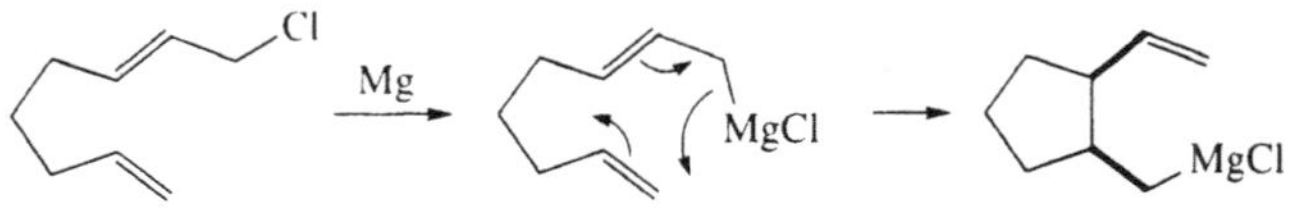

Note how the cis product is obtained in the metalla-ene reaction. This stereochemical result reflects the preference of this particular substrate for an endo transition state, just as in cycloadditions. In cycloadditions, though, the endo substituent of the dienophile is usually an electron-withdrawing substituent, whereas in this ene reaction the endo substituent of the enophile is simply the alkyl chain joining it to the ene.

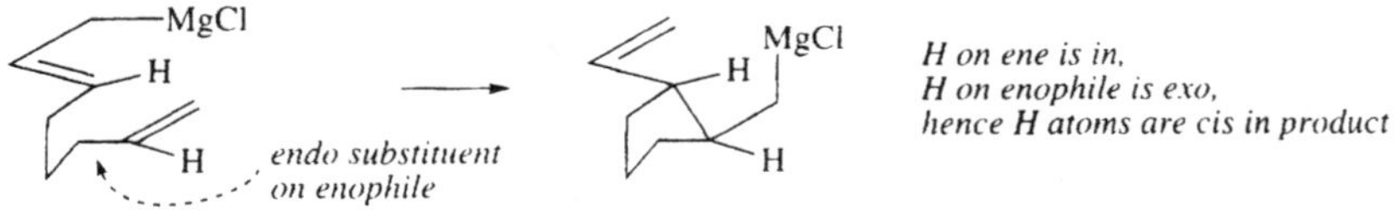

*H on ene is in,
H on enophile is exo,
hence H atoms are cis in product*

The ene reaction also occurs when heteroatoms replace C in either the ene or the enophile. The following hetero-ene reaction requires a much lower activation energy than normal because of the driving force of formation of the aromatic indole ring and because of the instability of ethyl glyoxylate, which has adjacent carbonyl groups.

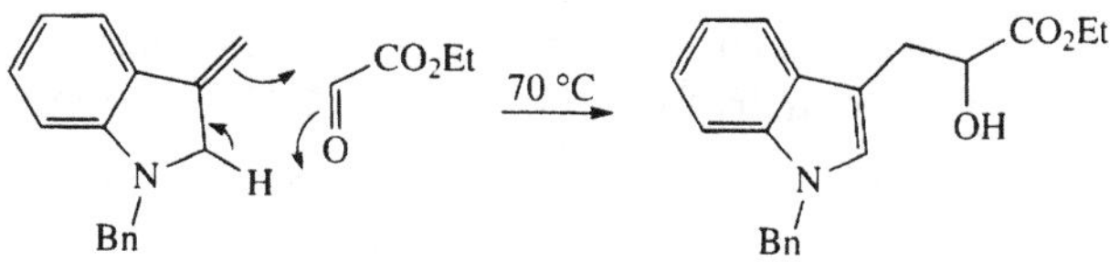

Ene reactions with enols as the ene component are driven in the forward direction by the favorable energy derived from the regeneration of the carbonyl group, just like in the oxy-Cope rearrangement. Very high temperatures may be required to generate a sufficiently high concentration of enol to make the reaction proceed.

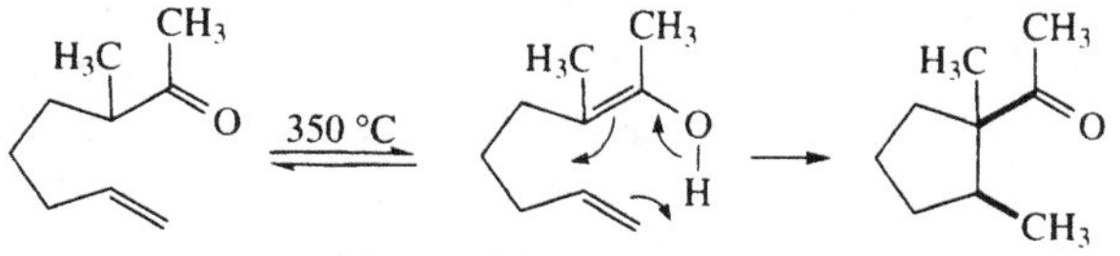

Selenium dioxide (SeO$_2$) is commonly used to hydroxylate alkenes in the allylic position. The mechanism of this transformation involves two sequential pericyclic reactions. The first reaction is an ene reaction. It gives a selenenic acid. Then a [2,3] sigmatropic rearrangement occurs, and the intermediate loses SeO to give the observed product. Note how two allylic transpositions result in no allylic transposition at all!

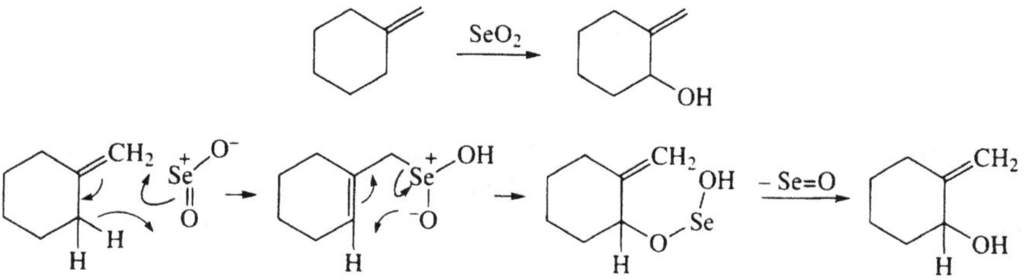

A number of synthetically useful elimination reactions proceed thermally, with no base or acid required. These elimination reactions proceed through a concerted *retro-ene* mechanism. (The mechanism is sometimes called E_i.) The thermal elimination of acetic acid from alkyl acetates and the elimination of RSCOSH from alkyl xanthates (the *Chugaev reaction*) are retro-ene reactions.

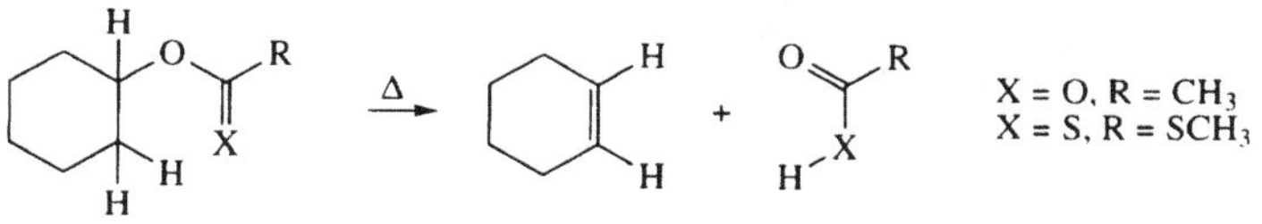

X = O, R = CH$_3$
X = S, R = SCH$_3$

Retro-ene reactions are partly driven by the gain in entropy. (Entropic contributions to $\Delta G°$ are much more important at the high temperatures required for ene reactions.) The formation of the C=O π bond at the expense of the C=S π bond provides an additional driving force for the Chugaev reaction. Acetic acid elimination is enthalpically unfavorable, and it requires much higher temperatures so that the entropic contribution becomes dominant. The formation of the C=O π bond and entropic gains also provide the driving force for the elimination of methyl formate from the following allylic MOM (methoxymethyl) ether.

Selenoxide elimination, a *retro-hetero-ene* reaction, is widely used for the oxidation of carbonyl compounds to the α,β-unsaturated analogs. An α-selenocarbonyl compound is oxidized to the selenoxide with one of a variety of oxidizing agents (H_2O_2, mCPBA, NaIO$_4$, O$_3$, etc.). The retro-hetero-ene reaction of the selenoxide is so facile that it occurs at room temperature within minutes. Amine oxides (*Cope elimination*, not to be confused with the Cope rearrangement) and sulfoxides also undergo the retro-hetero-ene reaction, but they require higher temperatures. Retro-hetero-ene reactions are driven by entropy, by charge neutralization, and sometimes by cleavage of a weak σ bond (e.g., C–Se) in favor of a C=C π bond. All retro-hetero-ene reactions involve a 1,2-dipole, usually $\bar{E}-\bar{O}$, where E is a heteroatom.

The factors that make ene and retro-ene reactions proceed are nicely illustrated by a synthesis of enantiopure, isotopically labeled acetic acid CH(D)(T)CO$_2$H, a useful compound for studying the mechanisms of enzyme-catalyzed reactions. One ene and one retro-ene reaction occur in this synthesis. The ene reaction is driven by formation of a new σ bond at the expense of a C≡C π bond; the retro-ene reaction is driven by the formation of a C=O π bond. Note that both pericyclic reactions proceed stereospecifically, even at the very high temperatures required for them to proceed!

Ene reactions make one new σ bond at the expense of one π bond, like electro-cyclic reactions, but they are fairly easy to distinguish from electrocyclic reactions otherwise. Look for allylic transposition of a double bond and transfer of an allylic H atom. In a retro-ene reaction, a nonallylic H is transferred to an allylic position.

Problem 4.20. The final step of the Swern oxidation involves elimination of Me_2S and H^+ from a sulfonio ether. An E2 mechanism for this elimination is reasonable, but a retro-hetero-ene mechanism is actually operative. Draw the retro-hetero-ene mechanism.

4.6   Summary

The Woodward–Hoffmann rules for all pericyclic reactions (Table 4.6) are as fol-lows. A pericyclic reaction involving an odd number of electron pairs must have an even number of antarafacial components under thermal conditions and an odd number of antarafacial components under photochemical conditions. A pericyclic reaction involving an even number of electron pairs must have an odd number of antarafacial components under thermal conditions and an even number of an-tarafacial components under photochemical conditions. In practice, of course, "even number of antarafacial components" means "no antarafacial components," and "odd number of antarafacial components" means "one antarafacial component."

TABLE 4.6. Woodward–Hoffmann Rules for
Pericyclic Reactions

Number of electron pairs	Number of antarafacial components	
	Δ	$h\nu$
Odd	even	odd
Even	odd	even

Pericyclic mechanisms are undoubtedly the hardest for students to draw. The superficial similarity of the mechanistic types, the way in which seem-ingly reasonable steps are disallowed by theoretical considerations, the si-multaneous formation of many bonds, the lack of a clearly reactive center—all these features of pericyclic reactions combine to make them anathema to many students. You can learn some useful techniques to help you work through pericyclic mechanism problems. Some of these techniques have already been mentioned in the course of the discussion, but they are reiterated here for emphasis.

General considerations when drawing pericyclic mechanisms:

• Draw dashed lines between the atoms where you are forming new bonds, and draw squiggles across bonds that break. Sometimes this procedure will help you identify the pericyclic reaction you need to draw.

• In many problems, a series of polar reactions proceeds to give a reactive intermediate that then undergoes a pericyclic reaction to give the product. Look at the product, and determine what pericyclic reaction could generate it. Then draw the immediate precursor to the final product. Don't be shy about drawing curved arrows on the product to help you draw the precursor with all its bonds in the correct position. This procedure often simplifies the problem considerably.

* **Common error alert:** *Students often draw fictional pericyclic steps such as* $A-B + C=D \rightarrow A-C-D-B$. A good preventive is to be sure that you can name any reactions that occur. If you can't name it, it ain't a reaction!

• As always, draw your by-products, label your atoms, make a list of bonds to make and break, and obey Grossman's rule!

Whether working forward or backward, look for some key substructures.

• The presence of a 1,3-butadiene or a cyclobutene in the starting material or the product may indicate a four-electron electrocyclic reaction.

• The presence of a 1,3-cyclohexadiene or a 1,3,5-hexatriene in the starting material or the product may indicate a six-electron electrocyclic reaction.

• Cyclopropyl cations and allylic cations can open and close, respectively, by two-electron electrocyclic reactions, as can the corresponding halides.

• A six-membered ring with two new σ bonds in a 1,3-relationship may indicate a [4 + 2] cycloaddition (Diels–Alder or hetero-Diels–Alder reaction). When the new ring is fused to an aromatic one, benzyne or an *o*-xylylene may have been a reactive intermediate.

• The formation of a five-membered heterocycle with two new σ bonds almost always indicates a [3 + 2] cycloaddition.

• The formation of a cyclobutanone almost always indicates a ketene–alkene [2 + 2] cycloaddition.

• Light and a cyclobutane often indicate a [2 + 2] cycloaddition.

* **Common error alert:** *A cyclobutane that lacks a ketone and that is produced in the absence of light is likely not formed by a [2 + 2] cycloaddition.*

• Loss of CO_2, N_2, CO, or SO_2 by cleavage of two σ bonds often indicates a retro [4 + 2] or [4 + 1] cycloaddition.

• A 1,5-diene, including a γ,δ-unsaturated carbonyl compound, often indicates a [3,3] sigmatropic rearrangement, i.e., a Cope or Claisen rearrangement.

• A δ,ε-unsaturated carbonyl compound is often the product of an oxy-Cope rearrangement. Oxy-Cope rearrangements are accelerated under strongly basic conditions (KH with or without 18-crown-6).

• The H atoms of cyclopentadienes undergo [1,5] sigmatropic rearrangements with great facility.

• Migration of an allylic group from a heteroatom to its next-door neighbor often indicates a [2,3] sigmatropic rearrangement.

• The formation of a new σ bond at the ends of two π bonds and the simultaneous migration of a H atom may indicate an ene reaction.

• The elimination of an acid such as AcOH or PhSeOH may indicate a retro-ene or retro-hetero-ene reaction.

• A requirement of light (hv) suggests that either an electrocyclic reaction (usually a ring closing), an even-electron-pair cycloaddition (e.g., a [2 + 2] cycloaddition), or a free-radical reaction has occurred.

Good luck!

PROBLEMS

1. Name the mechanism of each of the following reactions as specifically as possible, and predict whether the reaction proceeds thermally or photochemically.

(a)

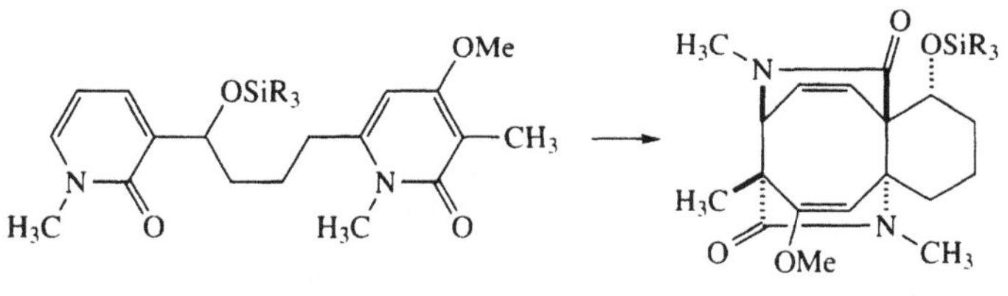

(b)

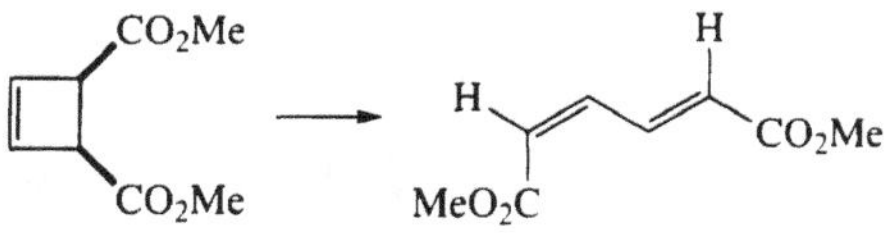

(c)

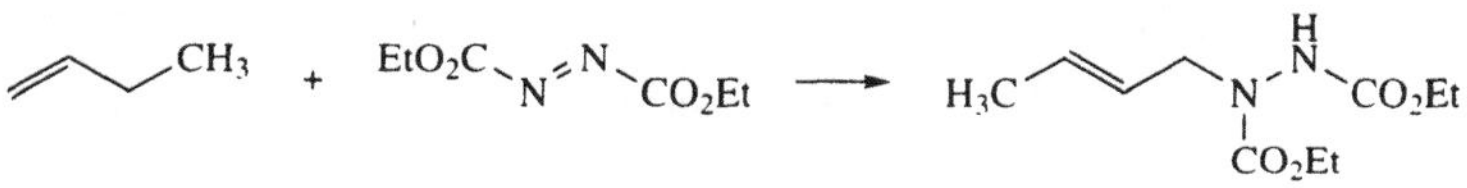

(d)

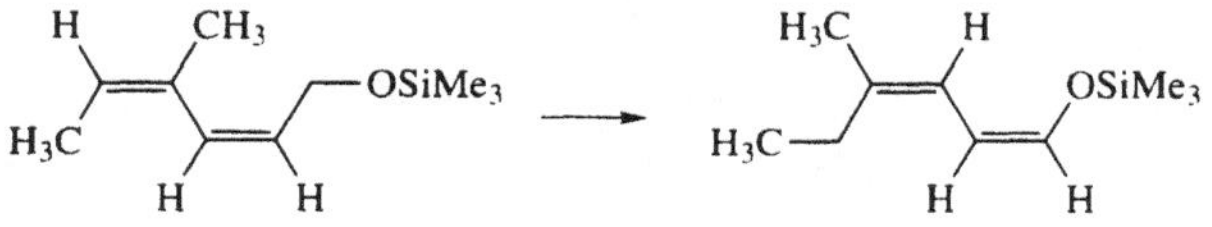

(e)

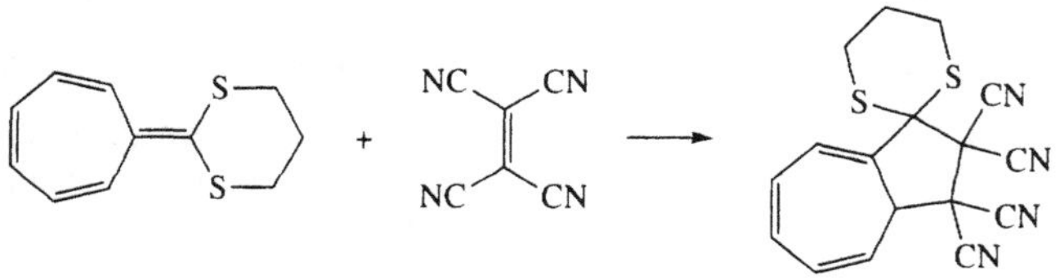

(f)

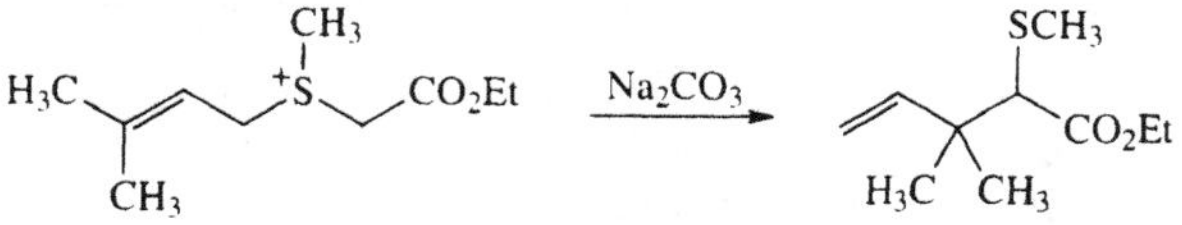

(g)

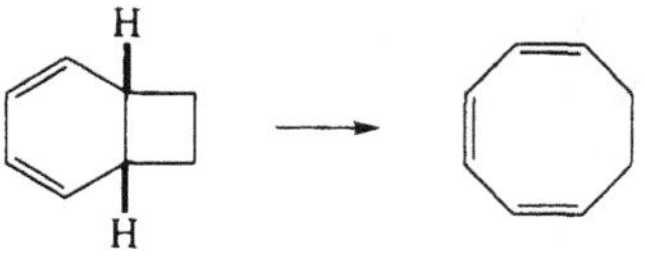

(h)

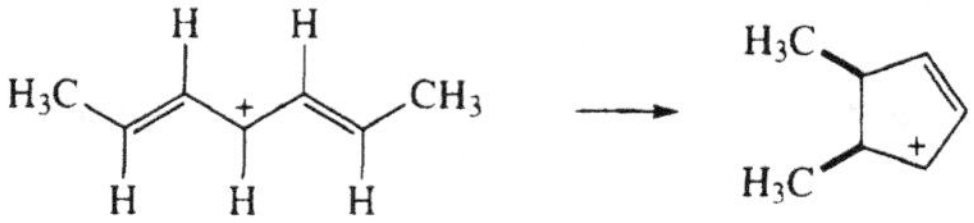

(i)

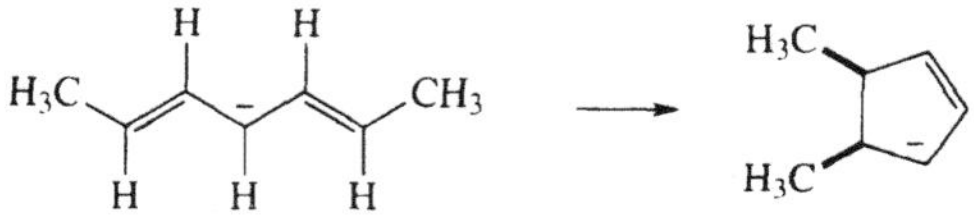

(j)

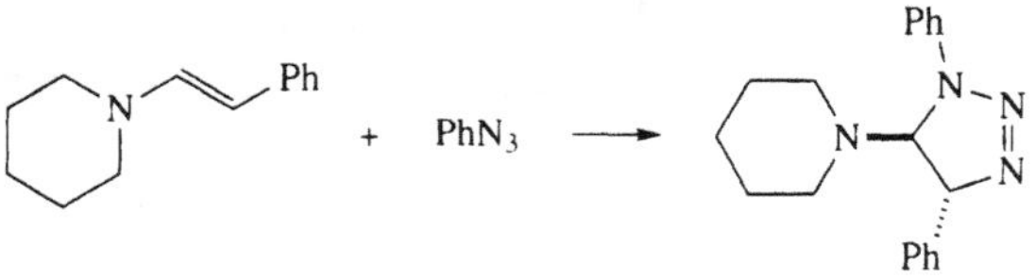

(k)

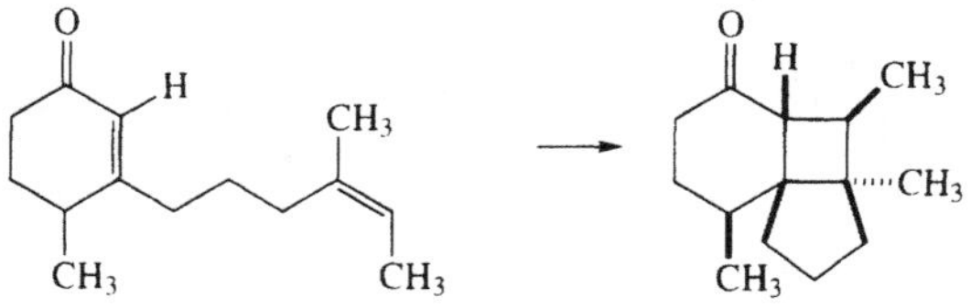

(l)

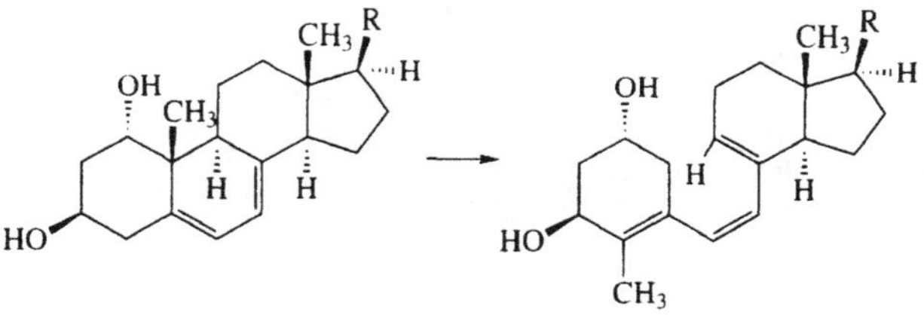

2. Predict the major product (regio- and stereoisomer) of each of the following cycloadditions. All reactions but (h) are [4 + 2] or [3 + 2] cycloadditions.

(a)

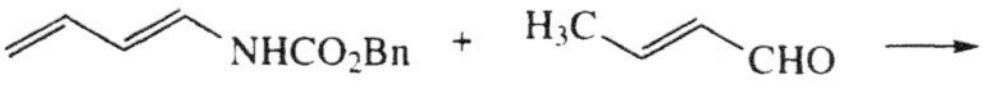

(b)

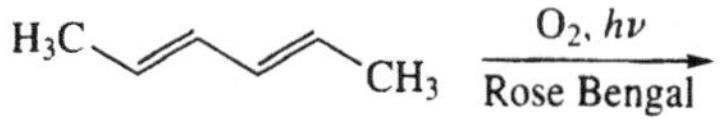

(c)

(d)

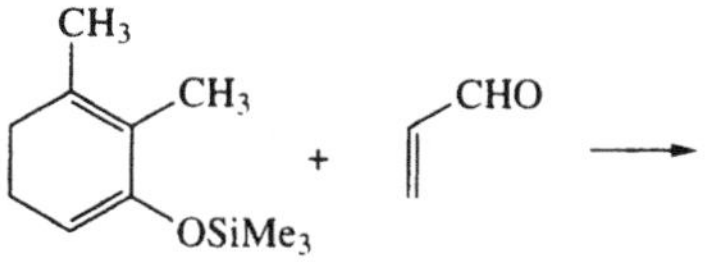

(e)

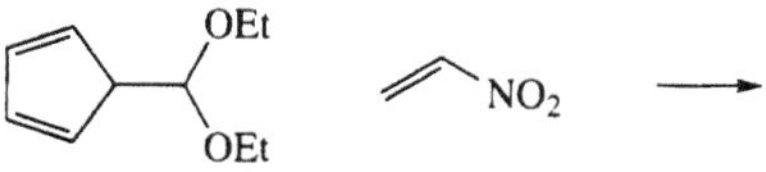

(f)

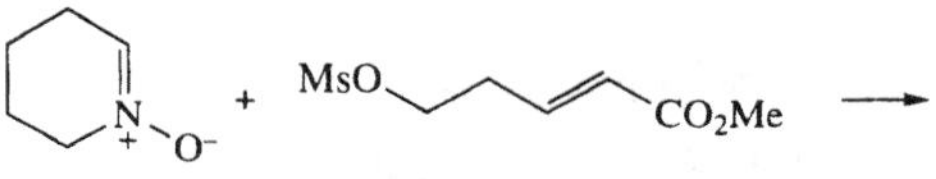

(g)

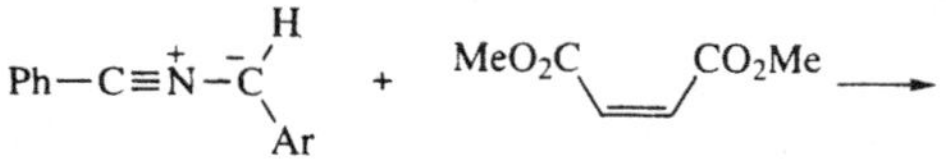

(h) The dots mark the locations where the polyene reacts.

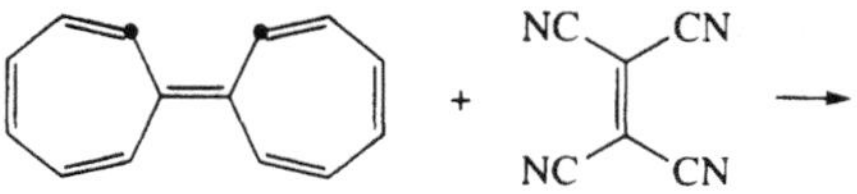

3. *all-cis*-1,3,5,7-Cyclononatetraene can theoretically undergo three different electrocyclic ring closures. Draw the product of each of these reactions, determine the stereochemistry of each when it is obtained under thermal conditions, and order the products by thermodynamic stability.

4. Draw the product of each of the following [3,3] sigmatropic rearrangements, including its stereochemistry. In some cases you may find it necessary to build molecular models in order to see the stereochemical result.

(a)

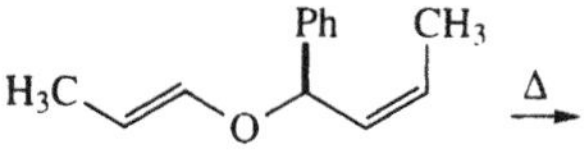

(b)

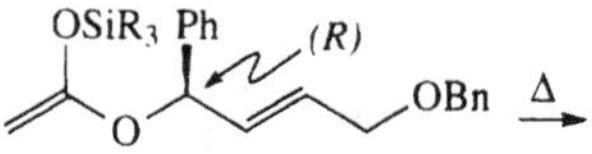

(c)

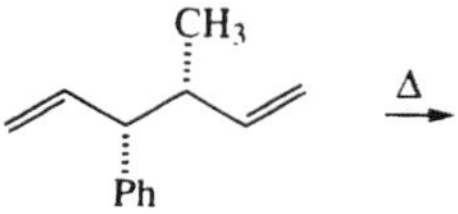

(d)

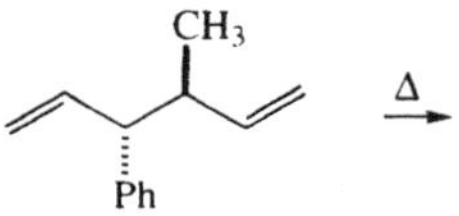

(e)

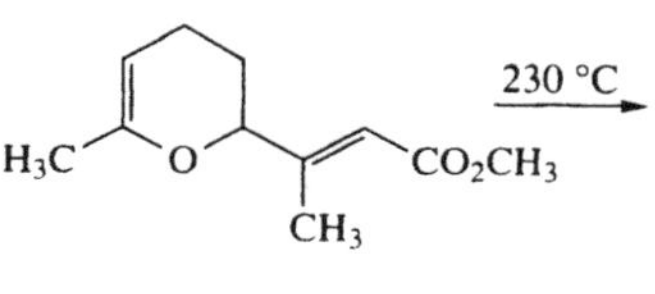

(f)

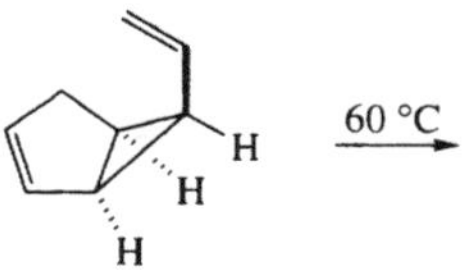

(g)

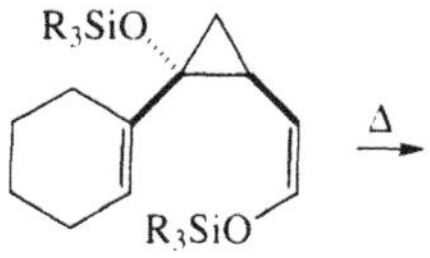

(h)

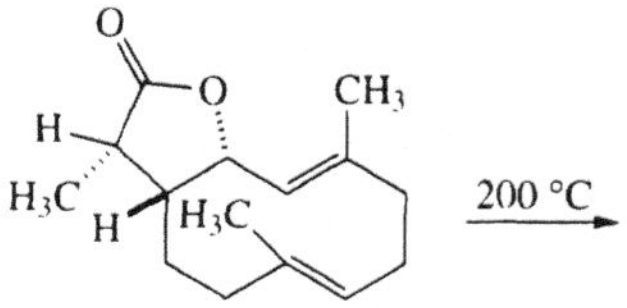

200 °C

5. When ketenes react with 1,3-dienes, bicyclic cyclobutanones are obtained. The mechanism of this reaction is usually described as a one-step [2 + 2] cycloaddition reaction, but the following two-step mechanism was recently proposed.

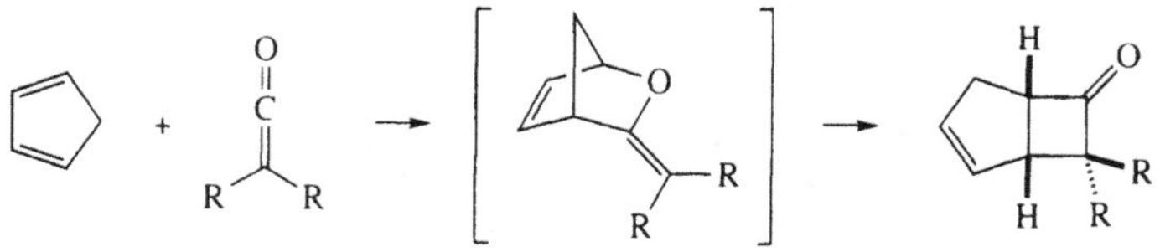

(a) Name the two steps of the proposed mechanism.

(b) Explain why the C=O and not the C=C bond of the ketene reacts with the 1,3-diene in the first step of the proposed mechanism.

(c) When an unsymmetrically substituted ketene ($(R_L)(R_S)C=C=O$) is used in this reaction, the larger group (R_L) is found in the more sterically hindered endo position in the ultimate product ("chemical masochism"), as shown here. Use the proposed mechanism to explain (in words and pictures) this phenomenon. You will need to look at the stereochemical result of both steps of the mechanism.

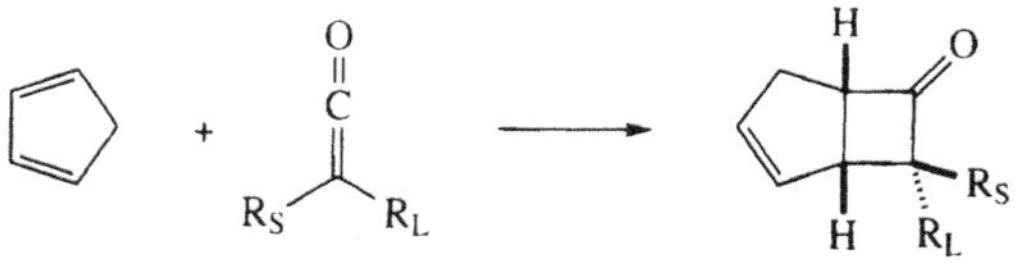

6. Draw mechanisms for the following reactions.

(a)

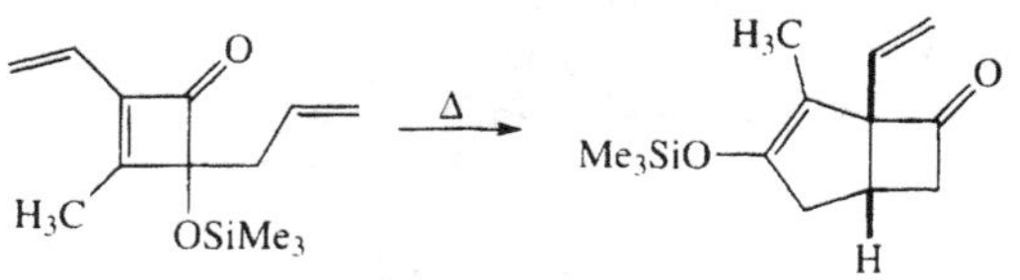

(b)

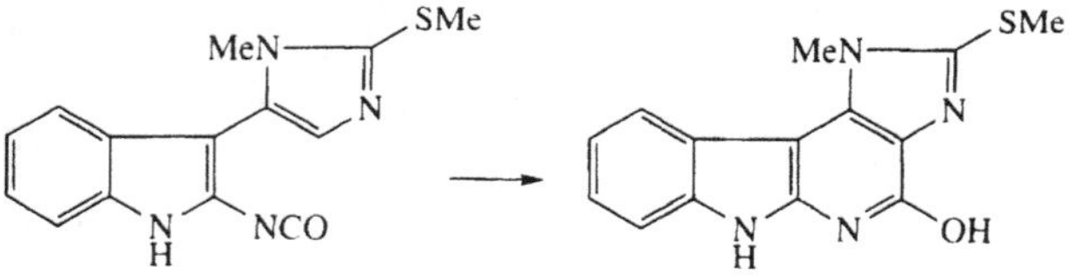

(c)

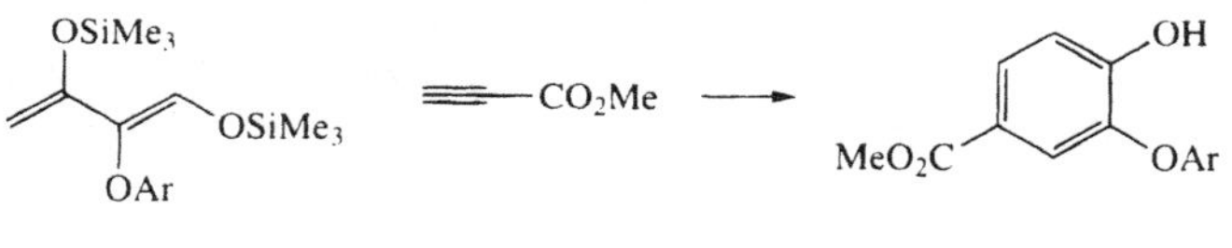

(d)

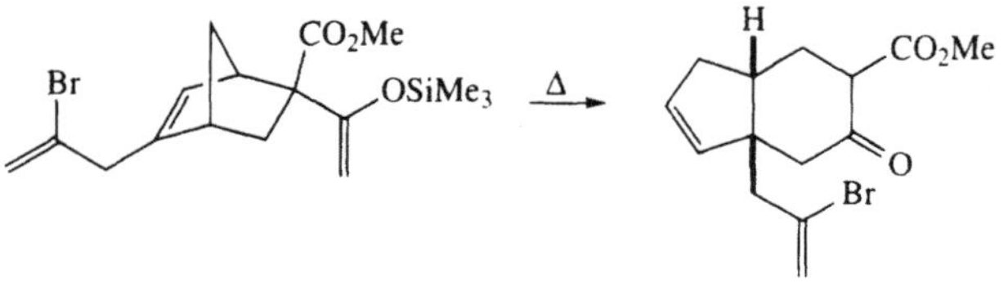

(e)

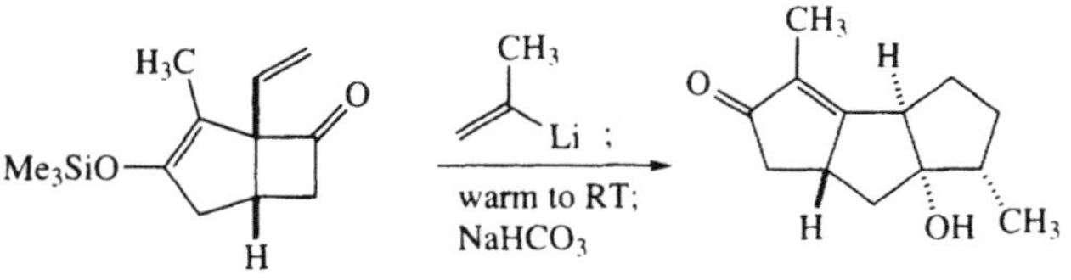

(f)

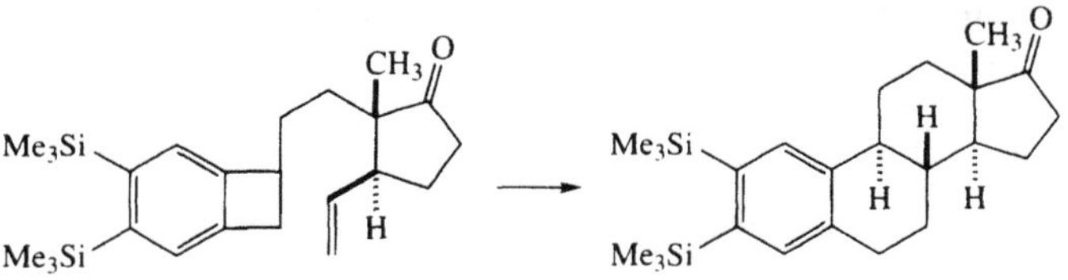

(g)

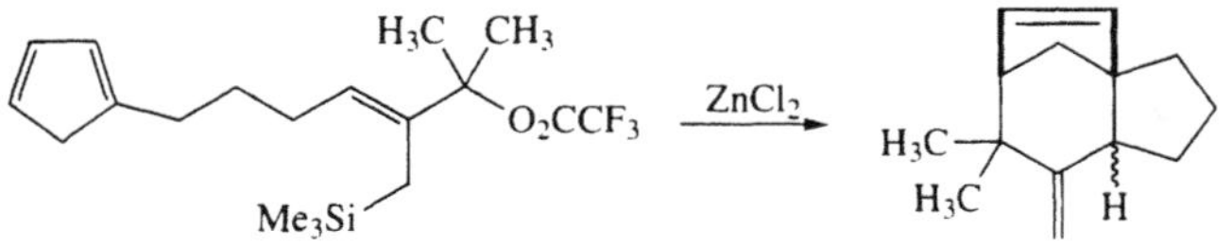

(h)

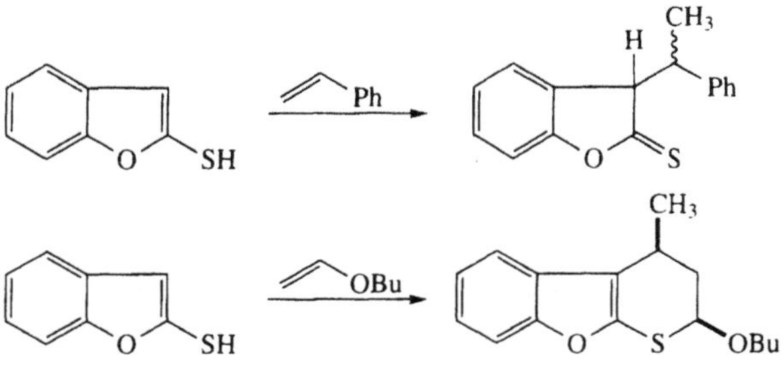

(i)

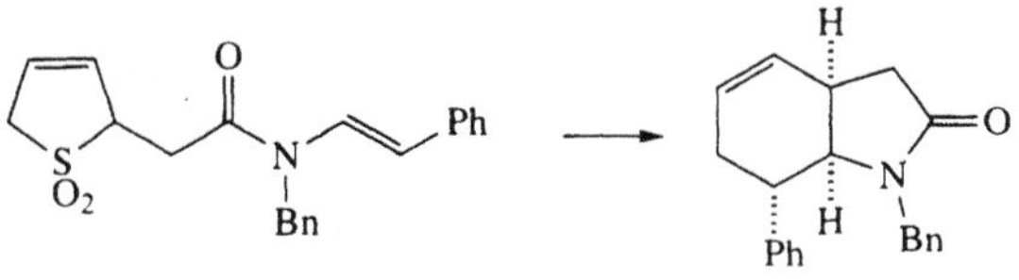

(j) Make an analogy between acyl chlorides and sulfonyl chlorides to solve this problem.

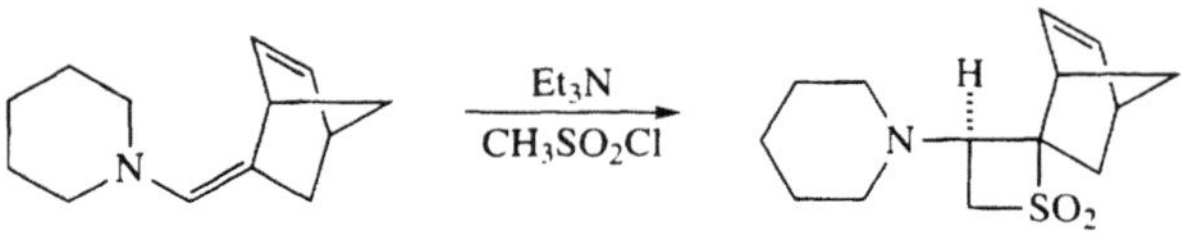

(k) NPhth = phthalimidoyl, an analog of succinimidoyl (as in NBS).

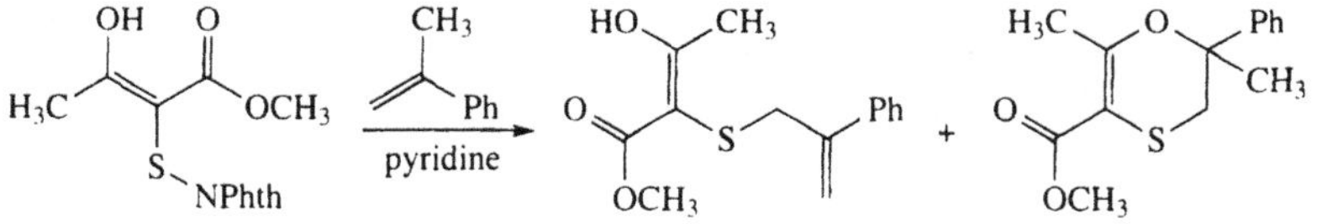

(l)

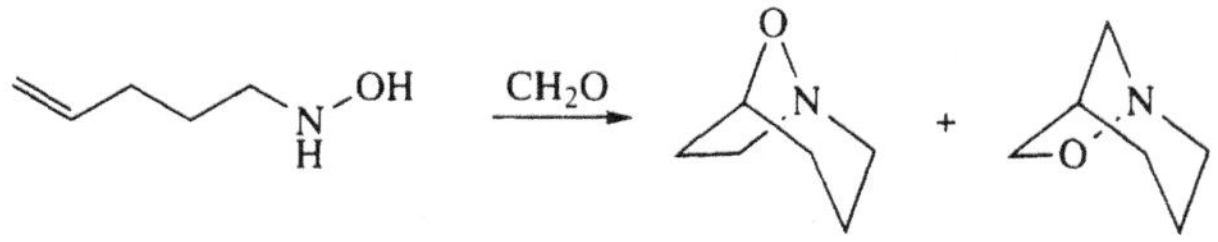

(m) The problem will be simpler if you number the atoms in the very first intermediate, rather than the starting material. *Hint*: The carbon-containing
by-product of the second reaction is HCO_2^-.

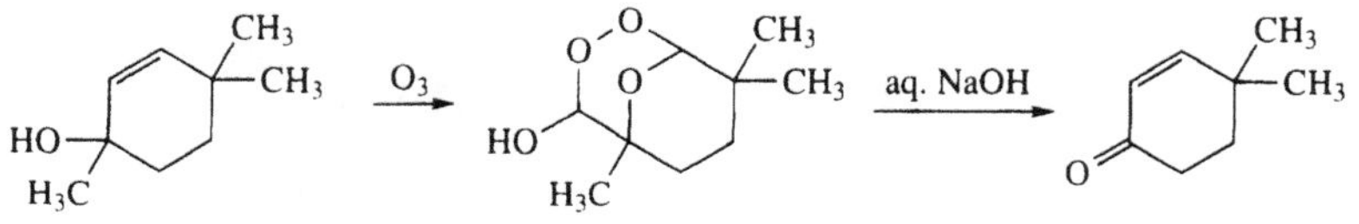

(n)

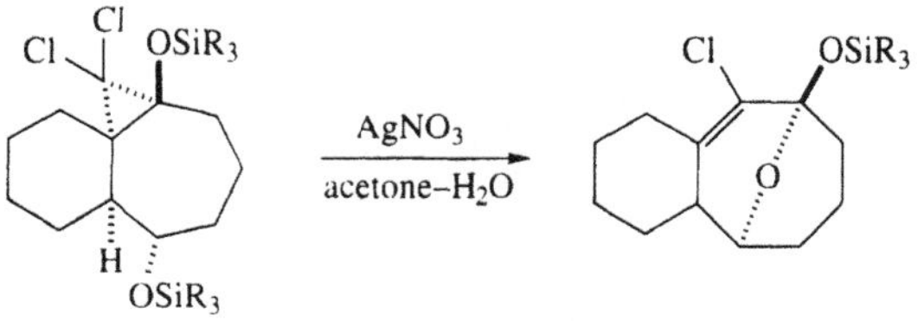

(o)

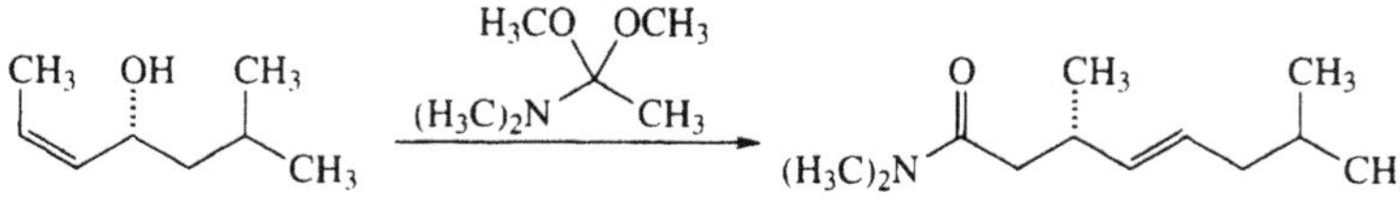

(p) What reaction could the second equivalent of the very strong, nonnucleophilic base LDA induce?

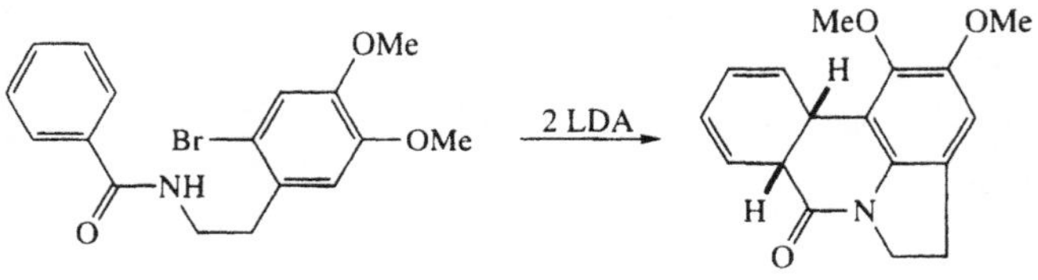

(q) The *Fischer indole synthesis*. Note the by-product!

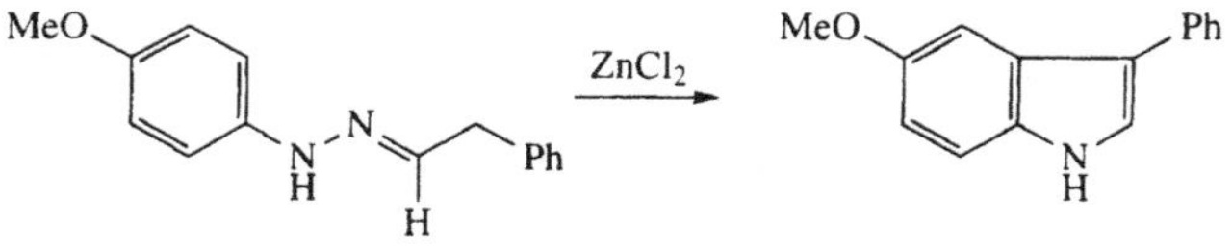

(r) *Two* equivalents of sodiomalonate are required for the following reaction to occur. *Hint*: What is the role of the second equivalent of malonate?

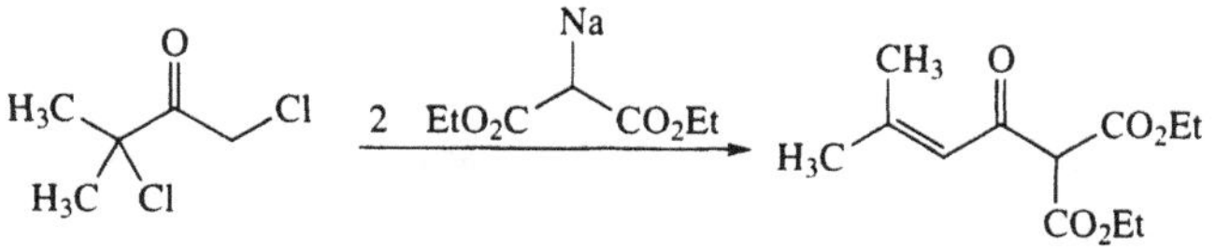

(s)

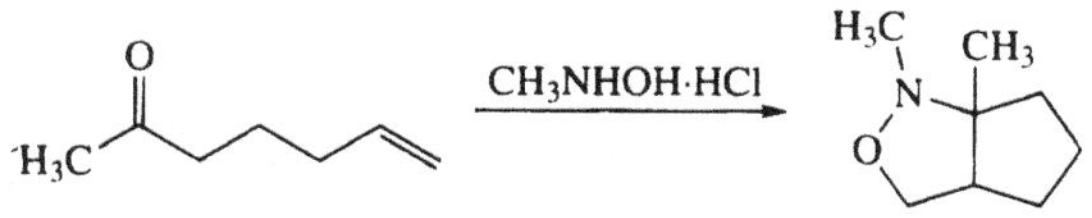

(t) The following problem is tough but doable. Number all your atoms and draw in the H's. Then ask, what is the role of the Rh?

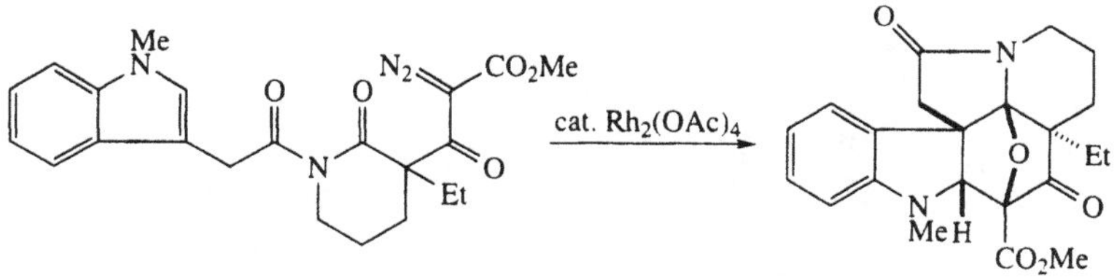

(u)

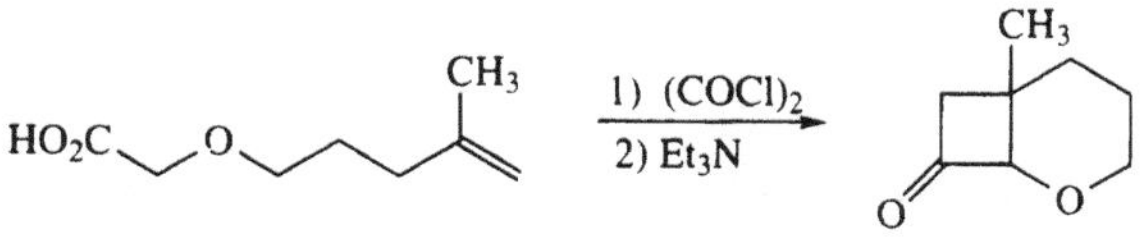

(v) The reaction is thermally allowed, but light is required to supply enough activation energy.

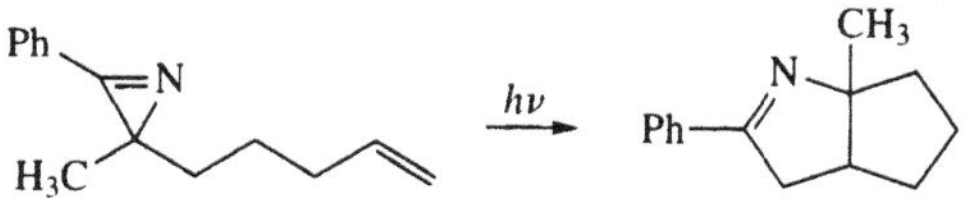

(w)

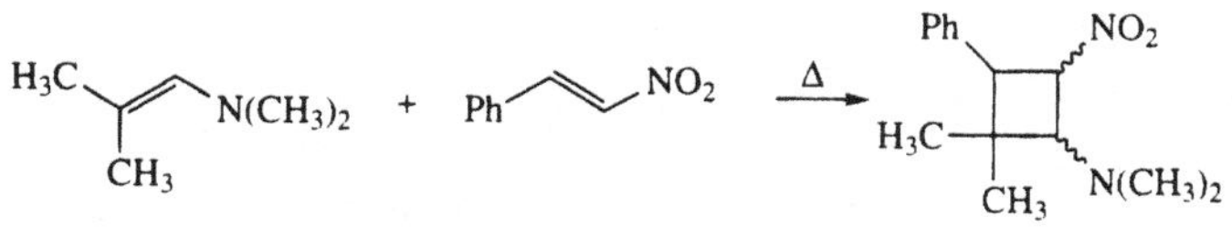

(x)

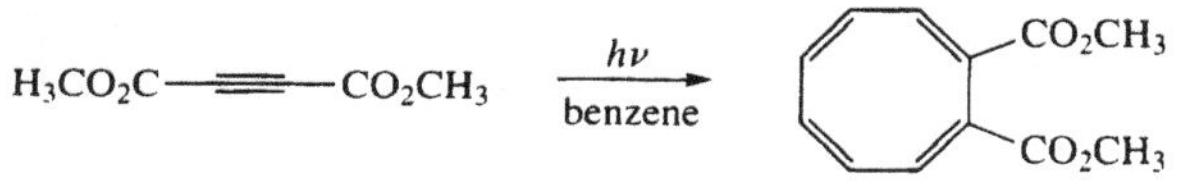

(y)

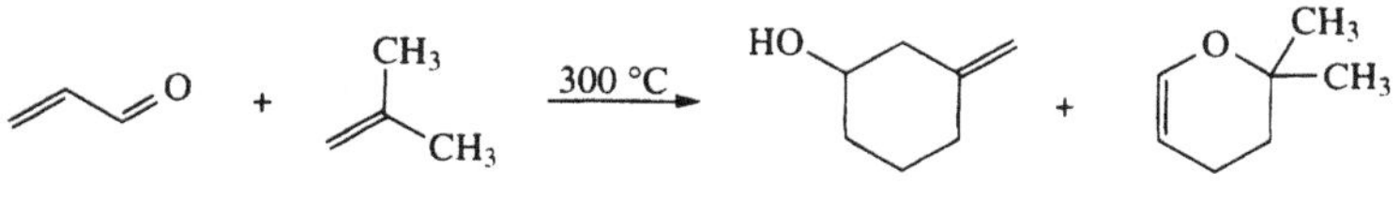

(z)

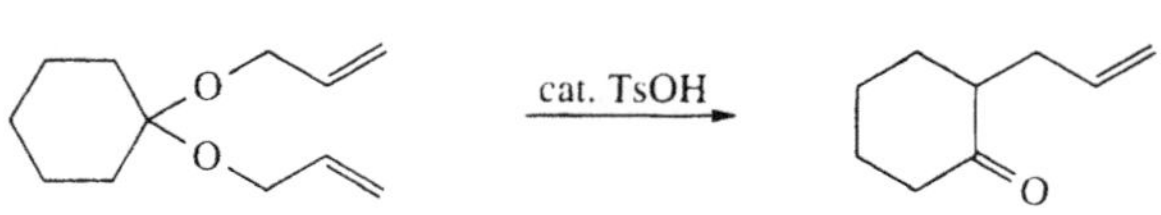

cat. TsOH

(aa) Be sure to account for the observed regio- and stereochemistry. *Hints:*
(1) The C−O bond is already present in **A**. (2) Na$_2$S is an excellent nucleophile.

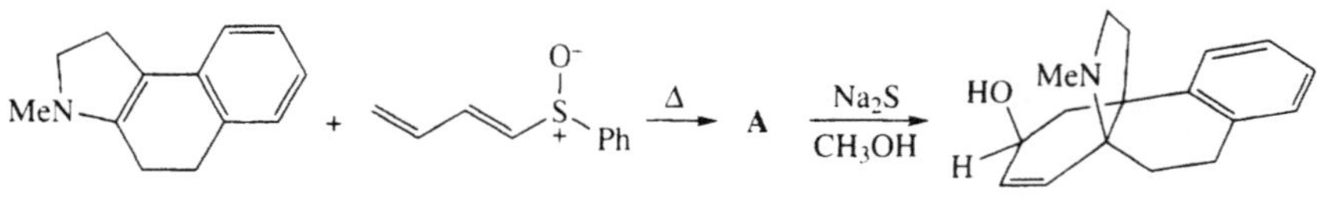

(bb)

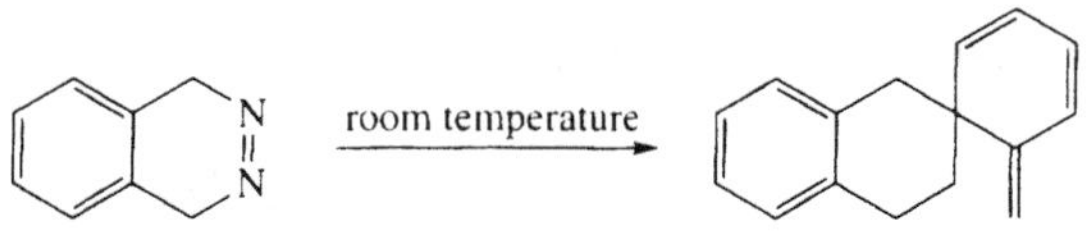

room temperature

(cc)

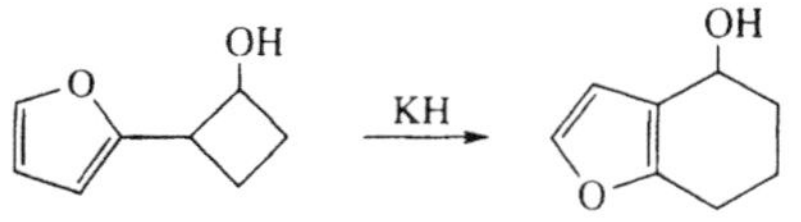

KH

(dd)

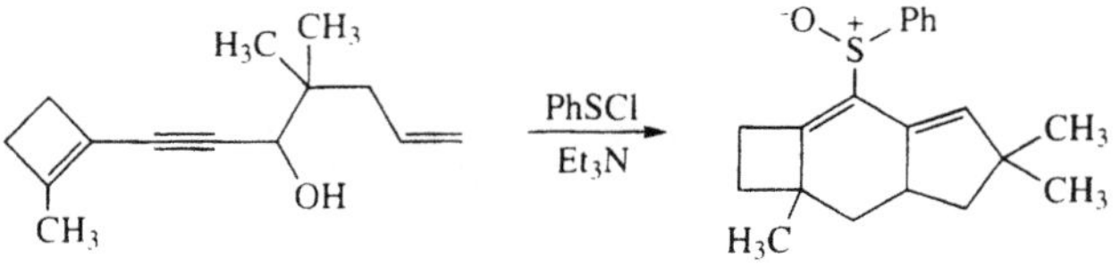

PhSCl
Et$_3$N

(ee)

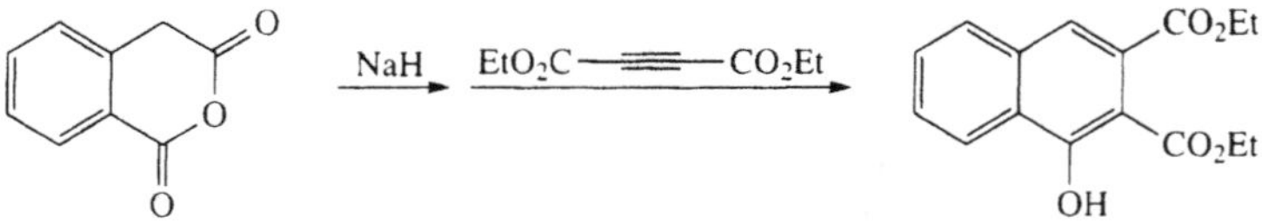

NaH   EtO$_2$C━━━CO$_2$Et

(ff)

110 °C

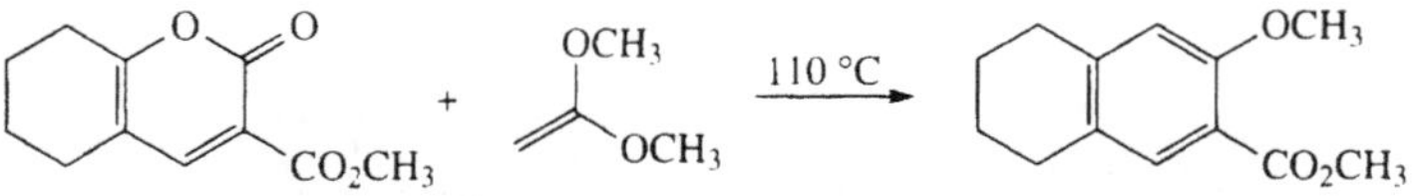

(gg) Here's a clue to rule out one of the possible mechanisms: Irradiation of
benzene under the same conditions does *not* give benzocyclooctatetraene,
the product of the following reaction.

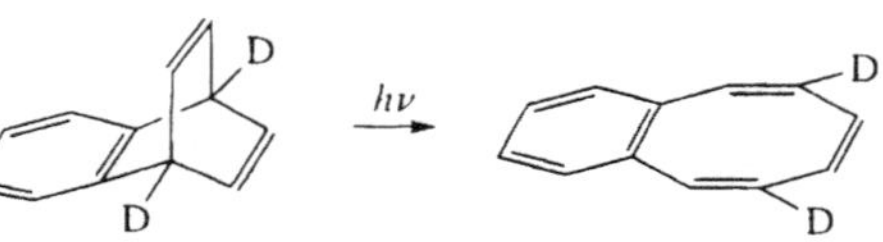

(hh) Atomic C has the reactivity of ±C±, a double carbene.

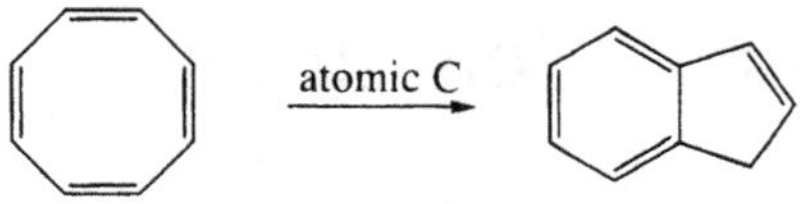

(ii) The key to this very difficult problem is to number the N atoms in the product correctly. Draw the azido groups in their RN–N$_2$ resonance structure, draw the by-products, and *then* number the atoms. Then, for the first step, think of a thermal rearrangement of organic azides that we have already learned.

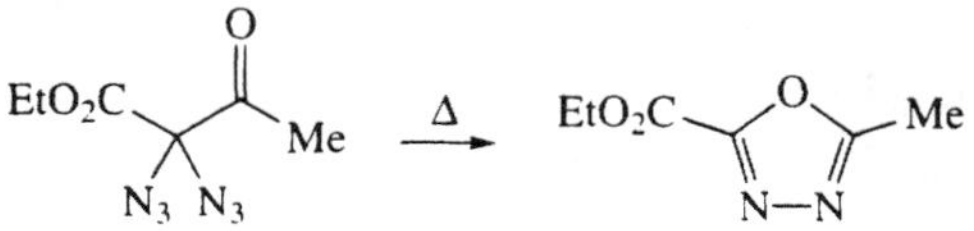

5

Free-Radical Reactions

5.1 Free Radicals

A free radical is a species containing one or more unpaired electrons. Free radicals are electron-deficient species, but they are usually uncharged, so their chemistry is very different from the chemistry of even-electron electron-deficient species such as carbocations and carbenes.

> The word *radical* comes from the Latin word for "root" (as in radish). The term *radical* was originally used to denote a fragment of a molecule that remained unchanged throughout a series of reactions. The term *free radical* was later coined to refer to a fragment of a molecule that wasn't bonded to anything. Today, *radical* and *free radical* are used interchangeably, although *radical* still retains its original meaning in certain contexts (e.g., in the use of R in organic structures).

5.1.1 Stability

Most of the chemistry that is discussed in this chapter involves alkyl radicals ($\cdot CR_3$). The alkyl radical is a seven-electron, electron-deficient species. The geometry of the alkyl radical is considered to be a shallow pyramid, somewhere between sp^2 and sp^3 hybridization, and the energy required to invert the pyramid is very small. In practice, one can usually think of alkyl radicals as if they were sp^2-hybridized.

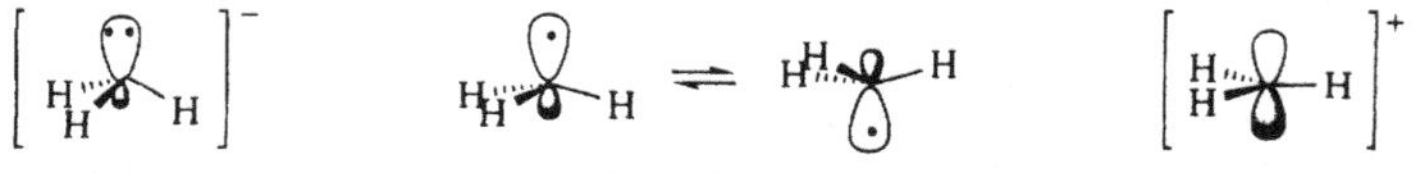

carbanion
pyramidal

alkyl radical
shallow pyramid

carbocation
trigonal planar

Both alkyl radicals and carbocations are electron-deficient species, and the structural features that stabilize carbocations also stabilize radicals. Alkyl radicals are stabilized by adjacent lone-pair-bearing heteroatoms and π bonds, just as carbocations are, and the order of stability of alkyl radicals is $3° > 2° > 1°$. However, there are two major differences between the energy trends in carbocations and alkyl radicals.

224

• A C atom surrounded by seven electrons is not as electron-deficient as a C atom surrounded by six electrons, so alkyl radicals are generally not as high in energy as the corresponding carbocations. Thus, the very unstable aryl and 1° alkyl carbocations are almost never seen, whereas aryl and 1° alkyl radicals are reasonably common.

• The amount of extra stabilization that adjacent lone pairs, π bonds, and σ bonds provide to radicals is not as great as that which they provide to carbocations. The reason is that the interaction of a filled AO or MO with an empty AO (as in carbocations) puts two electrons in an MO of reduced energy, whereas the interaction of a filled AO or MO with a half-filled AO (free radicals) puts two electrons in an MO of reduced energy and one electron in an MO of increased energy.

Orbital interaction diagram for radicals

Orbital interaction diagram for carbocations

one electron
raised in energy

p, π,
or σ C(p)   two electrons
lowered in energy

p, π,
or σ C(p)   two electrons
lowered in energy

Even though adjacent lone pairs, π bonds, and σ bonds do not stabilize radicals as much as they stabilize carbocations, the cumulative stabilizing effect of several such groups on a radical can be considerable. Benzylic radicals are particularly low in energy, as the radical center is stabilized by three π bonds. The triphenylmethyl (trityl) radical, a triply benzylic radical, was the first free radical to be recognized as such. This remarkably stable radical is in equilibrium with the dimer that results from combination of the methyl carbon of one radical with a para carbon on another radical. (The structure of the dimer was originally misidentified as hexaphenylethane.)

Neutral free radicals are electron-deficient, so radicals centered on less electronegative elements are lower in energy than radicals centered on more electronegative elements. As a result, the order of stability for first-row radicals is alkyl ($\cdot CR_3$) > aminyl ($\cdot NR_2$) > alkoxyl (RO$\cdot$), and for halogens it is I$\cdot$ > Br$\cdot$ > Cl$\cdot$ > $\cdot$F.

* **Common error alert:** *The hydroxyl radical ($\cdot OH$) is very high in energy.* Although it is involved in certain very important reactions in biological systems and in atmospheric chemistry, its intermediacy in a synthetic reaction should be viewed with skepticism.

* **Common error alert:** *The H$\cdot$ radical is also very high in energy and is hardly ever seen.*

Unlike carbocations, free radicals are stabilized both by electron-rich π bonds, like C=C π bonds, and by electron-poor π bonds, like C=O π bonds. The additional resonance provided by the carbonyl group offsets the destabilization of localizing a radical on O and making O electron-deficient. However, an electron-rich π bond is more stabilizing than one that is electron-poor. When a radical is substituted by both an electron-donating and an electron-withdrawing group, the total stabilization is greater than the sum of the parts in a phenomenon called the *captodative effect*.

Some captodatively stabilized radicals

The *nitroxyls* (a.k.a. *nitroxides*) are remarkably stable free radicals. Nitroxyls have two major resonance structures, one N-centered and one O-centered; the lone electron may also be considered to be in the π^* orbital of an N=O π bond. Nitroxyls are thermodynamically stable because dimerization would give a very weak N—N, N—O, or O—O bond. TEMPO (2,2,6,6-tetramethylpiperidin-1-oxyl), a commercially available nitroxyl, is further stabilized by steric shielding. Other thermodynamically stable free radicals include the small molecules O_2 (a 1,2-diradical, best represented as $\cdot O-O \cdot$) and nitric oxide ($\cdot N=O$), a "messenger molecule" in mammals that mediates smooth muscle contraction.

TEMPO, a stable nitroxyl

Tables of homolytic bond strengths (bond dissociation energies, BDEs) offer a good guide to the relative energies of radicals. Such tables can be found in most organic chemistry textbooks. Compare the BDEs of the two bonds H_3C-H and Me_3C-H. The BDE for the former is 104 kcal/mol, whereas the BDE for the latter is 91 kcal/mol. The smaller the BDE, the weaker the bond, and the lower energy the radical, so the $Me_3C\cdot$ radical is lower in energy than the $H_3C\cdot$ radical. If you compare the H_3C-Br bond (70 kcal/mol) with the Me_3C-Br bond (63 kcal/mol), you will see a similar trend. You should note, however, that the difference between the BDEs for H_3C-H and Me_3C-H (13 kcal/mol) is not the same as the difference between the BDEs for H_3C-Br and Me_3C-Br (7 kcal/mol). The difference between the ΔBDEs reflects the fact that the H· free radical is very high in energy. One must use BDEs with care!

Free radicals can also be stabilized kinetically by steric shielding of the radical centers. Such free radicals are called *persistent*. Examples of persistent free radicals include perchlorotrityl, galvinoxyl, and the radical derived from BHT (butylated hydroxytoluene, 2,6-di-*tert*-butyl-4-methylphenol), an antioxidant that is used as a food preservative.

Some persistent free radicals

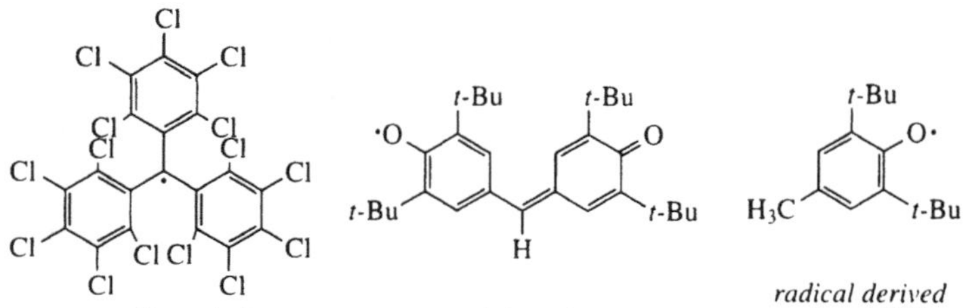

perchlorotrityl   *galvinoxyl*   *radical derived from BHT*

5.1.2 Generation from Closed-Shell Species

Most free radicals are not kinetically stable—they tend to react with one another to give electron-sufficient species—so free radicals that are to be used in reactions must usually be generated from closed-shell species in situ. Free radicals can be generated from closed-shell species in four ways: σ-bond homolysis, photochemical excitation of a π bond, one-electron reduction or oxidation, and cycloaromatization.

• Sigma-bond homolysis is a very common way to generate free radicals. The σ bond is usually a heteroatom–heteroatom bond such as N–O or Br–Br, but even σ bonds such as C–C or C–N that are normally very strong can homolyze if a very stable fragment is formed or if the bond is very strained. Light is sometimes used to induce σ-bond homolysis by promoting one of the electrons in the σ orbital to the $\sigma*$ orbital, but if the σ bond is sufficiently weak and the product radicals are sufficiently low in energy, heat may suffice to break it.

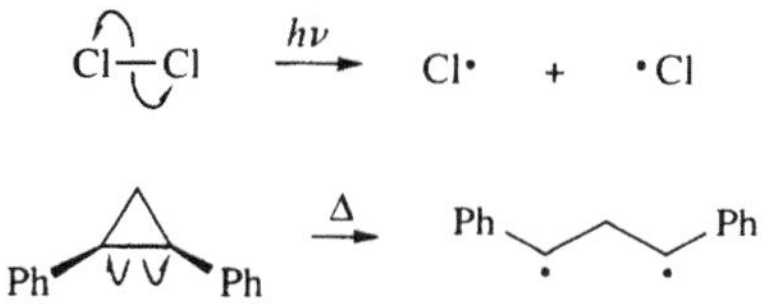

Benzoyl peroxide and AIBN are two of the compounds most widely used to generate free radicals in a reaction mixture. The O–O bond in benzoyl peroxide and the C–N bonds in AIBN homolyze under thermal or photochemical conditions.

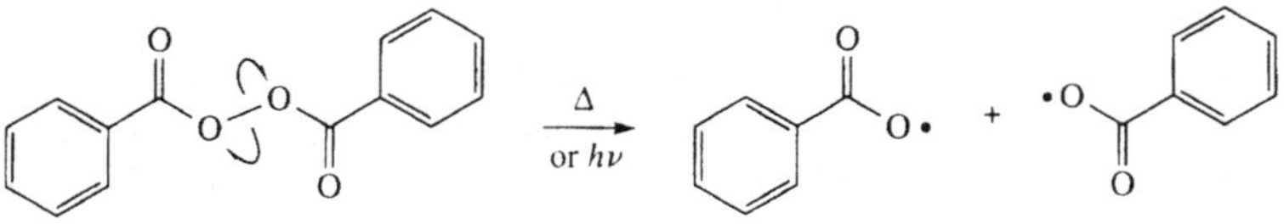

benzoyl peroxide

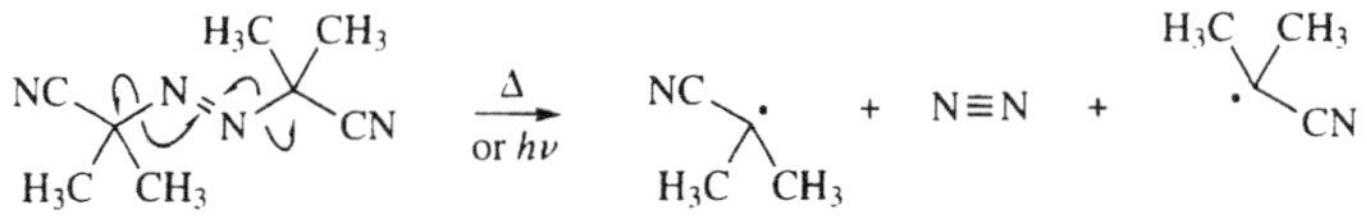

AIBN, a free radical initiator

The movement of unpaired electrons in free radical reactions is shown with *single-headed* curved arrows.

The likelihood of bond homolysis is directly related to the BDE for that bond. The BDE for the H−H bond is 104 kcal/mol, whereas the BDE for the Br−Br bond is 46 kcal/mol, so the likelihood of H−H homolysis is much smaller than the likelihood of Br−Br homolysis. Sigma bonds that are particularly prone to homolysis include N−O and O−O bonds, bonds between C and very heavy atoms like Pb and I, halogen–halogen bonds, and very strained bonds.

Upon photolysis of a diazo compound, the C−N bond cleaves in heterolytic fashion, N_2 is lost, and a *carbene* is generated, usually in the *triplet* form. Triplet carbenes have one electron in each of two orbitals and may be thought of as 1,1-diradicals. Alkyl and acyl azides also undergo loss of N_2 upon photolysis. The products are highly reactive *nitrenes*, the N analogs of carbenes.

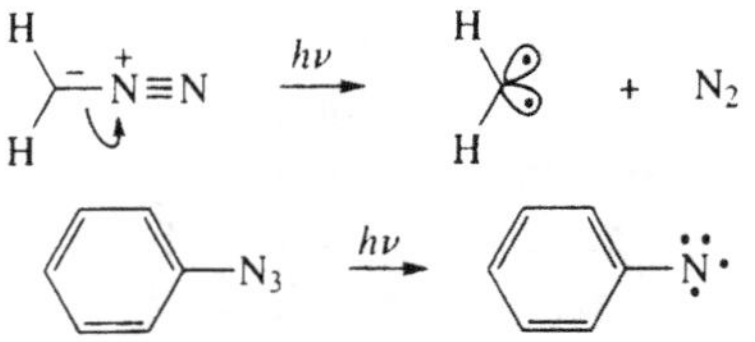

• When light of the appropriate wavelength is allowed to shine on a compound containing a π bond, an electron from the π orbital is promoted to the π^* orbital. The product can be considered to be a 1,2-diradical, and it undergoes reactions typical of free radicals. The C=O, C=S, and C=C π bonds can all be photoexcited in this manner. The weaker the π bond, the easier it is to photoexcite.

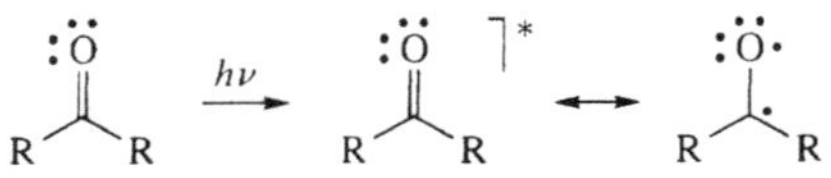

An asterisk (*) is often used to denote a compound in a photoexcited state, but this notation does not convey the reactivity of the compound. The 1,2-diradical structure is a better description.

In photoexcited alkenes, there is no π bond between the two radical centers, so free rotation about the C–C bond can occur. Alkenes can be isomerized from the cis to the trans form photochemically. The reaction does not usually proceed in the reverse direction, because the photoexcited cis alkene and the photoexcited trans alkene are in equilibrium, and the equilibrium favors the less sterically encumbered trans isomer.

The eyes of arthropods, mollusks, and vertebrates use the cis–trans isomerization reaction to detect light. When light enters the eye, it is absorbed by an imine of 11-*cis*-retinal, which isomerizes to the lower energy all-*trans*-retinal imine. The isomerization is detected by various enzymes that initiate an electrical impulse that enters the brain via the optic nerve. Meanwhile, the all-*trans*-retinal is transported to the liver (!), where the enzyme retinal isomerase uses acid catalysis and ATP to convert it back to the higher energy 11-*cis* form. The 11-*cis*-retinal is then sent back to the eye, ready to receive the next photon.

11-*cis*-retinal imine   all-*trans*-retinal imine

- A compound with an electron in a high-energy orbital may transfer the electron to a compound that has a lower energy orbital. Most often, the electron donor is a metal or reduced metal salt such as Li, Na, or SmI_2, but the donor may also be a lone-pair-bearing compound such as an amine or phosphine, especially if it is photoexcited. (Photoexcitation promotes an electron from a nonbonding orbital into a higher energy orbital, whence it is more likely to participate in electron transfer.) The orbital accepting the electron is usually a π^* orbital, most often one associated with an aromatic ring or a C=O π bond, but it may be a σ^* orbital. After the electron is accepted, the product is called a *radical anion*.

Aromatic radical anions are often drawn in their "toilet bowl" or "happy alien" resonance form, but it's easier to keep track of all of the electrons with the localized resonance structures, and they are therefore recommended. About 18 different resonance structures in which a pair of electrons and an unpaired electron are localized on two atoms of the π system can be drawn. The three electrons can be distributed on any two ortho or para atoms of the ring!

The radical anion derived from a ketone is called a *ketyl*. The dark blue, sterically and electronically stabilized benzophenone ketyl is widely used in solvent stills as a deoxygenating agent.

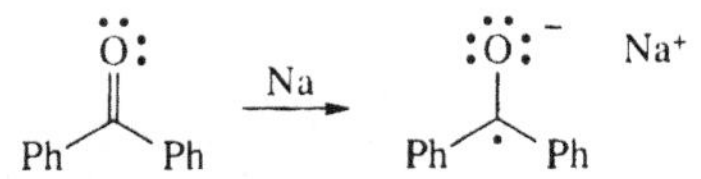

benzophenone ketyl

Electron transfer is also the first step of the $S_{RN}1$ substitution mechanism (Chapter 2). In reactions that proceed by the $S_{RN}1$ mechanism, the electron donor is usually the nucleophile. The nucleophile may be photoexcited to give its electron more energy and make it more prone to transfer.

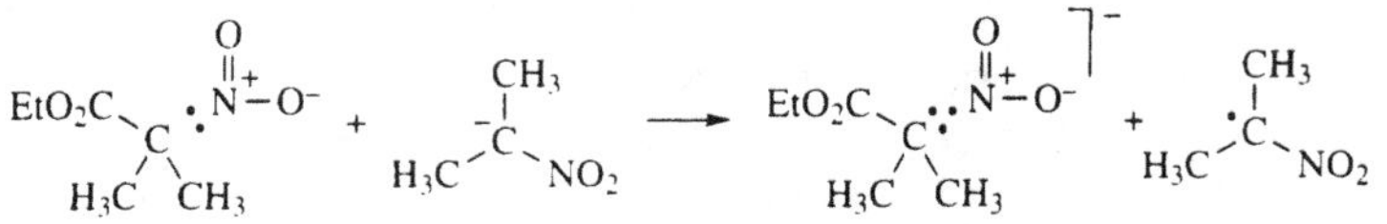

In this example, the radical anion is actually better described as the nitro analog of a ketyl, but the two-center, three-electron bond is more descriptive of the reactivity of the compound.

More descriptive of reactivity

Better description of ground state

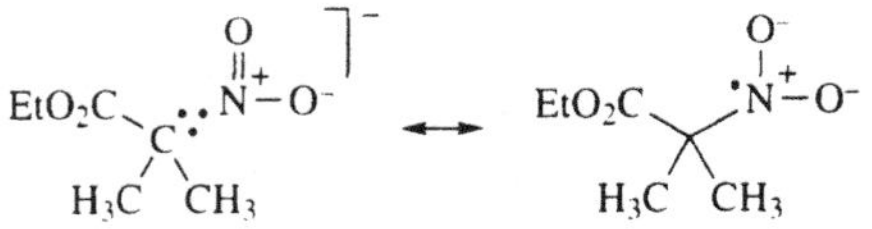

One-electron oxidation of organic substrates gives radical cations. Oxidation is much less common in synthesis than is reduction. Certain metal salts such as $Pb(OAc)_4$, $Mn(OAc)_3$, and $(NH_4)_2Ce(NO_3)_6$ (ceric ammonium nitrate, CAN) are one-electron oxidizing agents.

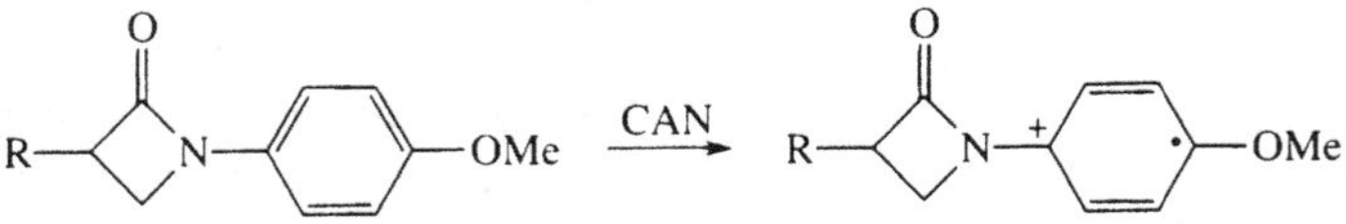

$Mn(OAc)_3$ is usually used to remove H· from the α-carbon atoms of carbonyl compounds. The Mn(III) exchanges with H^+ of the enol, and Mn–O σ-bond homolysis generates Mn(II) and the enoxy radical.

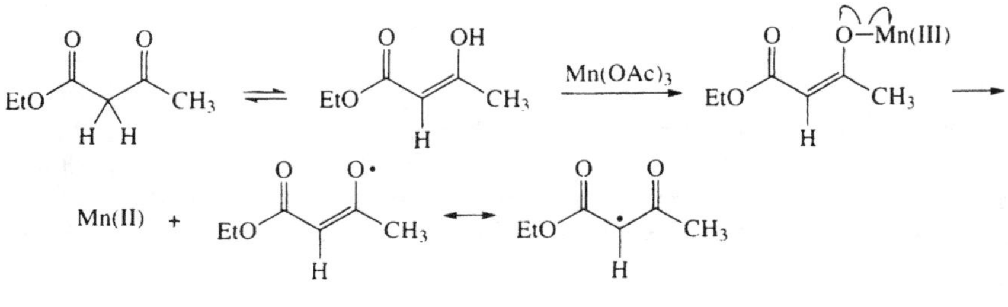

Quinones such as DDQ (2,3-dichloro-5,6-dicyano-1,4-benzoquinone) and chloranil can also be used to remove one electron from a substrate.

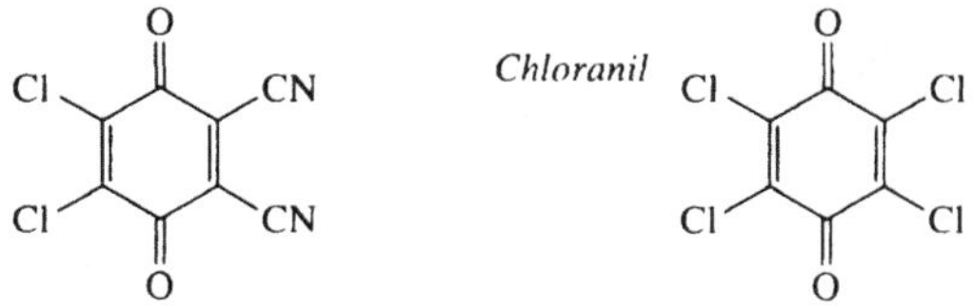

• The *cycloaromatization* of certain highly unsaturated organic compounds can give diradicals. These reactions can be thought of as six-electron electrocyclic ring closings (Chapter 4), but whether the p orbitals of C3 and C4 of the six-atom system actually participate in the reaction is unclear. The best-known cycloaromatization reaction is the *Bergman cyclization* of an enediyne such as 3-hexene-1,5-diyne to give a 1,4-phenylene diradical. Allenylenynes (1,2,4-trien-6-ynes) undergo cycloaromatizations, too.

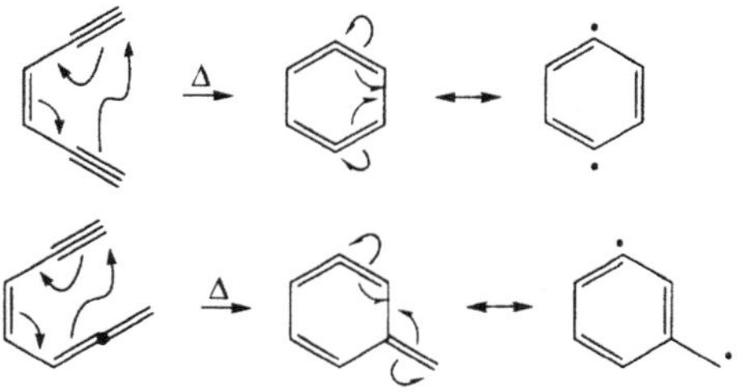

The Bergman cyclization was discovered in the late 1960s and studied for its academic interest, but it suddenly became a hot topic of research in the mid-1980s when certain naturally occurring antitumor antibiotics that used the Bergman cyclization to attack DNA were discovered. In calicheamycin γ_1, the aryl and sugar groups bind to DNA, and the polyunsaturated "warhead" portion of the molecule (the left-hand portion in the following drawing) damages the DNA.

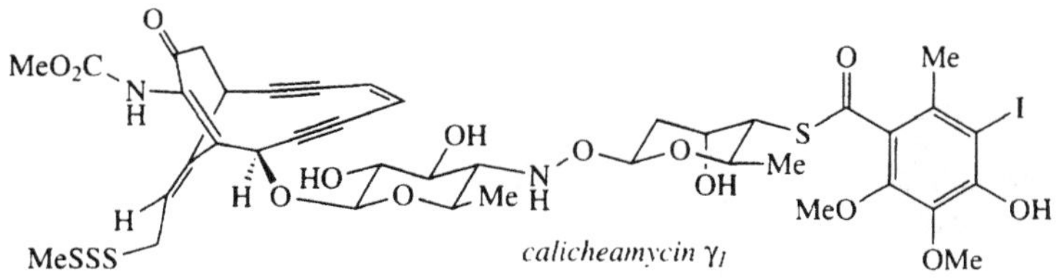

calicheamycin γ_1

The warhead works as follows. Reduction of an S–S bond in the warhead portion of the molecule by endogenous substances in the cell nucleus generates a thiolate, which adds to the enone in an intramolecular Michael reaction. Rehybridization of the β-carbon of the enone from sp^2 to sp^3 upon Michael reaction brings the two ends of the enediyne close enough that they can undergo Bergman cyclization. The 1,4-diradical species that is thereby obtained abstracts H· atom from the nearby DNA, causing damage and eventual cell death.

Other enediyne antitumor antibitotics (neocarzinostatin, esperamycin, dynemicin) may have similar or different mechanisms of activation, but they all use cycloaromatization as the key step in generating the DNA-damaging species.

5.1.3 Typical Reactions

Free radicals undergo eight typical reactions: addition to a π bond, fragmentation, atom abstraction (reaction with a σ bond), radical–radical combination, disproportionation, electron transfer, addition of a nucleophile, and loss of a leaving group. *The first three are by far the most important.* Atom abstraction and addition reactions involve the reaction of a free radical with a closed-shell species, and fragmentation is the conversion of a free radical to a new free radical and a closed-shell species. Addition and fragmentation reactions are the microscopic reverse of one another. In general, the propagation part of a chain mechanism consists exclusively of addition, fragmentation, and atom abstraction steps.* Radical–radical combination, disproportionation reactions, and electron transfers are found in the termination part of chain mechanisms and in nonchain mechanisms. Chain mechanisms can be initiated by electron transfer, too. Addition of a nucleophile and loss of a leaving group are both two-electron reactions.

The free-radical chain mechanism was discussed in Chapter 1. It is *essential* that you be familiar with the rules for drawing a chain mechanism!

• A free radical can *add to a π bond* of a closed-shell species to give a new radical. The radical (X·) adds to a Y=Z π bond to give X–Y–Z·, in which the new X–Y σ bond is made up of one electron from the radical and one electron from the former π bond. The Y=Z π bond may be polarized or nonpolarized. Intermolecular additions of radicals to π bonds usually occur in such a way that the lowest energy radical is formed, but intramolecular additions are often subject

---

*There is a radical–radical combination step in the propagation part of the chain mechanism for the autoxidation reaction. The propagation part of the $S_{RN}1$ mechanism (Chapter 2) consists of loss of a leaving group, addition of a nucleophile, and electron transfer.

to stereoelectronic requirements that may cause the higher energy radical to be formed. The addition of free radicals to π bonds is mechanistically similar to the addition of π bonds to carbocations.

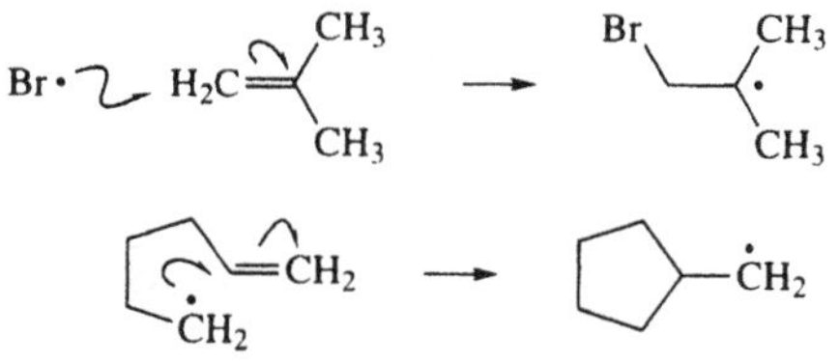

In this book, single-headed arrows are often shown moving *in one direction only* for free-radical reactions other than bond homolysis. This convention is used in order to avoid cluttering the drawings with arrows. You may show the movement of electrons in all directions if it helps you to see the reactions better. In fact, most chemists show arrows moving in both directions.

A free radical does not always add to a polarized π bond in the same way that a nucleophile would add. For example, the $Bu_3Sn\cdot$ radical adds to the S atom of $C=S$ π bonds, not to the C atom.

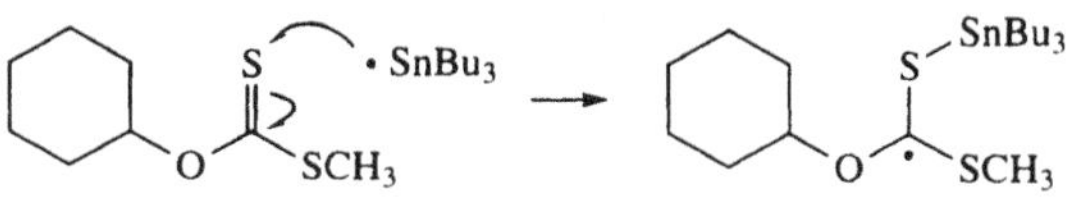

Free radicals can also add to stable "carbenes" such as CO and isocyanides (RNC). The carbene C becomes a radical center in the product.

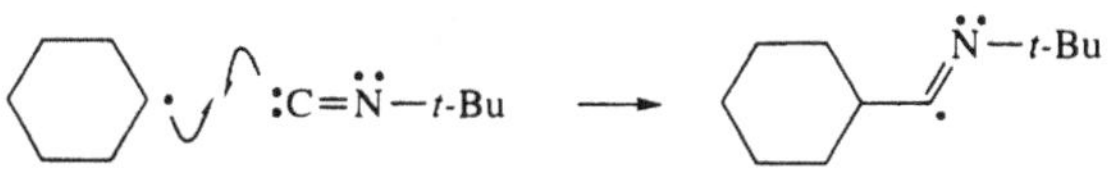

The addition of free radicals to nitrones and nitroso compounds gives very stable nitroxyls. In fact, nitrones are commonly used as *spin trapping reagents* for the study of free radicals by electron paramagnetic resonance (EPR), the electronic analog of NMR. Free-radical intermediates in reactions have too fleeting a lifetime to be examined directly, but their nitroxyl derivatives are long-lived and can be studied by EPR.

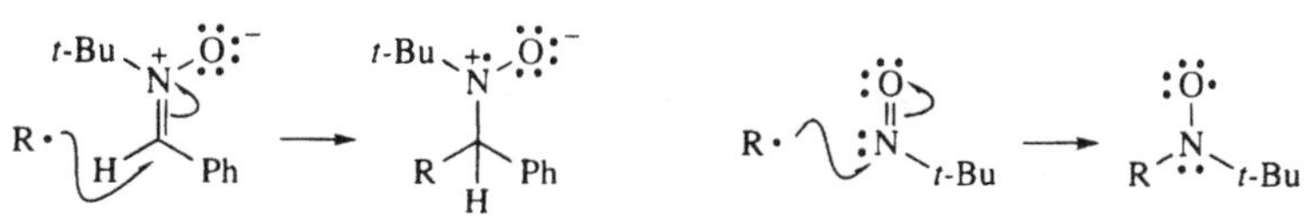

• *Fragmentation* is the microscopic reverse of addition to a π bond. A σ bond that is *adjacent* (β) to a radical center homolyzes, and one of the electrons of that σ bond and the former unshared electron together form a new π bond at the former radical center. The fragmentation of a radical is mechanistically similar to the fragmentation of a carbocation.

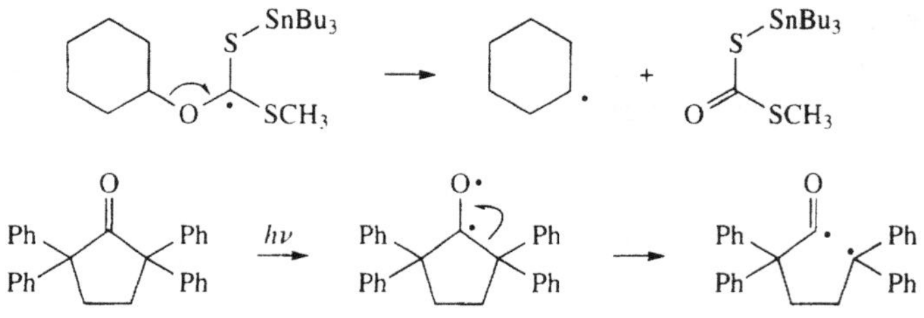

CO can be lost from an acyl radical by a fragmentation reaction. In this case, the bond that breaks is *directly attached* (α) to the radical center.

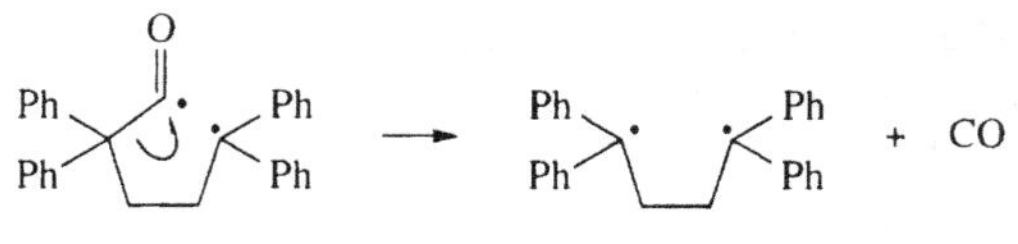

The combination of Et_3B and O_2 is widely used to initiate radical reactions. The mechanism involves two unusual radical reactions. Addition of O_2 to the empty orbital of BEt_3 provides a one-electron bond between B and O, and a fragmentation follows to provide Et· and Et_2BOO·.

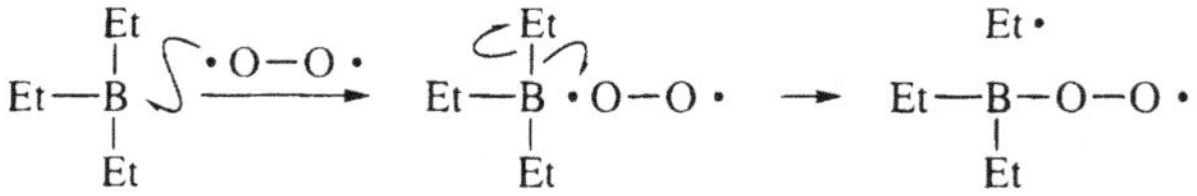

• In an *atom abstraction reaction*, a radical X· attacks a Y−Z σ bond to give a new closed-shell species X−Y and a new radical ·Z. One of the electrons in the old σ bond goes to form a bond to X, and the other one ends up on Z. The transferred atom, Y, is usually H or halogen, but not always.

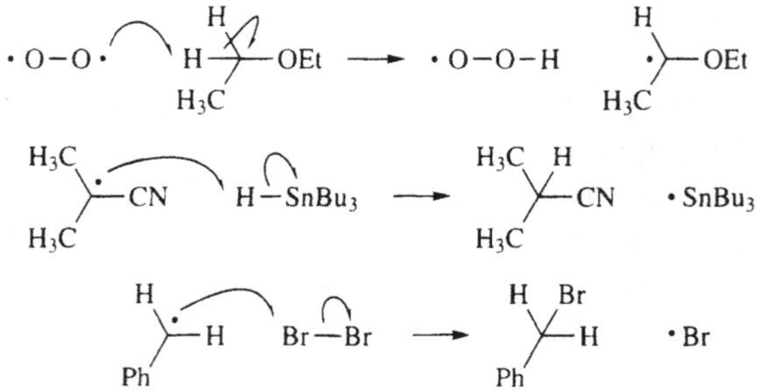

* **Common error alert:** *The abstraction of hydrogen atom H·, which has a low activation energy, is* not *to be confused with the abstraction of hydride anion H^-, which is a very high energy process.*

The tributyltin radical (Bu_3Sn·) is often used to abstract heavy atoms X from C−X bonds. The atom X is usually Br or I, but it may be Se or even S. The Sn−Br bond is strong enough that Bu_3Sn· can abstract Br· from aryl bromides to generate very high energy aryl radicals. Bu_3Sn· can itself be generated by H· abstraction from Bu_3SnH.

Food containing BHT is preserved from oxidative degradation because O_2 abstracts H·
from the phenol group of BHT instead of from the food. The resulting phenoxy radical is persistent and unreactive due to steric encumbrance and resonance stabilization.

* **Common error alert:** *Abstraction of C (e.g., X· + $R_3C-Y \rightarrow X-CR_3 + \cdot Y$) does
not occur in free-radical reactions.* An addition–fragmentation mechanism can
be drawn for an apparent atom abstraction of C(sp²) or C(sp). If your mechanism requires an abstraction of C(sp³), it is almost certainly incorrect. Even abstraction of O is very rare.

*Direct abstraction of
CN by the aryl radical
does not occur!*

*An addition–fragmentation
mechanism is much more
reasonable.*

BDEs offer a good guide to the likelihood of atom abstraction. The smaller
the BDE, the weaker the bond, and the more likely the bond will be broken by
atom abstraction.

* **Common error alert:** *Homolytic bond strengths are very different from heterolytic bond strengths.* In polar mechanisms, the likelihood of bond cleavage is
usually related to the heterolytic bond strength. In radical reactions, on the other
hand, the likelihood of bond cleavage is usually related to the homolytic bond
strength. For example, under basic conditions, the RO−H bond is very likely to
be cleaved in a heterolytic fashion (by deprotonation), whereas the Me₃C−H bond
is not. However, the RO−H bond has a much larger BDE than the Me₃C−H bond,
so abstraction of H· from RO−H by a free radical is much less favorable than
abstraction of H· from Me₃C−H by the same free radical.

The *rate* of the atom abstraction reaction, X· + Y−Z → X−Y + ·Z, is dependent not
only on the strength of the X−Y and Y−Z bonds but also on the strength of the X−Z
bond: The weaker the X−Z bond, the faster the abstraction. This phenomenon explains
why the abstraction of H· from ArOH by O_2 is so fast: the weakness of the O−O bond
causes the reaction to proceed very quickly.

The solvents of choice for free-radical reactions (CH_3OH, H_2O, benzene) share
the feature that their X−H bonds have very large BDEs, so that the solvents are
unlikely to participate in the reaction. Free-radical reactions executed in ether,
THF, toluene, CH_2Cl_2, and $CHCl_3$ are often complicated by atom abstraction
from the solvent. Acetone is sometimes used as a solvent for free-radical reac-

tions, because it can be photoexcited to act as an initiator, but it can also com-
plicate free-radical reactions by giving up H· to another radical to afford the relatively stable 2-oxopropyl (acetonyl) radical.

Unlike radical addition and fragmentation reactions, atom abstraction has no
common counterpart in carbocation chemistry.

• Two radicals can react with one another in one of two ways: *radical–radical combination* or *disproportionation*. Radical–radical combination is the opposite of bond homolysis. Each of the two radical centers contributes its lone
electron to form a new σ bond. Radical–radical combination is also known as
homogenesis (opposite of homolysis). Radical–radical combination reactions are
usually very fast and favorable. Most free radicals are very unstable because they
undergo very rapid radical–radical combination.

In *disproportionation*, one radical X· abstracts H· atom from another radical
H–Y–Z· to give the two electron-sufficient species X–H and Y=Z. The abstracted
atom is almost always H, and it is always β to the radical center Z. The two radicals may be the same or different. Disproportionation, like radical–radical combination, is a very fast and favorable process.

The reaction is called *disproportionation* because two radicals with the same formula
become two products with different formulas. (In the preceding example, C_8H_9 +
$C_8H_9 \rightarrow C_8H_8 + C_8H_{10}$.)

In chain mechanisms, radical–radical combination and disproportionation reactions are seen only in the termination part of the mechanism. The one exception is the combination of O_2 with a free radical, one of the propagation steps in
the autoxidation chain mechanism.

Nonchain mechanisms, though, often involve radical–radical combinations
and disproportionations. Nitroxyls can be used to trap alkyl radicals that may be
present in a reaction medium by a radical–radical combination reaction, giving
an electron-sufficient species that is more easily studied than the free radical. The
photochemical reactions of carbonyl compounds often involve radical–radical
combination and disproportionation steps.

• A radical can undergo *one-electron transfer* (oxidation or reduction) in the presence of an oxidizing or reducing agent to give an even-electron species. In nonchain mechanisms, a radical that undergoes electron transfer is itself usually generated by an electron transfer. A chain mechanism can be initiated or terminated by electron transfer. The initiation step and one of the propagation steps in the $S_{RN}1$ mechanism (Chapter 2) are electron transfers.

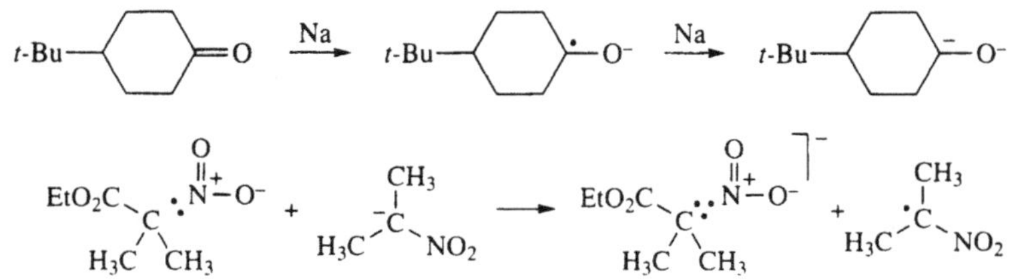

A curved arrow is *not* used to show an electron transfer.

• A radical anion can lose an anionic leaving group to give a neutral free radical, and in the reverse direction, a neutral free radical can combine with an anionic nucleophile to give a new radical anion. A radical cation can also combine with a nucleophile. Such steps occur in the $S_{RN}1$ mechanism (Chapter 2), in dissolving metal reductions, and in oxidative deprotections. These two-electron reactions have obvious counterparts in carbocation chemistry.

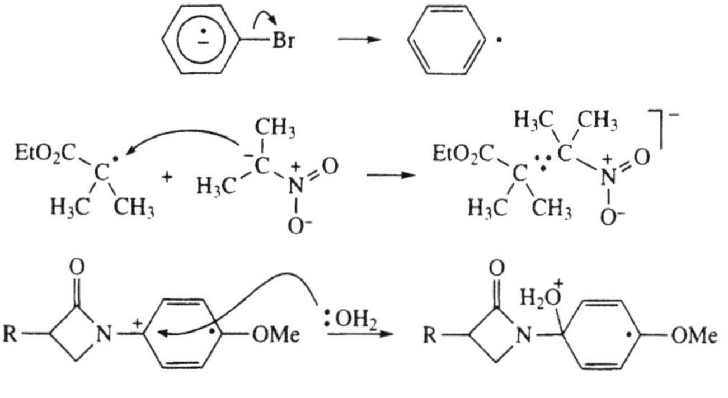

Analogies between free radicals and carbocations have been drawn several times, but one of the typical reactions of carbocations, the concerted 1,2-shift, does not occur in free radicals. A concerted 1,2-radical shift is allowed only when one of the components can react antarafacially. The 1,2-hydrogen atom shift is geometrically impossible, and the geometric requirements of the 1,2-alkyl radical shift are so stringent that it is not observed.

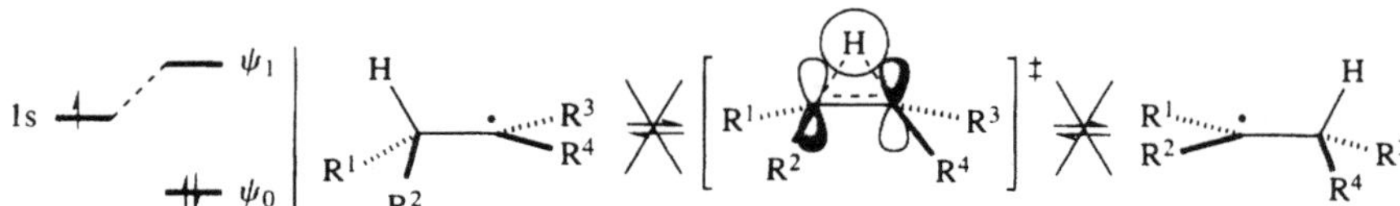

[1,2] hydrogen atom shift

Unsaturated groups, however, can shift by a two-step *addition–fragmentation* mechanism via a discrete cyclopropane radical intermediate. Phenyl groups are especially good at undergoing radical 1,2-shifts; 1,2-shifts of acyl, alkenyl, and, to a lesser extent, alkynyl and cyano groups are also seen.

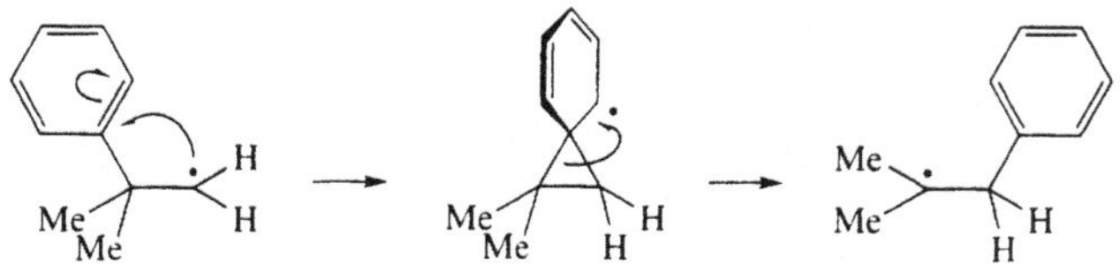

Note that the lowest energy radical that could be obtained would result from a 1,2-methyl shift, but this is not the shift that occurs, because 1,2-methyl shifts cannot occur in free radicals.

Problem 5.1. Draw a mechanism for the following radical rearrangement.

Heavy atoms such as Br and Se can also undergo a radical 1,2-shift by an addition–fragmentation mechanism. An intermediate that features a heavy atom with a *nonet* of electrons intervenes. First-row elements cannot accommodate a nonet of electrons like heavy elements can, so the addition–fragmentation mechanism is disallowed for them.

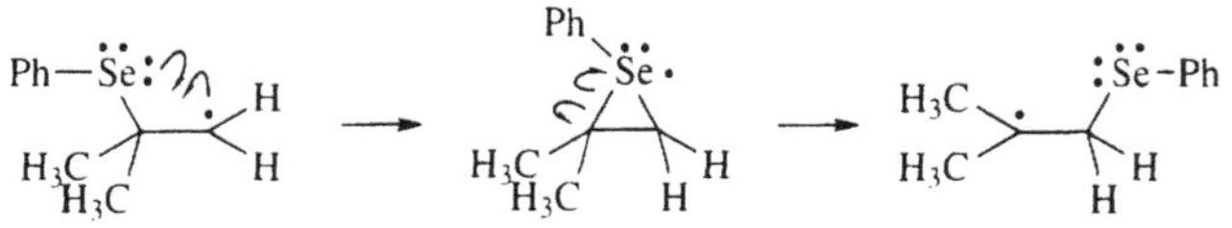

nine-electron Se intermediate!

5.1.4   Chain vs. Nonchain Mechanisms

A free-radical reaction may proceed by a chain or a nonchain mechanism. There are many experimental methods for determining whether a chain or a nonchain mechanism is operative in a reaction, but these methods don't help much in pencil-and-paper problems such as the ones in this book. Luckily, the reagents or reaction conditions will usually indicate which type of mechanism is operative.

• Reactions involving stoichiometric amounts of one-electron reducing or oxidizing agents such as Li, Na, SmI_2, or $Mn(OAc)_3$ always proceed by *nonchain mechanisms*, as free electrons are available throughout the reaction to quench the intermediate radicals. Reactions that generate two radicals simultaneously and in close proximity, such as cycloaromatizations and many photochemical reactions, also usually proceed by nonchain mechanisms.

• All chain reactions require an initiator and there are only a few widely used initiators (O_2, peroxides, AIBN, $h\nu$), so the presence of these initiators is a good sign of a *chain mechanism*. Be aware, though, that the (apparent) absence of an initiator does not rule out a chain mechanism. For example, the O_2 in the ambient atmosphere can act as an initiator, and many 3° alkyl iodides can undergo spontaneous C–I bond homolysis to initiate a chain reaction upon gentle heating. If a free-radical reaction involves a tin compound (e.g., Bu_3SnH), it proceeds by a chain mechanism. However, not all reactions involving Bu_3SnH are free-radical reactions. (A few transition-metal-catalyzed reactions also use Bu_3SnH.)

You may have noticed that $h\nu$ may indicate either a nonchain or a chain mechanism. A good rule of thumb for distinguishing light-initiated nonchain and chain mechanisms is that unimolecular rearrangements or eliminations usually proceed by nonchain mechanisms, whereas addition and substitution reactions (and especially intermolecular ones) almost always proceed by chain mechanisms. However, there are a few exceptions to this rule (photochemical pinacol reaction, Barton reaction). Of course, many pericyclic reactions require light, too (Chapter 4)!

5.2   Chain Free-Radical Reactions

5.2.1   Substitution Reactions

Probably the best-known example of a free-radical reaction is the halogenation of alkanes with Br_2 or NBS. This chain reaction is initiated by homolytic cleavage of Br_2 induced by light. The propagation part consists of two atom abstraction reactions.

Example

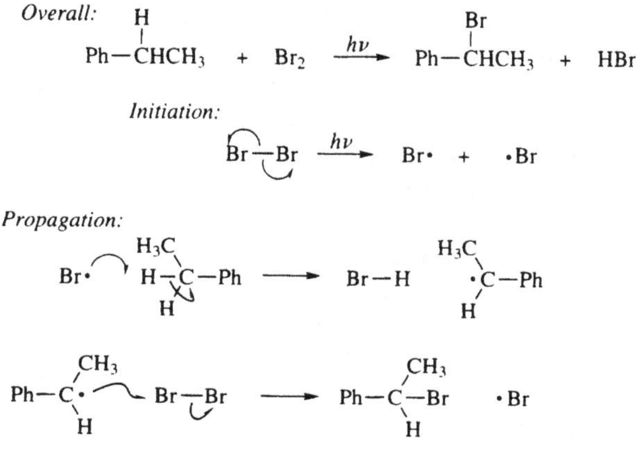

Termination:

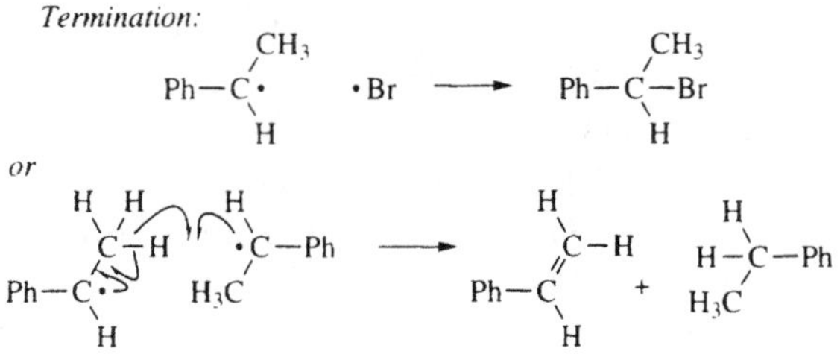

or

or *others.*

The termination parts of free-radical chain mechanisms are almost always the same: two free radicals from the propagation part react with each other either in a radical–radical combination or disproportionation reaction. The termination parts of chain mechanisms will not routinely be drawn in this text.

Free-radical halogenation is most commonly applied to allylic or benzylic halogenation, as the radicals formed at these positions are lowest in energy. Of course, in allylic halogenation, transposition of the double bond can easily occur.

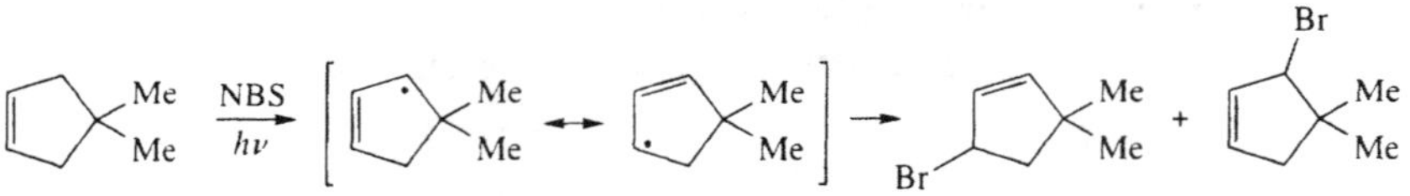

In brominations with NBS, it is thought that Br_2 is the actual halogenating agent. The Br_2 is generated by reaction of HBr (the by-product of halogenation) with NBS. This polar reaction intervenes between the H· abstraction and Br· abstraction steps of the propagation part of the chain mechanism.

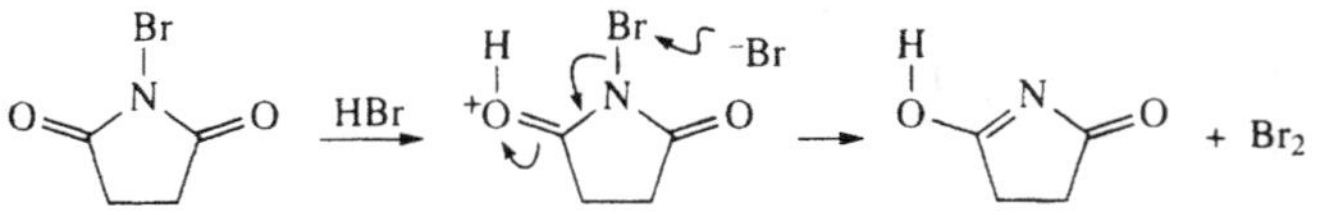

Elemental chlorine can be used in free-radical halogenation reactions, too, but these reactions are less easily controlled, because the Cl· radical is more reactive than the Br· radical and hence less selective. The reagents t-BuOCl and SO_2Cl_2 are used as alternative chlorinating agents. The F· radical is so reactive, and the reaction $F–F + C–H \rightarrow H–F + C–F$ is so exothermic, that free-radical fluorinations result in violent and uncontrollable exotherms (explosions). At the other extreme, free-radical iodinations of alkanes do not work well at all, as the H· abstraction step is too endothermic.

Free-radical *dehalogenations* are common, too. In these reactions, a C–X bond is replaced with a C–H bond. The reducing agent is most commonly Bu_3SnH, although $(Me_3Si)_3SiH$ or a combination of a catalytic amount of Bu_3SnCl and a stoichiometric amount of $NaBH_4$ may be used instead as environmentally more friendly reagents. A catalytic amount of AIBN or $(BzO)_2$ is commonly used as a free-radical initiator. Initiation involves abstraction of H·

from Bu_3SnH by a radical derived from the initiator. In the propagation part, $Bu_3Sn\cdot$ abstracts $X\cdot$ from the C–X bond to give Bu_3SnX and an alkyl radical, which then abstracts $H\cdot$ from Bu_3SnH to give the alkane and regenerate $\cdot SnBu_3$.

Example

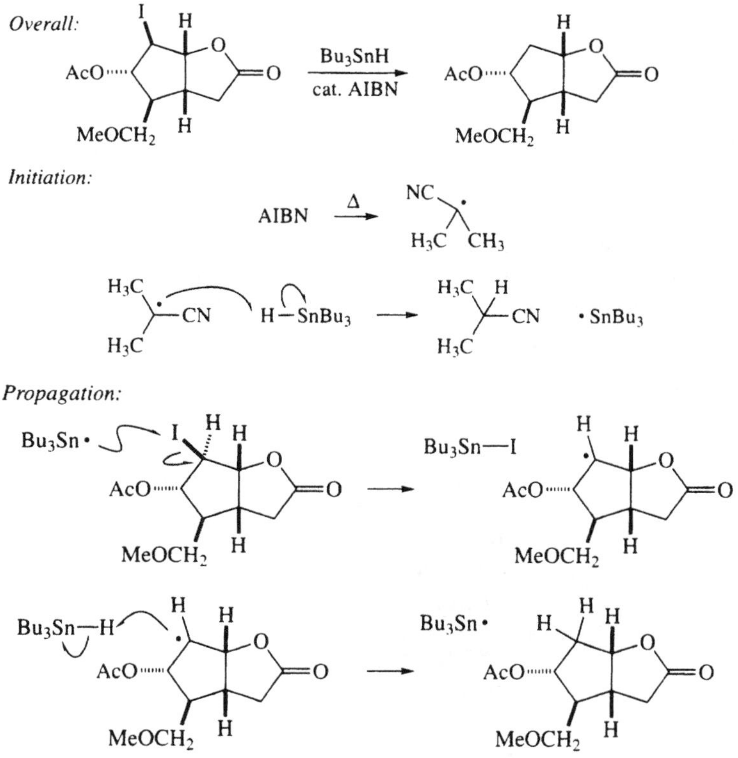

Overall:

Initiation:

Propagation:

Alcohols are deoxygenated in the *Barton–McCombie reaction*. The alcohol is first converted into a xanthate ($ROCS_2CH_3$) or another thiocarbonyl compound, and then the entire functional group is removed by Bu_3SnH and replaced with H. In the propagation part of the mechanism, $Bu_3Sn\cdot$ adds to S of the S=C bond, and the resulting radical fragments to give the alkyl radical and a carbonyl compound. The alkyl radical then abstracts $H\cdot$ from Bu_3SnH to regenerate the chain-carrying radical $\cdot SnBu_3$. Formation of the strong C=O π bond provides the driving force for fragmentation of the C–O σ bond.

Example

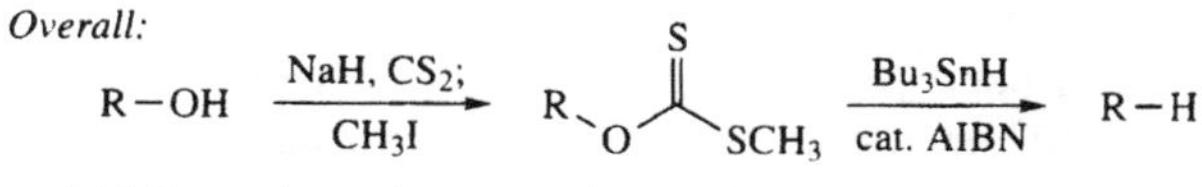

Overall:

Initiation as in previous examples.

Propagation:

$$R_O \overset{S \cdot SnBu_3}{\underset{SCH_3}{\bigg|\bigg|}} \longrightarrow R_O \overset{S \cdot SnBu_3}{\underset{SCH_3}{\cdot}}$$

$$R_O \overset{S \cdot SnBu_3}{\underset{SCH_3}{\cdot}} \longrightarrow R \cdot \; + \; \underset{SCH_3}{\overset{S \cdot SnBu_3}{\bigg|}}_O$$

$$R \cdot \overset{\frown}{} H \!-\! SnBu_3 \longrightarrow R\!-\!H \; + \; \cdot SnBu_3$$

The substitution of H with OOH in alkanes is called *autoxidation.* ("Autoxidation" is a misnomer because the substrate is not oxidizing itself; O_2 is doing the oxidizing!) Autoxidation proceeds by a free-radical chain mechanism. Note that the mechanism for oxidation includes a very rare radical–radical combination step in the propagation part. The radical–radical combination step doesn't terminate the chain in this particular reaction because O_2 is a 1,2-diradical.

Example

Overall:

$$EtO-\overset{H}{\underset{}{C}}HCH_3 \; + \; O_2 \longrightarrow EtO-\overset{OOH}{\underset{}{C}}HCH_3$$

Initiation:

$$\cdot O\!-\!O\cdot \overset{\frown}{} H\!-\!\overset{H_3C}{\underset{H}{C}}\!-\!OEt \longrightarrow \cdot O\!-\!O\!-\!H \quad \overset{H_3C}{\underset{H}{\cdot C}}\!-\!OEt$$

Propagation:

$$EtO\!-\!\overset{CH_3}{\underset{H}{C}}\!\cdot \overset{\frown}{} \cdot O\!-\!O\cdot \longrightarrow EtO\!-\!\overset{CH_3}{\underset{H}{C}}\!-\!O\!-\!O\cdot$$

$$EtO\!-\!\overset{CH_3}{\underset{H}{C}}\!-\!O\!-\!O\cdot \overset{\frown}{} H\!-\!\overset{H_3C}{\underset{H}{C}}\!-\!OEt \longrightarrow EtO\!-\!\overset{CH_3}{\underset{H}{C}}\!-\!O\!-\!O\!-\!H \quad \overset{H_3C}{\underset{H}{\cdot C}}\!-\!OEt$$

Ethers are most prone to autoxidation because O can stabilize an adjacent alkyl radical, but aldehydes are also quite prone to autoxidation. In the latter case, the immediate products, carboxylic peracids, react with the starting aldehydes to give carboxylic acids by a Baeyer–Villiger reaction (Chapter 2).

$$\underset{Ph}{\overset{O}{\big\|}}\!H \xrightarrow{O_2} \underset{Ph}{\overset{O}{\big\|}}\!OOH \xrightarrow{PhCHO} 2\; \underset{Ph}{\overset{O}{\big\|}}\!OH$$

Autoxidation is one of the key steps in the industrial synthesis of phenol and acetone from benzene and propylene. In the second step of this synthesis, cumene (isopropylbenzene) is autoxidized to give cumyl hydroperoxide.

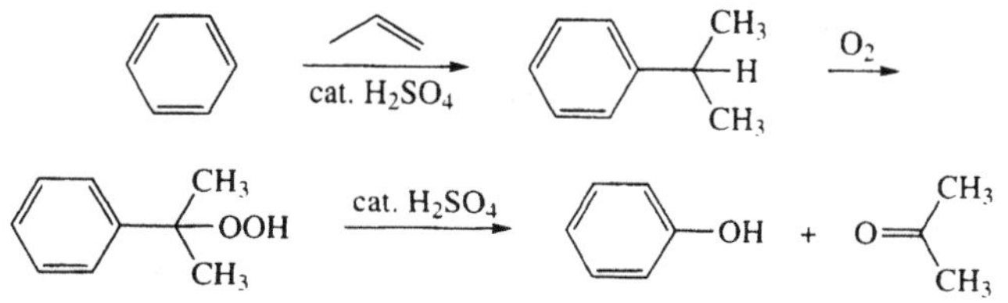

Problem 5.2. Draw mechanisms for all three steps of the phenol synthesis.

Autoxidation is one of the most important abiotic processes by which degradation of organic compounds occurs. Autoxidation of ethers like diethyl ether, THF, and other common solvents leads to the formation of hydroperoxides, which are explosively heat- and shock-sensitive. Diisopropyl ether is so prone to autoxidation that it is necessary to dispose of it *immediately* after opening a bottle of it. Autoxidation of biological compounds may be partly responsible for the aging process. The autoxidation of oily rags in garages releases heat, accelerating the autoxidation process, leading more heat to be released and eventually causing a fire.

In the *Hofmann–Loeffler–Freytag* reaction, an *N*-chloroammonium ion is converted by a free-radical substitution reaction into a 4-chloroalkylammonium ion, which then undergoes intramolecular S$_N$2 substitution to give a pyrrolidine. In the free-radical substitution mechanism, the abstraction step occurs in an intramolecular fashion. Entropic and stereoelectronic factors make the regioselectivity very high for the C4 hydrogen.

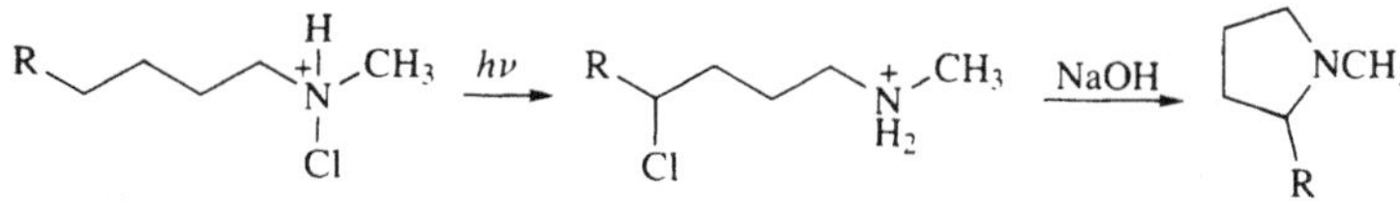

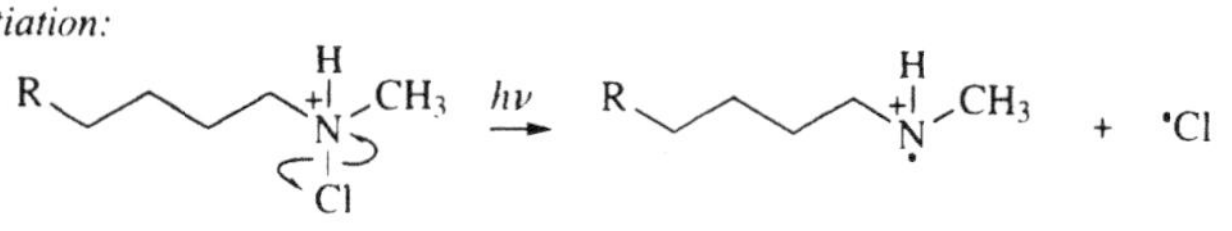

Example

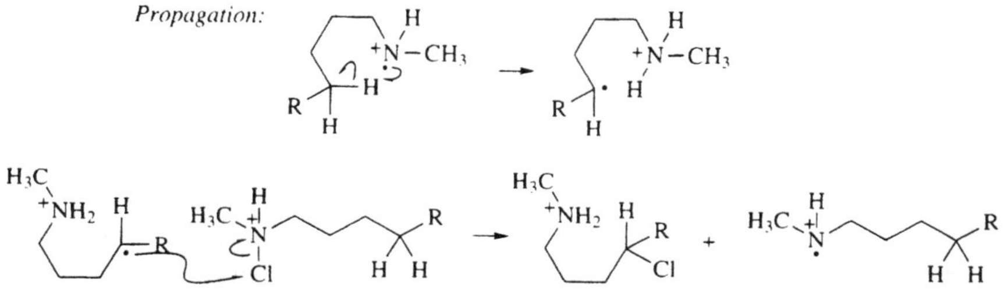

5.2.2   Addition and Fragmentation Reactions

5.2.2.1   Carbon–Heteroatom Bond-Forming Reactions

The anti-Markovnikov addition of HBr to alkenes was probably the first free-radical addition reaction to be discovered. The discovery was inadvertent; around the turn of the twentieth century, scientists studying the regiochemistry of addition of HBr to alkenes found that the proportion of Markovnikov to anti-Markovnikov addition products varied inexplicably from run to run. Eventually, it was discovered that impurities such as O_2 and peroxides greatly increased the amount of anti-Markovnikov addition product. The results were later explained by a free-radical addition mechanism. The anti-Markovnikov regiochemistry derives from the addition of the Br· radical to the less substituted C of the alkene (steric reasons) to give the lower energy, more substituted radical (electronic reasons). In a polar reaction, Br· would add to the more substituted C of the alkene.

Example

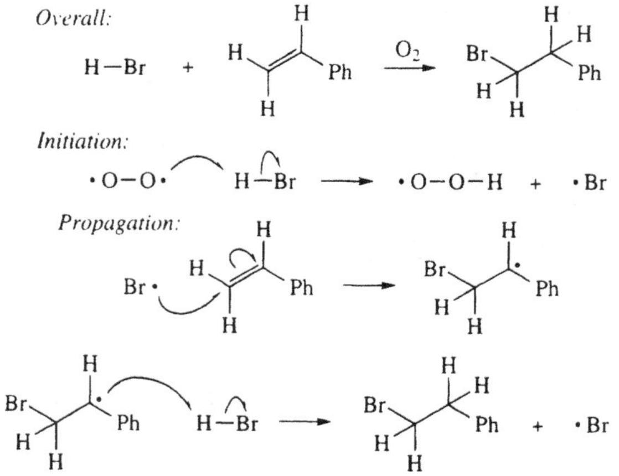

In contrast to HBr, the acids HCl and HI do not undergo free-radical addition to alkenes, even in the presence of peroxides or O_2. Abstraction of H· from HCl is too endothermic, and addition of I· to an alkene is too endothermic. However, thiols (RSH) add to alkenes by a free-radical mechanism exactly analogous to the addition of HBr. The initiator is usually AIBN or $(BzO)_2$. The alkene may be electron-rich or electron-poor. Note that the conjugate addition of thiols to electron-poor alkenes can occur either by a free-radical mechanism or by a polar, nucleophilic mechanism.

Problem 5.3. Draw a free-radical mechanism for the following addition reaction.

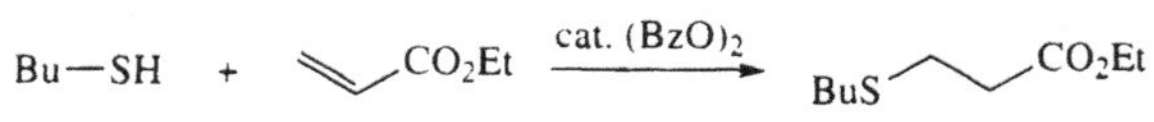

Tributyltin hydride adds across C=C π bonds by a free-radical mechanism. Addition of Bu_3SnH across alkynes is one of the best ways of making alkenyltin compounds, which are useful reagents in organic synthesis. The mechanism is exactly the same as the one for addition of HBr across an alkene. The intermediate alkenyl radical is stabilized by hyperconjugation with the relatively high energy C–Sn bond.

Problem 5.4. Draw a free-radical mechanism for the following addition reaction.

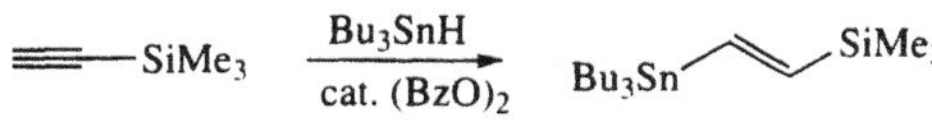

5.2.2.2   Carbon–Carbon Bond-Forming and Bond-Cleaving Reactions

The most important free-radical chain reaction conducted in industry is the free-radical polymerization of ethylene to give polyethylene. Industrial processes usually use $(t\text{-BuO})_2$ as the initiator. The t-BuO· radical adds to ethylene to give the beginning of a polymer chain. The propagation part has only one step: the addition of an alkyl radical at the end of a growing polymer to ethylene to give a new alkyl radical at the end of a longer polymer. The termination steps are the usual radical–radical combination and disproportionation reactions.

> ***Example***
>

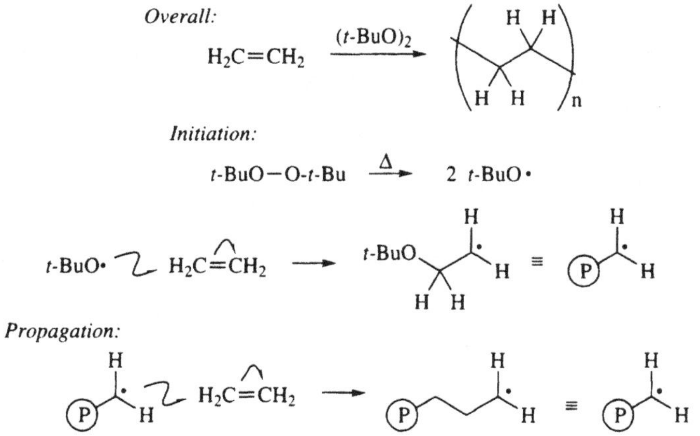

Polyethylene is used for everything from sandwich wrap to soda bottles to park benches. Sitting on a park bench does not feel the same as wrapping one's buttocks in sandwich wrap, though, so it is apparent that the physical properties of polyethylene can vary tremendously. The physical properties of a particular batch of polyethylene depend partly on the amount and kind of branching in the polymer chains. Polyethylene can have either short or long branches. A short branch in a polymer chain is created when the radical at the end of a growing polymer abstracts H· from a C atom four or five atoms back in the *same* chain. The radical thus generated continues to polymerize, giving a four- or five-carbon side chain.

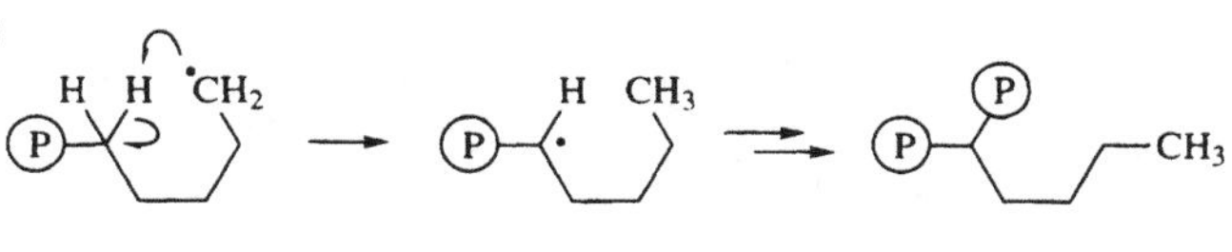

By contrast, a long branch in a polymer chain is created when the radical at the end of a growing polymer abstracts H· from the middle of *another* polymer chain. The radical thus generated continues to polymerize, giving a polymeric side chain.

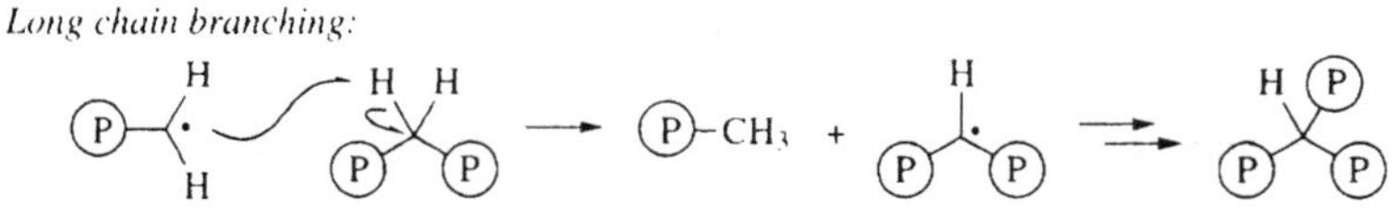

Long chain branching:

Reaction conditions (ethylene density, polymer density, temperature, initiator concentration, pressure) during the polymerization can be altered to provide more or less short and long branching. Other factors such as average molecular weight and polydispersity (the degree of variation of chain length among the molecules in the sample) also affect the properties of polyethylene.

In the laboratory, free-radical addition reactions often use $Bu_3Sn·$ radicals as key chain carriers. The initiation part of such reactions usually has a radical initiator (often derived from AIBN) abstract H· from $HSnBu_3$ to give $Bu_3Sn·$. The propagation part of the mechanism usually involves (1) abstraction of ·Br, ·I, or ·SeR by $Bu_3Sn·$ to give an alkyl radical, (2) addition of an alkyl radical to a π bond, and (3) abstraction of H· from $HSnBu_3$ by an alkyl radical to give the product and regenerate ·$SnBu_3$.

Example

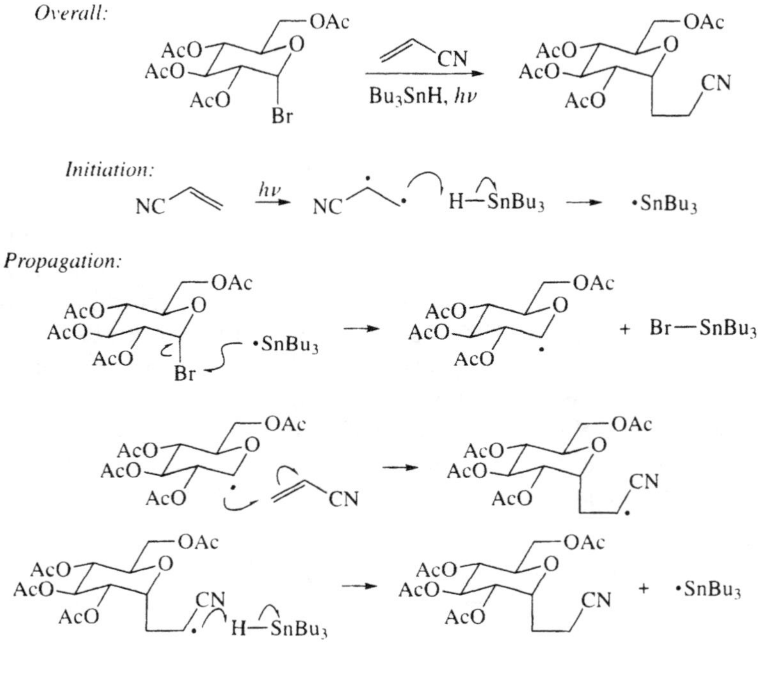

Overall:

Initiation:

Propagation:

Free-radical cyclization reactions (i.e., the intramolecular addition of an alkyl radical to a C=C π bond) have emerged as one of the most interesting and widespread applications of free-radical chemistry to organic synthesis. Free-radical cyclizations are useful because they are so fast. The cyclization of the 5-hexenyl radical to the cyclopentylmethyl radical is very fast, occurring at a rate of about 1.0×10^5 s^{-1}. In fact, the rate of formation of the cyclopentylmethyl radical is much faster than the rate of cyclization to the lower energy cyclohexyl radical. This *stereoelectronic* effect is derived from the fact that the overlap between the p orbital of the radical and the $\pi*$ MO of the double bond is much better when C1 attacks C5 than when it attacks C6. The relative rates of 5-exo and 6-endo ring closures are strongly dependent on the nature of the substrate and especially on the amount of substitution on the π bond. Cyclization of the 6-heptenyl radical in the 6-exo mode is also very favorable.

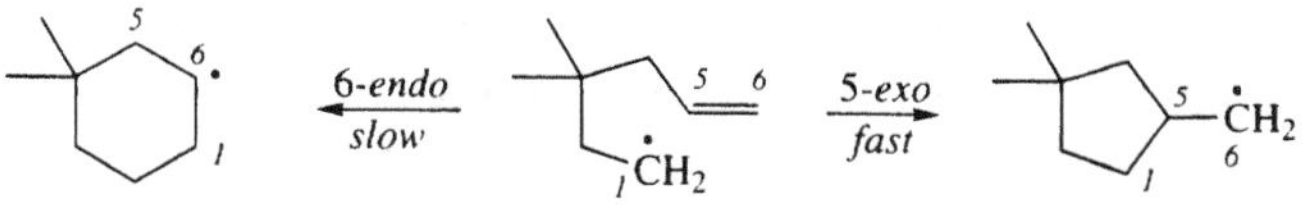

The mechanisms of intramolecular free-radical cyclization reactions are no different from their intermolecular counterparts.

Example

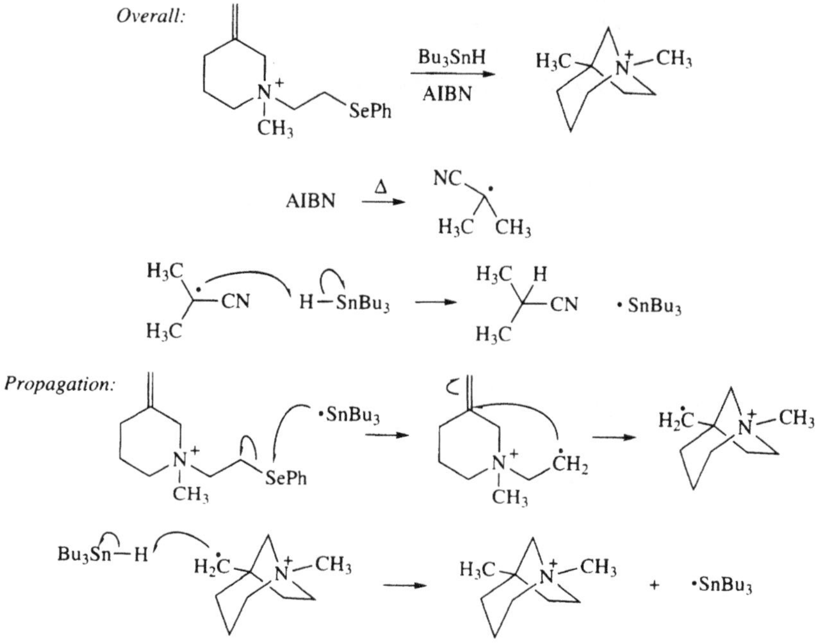

A two-step addition–fragmentation process involving a nine-electron Se atom intermediate could be drawn in place of the atom abstraction reaction.

Problems 5.5. Draw mechanisms for the following free-radical cyclization reactions. All cyclization steps should give 5- or 6-membered rings, if possible.

(a)

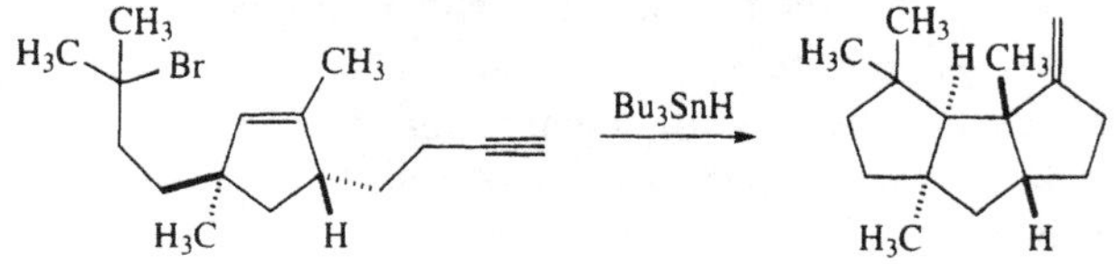

Bu₃SnH

(b)

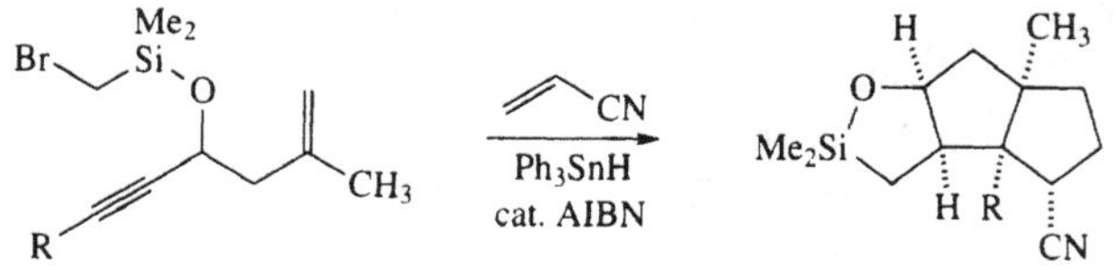

A free radical can add to CO or an isocyanide (RNC) in the course of a free-radical cyclization reaction, too, to give an acyl radical (RĊ=O) or an *iminyl* radical (RĊ=NR), either of which can undergo further reactions. In the following example, an alkyl radical adds to the terminal C of *t*-BuNC to give an iminyl radical. The iminyl radical then fragments to give *t*-Bu· and an alkyl cyanide :N≡CR. In a different substrate, the iminyl radical may undergo an addition or an abstraction reaction instead.

A second interesting feature of the following reaction is that a catalytic amount of Bu₃SnCl and a stoichiometric amount of NaBH₃CN are added to the reaction mixture. The small amount of Bu₃SnCl that is added to the reaction mixture is reduced to Bu₃SnH with a small amount of NaBH₃CN. After each catalytic cycle, the Bu₃SnH is converted to Bu₃SnI, and the stoichiometric NaBH₃CN reduces it back to Bu₃SnH. The technique allows one to minimize the quantity of malodorous and toxic Sn compounds used in the reaction.

Example

Overall:

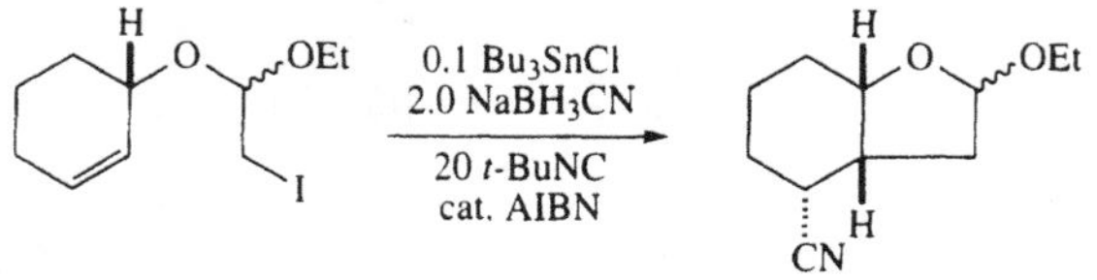

Propagation:

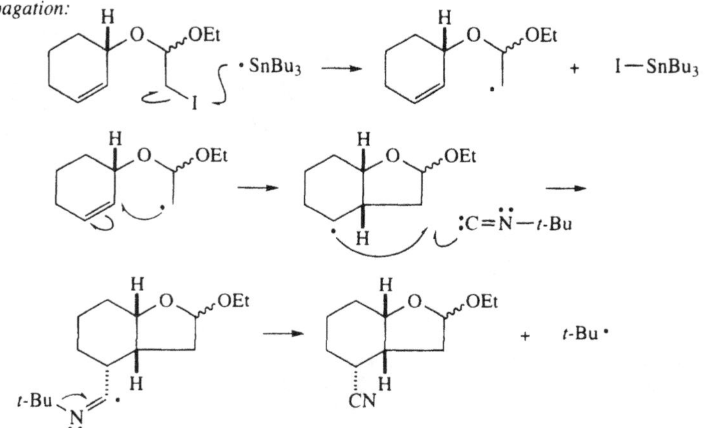

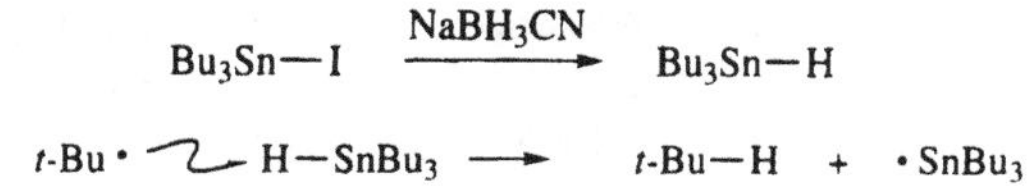

Bu$_3$Sn—I $\xrightarrow{\text{NaBH}_3\text{CN}}$ Bu$_3$Sn—H

t-Bu$\cdot$ ⌒ H—SnBu$_3$ $\longrightarrow$ t-Bu—H + $\cdot$SnBu$_3$

Problem 5.6. Draw a mechanism for the following free-radical cyclization reaction.

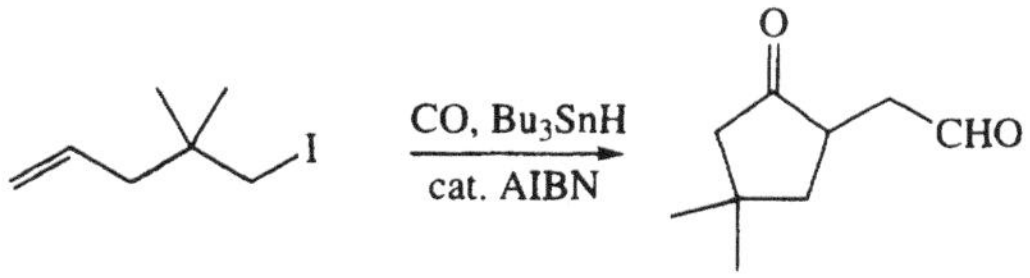

Some free-radical cyclization reactions require only a catalytic amount of Bu$_3$SnH or Bu$_3$SnSnBu$_3$. Such a reaction, instead of replacing a C—X bond with a C—H bond, simply relocates the X atom. These *atom transfer cyclizations* differ from more conventional free-radical cyclizations in that the last step in the propagation involves abstraction of X$\cdot$ from a C—X bond not by $\cdot$SnBu$_3$ but by the cyclized starting material. The last step is kinetically viable only when a stronger bond is made from a weaker one. In the following example, a Si—I bond is made at the expense of a C—I bond.

Example

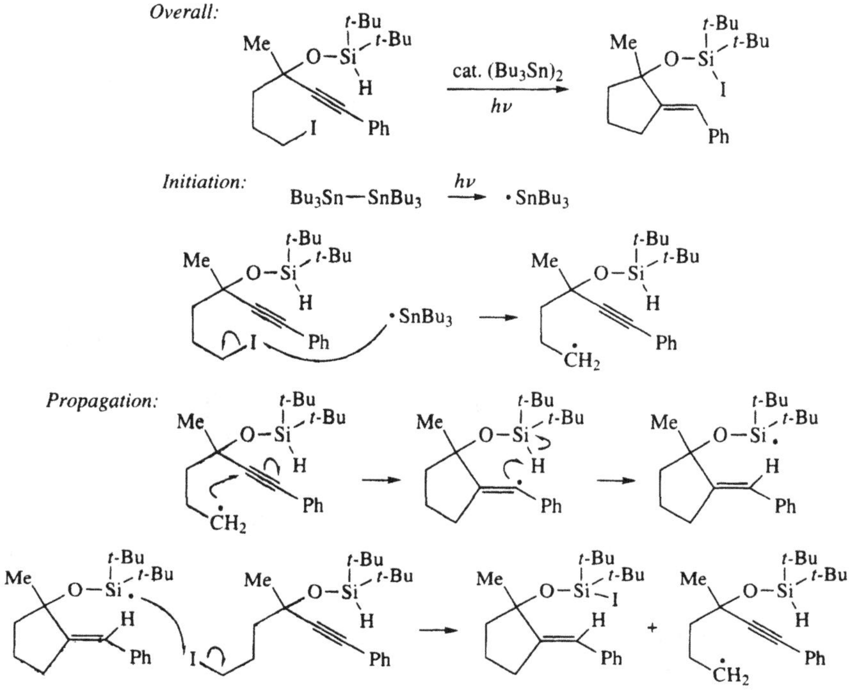

Compounds with weak C—H bonds can add to alkenes by a free-radical chain mechanism. Compounds that can add to alkenes in this way include RCHO

compounds (aldehydes, formates, etc.) and 1,3-dicarbonyl compounds. In the initiation part of the mechanism, an initiator radical abstracts H· from the weak C–H bond to give an alkyl radical. In the propagation part of the mechanism, the alkyl radical adds across the C=C π bond, and then the new radical abstracts H· from the weak C–H bond to give the product and regenerate the first alkyl radical.

Example

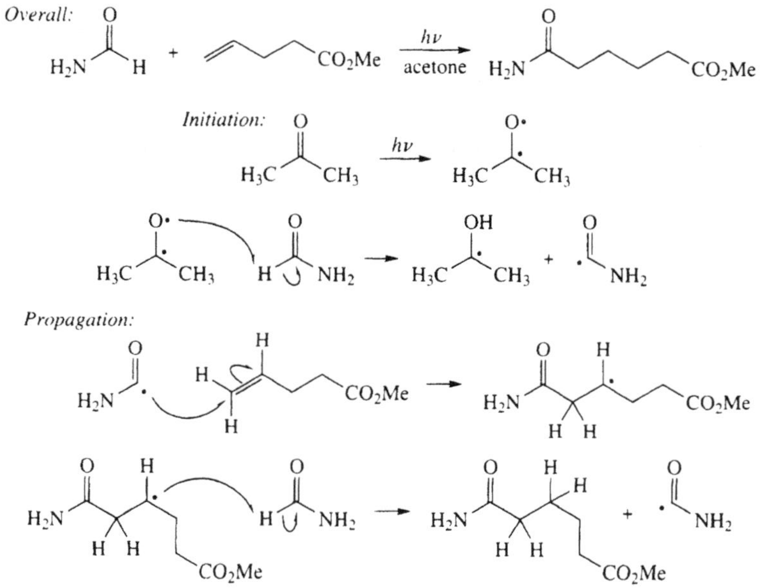

Overall:

Initiation:

Propagation:

Problem 5.7. Draw free-radical mechanisms for the following addition reactions.

(a)

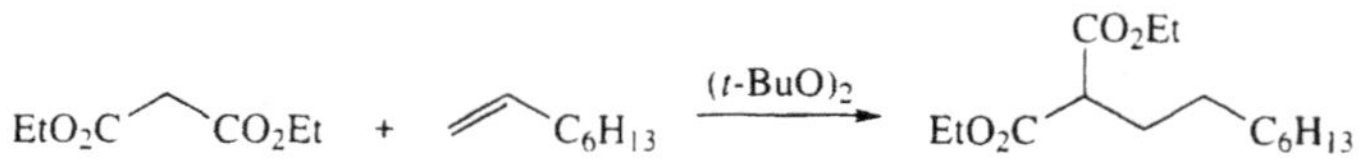

(b)

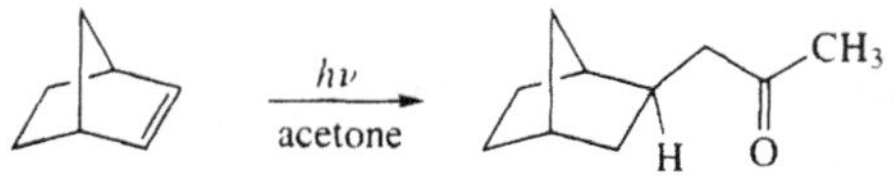

The addition of a free radical to a π bond is sometimes followed directly by fragmentation of the new free radical. For example, alkyl halides and selenides react with (allyl)SnBu$_3$ in a *free-radical allylation reaction* to give a product in which the halogen or selenium is replaced with an allyl group. In the propagation part of the mechanism, Bu$_3$Sn· abstracts X· from the starting material to generate an alkyl radical, which adds to the terminal C of the allyl group of the stannane. Fragmentation then occurs to give the product and to regenerate ·SnBu$_3$.

Example

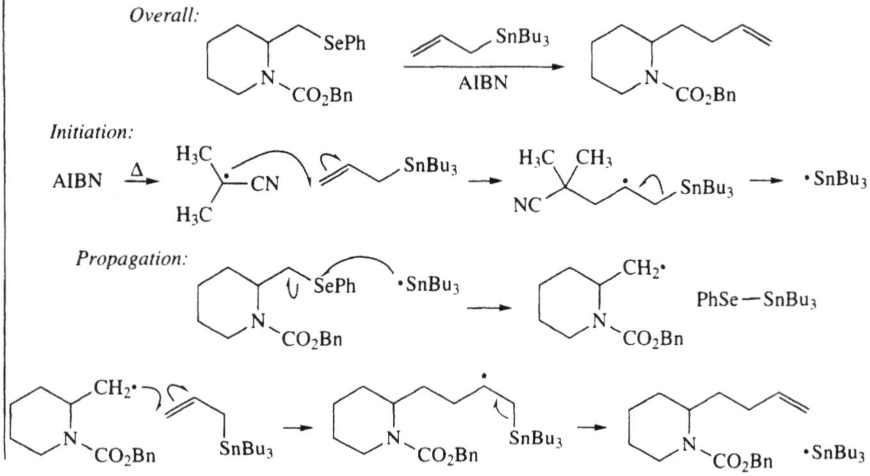

Overall:

Initiation:

Propagation:

Problem 5.8. Draw an addition–fragmentation mechanism for the following reaction.

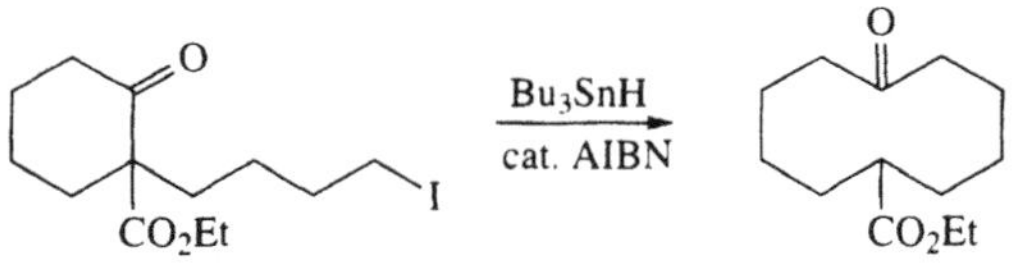

The fragmentation of cyclopropylmethyl radicals to give 3-butenyl radicals is one of the fastest reactions known, occurring at a rate of 2.1×10^8 s^{-1}! Phenyl-substituted cyclopropylmethyl groups open at even faster rates. These *radical clocks* can be used to probe the mechanism of very fast reactions, especially those in enzymes. If a substrate undergoes a reaction that involves a radical intermediate, and the lifetime of the intermediate is longer than about 10^{-8} s, then an analog of that substrate that has a cyclopropyl group attached to the putative radical center will give a product in which the cyclopropyl ring has opened. Lack of ring opening does not mean that there was no radical intermediate; instead, it simply means that the lifetime of the radical intermediate was shorter than the lifetime of the cyclopropyl-methyl radical. However, if the lifetime of a radical intermediate is shorter than the fastest reaction known, then, practically speaking, one can say that it does not exist.

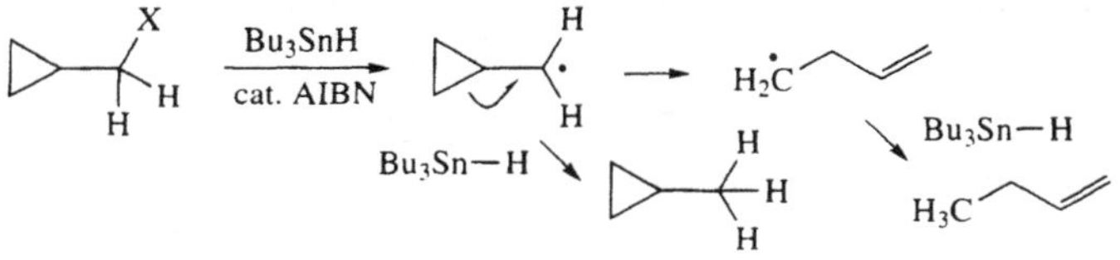

In the *Hunsdiecker reaction*, the silver salt of a carboxylic acid (RCO$_2$Ag) is treated with Br$_2$ to give an alkyl bromide RBr with one fewer C atom. The reaction does not work well with aromatic acids, suggesting that a free-radical mechanism is involved. The carboxylate and bromine react to give an acyl hypobromite, which decomposes by a free-radical chain mechanism.

Example

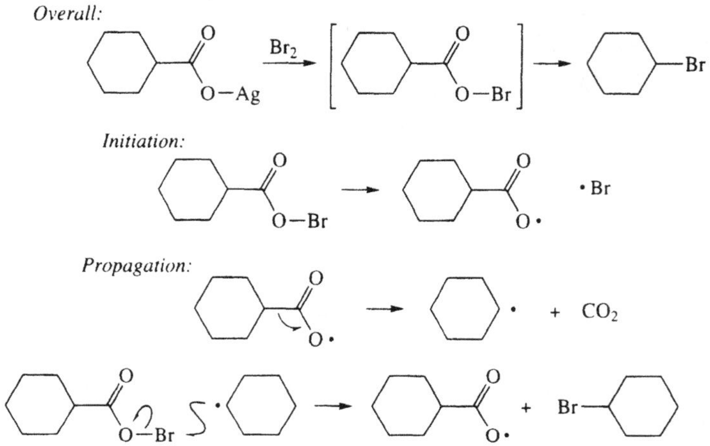

Overall:

Initiation:

Propagation:

5.3   Nonchain Free-Radical Reactions

The major classes of nonchain free-radical reactions are photochemical reactions, reductions and oxidations with metals, and cycloaromatizations.

5.3.1   Photochemical Reactions

The photoexcitation of a carbonyl compound to give a 1,2-diradical is often followed by a fragmentation reaction. The fragmentation can take one of two courses, called *Norrish type I* and *Norrish type II* cleavages. In Norrish type I cleavage, the bond between the α-carbon and the carbonyl carbon cleaves to give an acyl radical and an alkyl radical. These radicals can subsequently undergo a variety of reactions. For example, the acyl radical can decarbonylate to give a new alkyl radical, which can undergo radical–radical combination with the alkyl radical generated in the Norrish cleavage. Alternatively, the acyl and alkyl radicals can undergo disproportionation to give an aldehyde and C=C π bond. A third posibility, radical–radical recombination, simply regenerates the starting material.

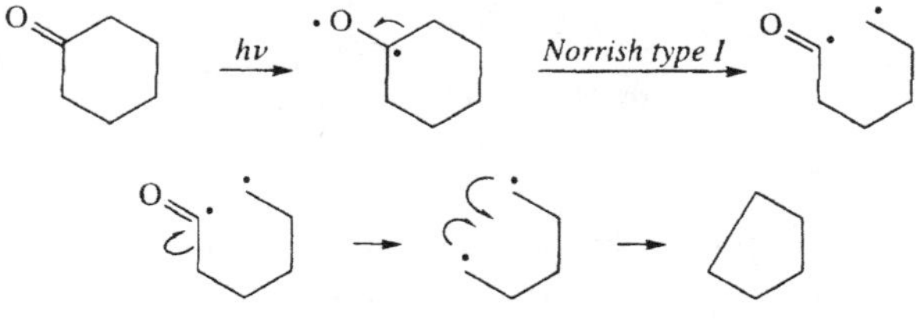

In Norrish type II cleavage, the O radical abstracts H from the γ-carbon in a six-membered TS, and the 1,4-diradical then fragments to give an alkene and an enol, the latter of which tautomerizes to the ketone. Sometimes, the 1,4-diradical undergoes radical–radical combination to give a cyclobutane, instead. The Norrish type II cleavage is closely related to the *McLafferty rearrangement* that is often seen in the mass spectra of carbonyl compounds.

Problem 5.9. The 2-nitrobenzyl group is used as a base- and acid-stable but photolabile protecting group for alcohols. Draw a Norrish mechanism for the release of an alcohol from a 2-nitrobenzyl ether upon photolysis.

Photoexcited ketones can undergo other reactions typical of free radicals. Benzophenone dissolved in isopropyl alcohol undergoes a *pinacol coupling* merely upon exposure to sunlight. The *i*-PrOH acts as a hydrogen atom donor.

In the *Barton reaction*, an alkyl nitrite is converted into an alcohol–oxime. The nitrite is photoexcited to a 1,2-diradical. It fragments to NO and an alkoxy radical (RO·), and the latter abstracts H· from the nearest C–H bond. The resulting alkyl radical then combines with NO to give a nitroso compound, which then tautomerizes to the oxime, probably by a polar stepwise mechanism. The Barton reaction has been used for the remote functionalization of hydrocarbons, especially steroids.

Example

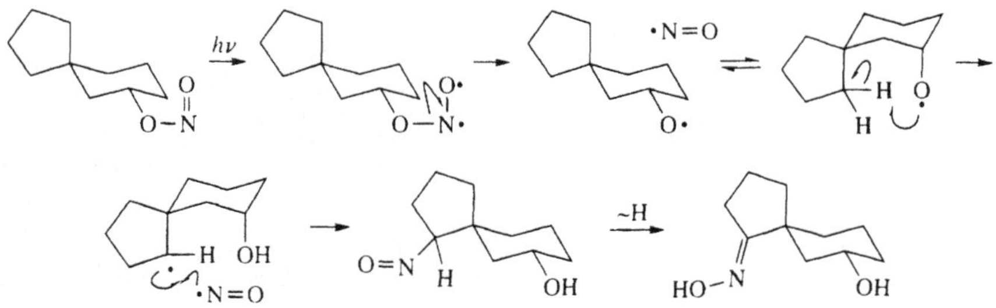

One could write a perfectly reasonable chain mechanism for this reaction, but a non-chain mechanism is appropriate because NO is a stable free radical, and it hangs around until it is able to combine with the alkyl radical to form a strong C–N bond.

Finally, it should be mentioned that the photochemically allowed [2 + 2] cycloaddition reaction of alkenes can be considered to be a radical-mediated process. Photoexcitation of an alkene gives a 1,2-diradical. The radical at C1 adds to one terminus of the other alkene to give a 1,4-diradical, which then cyclizes to give the observed product. Spectroscopic measurements at the femtosecond time scale have recently proven that the 1,4-diradical is a true intermediate along the reaction pathway of [2 + 2] cycloadditions. However, the lifetime of the 1,4-diradical is shorter than the rate of rotation about C–C σ bonds, as [2 + 2] cycloadditions are stereospecific.

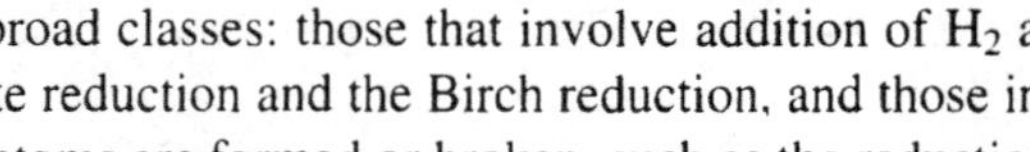

5.3.2   Reductions and Oxidations with Metals

Reductions for which group 1 metals are used in organic chemistry can be divided into two broad classes: those that involve addition of H_2 across a π bond, such as conjugate reduction and the Birch reduction, and those in which σ bonds between non-H atoms are formed or broken, such as the reduction of C–X bonds and acyloin and pinacol condensations.

5.3.2.1   Addition of H_2 across π Bonds

Reductions involving addition of H_2 across a π bond follow one or both of the following two sequences of events.

1. An electron is transferred from the metal to the substrate to give a radical anion.
2. A second electron transfer to the radical anion occurs to give a closed-shell dianion.
3. The dianion is protonated to give a closed-shell anion.

or

1. An electron is transferred from the metal to the substrate to give a radical anion.

2. The radical anion is protonated to give a neutral radical.

3. A second electron transfer occurs to give a closed-shell anion.

The solvent is usually liquid ammonia, in which the electrons from the metal are solvated, giving a dark blue solution. The proton source is usually an alcohol or ammonia itself.

Dissolving metals have been used for the *conjugate reduction* of α,β-unsaturated ketones for a very long time, and the method is still one of only a few good ones for accomplishing this transformation. In conjugate reduction, two electron transfers occur to give a closed-shell dianion. Protonation of the dianion occurs to give the thermodynamically more stable (usually) enolate. The enolate may be converted to the ketone by an aqueous workup, to the silyl enol ether by treatment with Me$_3$SiCl, or to an alkylated ketone by addition of an alkyl halide.

Example

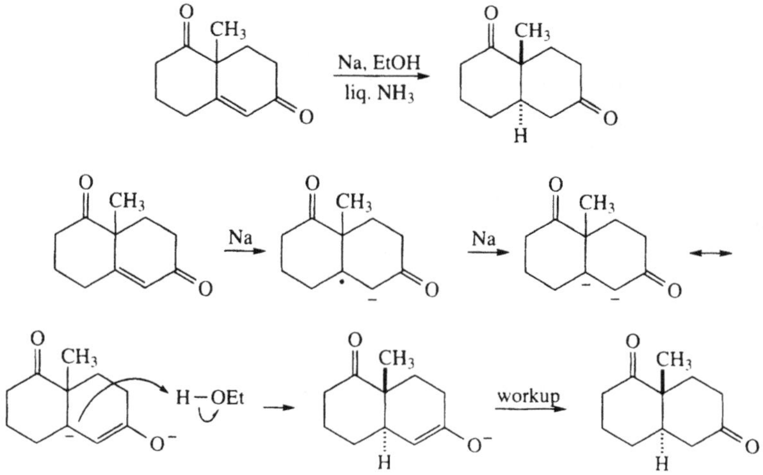

Conjugate reduction is also useful for the regiospecific generation of enolates. Deprotonation of a 3-substituted cyclohexanone gives a nearly equimolar ratio of regioisomeric enolates, but either enolate can be regiospecifically generated by conjugate reduction of the appropriate cyclohexenone.

Deprotonation leads to a mixture of regioisomeric enolates...

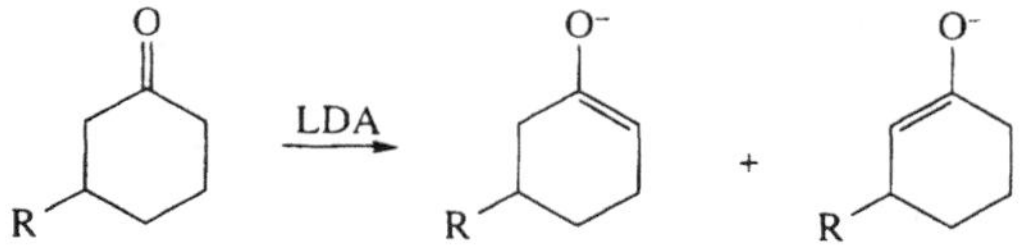

... but conjugate reduction leads to regiospecific enolate formation.

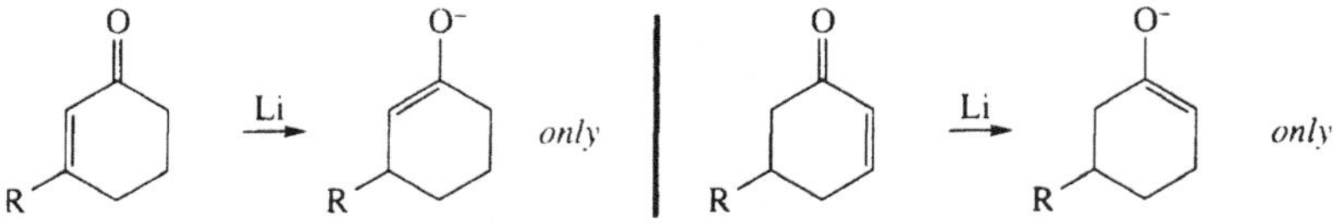

Saturated ketones are reduced to the thermodynamically more stable alcohols in the *Bouveault–Blanc reduction* by an almost identical mechanism. This old reaction has been largely superseded by the advent of complex metal hydrides like NaBH₄ and LiAlH₄, but it is still useful when the metal hydride gives the undesired, nonthermodynamic product.

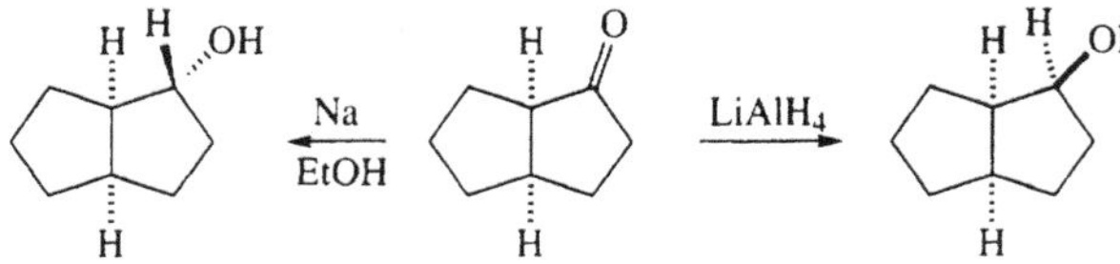

Problem 5.10. Draw a mechanism for the Bouveault–Blanc reduction of the preceding ketone.

In the reduction of alkynes to trans alkenes, the second electron transfer occurs *after* the first protonation. The stereochemistry-determining step is protonation of the carbanion obtained after the second electron transfer. The thermodynamically more stable trans product is obtained.

Example

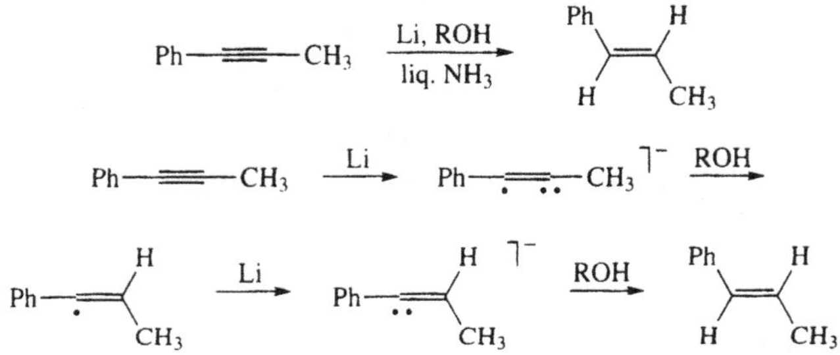

The *Birch reduction* converts arenes to 1,4-cyclohexadienes. The mechanism, again, is electron transfer to the ring, protonation of the radical anion, a second electron transfer to give a carbanion, and protonation to give the neutral product. The reaction is highly regioselective. The protonation steps determine the regiochemistry of the reduction. They occur para and ipso to electron-withdrawing groups, and ortho and meta to electron-donating groups. Thus, in the product, electron-donating groups such as $-OMe$ are conjugated to the π bonds, and electron-withdrawing groups such as $-CO_2H$ are placed at the sp^3-hybridized positions.

Example

When a benzoic acid derivative is reduced, the final protonation step does not occur, and a carboxylate enolate is obtained. The enolate can be protonated upon workup to give the usual product, or it can be alkylated by addition of an electrophile. The proton or the electrophile adds ipso to the carboxylate group exclusively. This position has the largest coefficient in the HOMO of the pentadienyl anion.

Birch reductions are highly intolerant of functionality. All benzylic ethers, alcohols, and carbonyl groups are reduced away and replaced with H, as are halides anywhere in the compound. Nonbenzylic carbonyls may or may not be reduced to alcohols, but alkynes and conjugated alkenes are reduced to simple alkenes.

The reductive cleavage of benzylic ethers makes the benzyl group a useful protecting group for alcohols. It is stable to many acidic and basic conditions, and it can be removed by reduction with Li in liquid NH_3 or by Pd-catalyzed hydrogenolysis (Chapter 6). The products of the reductive cleavage are the desired alcohol plus toluene or methylcyclohexadiene (either of which is easily removed by evaporation).

(product of interest)

Problem 5.11. Draw a mechanism for the dissolving metal reduction of acetophenone (PhCOMe) to ethylbenzene and thence to 1-ethyl-1,4-cyclohexadiene.

5.3.2.2   Reduction of C–X Bonds; Reductive Coupling

Leaving groups on the α-carbon of carbonyl compounds (e.g., Br and OR) are reduced away by one-electron reducing agents. α-Bromocarbonyl compounds are reduced to the corresponding enolates by Zn in the *Reformatsky reaction* (Chapter 2). After the initial electron transfer, several pathways are possible, but all lead to the enolate. The enolate is usually allowed to react immediately with an electrophile such as another carbonyl compound. Before the advent of strong, non-nucleophilic bases, the Reformatsky reaction was the only way to prepare enolates of simple carbonyl compounds quantitatively.

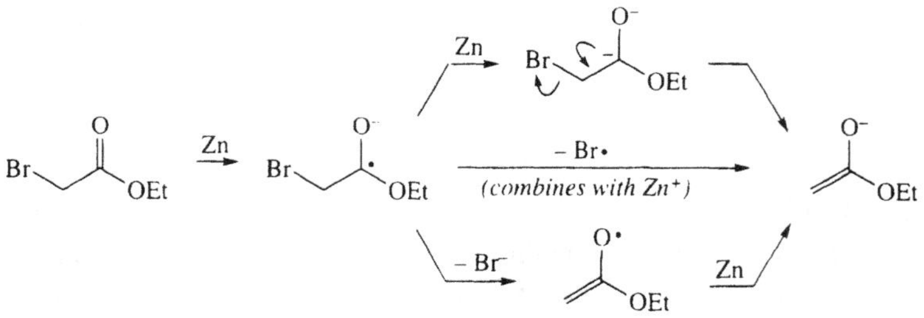

Samarium diiodide (SmI$_2$), an oxophilic, one-electron reducing agent, is particularly useful for the α-deoxygenation of ketones. The metal in SmI$_2$ is in the $+2$ oxidation state, but it prefers to be in the $+3$ oxidation state. Therefore, it transfers one electron to a C=O π bond to give the ketyl (R$_2$Ċ–O–SmI$_2$ $\rightleftharpoons$ R$_2$Ċ–O$^-$ $^+$SmI$_2$). The ketyl can do any of its usual chemistry, including adding to pendant π bonds. When there is a hydroxy or alkoxy group α to the carbonyl, though, a second equivalent of SmI$_2$ coordinates to it. Sigma-bond homolysis then gives ROSmI$_2$ and the deoxygenated carbonyl compound as its enolate, which can be protonated in situ or upon workup.

Example

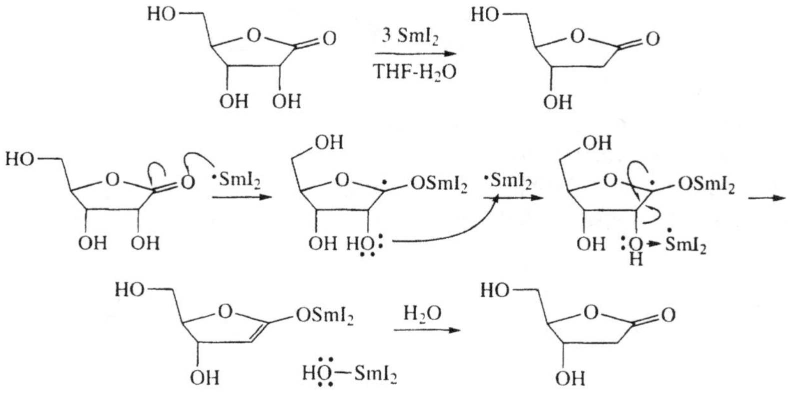

In this example, three equivalents of SmI$_2$ are used in order to provide an excess of the reagent.

In the *pinacol coupling*, two ketones are reductively coupled to give a 1,2-diol. (Compare the photochemical pinacol coupling discussed in Section 5.3.1.) The two ketones are usually identical, but intramolecular dimerizations can give unsymmetrical 1,2-diols. The reaction proceeds by electron transfer to the ketone to give a ketyl radical anion. This compound dimerizes to give the 1,2-diol.

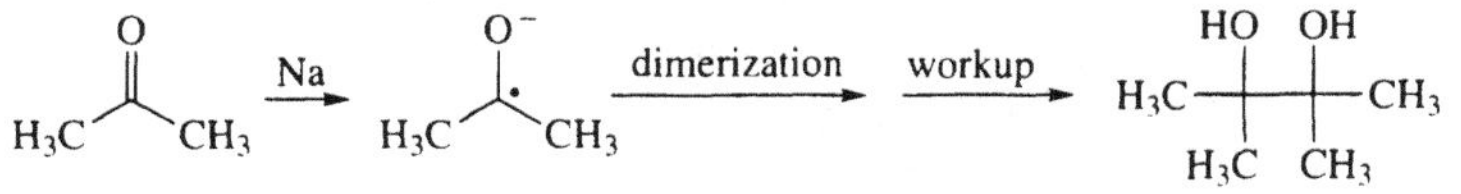

The *McMurry coupling* (Chapter 6) is closely related to the pinacol coupling.

The *acyloin condensation* converts two esters to an α-hydroxyketone (an acyloin), often in an intramolecular fashion. The reaction proceeds by a mechanism very similar to the pinacol coupling, except that after the radical–radical combination step there are two elimination steps and two more electron transfer steps. The intramolecular reaction works well for a wide variety of ring sizes.

Example

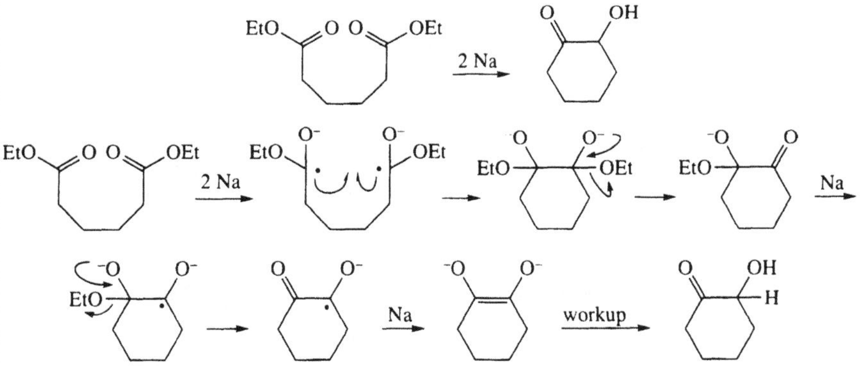

Modern acyloin condensations are usually executed in the presence of Me₃SiCl, and a bis(silyloxy)alkene is obtained as the immediate product. The bis(silyloxy)alkene may then be hydrolyzed to the acyloin upon workup. The yield of the acyloin condensation is greatly improved under these conditions, especially for intramolecular cyclizations. The Me₃SiCl may improve the yield by reacting with the ketyl to give a neutral radical, which can undergo radical–radical combination more easily with another ketyl radical due to a lack of electrostatic repulsion.

Example

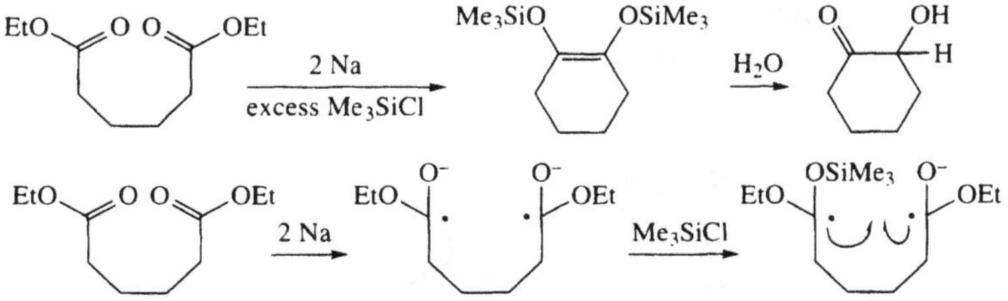

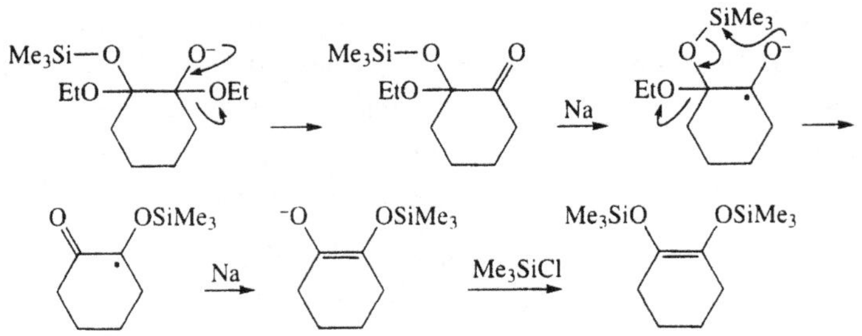

5.3.2.3   One-Electron Oxidations

The *p*-methoxyphenyl group can be removed from N or O by a one-electron oxidation promoted by CAN or DDQ. The aryl group is further oxidized to the quinone after it is cleaved from the substrate, so at least two equivalents of oxidant is required.

Example

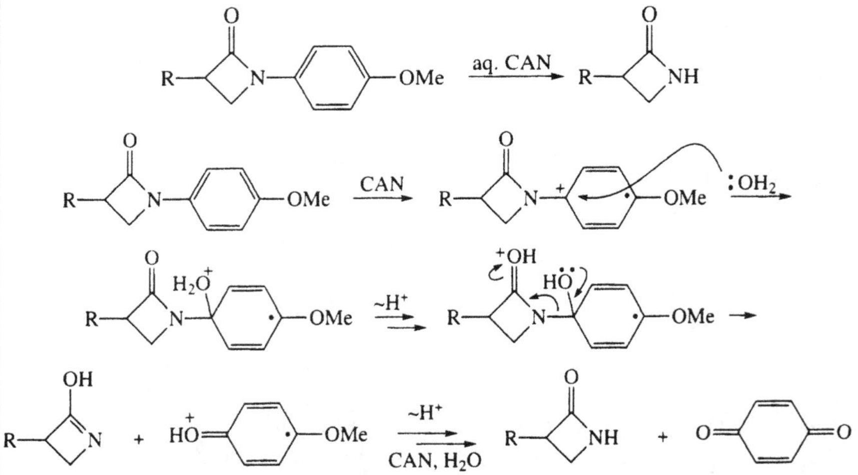

The enzyme cytochrome P450 metabolizes tertiary amines in the liver by a one-electron transfer mechanism. An electron is transferred from N to a heme group featuring an Fe=O bond to give an aminium radical cation and the radical anion [F̄e]–O·. An α-hydrogen atom is then abstracted by the oxygen-based radical to give an iminium ion. Hydrolysis of the iminium ion affords a secondary amine and an aldehyde, both of which are further oxidized into water-soluble compounds and then excreted. The oxidation of tertiary amines to secondary amines can also be executed in the laboratory.

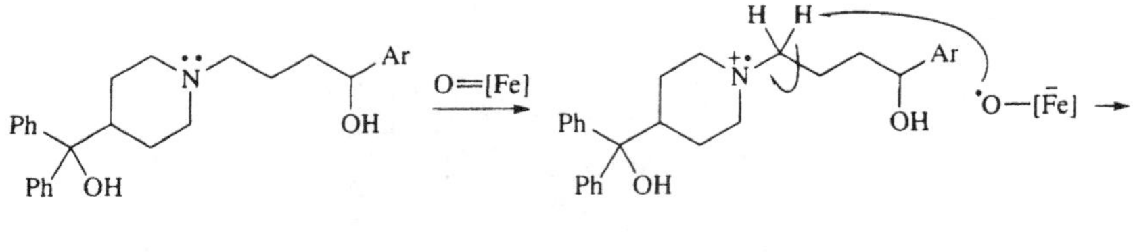

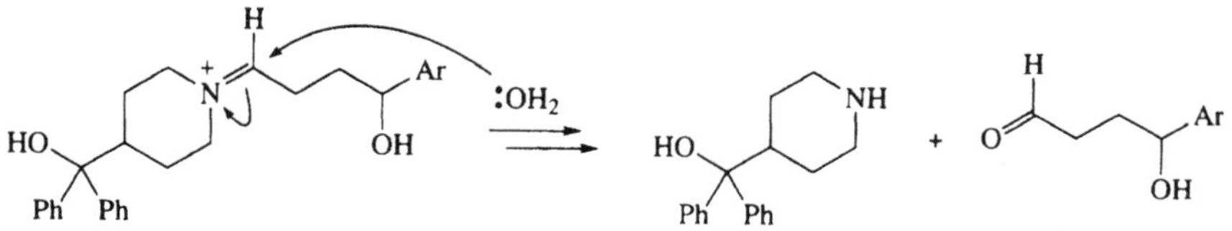

Problem 5.12. Draw a reasonable mechanism for the following reaction. Why aren't any of the C atoms adjacent to the ring N oxidized?

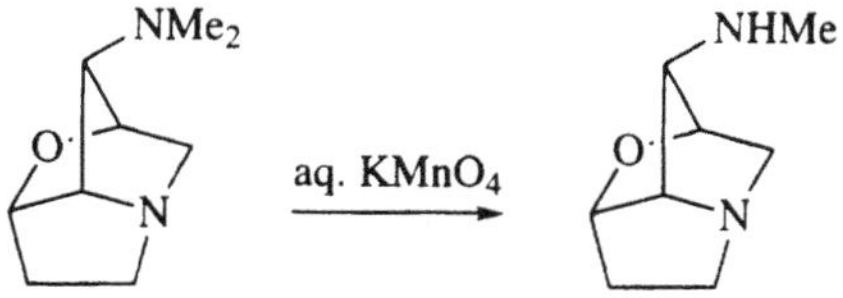

5.3.3   Cycloaromatizations

The synthetic potential of cycloaromatizations has only just begun to be explored. The two aryl radicals that are generated in the cycloaromatization are usually trapped by abstraction of H· from a compound such as 1,4-cyclohexadiene. Alternatively, one of the aryl radicals may add to a nearby π bond to give a new radical, which may itself be quenched by H· abstraction.

Example

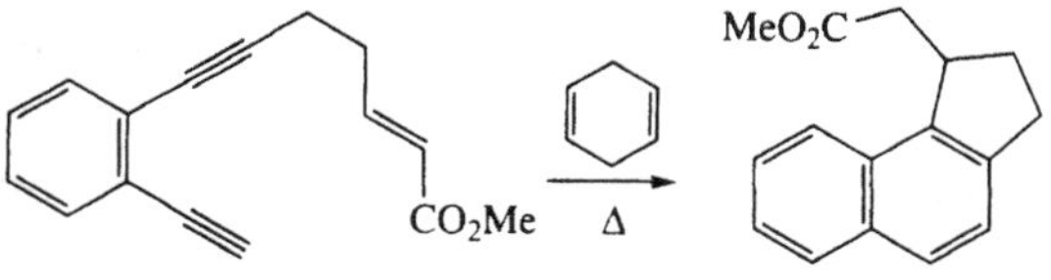

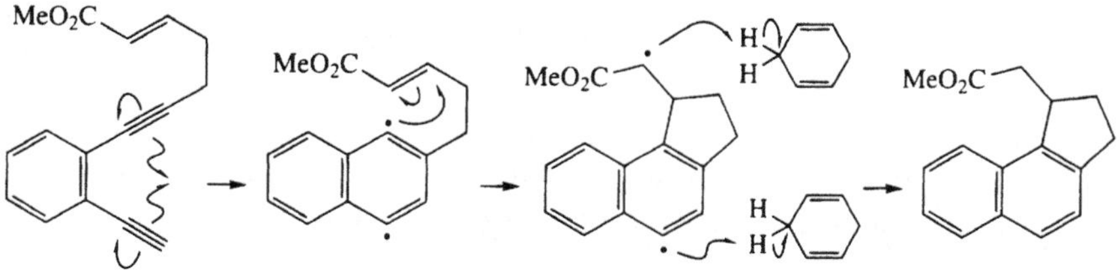

5.4   Miscellaneous Radical Reactions

5.4.1   1,2-Anionic Rearrangements; Lone-Pair Inversion

The Stevens rearrangement and the Wittig rearrangement (nonallylic version) (Chapter 4) can be classified as four-electron [1,2] sigmatropic rearrangements. The Woodward–Hoffmann rules state that for a four-electron sigmatropic rearrangement to be allowed, one of the components must be antarafacial, yet it is

geometrically impossible for either component in the Stevens and Wittig rearrangements to react antarafacially. How can this be?

Stevens rearrangement

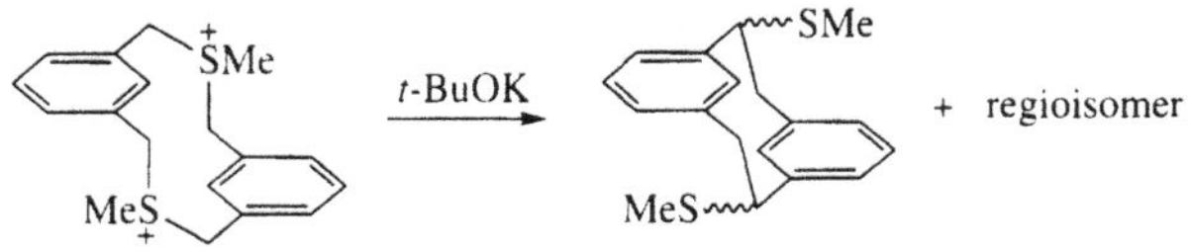

Wittig rearrangement

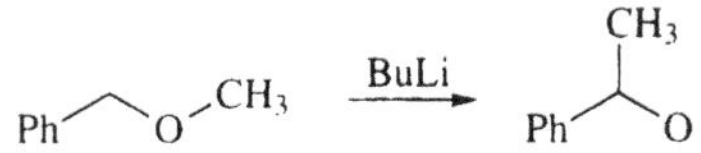

One of the conditions of the Woodward–Hoffmann rules is that they are applicable only if the reaction proceeds in a concerted fashion. The fact that the
Stevens and the Wittig rearrangements proceed can be explained by a two-step,
nonconcerted mechanism involving free radicals. After deprotonation of the position α to the heteroatom, homolysis of the bond between the heteroatom and
the migrating group occurs to give a radical and a radical anion. The resonance
structure of the radical anion shows that it is a ketyl. Radical–radical combination of the migrating group with the ketyl C gives the rearranged product.

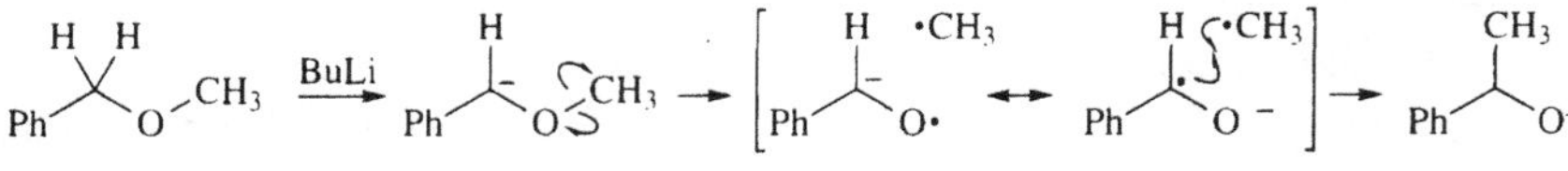

Problem 5.13. Draw the mechanism for the Stevens rearrangement.

The lone pairs on amines are nonstereogenic; that is, a compound such as
MeṄ(Et)(*i*-Pr) is pyramidalized and chiral, but the rapid inversion of the N lone
pair (the "umbrella effect") at room temperature makes it impossible to isolate the
enantiomers of all but a few amines. By contrast, lone-pair-bearing heavy atoms
such as S and P do not undergo inversion at room temperature, and trivalent S and
P compounds are configurationally stable at normal temperatures (<100 °C).
Sulfoxides do racemize when they are heated to sufficiently high temperatures. The
mechanism of racemization is thought not to involve inversion of the lone pair on
S. Instead, the racemization is thought to proceed by homolysis of a S–C bond.
The divalent S radical no longer has a configuration. Radical–radical recombina-
tion can then occur to give either enantiomer of the starting material.

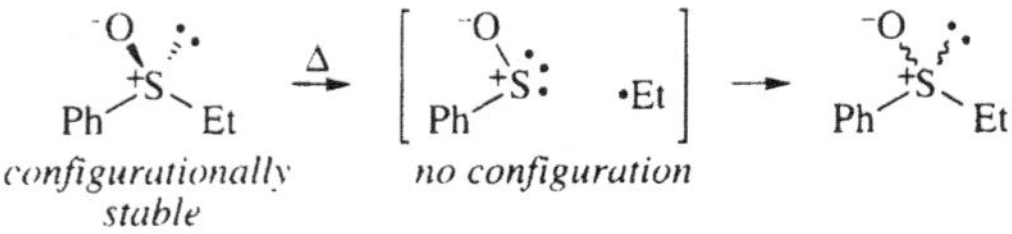

5.4.2   Triplet Carbenes and Nitrenes

Upon photolysis of a diazo compound or an azide, a triplet carbene or nitrene is
generated. The two unshared electrons in triplet carbenes have aligned spins and

reside in different orbitals. Triplet carbenes and nitrenes undergo all the typical reactions of singlet carbenes and nitrenes (Chapter 2) but with one important difference: the reactions are not stereospecific. The lack of stereospecificity in the reactions of triplet carbenes and nitrenes is due to their 1,1-diradical character. For example, consider a cyclopropanation reaction. The two electrons in the π bond have opposite spins, but the two electrons in the triplet carbene have aligned spins. For the cyclopropanation to be concerted, all four electrons have to flow smoothly into two new σ bonding orbitals, but this is not possible when three electrons are aligned opposite the fourth. The reaction therefore proceeds in a stepwise fashion: one bond forms, the two lone electrons of the 1,3-diradical lose contact with each other and randomize their spins, then the second bond forms. The lifetime of the 1,3-diradical intermediate is long enough that rotation about σ bonds can occur, causing a loss of stereochemical purity about the former π bond.

Example

21%   78%

The symbol $R_2C\pm$ should not be used to describe triplet carbenes. The "$-$" in "$\pm$" represents a pair of electrons, but the two electrons in triplet carbenes are unpaired. The symbol $R_2C:$ can be used to describe either a singlet or triplet carbene.

Triplet carbene and nitrene C–H insertion reactions occur with loss of stereochemical purity around the C from which the H is abstracted, in contrast to C–H insertion reactions of singlet carbenes and carbenoids (Chapter 2). The triplet carbene or nitrene, a 1,1-diradical, abstracts H· from the substrate to give two radical intermediates. After the lone electrons lose track of each other's spins, radical–radical combination gives the product. The nitrene insertion reaction is the basis of *photoaffinity labeling*, in which the ligand of a biological receptor is modified with an azido group and a radioactive or fluorescent tag, allowed to bind to the protein, and then photolyzed to install the label covalently and specifically in the binding pocket.

The singlet form of carbenes substituted with a carbonyl group is much lower in energy than the triplet, so the photolysis of α-diazocarbonyl compounds gives singlet carbenes. The photo-Wolff rearrangement (Chapter 2) of α-diazoketones occurs upon photolysis of these compounds.

The reactions of carbenes are complicated by the fact that *intersystem crossing* (i.e., interconversion of the triplet and singlet states by a spin flip of an electron) can occur readily, as in the photo-Wolff rearrangement. Which of the two states of a particular carbene is lower in energy is also strongly dependent on the nature of the substituents. In general, singlet carbenes are more useful synthetically because they react stereospecifically. Most carbene-generating techniques give singlet carbenes.

5.6   Summary

Free radicals are seven-electron, electron-deficient species. Many free radicals proceed by chain mechanisms, but photochemical rearrangements, metal-mediated oxidations and reductions, and cycloaromatizations do not. Free radicals undergo eight typical reactions, but three are most important: atom abstraction, addition to a π bond, and fragmentation. (The other five are radical–radical combination, disproportionation, electron transfer, addition of a nucleophile, and loss of a leaving group.) *Carbocations* are six-electron, electron-deficient species. They undergo three typical reactions: addition of a nucleophile, fragmentation, and rearrangement by a 1,2-shift. *Carbenes* are also six-electron, electron-deficient species. They undergo four typical reactions: cyclopropanation, insertion into a C–H bond, addition of a nucleophile, and rearrangement by a 1,2-shift. The similarities that exist among free radicals, carbocations, and carbenes are apparent. The differences among them are due to their different electron counts and the presence or absence of an unshared pair of electrons.

PROBLEMS

1. Methyl *tert*-butyl ether, MTBE, and ethyl *tert*-butyl ether, ETBE, are added to gasoline in order to increase the efficiency of gasoline combustion, which reduces the quantity of volatile organic compounds (VOCs) that escape into the atmosphere and cause smog. The chemical industry is very interested in using MTBE as a substitute for THF and diethyl ether, because autoxidation of THF and diethyl ether is a major safety problem for chemical companies. The industry is less interested in ETBE as a solvent substitute.

   (a) Why is MTBE less prone to autoxidize than ether and THF?

(b) Why is ETBE of less interest than MTBE as a substitute for diethyl ether and THF?

(c) Draw a mechanism for the synthesis of MTBE from methanol and isobutylene. What kind of conditions are required?

(d) Senators from farm states are pushing EPA to require a certain percentage of ETBE (but not MTBE) in gasoline oxygenates. Why? (*Hint:* What starting materials are used to prepare ETBE, and what do they have to do with farming?)

(e) MTBE has been banned in California because it has been contaminating groundwater, imparting a very bad odor to it. Its presence there derives from leaking underground gasoline storage tanks. Ironically, gasoline leakage from the same tanks is not causing a groundwater contamination problem. Provide two reasons why MTBE is much more prone to contaminate groundwater.

2. The production of chlorofluorocarbons, or CFCs, has been banned by international treaty because of their deleterious effect on the ozone layer. The ozone layer absorbs much of the Sun's dangerous UV radiation before it reaches the Earth's surface. CFCs are extremely stable in the lower atmosphere (one reason why they are so useful), but when they reach the stratosphere they decompose, producing potent catalysts of ozone destruction. Ozone destruction is most evident above Antarctica during the spring, when this region is exposed to the Sun for the first time in months. Dichlorodifluoromethane (CF_2Cl_2) is a typical CFC.

   Hydrochlorofluorocarbons, or HCFCs, are being promoted as temporary replacements for CFCs. HCFCs, unlike CFCs, have at least one C–H bond. HCFCs are not less prone than CFCs to decompose in the stratosphere, but there is a pathway by which HCFCs can decompose in the *lower* atmosphere (where they cannot damage ozone) that is not accessible to CFCs. 2,2,2-Trichloro-1,1-difluoroethane, CHF_2CCl_3, is a typical HCFC.

(a) What is the first step in the decomposition of CFCs in the stratosphere?

(b) Propose a first step of a reaction by which HCFCs decompose in the lower atmosphere.

3. Draw mechanisms for each of the following reactions.

(a)

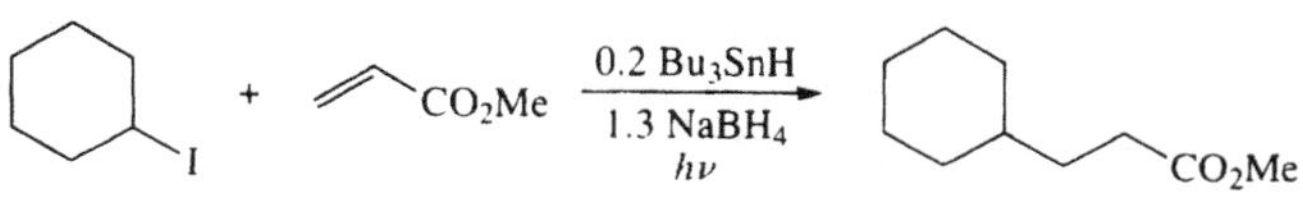

(b) Can you explain the regioselectivity of the first reaction?

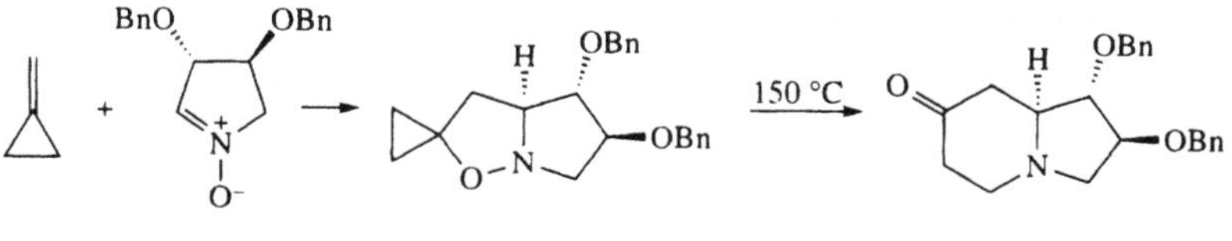

(c)   An alkylative Birch reduction.

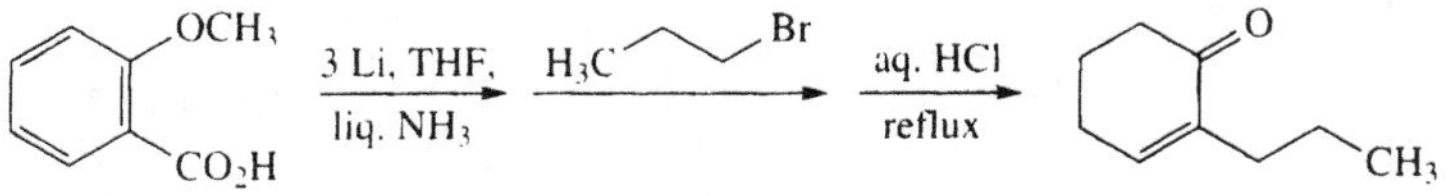

(d)   The starting material can react by two pathways. Both pathways give
      the four-membered ring shown. One pathway gives the ketene as the by-
      product; the other gives PhCHO and CO (1:1).

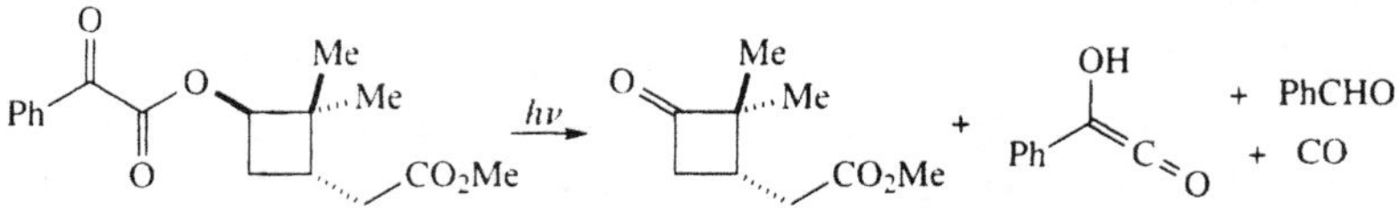

(e)

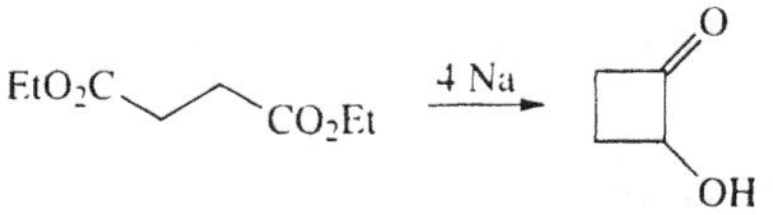

(f)

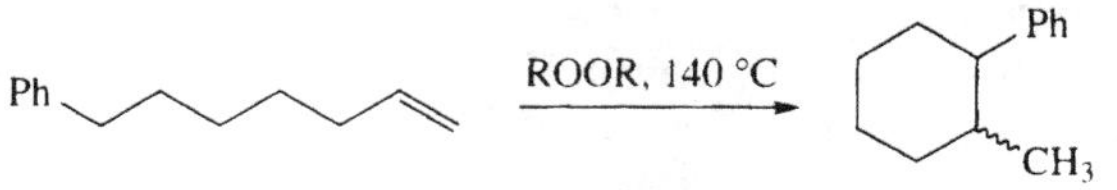

(g)

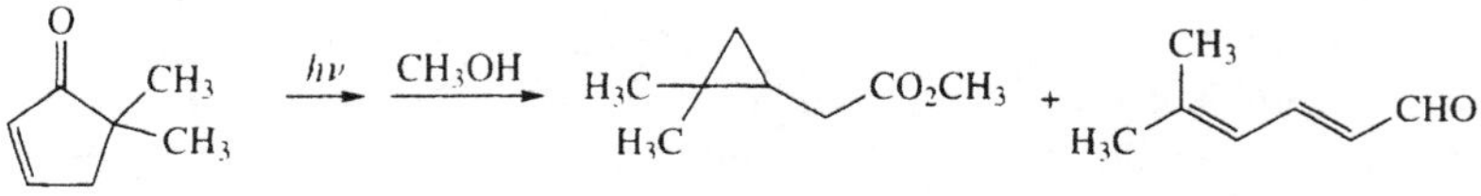

(h)

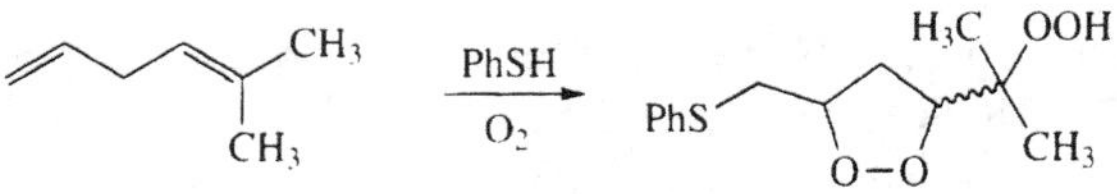

(i)

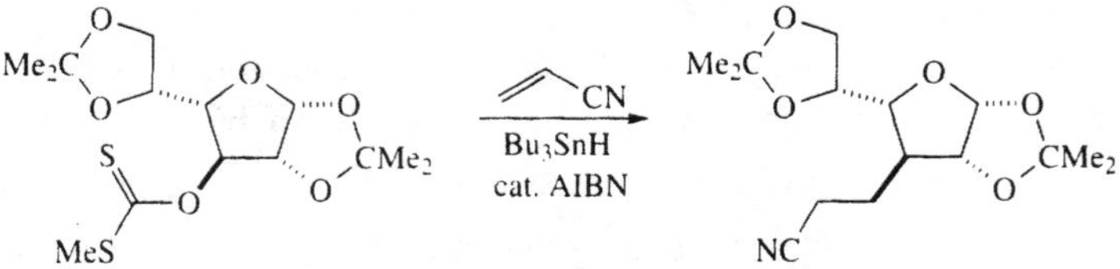

(j)

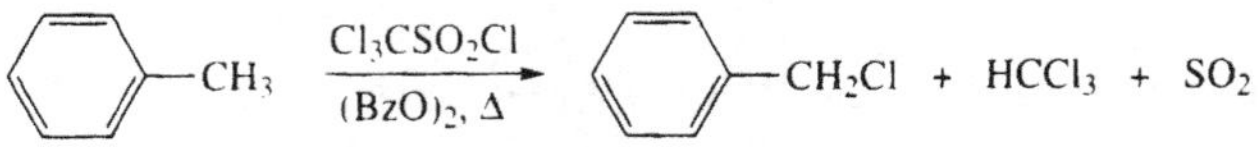

(k)

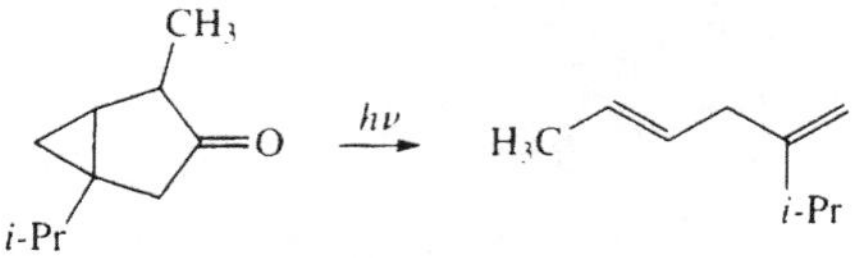

(l)   Note the use of *catalytic* quantities of Bu$_3$SnH.

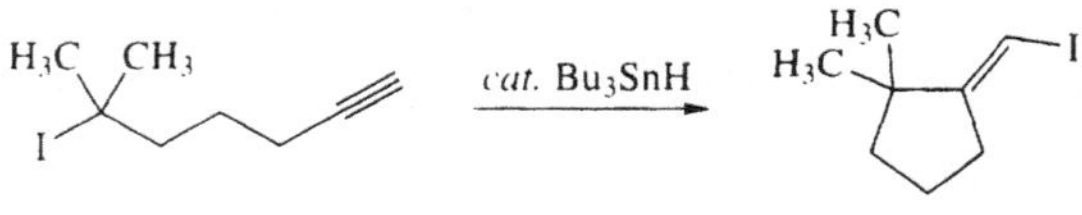

(m)

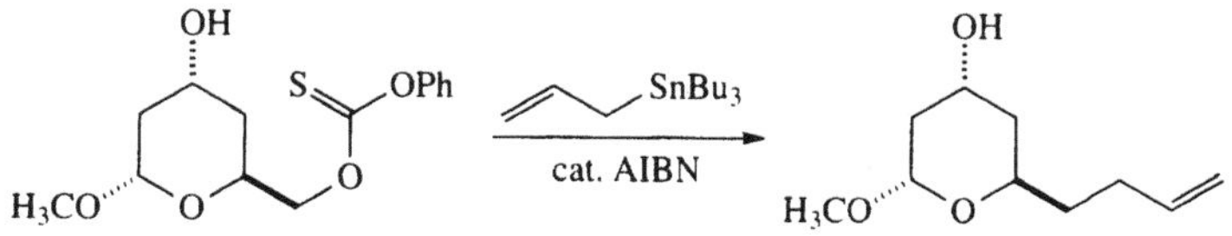

(n)

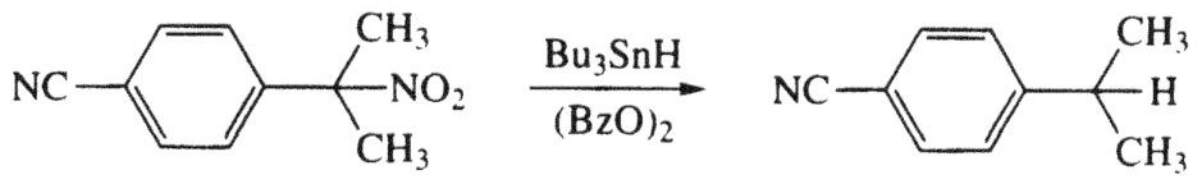

(o)   There is no typo in the following reaction.

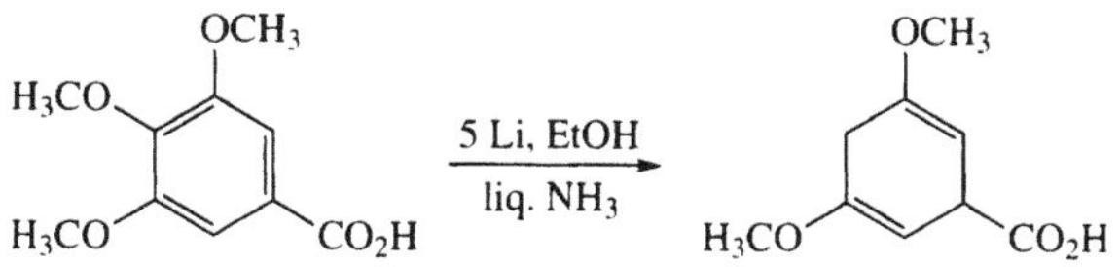

(p)

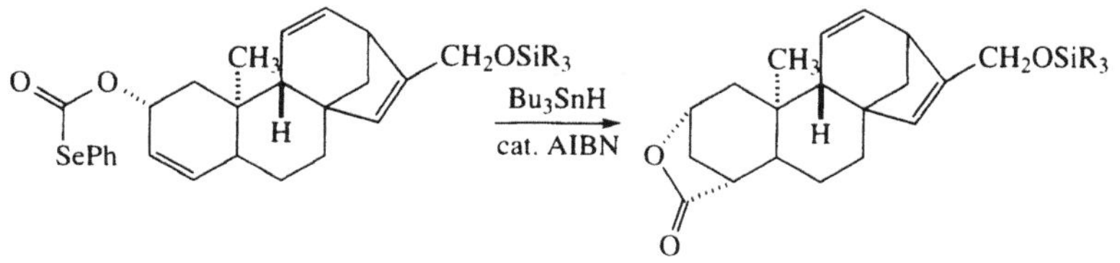

(q)

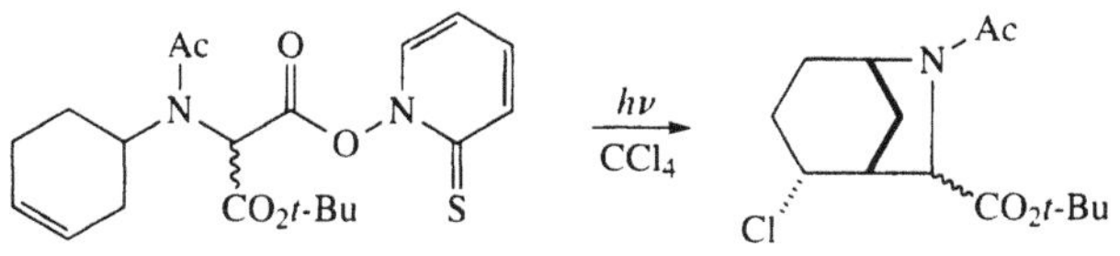

(r)

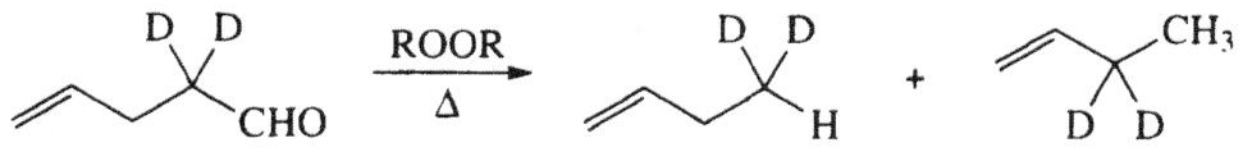

(s)

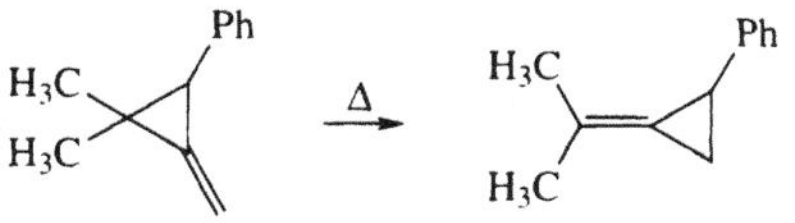

(t)

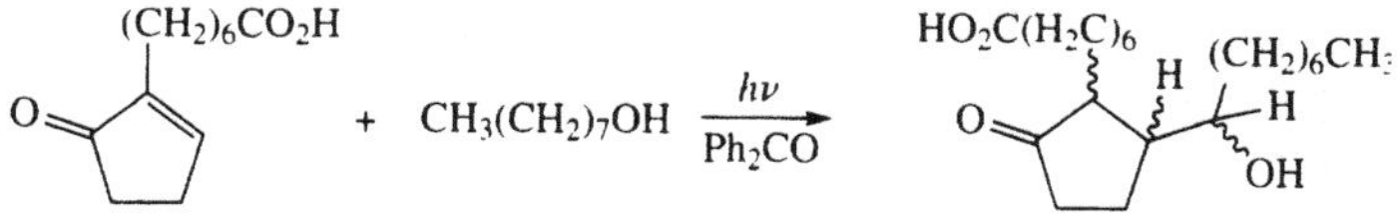

(u)

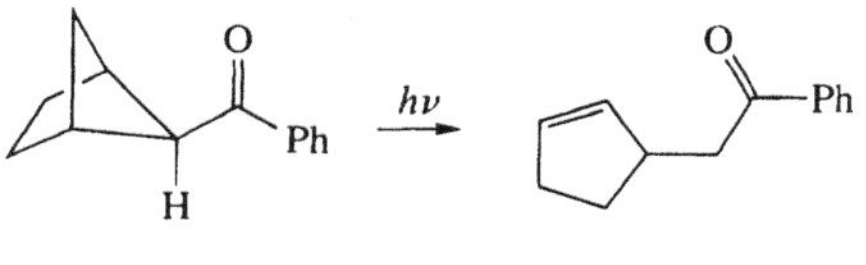

(v)

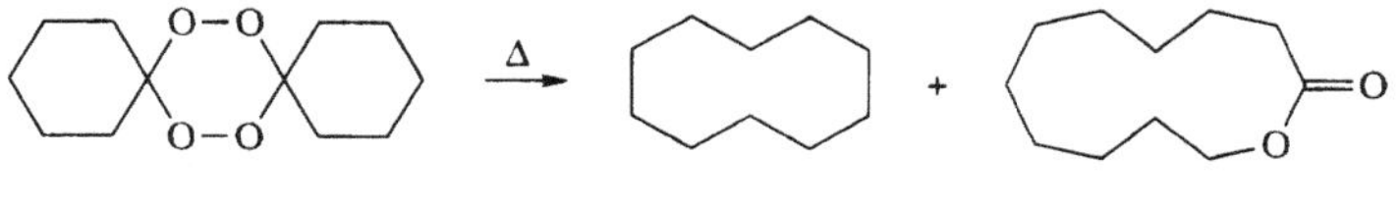

(w)

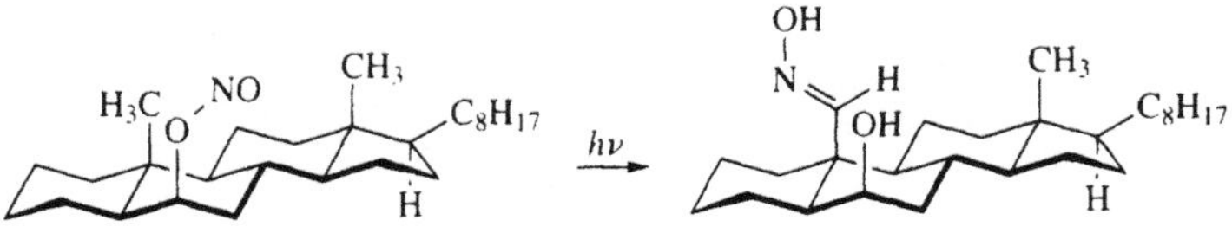

(x)

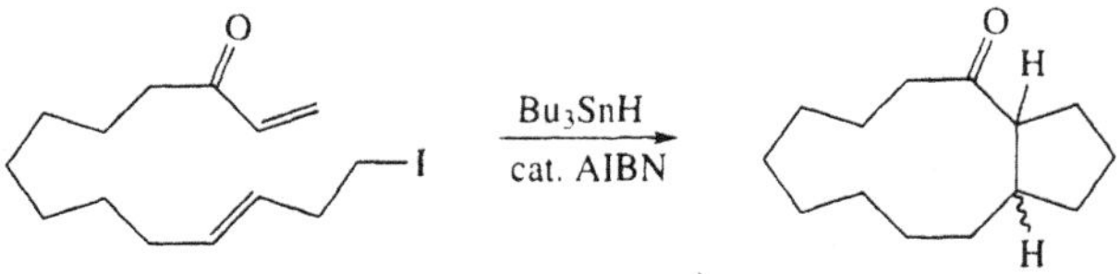

$$Bu_3SnH \atop cat.\ AIBN$$

(y)

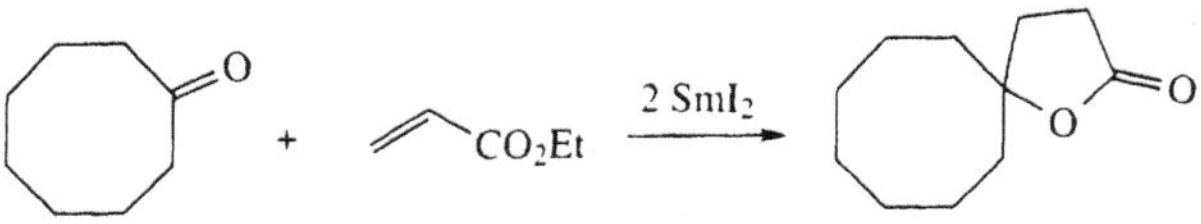

$$2\ SmI_2$$

(z)

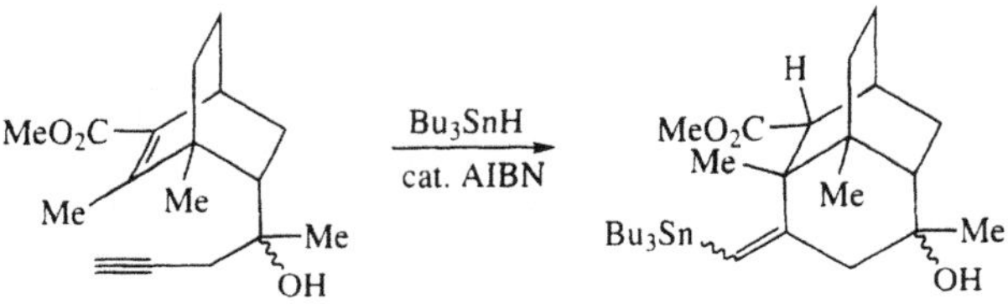

$$Bu_3SnH \atop cat.\ AIBN$$

(aa)

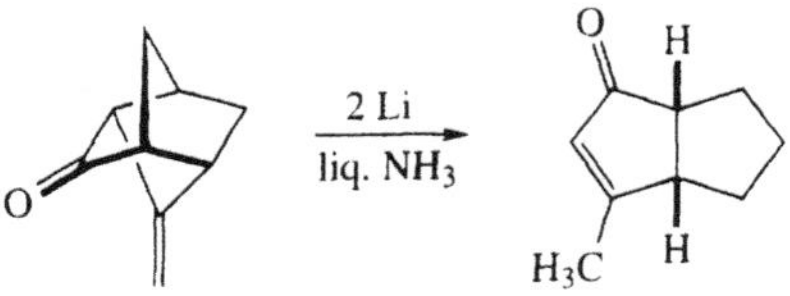

$$2\ Li \atop liq.\ NH_3$$

(bb)

$$(Me_3Si)_3SiH \atop cat.\ AIBN$$

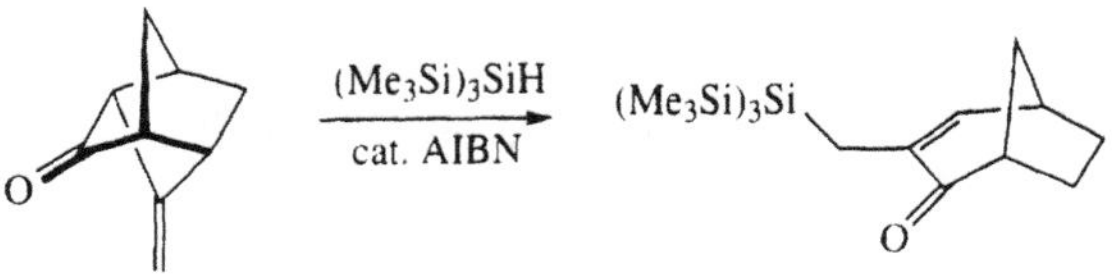

(cc)

$$\Delta$$

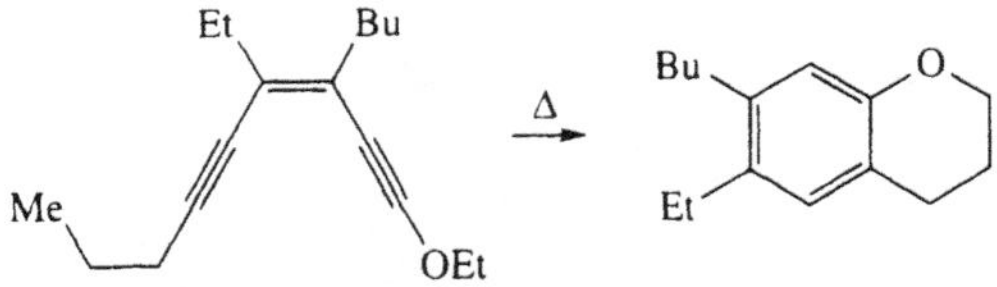

(dd) Fe(NO$_3$)$_3$ has the same reactivity as CAN.

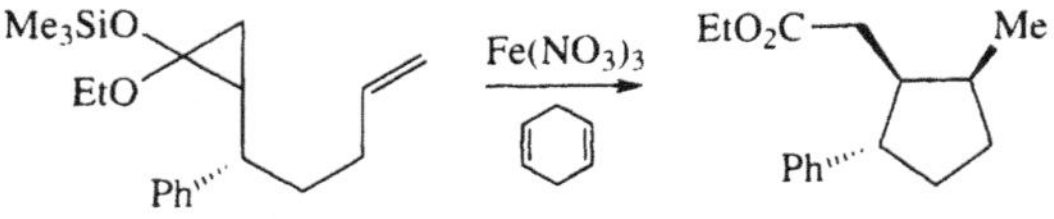

$$Fe(NO_3)_3$$

(ee)

$$Mn(OAc)_3 \atop Cu(OAc)_2$$

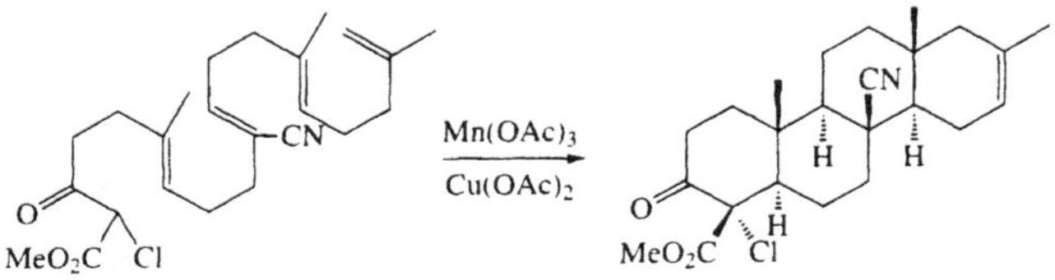

4. The following reaction of C_{60} appeared recently in a respected journal. The reaction requires O_2.

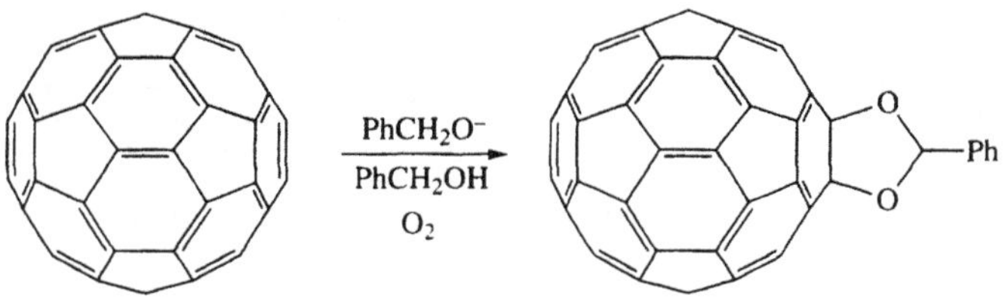
PhCH₂O⁻
PhCH₂OH
O₂
O
O
Ph

The authors proposed the following mechanism. C_{60} is an electrophilic compound, and both $RC_{60}\cdot$ radicals and RC_{60}^- anions are quite stable and long-lived species, so the first two steps are perfectly reasonable.

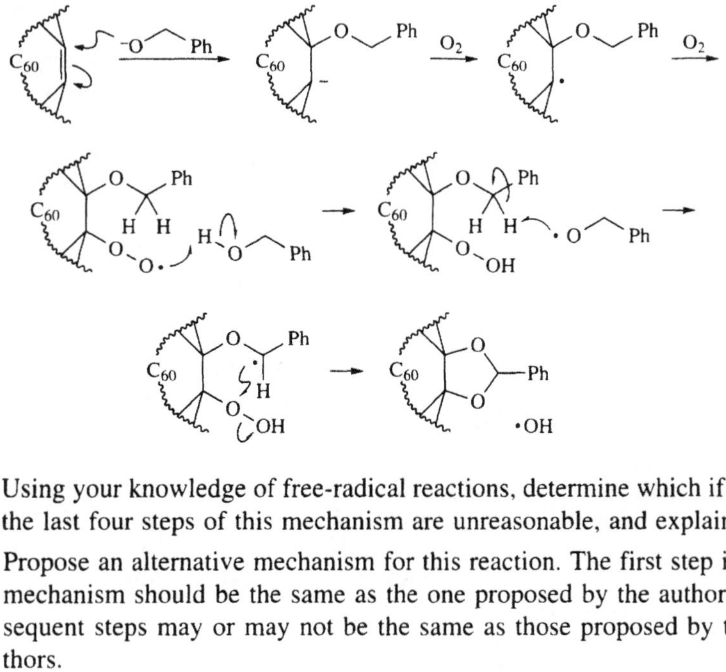
C60
⁻O
Ph
O
O₂
H
OH
•OH

(a) Using your knowledge of free-radical reactions, determine which if any of the last four steps of this mechanism are unreasonable, and explain why.

(b) Propose an alternative mechanism for this reaction. The first step in your mechanism should be the same as the one proposed by the authors; subsequent steps may or may not be the same as those proposed by the authors.

6

Transition-Metal-Catalyzed and -Mediated Reactions

6.1   Introduction to the Chemistry of Transition Metals

Many widely used organic reactions require the use of transition metals. You have probably already learned about metal-catalyzed hydrogenations of alkenes and alkynes, dihydroxylation of alkenes using OsO_4, and the use of lithium dialkylcuprates (Gilman reagents) as "soft" nucleophiles. Although the mechanisms of many of these reactions appear mysterious, in fact they are quite easy to understand (in most cases) using some very basic principles. This chapter discusses some of the typical reactions of transition metals. These principles will then be applied to understanding some of the organic transformations mediated by these metals.

> The reactions in this chapter are organized according to the overall transformation achieved. An organization by mechanistic type (insertions, transmetallations, etc.) was considered and rejected. The purpose of this text is to teach the student how to go about drawing a mechanism for an unfamiliar reaction. A student who does not already know the mechanism for a reaction will find it easier to narrow down the possible mechanisms by considering the overall transformation rather than by trying to determine the mechanistic type. An organization by metal (early, middle, or late) was also considered and rejected, because it would obscure the important similarities between the reactions of different metals.
>
> No attempt is made to cover every aspect of organometallic chemistry. Many interesting organometallic reactions, such as the reactions of Cr arene complexes, have been discovered and even investigated extensively but have not been widely adopted by organic chemists for one reason or another. These reactions are not discussed. Moreover, as in previous chapters, questions of stereochemistry are deemphasized in favor of concentration on electron-pushing. The reader is urged to consult any of the excellent textbooks on organometallic chemistry for further coverage of these important topics.

6.1.1.   Conventions of Drawing Structures

The conventions for drawing organometallic and inorganic compounds differ in subtle ways from those used to draw "ordinary" organic compounds. The most important difference is the way in which bonds are drawn. In organic compounds, one does not use a line to connect a bond to an atom. In organometallic and inorganic compounds, however, a line sometimes connects an atom and a σ or π bond. In this case, the line indicates that the pair of electrons in the σ or π bond is shared with the metal also.

This bond represents two electrons ...

*... and this bond represents the **same** pair of electrons used in the C=C π bond!*

An even more confusing situation arises in complexes in which the electrons in a π system spread over three or more atoms are used to make a bond to a metal. In this case, the usual convention is to use a curved line to indicate the π system and a single line to connect the π system to the metal, regardless of the number of electrons in the π system (the organometallic chemists' convention). However, sometimes the curved line is omitted and single lines are used to connect the metal to each of the atoms in the π system (the crystallographers' convention). The representation that would make the most sense to organic chemists, in which a single line represents a two-electron σ bond and a dative bond is used to show two-electron bonds between each individual C=C π bond and the metal, is simply never used.

organometallic chemists' convention	crystallographers' convention	organic chemists' convention	number of electrons in bond between metal and π system
			four
			six

Formal charges are usually omitted in inorganic and organometallic complexes. Only the overall charge on the complex is indicated. For example, formal charges are usually not assigned in Lewis acid–base complexes involving transition metals. The acid–base bond is sometimes indicated by an arrow pointing from the ligand to the metal, but more often it is indicated by an ordinary line.

usual *occasional* *rare*

6.1.2   Counting Electrons

The chemistry of any particular metal complex can be understood by examining its *total electron count*, its *number of d electrons*, and the metal's *oxidation state*.

6.1.2.1   Typical Ligands; Total Electron Count

Main group elements like C and S have 4 valence AOs, one s and three p, and they follow the octet rule (although heavier main group elements can extend their octet). Transition metals, by contrast, have 10 valence AOs—one s, five d, and three p, in that order—and they follow the *18-electron* rule. The 18-electron rule is much less rigorous for transition metals than the octet rule is for main-group elements. First, it can be difficult to surround a metal, especially an early metal, with sufficient numbers of substituents to provide 18 electrons to the metal. Second, the valence orbitals of metals are sufficiently extended from the nucleus that the nucleus doesn't care much about what's going on in its valence shell.

It's more difficult to obtain a count of the total number of electrons around a metal than it is around a main-group element. Unlike in main-group compounds, where substituents almost always bring one or two electrons to an atom, substituents (or *ligands*) attached to metals can provide anywhere from one to six electrons to the metal. However, just a few classes of ligands are widely used in organometallic chemistry, and it's not too hard to remember how many electrons each one donates.

• Monovalent groups such as alkyl, alkoxy, H, R_2N, halogen, etc., that are singly bound to metals are one-electron donors.

• Divalent fragments such as $R_2C=$ (carbenes or alkylidene groups), $RN=$ (imido groups), or $O=$ (oxide) that are doubly bound to a metal are two-electron donors.

• Lewis bases such as R_3N, $RC\equiv N$, H_2O, and R_3P are also two-electron donors, as are carbon monoxide ($:C\equiv O$) and the isoelectronic isocyanides ($:C\equiv NR$). (When CO is attached to a metal it is called a *carbonyl* group.) The bond between a Lewis basic ligand and the Lewis acidic metal is sometimes called a *dative* bond.

• Alkenes and alkynes can also act as Lewis bases toward metals by using the two electrons in their π bonds. Such an interaction is called a π *complex*. The metal–alkene bond is a σ bond due to its spherical symmetry around the axis formed by the metal and the midpoint of the C=C π bond. Even σ bonds such as the H–H and C–H bonds can act as two-electron donors to metals in this manner, and these compounds are called σ *complexes*. When a σ complex is formed in an intramolecular fashion, the bond is called an *agostic bond* or *agostic interaction*. Both π and σ complexes can be thought of as two-electron, three-center bonds in which three contributing orbitals comprise the occupied, bonding orbital.

π complex σ complex

- Trivalent fragments such as RC≡ (carbyne or alkylidyne group) and N≡ (nitride) that are triply bound to a metal are three-electron donors.

- The allyl group can be either a one- or a three-electron donor. Imagine a three-electron allyl group as being a one-electron donor through the C–M σ bond and a two-electron donor from its π bond.

allyl group (three-electron donor)

- Dienes are four-electron donors. Alkynes are four-electron donors when they use one π bond to form a σ bond and the other π bond to form a π bond to metal orbitals of the proper symmetry. When only two of the three π bonds in a benzene ring interact with a metal, the ring is a four-electron donor.

- The cyclopentadienyl group (C_5H_5 or Cp) is a five-electron donor. Conceptually, one electron is derived from a covalent bond to C, and each of the two π bonds of the group acts as a two-electron donor ligand. In fact, though, the five C atoms of the Cp ring are indistinguishable. Occasionally, the Cp group can "slip" and decoordinate one π bond to turn itself into a three-electron donor. Acyclic pentadienyl groups are five-electron donors also.

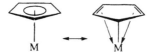

Cp group (five-electron donor)

- Benzene rings can act as six-electron donors.
- Ligands with lone pairs such as R, R_2N, X, and O groups can use their lone pairs to make π bonds to the metal, as in M–ÖR ⟷ M=OR. (Again, formal charges are usually not drawn.) Thus, these groups can be three- (R, R_2N, X), four- (O), or even five-electron donors (RO, X), if the metal needs the extra electron density and it has orbitals of the right symmetry to overlap with the donor orbitals.

To calculate the total electron count around a metal, add the number of valence electrons that the metal brings to the number of electrons that each ligand donates, and subtract or add electrons for total charges on the complex. The number of valence electrons brought by a metal is given by its Arabic group number or by its Roman group number plus 2. Thus, Ti is a group 4, a IIA, or a IIB metal, depending on which nomenclature system you use, but it always has four valence electrons.

> Another method for determining total electron count treats odd-electron donor ligands (alkyl, Cp, etc.) as anionic, *even-electron* donors. The number of electrons contributed by the metal is calculated from the metal's oxidation state. Both methods give the same answer for total electron count.

Alkenes and alkynes have special characteristics as Lewis base donors. The face-on interaction of the π bond of an alkene with a p or d orbital on the metal produces a σ bond, as shown in the first of the following structures. This interaction is a simple Lewis acid–base interaction, no different from a lone pair on PR₃ donating to the same orbital. However, if the metal has a pair of electrons in another d orbital, the d orbital can overlap with the empty π* orbital of the alkene, producing a second, *back-bonding* interaction as shown in the second structure. This second bonding interaction weakens the C=C π bond (by dumping electron density into the π* orbital), but it also strengthens the M–C interactions (by increasing the number of electrons residing in delocalized orbitals from two to four). Thus, two resonance structures for alkene–metal complexes can be written, as shown on the right, in one of which (a *metallacyclopropane*) the π bond is gone and there are *two* M–C σ bonds. The total electron count around a metal in an alkene–metal complex is the same regardless of whether the dative resonance structure or the metallacyclopropane resonance structure is drawn.

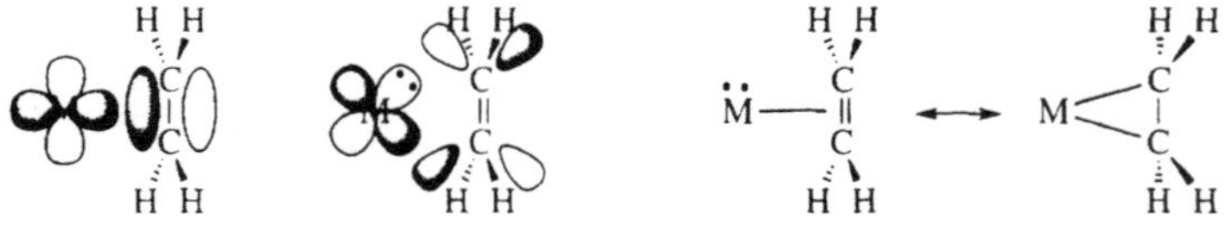

The preceding pair of resonance descriptions is called the *Dewar–Chatt–Duncanson* model of π bonding. Similar pictures can be drawn for metal–alkyne complexes, regardless of whether the second π bond is acting as a π donor, and for metal–1,3-diene complexes.

* **Common error alert:** *The metallacyclopropane resonance structure is valid only if the metal has at least two valence electrons available for back-bonding. It should not be drawn for metals with no valence electrons.*

A similar picture can be drawn for σ complexes. H_2 can use the two electrons in its σ bond to form a σ complex with a metal, $M-(H_2)$. If the metal has at least two valence electrons, it can back-bond to the σ^* orbital, lengthening and weakening the $H-H$ bond until the complex is better described as a dihydride complex $H-M-H$. In fact, a continuum of H_2 complexes of metals has been discovered from one extreme to the other.

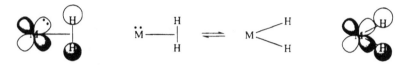

CO and RNC ligands can also interact with lone pairs on a metal. A metal carbonyl can thus be thought of as a metallaketene.

$$\ddot{M}-C\equiv O\colon \quad\longleftrightarrow\quad M=C=\ddot{\underset{\cdot\cdot}{O}}$$

You might imagine that a metal with a total electron count of less than 18 displays electrophilic character, but this is not always true. The 18-electron complex $Pd(PPh_3)_4$ is used as a nucleophilic catalyst, but it is actually the 16- or 14-electron complex $Pd(PPh_3)_3$ or $Pd(PPh_3)_2$ that does the chemistry. (Yes, metals can be nucleophilic!) The Ph_3P groups are too bulky to allow $Pd(PPh_3)_4$ to undergo any reactions, and one or two ligands must dissociate before the complex can react.

6.1.2.2 Oxidation State and d Electron Count

A second important determinant of the reactivity of a metal is its *oxidation state*. The assumption in calculating a metal's oxidation state is that all groups attached to the metal are more electronegative than the metal, and so all σ bonds are ionic bonds. If both electrons in every σ bond "belong" to the ligand, the metal is left with a charge, and that is its oxidation state. In practice, you can determine the oxidation state by counting the number of *covalent* bonds to the metal. Add 1 to the oxidation state for each R, H, RO, R_2N, X, allyl, and Cp group, add 2 for groups like $R_2C=$, $RN=$, and $O=$, and add 3 for $N\equiv$ and $RC\equiv$, but do not add anything for ligands such as R_3P, CO, H_2O, π bonds, dienes, and arenes, all of which are Lewis bases and form dative bonds to the metal. Also add the total charge on the complex to the oxidation state.

> Oxidation states may be written in three ways: Fe^{3+}, Fe^{III}, or Fe(III). The first way is not really appropriate for nonionic complexes, so it will not be used in this book. Either of the other ways is fine.

The *d electron count* of the metal is calculated by subtracting the metal's oxidation state from the number of valence electrons (including the two s electrons) in its elemental state. The "d electron count" is an inorganic chemistry term for "unshared valence electrons." The d electron count of a metal has important ramifications for reactivity. For example, metallocyclopropane resonance structures cannot be drawn for alkene complexes of d^0 metals.

* **Common error alert:** *Do not confuse total electron count, d electron count, and oxidation state with one another.* All three characteristics are important to the reactivity of the metal.

> The d electron count is written as a superscript (e.g., a d^2 complex). The two valence s electrons are always counted toward the d electron count. Thus, Pd(0) is said to be d^{10}, even though it has eight 3d electrons and two 4s electrons.

In alkene–metal complexes, the oxidation state of the metal in the metallacyclopropane resonance structure is 2 higher than in the dative resonance structure. In the following example, the metal has two covalent bonds (to the Cp ligands) in the dative resonance structure, but has *four* covalent bonds in the metallacyclopropane resonance structure.

6.1.3 Typical Reactions

Almost all of the reactions of metals can be classified into just a few typical reactions, and the reactions that metals promote in organic chemistry are simple combinations of these typical reactions. If you learn these typical reactions, you will have no trouble drawing metal-mediated mechanisms. The typical reactions of metal complexes are ligand addition/ligand dissociation/ligand substitution, oxidative addition/reductive elimination, insertion/β-elimination, α-insertion/ α-elimination, σ-bond metathesis (including transmetallations and abstraction reactions), [2 + 2] cycloaddition, and electron transfer.

• *Ligand addition* and *dissociation* reactions are simple Lewis acid–base reactions, either in the forward direction to make a bond (addition) or in the reverse direction to break a bond (dissociation). Many metal catalysts undergo a dissociation or association reaction to form the actual active species, as in Pd(PPh$_3$)$_4$ → Pd(PPh$_3$)$_3$ + PPh$_3$. In an addition (or dissociation), the total electron count of the metal increases (or decreases) by 2, but the d electron count and the oxidation state do not change. Metals also easily undergo *ligand substitution* reactions. These usually proceed in two steps by addition–dissociation or dissociation–addition, rather than in a concerted fashion like an S_N2 reaction.

> Curved arrows may or may not be used to show association and dissociation reactions, depending on your taste. The word "association" or "dissociation" below the arrow often suffices.

The deprotonation of M–H to give M⁻ and H⁺ is one quite common "disso-ciation" reaction in which the pair of electrons of the bond goes to the *metal*. The metal doesn't change its total electron count, and its oxidation state decreases by 2. It's strange to think of a deprotonation as causing a reduction of the metal. The conundrum arises because of the oddities of the language that is used to de-scribe metal complexes. The oxidation state of a complex with a M–H bond is calculated as if the bond were polarized toward H (i.e., M⁺ H⁻). Thus, when metal hydrides are deprotonated, it seems as if the metal is gaining electrons (from the hydride) that it did not have previously.

$$\underset{\underset{PPh_3}{|}}{\overset{\overset{PPh_3}{\|}}{Cl-Pd}}\overset{II}{\underset{2}{-}}H \quad \curvearrowright :NEt_3 \quad \xrightarrow{} \quad \underset{\underset{PPh_3}{|}}{\overset{\overset{PPh_3}{|}}{Cl-Pd}}\overset{0}{-} \quad \xrightarrow{dissociation} \quad \underset{\underset{PPh_3}{|}}{\overset{\overset{PPh_3}{|}}{Pd}}{}^0$$

Of course, the reverse reaction can also occur: protonation of a metal gives a metal hydride in a higher oxidation state.

It is often not clear when or whether ligand addition or dissociation occurs in the course of a metal-mediated or -catalyzed reaction, and in fact knowledge of such details is often not necessary to draw a reasonable mechanism. For this rea-son, the metal center and its associated ligands are often indicated merely as L_nM or [M].

• *Oxidative addition* and *reductive elimination* are the microscopic reverse of each other. In oxidative addition, a metal inserts itself into an X–Y bond (i.e., M + X–Y → X–M–Y). The X–Y bond is broken, and M–X and M–Y bonds are formed. The reaction is an oxidation, because the metal's *oxidation state* increases by 2 (and its d electron count decreases by 2), but the metal also increases its *total electron count* by 2, so it becomes less electron-deficient. The apparent paradox that the "oxidation" of a metal results in a larger electron count is an ar-tifact of the language with which compounds are described.

Oxidative additions can occur at all sorts of X–Y bonds, but they are most commonly seen at H–H (also H–Si, H–Sn, or other electropositive elements) and carbon-halogen bonds. In the oxidative addition of H_2 to a metal, a d orbital containing a lone pair interacts with the σ^* orbital of H_2, lengthening and break-ing the H–H σ bond. The two pairs of electrons from the metal and the H_2 σ bond are used to form two new M–H bonds. C–H bonds can oxidatively add to metals in the same way, and this process, called *C–H bond activation*, is of enor-mous current interest for its potential applications to petrochemical processing and "green" chemistry. Si–H, Si–Si, B–B, and other bonds between elec-tropositive elements undergo oxidative addition, too.

Alkyl halides, pseudohalides like triflates (R–OTf) and phosphates (R–OP(O)(OEt)$_2$), and other C–X species readily undergo oxidative addition. Oxidative addition is usually fastest for X = I, but not always. Oxidative additions at C–X bonds are fundamentally different from oxidative additions at H–H and C–H bonds. For C(sp^3)–X bonds, the same substitution mechanisms that operate in purely organic systems—S$_N$2, S$_{RN}$1, and S$_N$1—also operate in organometallic systems. For example, Pd(PPh$_3$)$_4$ undergoes oxidative addition to CH$_3$I to give (Ph$_3$P)$_2$Pd(CH$_3$)I. After dissociation of PPh$_3$ to give a 16-electron, d^{10} Pd(0) complex, Pd reacts with CH$_3$I in what is probably a straightforward S$_N$2 reaction to give the 16-electron, d^8 Pd(II) complex (Ph$_3$P)$_3$Pd$^+$–CH$_3$ and I$^-$. The I$^-$ then displaces another PPh$_3$ group to give the final neutral, 16-electron, d^8 Pd(II) complex (Ph$_3$P)$_2$Pd(CH$_3$)I.

$18\ e^-,\ d^{10},\ Pd(0)$ $16\ e^-,\ d^{10},\ Pd(0)$ $16\ e^-,\ d^8,\ Pd(II)$ $14\ e^-,\ d^8,\ Pd(II)$ $16\ e^-,\ d^8,\ Pd(II)$

When stereochemically pure 2° alkyl halides undergo oxidative addition to Pd(0), inversion of configuration at C is observed, confirming the S$_N$2 mechanism.

Oxidative addition can also occur at C(sp^2)–X bonds (i.e., at vinyl and aryl halides). It always occurs with retention of configuration about the double bond. Oxidative addition at C(sp^2)–X obviously cannot proceed by an S$_N$2 mechanism. The S$_{RN}$1 mechanism is a possibility. Another possibility is that the Pd coordinates first to the π bond of, say, vinyl iodide, to form a π complex. The metallacyclopropane resonance structure of the π complex can be drawn. An electrocyclic ring opening of the metallacyclopropane (Chapter 4) can then occur. One Pd–C bond breaks, and I$^-$ leaves to give a new cationic Pd-vinyl complex. When I$^-$ associates with the Pd, the overall result is oxidative addition.

Any metal with a d electron count of d^2 or greater can undergo oxidative addition, but 18-electron complexes do not undergo oxidative addition reactions. Oxidative addition is very common for late metals like Pd, Pt, Ir, Rh, and the like.

* **Common error alert:** *d^0 Metals cannot undergo oxidative addition reactions.*

Reductive elimination (X–M–Y → M + X–Y) is the microscopic reverse of oxidative addition. This reaction is usually most facile when the X–Y bond is strong (e.g., H–Ti–Bu → Bu–H + Ti). Not as much is known about the mechanism of reductive elimination as is known about oxidative addition. It is known that the two groups must be adjacent to each other in the metal's coordination sphere. In square planar Pd complexes ((R$_3$P)$_2$PdR$_2$), if the PR$_3$ groups are forced to be trans to one another, reductive elimination does not take place. Reductive elimination is accompanied by a decrease in the total electron count around the metal, a decrease in the oxidation state, and an increase in the d electron count.

16-electron, d^8, Pd(II)   reductive elimination   14-electron, d^{10}, Pd(0)

• *Insertions* and β-*eliminations* are also the microscopic reverse of each other. In an *insertion*, an A=B π bond inserts into an M−X bond (M−X + A=B → M−A−B−X). The M−X and A=B bonds are broken, and M−A and B−X bonds are formed. Insertion is usually preceded by coordination of the A=B π bond to the metal, so it is sometimes called *migratory insertion*. In an insertion, an M−X bond is replaced with an M−A bond, so there is no change in oxidation state, d electron count, or total electron count. However, a new σ bond is formed at the expense of a π bond. The nature of the reaction requires that the new C−M and C−H bonds form to the same face of the A=B π bond, resulting in syn addition. The reaction of a borane (R₂BH) with an alkene to give an alkylborane is a typical insertion reaction that you have probably seen before.

Both early- and late-metal complexes undergo insertions. The likelihood of insertion, though, depends on the nature of X, A, and B. When X = H, as in metal hydrides such as B−H, Zr−H, and Pd−H, insertion is very facile. Insertions of C=C π bonds into M−C and M−H bonds are very important, as will be seen soon. On the other hand, insertions where X = halogen and alkoxy are much rarer. The A=B π bond may be C=C, C=N, C=O, or any other.

β-*Elimination* is the microscopic reverse of insertion. Just as insertion does not, a β-elimination causes no change in the oxidation state, d electron count, or total electron count of the metal. By far the most common β-elimination is the β-*hydride elimination*, in which M−A−B−H → M−H + A=B. The β-hydride elimination is the bane of the organometallic chemist's existence, as it causes many metal–alkyl bonds to be extremely labile. β-Alkoxy and β-halide eliminations are also known, as in the reaction of $BrCH_2CH_2Br$ with Mg.

Insertions are generally thermodynamically favorable, providing a new σ bond at the expense of a π bond. β-Eliminations are very fast reactions, however, so insertions are readily reversible even when they are thermodynamically favorable. For example, when Cp$_2$Zr(H)Cl, a 16-electron d^0 complex, is allowed to react with an internal alkene, the alkyl group in the product is 1°, not 2°! A rapid series of insertions and β-eliminations takes place until the 1° alkyl–Zr bond, which is lowest in energy, is formed. None of the intermediates is observed. Alkylboranes can undergo this reversible series of reactions, too, albeit at much higher temperatures.

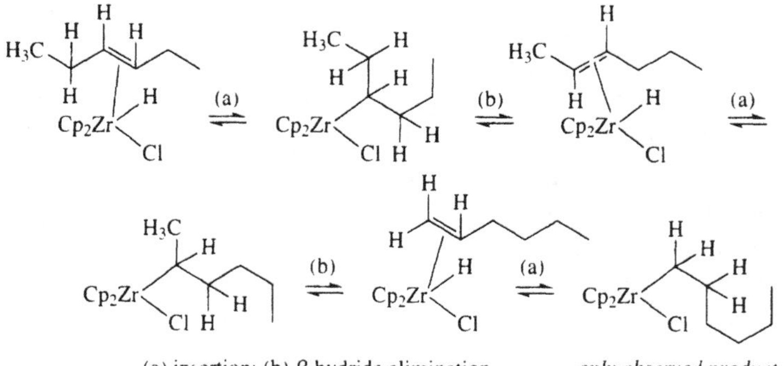

(a) insertion; (b) β-hydride elimination.   *only observed product*

Carbenoid species such as CO and RNC also undergo insertions and eliminations (e.g., M–X + CO ⇌ M–C(O)–X). These are 1,1-insertions, as opposed to the more common 1,2-insertions. Again, there is no change in total electron count or oxidation state of the metal, except transiently as CO coordinates to the metal before insertion. The insertion of CO into a M–C bond is a key step in many important reactions. Again, the insertion is reversible.

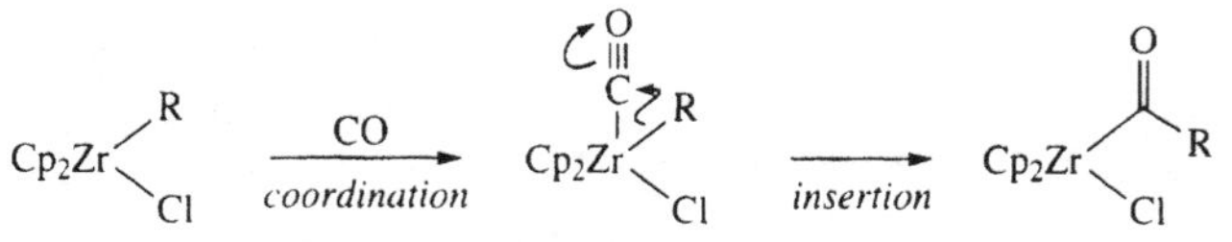

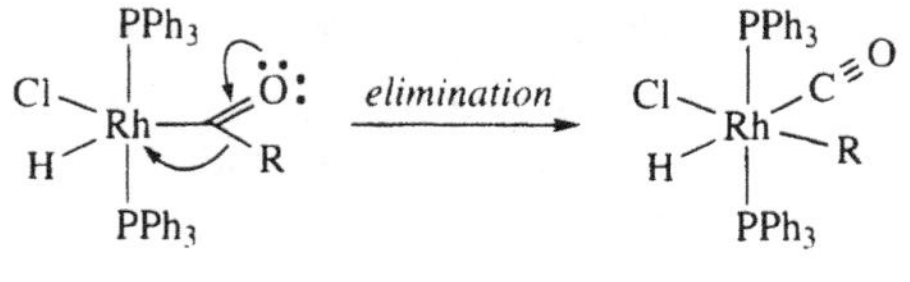

• α-*Insertions* and α-*eliminations* are considerably less common than migratory insertions and β-eliminations. In an α-insertion, a ligand on a metal migrates to an adjacent atom that is doubly bound to the metal, and the electrons of the π bond migrate to the metal (i.e., X–M=Y ⇌ M–Y–X). In an α-insertion, in contrast to its β-relative, the total electron count around the metal decreases by 2, the oxidation state decreases by 2, and the d electron count increases by 2. The opposite occurs in an α-elimination. α-Insertions and α-eliminations are most commonly seen in third-row transition metals, which are more likely to form multiple bonds.

16 electrons, Os(VIII), d^0   α-*insertion* →   *14 electrons, Os(VI), d^2*

• *Sigma-bond metathesis* reactions involve the swapping of M−X and Y−Z σ bonds to give M−Z and X−Y (or M−Y and X−Z). The reaction is concerted, involving a four-center transition state. No change in oxidation state or total electron count occurs, and for this reason the reaction is seen especially often among early, d^0 transition metals. (In metals with a d^2 or greater electron count, a two-step oxidative addition–reductive elimination pathway can be invoked to explain the same overall reaction.) The group that is transferred to the metal is often H, because its nondirectional s orbital provides better overlap in the four-center transition state, although it may also be Si or another element. The group that is transferred from the metal may be almost anything, including C and O.

Transmetallations, in which M−X and M′−Y swap partners to give M−Y and M′−X, are a special kind of σ-bond metathesis reaction. Sometimes, transmetallations are quite obviously thermodynamically favorable, as in 2 BuLi + $Cp_2ZrCl_2 \rightarrow Cp_2ZrBu_2$ + 2 LiCl, but sometimes not, as in $(Ph_3P)_2Pd(Ph)I$ + R−SnBu$_3 \rightarrow (Ph_3P)_2Pd(Ph)R$ + I−SnBu$_3$. Transmetallations involving relatively electronegative metals like Sn are not very well understood, but transmetallations from electropositive elements such as Li or Mg to transition-metal halides can be thought of as simple S_N2 displacements of the halides.

Abstraction reactions are four-center, concerted processes that are closely related to σ-bond metathesis. β-*Abstraction* occurs most often in d^0 metal dialkyl complexes such as Cp_2ZrEt_2. In this reaction, one Et group uses the electrons in the Zr−Et bond to make a new bond to the β-hydrogen on the other Et group. The electrons in the breaking C−H bond are used to form a new bond from C of the breaking bond to Zr. The products are an alkane and a metallacyclopropane, the latter of which can also be described as an alkene–metal π complex. No change in oxidation state occurs if the product is considered to be in its metallacyclopropane oxidation state, although, as discussed earlier, the metals in metallacyclopropanes can be considered to be in a lower oxidation state in their alternative, dative resonance structures. The products of β-abstraction are the same as the products of a two-step, β-hydride elimination–reductive elimination sequence, and experimental evidence is required to distinguish between the two mechanistic pathways. γ-Abstractions have also been observed, although they are much rarer.

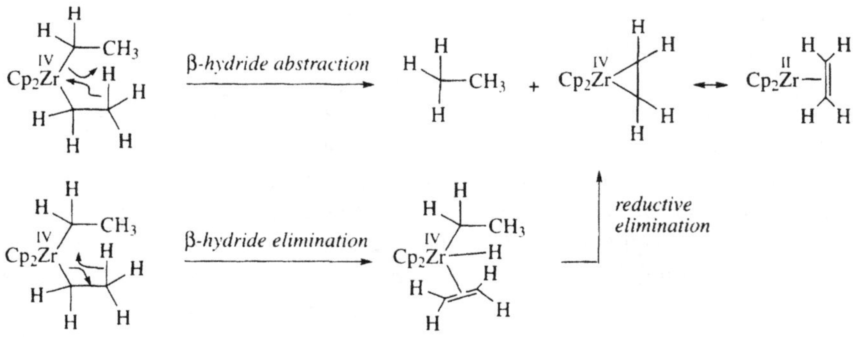

α-Abstraction reactions are especially common among third-row transition metals, which readily form multiple bonds to C. A dialkylmetal complex ($R-M-CH_2R'$) loses alkane RH to give an alkylidene complex ($M=CHR'$) in a concerted, four-center mechanism. No change in metal oxidation state occurs, and the metal may have any number of d electrons.

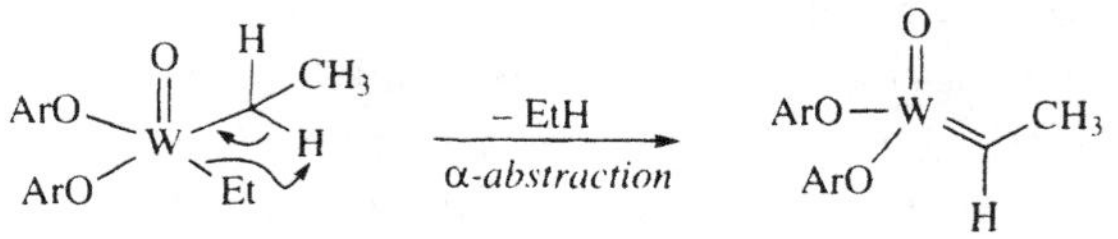

• The *[2 + 2] cycloaddition* (Chapter 4) is another typical reaction of metals. The reaction is quite common among both early and late metals that feature M=C, M=N, or M=O double bonds. The reaction may be preceeded by coordination of the alkene π bond to the metal. No change in oxidation state occurs in [2 + 2] cycloadditions. [2 + 2] Retro-cycloadditions are also seen.

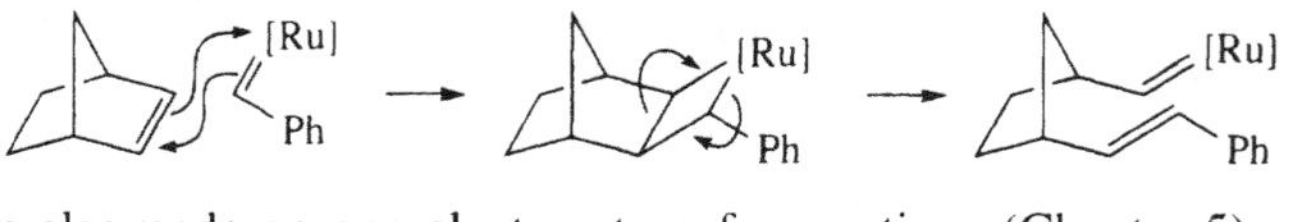

• Metals also undergo one-electron transfer reactions (Chapter 5).

6.1.4. Stoichiometric vs. Catalytic Mechanisms

Whether a reaction requires stoichiometric or catalytic quantities of a transition metal has a strong bearing on how one draws a mechanism for the reaction. The mechanism of a reaction that requires stoichiometric quantities of the metal can be written in a linear fashion like a polar or pericyclic mechanism. However, the mechanisms of metal-catalyzed reactions are usually drawn in a circular fashion, showing how the original metal species is regenerated at the end of each catalytic cycle. Whether a mechanism is catalytic or stoichiometric in metal is usually clear from the reaction conditions.

A stoichiometric reaction mechanism

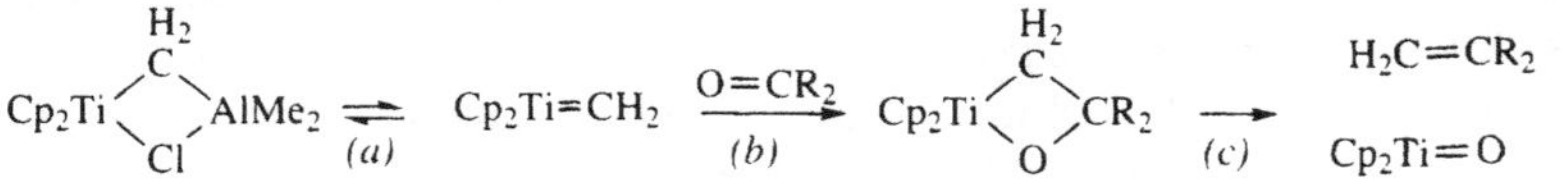

(a) dissociation; (b) [2 + 2] cycloaddition; (c) [2 + 2] retro-cycloaddition.

A catalytic reaction mechanism

(a) oxidative addition;
(b) coordination;
(c) insertion;
(d) reductive elimination.

Because it is not the case that every line represents exactly two electrons in a drawing of an organometallic or inorganic compound, it follows that the curved-arrow convention for showing the movement of electrons cannot be applied unambiguously when reaction mechanisms inolving transition metals are drawn. For this reason, in mechanisms involving transition metals, the name of each individual mechanistic step (insertion, transmetallation, oxidative addition, etc.) is indicated in place of curved arrows. You may use curved arrows to show electron movement in some steps if you wish, but it is more important for you to name every step.

Even more than in traditional areas of organic chemistry, it is often possible to draw more than one mechanism for transition-metal-mediated and -catalyzed reactions. One reason is that the mechanisms of transition-metal-mediated and -catalyzed reactions have not been investigated as closely or for as long a time as those of purely organic reactions, so less is known about them. Another reason is that because multiple oxidation states and coordination levels are tolerated by transition metals, more than one mechanism often operates in a single flask, and seemingly insignificant changes in solvent or ligand can sometimes alter the mechanism.

Organometallic and inorganic catalysts are often classified as *homogeneous* or *heterogeneous*. These two terms simply mean *soluble* or *insoluble*. Homogeneous catalysts are easier to study and often have more predictable behavior than heterogeneous catalysts. Heterogeneous catalysts tend to be used more in large-scale industrial synthesis. The first catalysts to be discovered for many of the reactions described in this chapter were heterogeneous. Afterward, homogeneous catalysts were developed and used both to clarify the mechanism of the reaction and to increase the scope of the reaction.

6.2   Addition Reactions

6.2.1   *Late-Metal-Catalyzed Hydrogenation and Hydrometallation (Pd, Pt, Rh)*

One of the first metal-catalyzed reactions learned by organic chemistry students is the Pd-catalyzed hydrogenation of alkenes and alkynes. The reaction is stereospecifically syn (i.e., the two H atoms add to the same face of the π bond). A wide variety of catalysts can be used for this reaction: Pd metal supported on activated C, Pd poisoned by Pb or Ba salts (Lindlar catalyst), $Pd(OH)_2$, PtO_2, Rh metal supported on alumina, $(Ph_3P)_3RhCl$ (Wilkinson's catalyst), and Ru–phosphine complexes are just some examples. For a long time, the mechanisms of the heterogeneous hydrogenations, those reactions in which the metal did not dissolve in the solution, were considered quite mysterious, but with the advent of homogeneous catalysts like Wilkinson's catalyst, it became possible to study the mechanisms in detail.

Consider the reaction of 2,3-dimethyl-2-butene with H_2 catalyzed by Pd/C. Two new C–H bonds are made, and a C=C π bond breaks. The fact that the addition is stereospecifically syn suggests that an insertion reaction is occurring. The Pd metal is in the (0) oxidation state (i.e., d^{10}), so it can react with H_2 by an oxidative addition to give two Pd–H bonds. At this point Pd is in the (II) oxidation state. Coordination and insertion of the alkene into one of the Pd–H bonds gives Pd–C and C–H bonds. Finally, reductive elimination gives the product and regenerates Pd^0, which begins the catalytic cycle anew.

(a) oxidative addition;
(b) coordination;
(c) insertion;
(d) reductive elimination.

Note that curved arrows are typically *not* used to show electron flow in each mechanistic step. However, the catalytic cycle is explicitly drawn, and every step is one of the typical reactions of metals.

Every late-metal hydrogenation catalyst, whether homogeneous or heterogeneous, probably uses exactly the same catalytic cycle, although some catalysts, particularly sterically encumbered ones such as Wilkinson's catalyst, require that ligand dissociation or substitution occur before the catalytic cycle gets underway. In the case of metals supported on solids such as activated C, silica, and alumina, the support may participate in the reaction in ways that need not concern you here.

In the hydrogenation of unsymmetrical π bonds like $R_2C=O$ (or $R_2C=NR$), insertion into the M–H bond can occur to give either $H–M–O–CHR_2$ or $H–M–CR_2–OH$. Either complex can then undergo reductive elimination to give the alcohol product. On first glance the $M–O–CHR_2$, with its metal-oxygen bond, seems more favorable, but it happens that the strengths of bonds of carbon and oxygen to late metals like Ru and Pd are about the same. Either mechanism may be operative in any particular reaction.

Formic acid and 1,4-cyclohexadiene are sometimes used as alternative H_2 sources in hydrogenation and hydrogenolysis reactions. Protonation of Pd(0) by formic acid and association of the formate counterion gives the oxidative addition product H−Pd(II)−O₂CH. β-Hydride elimination then gives H−Pd(II)−H and CO₂. Similarly, oxidative addition of Pd(0) to the allylic C−H bond of 1,4-cyclohexadiene gives cyclohexadienyl−Pd(II)−H, and β-hydride elimination from this compound gives benzene and H−Pd(II)−H. In both cases, β-hydride elimination may in principle take place before or after insertion of the alkene into the Pd−H bond. Certain classes of hydrogenations proceed better using these H_2 sources.

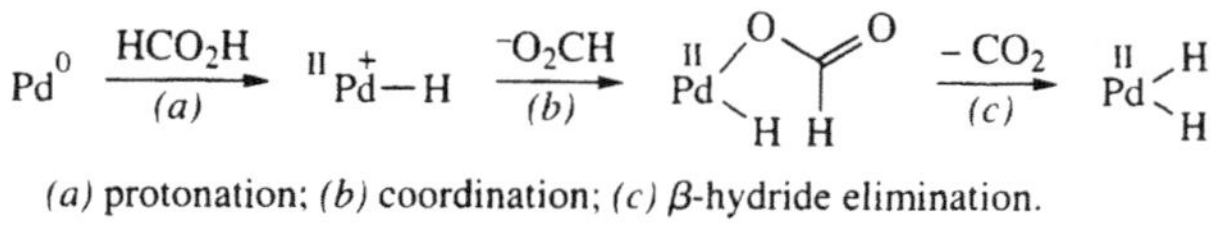

(a) protonation; (b) coordination; (c) β-hydride elimination.

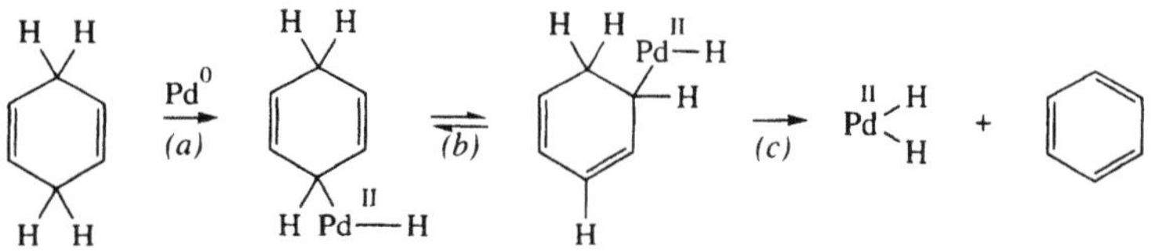

(a) oxidative addition; (b) allylic migration; (c) β-hydride elimination.

Late-metal complexes of Pd, Pt, and Rh can also catalyze hydrosilylation, hydrostannylation, hydroboration, and diborylation reactions of π bonds. Both C=C and C=O bonds may be hydrosilylated or hydroborated, whereas hydrostannylation is usually carried out only on C=C bonds. (Some boranes add to C=O and C=C bonds in the absence of catalyst, but less reactive ones, such as catecholborane ((C₆H₄O₂)BH), require a catalyst. Moreover, the metal-catalyzed reactions sometimes display different selectivities from the uncatalyzed variants.) The mechanisms of all these reactions are the same as hydrogenation, except that oxidative addition of H−H is replaced by oxidative addition of a R₃Si−H (R₃Sn−H, R₂B−H, R₂B−BR₂) bond.

Problem 6.1. Draw a mechanism for the following diborylation reaction.

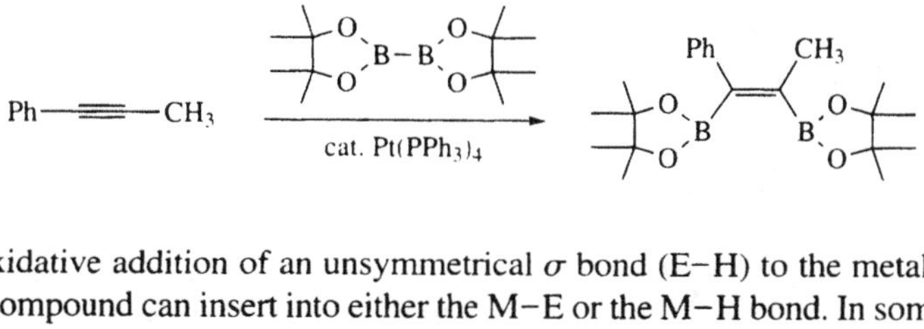

After oxidative addition of an unsymmetrical σ bond (E−H) to the metal, an unsaturated compound can insert into either the M−E or the M−H bond. In some cases, such as the hydrosilylation of carbonyl compounds, the π bond of the substrate inserts into the M−E bond, whereas, in others, insertion of the π bond of the substrate into the M−H bond occurs faster. In any case, either pathway gives the same product after reductive elimination.

(a) oxidative addition;
(b) coordination;
(c) insertion;
(d) reductive elimination.

Problem 6.2. Draw the mechanism for the hydrosilylation of acetophenone in which insertion of the C=O π bond into the Rh–H σ bond (instead of the Rh–Si bond) occurs. Your mechanism should give the same product as the one above.

Many metal complexes with chiral ligands such as phosphines, phosphites, or imidates will catalyze asymmetric hydrogenations, hydrosilylations, and the like. Extremely high enantioselectivities have been achieved with a wide variety of substrates.

All of the steps in the catalytic cycles just illustrated are reversible. In principle, a M–H species can undergo a series of insertions and β-hydride eliminations to give a product whose π bond has been isomerized to a different position. In fact, this is an occasional side reaction in Pd-catalyzed hydrogenations. The partial hydrogenation of fatty acids containing cis double bonds gives a small amount of trans fatty acids by this very mechanism.

6.2.2 Hydroformylation (Co, Rh)

Rh and Co complexes catalyze *hydroformylation*, the addition of H_2 and CO to a C=C π bond to give an aldehyde (H–C–C–CHO). Hydroformylation was one of the first organometallic reactions to be used industrially. The hydroformylation of propene is used to make butyraldehyde, which is hydrogenated to give butanol, a widely used solvent. The original reaction was carried out using a Co catalyst in molten $Ph_3P(!)$, but today the reaction is carried out under much milder, Rh-catalyzed conditions. The mechanism of hydroformylation is nearly the same as the mechanism of hydrogenation that was discussed earlier. Oxidative addition of H_2 to the metal(I) complex occurs to give a metal(III) dihydride. Insertion of an alkene occurs next to give an alkylmetal(III) complex. A step not found in the hydrogenation mechanism occurs next: CO inserts into the M–C bond to give an acylmetal(III) compound. Finally, reductive elimination gives the product and regenerates the metal(I) complex.

(a) oxidative addition; (b) coordination; (c) insertion; (d) reductive elimination.

Insertion of propene into the M–H bond can give two isomers, one with a 1° alkyl–metal bond (as shown) and one with a 2° alkyl–metal bond. The first isomer leads to *n*-butyraldehyde, whereas the second isomer leads to isobutyraldehyde. The first isomer predominates under all conditions, but the isomeric ratio is dependent on the metal, the temperature, and the ligands. As the technology has improved, higher and higher ratios in favor of *n*-butyraldehyde have been obtained.

The mechanism of the *silylformylation* reaction, in which silanes (R_3Si–H) are substituted for H_2, is exactly the same as the mechanism for hydroformylation.

Problem 6.3. Draw a mechanism for the following silylformylation reaction:

$$\text{OSiPh}_2\text{H, } i\text{-Pr} \xrightarrow[\text{1000 psi CO}]{\text{Rh(acac)(CO)}_2} \text{O–SiPh}_2, i\text{-Pr, O, H}$$

6.2.3 Hydrozirconation (Zr)

In hydroboration, a boron hydride (R_2BH) adds across an alkene ($R'CH=CH_2$) to give $R'CH_2CH_2BR_2$. The 16-electron, d^0 Zr(IV) complex $Cp_2Zr(H)Cl$, popularly known as *Schwartz' reagent*, undergoes a closely related reaction. The mechanism involves coordination of an alkene to the electrophilic Zr center followed by migratory insertion of the alkene into the Zr–H bond. The reaction proceeds for alkynes also, by exactly the same mechanism.

The alkylzirconium(IV) complexes produced by hydrozirconation are usually used as reagents in further transformations. The C–Zr bond is moderately nucleophilic at C, so hydrolysis cleaves the C–Zr bond and gives an alkane, whereas treatment with an electrophilic halogen such as NBS or I_2 gives an alkyl halide. Addition of CO or an isocyanide (RNC) results in the insertion of a one-carbon unit into the Zr–C bond. Hydrolysis then gives the corresponding aldehyde or imine, and halogenation gives the acyl halide or nitrile. Transmetallation of the C–Zr bond to another metal such as Cu, Zn, Pd, etc., may also occur, and these alkyl- or alkenylmetal compounds can be used to carry out Pd-catalyzed cross-coupling reactions with organic halides (*vide infra*) and the like. All of these reactions (except CO insertion) are usually drawn as σ-bond metathesis processes, although not much mechanistic work has been conducted.

(a) σ-bond metathesis; (b) insertion; (c) transmetallation.

As mentioned earlier, hydrozirconation of *internal* alkenes gives *terminal* alkylzirconium compounds. The isomerization occurs so quickly that no intermediates are observed. The mechanism of the reaction is a series of insertions and β-hydride eliminations.

Problem 6.4. When internal alkynes are hydrozirconated, the Cp_2ZrCl group does not migrate to the terminus of the chain. Why not?

6.2.4   Alkene Polymerization (Ti, Zr, Sc, and Others)

The polymerization of ethylene to give polyethylene is one of the most important industrial reactions in the world. A free-radical process is most commonly used to execute the polymerization of ethylene. A free-radical process can also be used to execute the polymerization of higher alkenes, but the yields are not as good, and questions of stereoselectivity and regioselectivity arise. The polymerization of higher alkenes such as propene, styrene, and butadiene is instead usually carried out using early-metal catalysts called *Ziegler–Natta catalysts*. The first Ziegler–Natta catalysts were heterogeneous catalysts that consisted of Ti, Zr, or other early metal catalysts supported on surfaces. Later, homogeneous catalysts were developed that allowed chemists to carry out studies on the mechanism of polymerization. More recently, the efficiency of the homogeneous catalysts has improved to the point to which they are now competitive with the heterogeneous catalysts in terms of cost, at least for the higher-value polymers.

The most widely studied homogeneous Ziegler–Natta catalysts are the group 4 metallocenes, especially zirconocene. Zirconocene dichloride (Cp_2ZrCl_2) is an air- and moisture-stable 16-electron, d^0 complex. Its derivative Cp_2ZrMe_2 is a precatalyst for the polymerization of alkenes. The precatalyst is made active by the addition of a Lewis acid; methylalumoxane, $(MeAlO)_n$, prepared from the addition of one equivalent of water to Me_3Al, is commonly used, as is $B(C_6F_5)_3$. The Lewis acid serves to abstract Me^- from Cp_2ZrMe_2 to give the 14-electron, d^0 complex Cp_2ZrMe^+, the active catalyst.

The catalytic cycle for alkene polymerization, the *Cossee mechanism*, is extremely simple. Coordination of the alkene to Cp_2ZrR^+ is followed by insertion to give a new complex Cp_2ZrR^+. No change in the oxidation state of Zr occurs in either of these steps.

$$\underset{\underset{\displaystyle \text{III}}{|||}}{\overset{IV}{Cp_2\overset{+}{Z}r}-R} \quad \xrightarrow[\text{coordination}]{H_2C=CH_2} \quad \overset{IV}{Cp_2}\overset{+}{Z}r\diagdown_R$$

$$\overset{IV}{Cp_2}\overset{+}{Z}r\diagdown\diagup R \quad \xleftarrow{\text{insertion}}$$

There are several mechanisms for termination of the growing polymer chain. One common chain termination step is β-hydride elimination to give Cp_2ZrH^+ and the polymer with a terminal double bond. When the polymerization is carried out under H_2, a σ-bond metathesis can take place to give the saturated polymer and Cp_2ZrH^+.

$$\overset{IV}{Cp_2}\overset{+}{Z}r\diagdown\diagup\underset{H\ \ H}{\diagup P} \quad \xrightarrow{\textit{\beta-hydride elimination}} \quad \overset{IV}{Cp_2}\overset{+}{Z}r-H \quad + \quad \diagdown\diagup\diagdown P$$

$$\underset{H-H}{\overset{IV}{Cp_2}\overset{+}{Z}r-CH_2P} \quad \xrightarrow{\textit{\sigma-bond metathesis}} \quad \underset{H-CH_2P}{\overset{IV}{Cp_2}\overset{+}{Z}r-H}$$

$$P = \text{polymer chain}$$

Group 4 metallocene complexes in which the Cp groups are substituted with alkyl or other groups are also effective in the reaction. The Cp groups can even be replaced with other five-electron donors such as RO groups (using their two lone pairs to make π bonds to the metal) and even three-electron donors like R_2N groups. Group 3 and lanthanide complexes are also widely used to study and carry out alkene polymerization reactions. Polymerizations using any of these d^0 metals proceed by the mechanism shown for Cp_2ZrMe^+.

An alternative mechanism for alkene polymerization, the *Green mechanism*, has been demonstrated for the 12-electron d^2 Ta(III) complex $(Me_3P)_2I_2TaCH_2R$. α-Hydride elimination gives a Ta(V) complex $(H-Ta=CHR)$. $[2 + 2]$ Cyclo-addition of the alkene with this alkylidene complex gives a Ta(V) tantalacyclobutane, and reductive elimination gives the starting catalyst with a chain extended by two C atoms. The chain termination steps are the same as for the Cossee mechanism. The Green mechanism requires an increase in the oxidation state of the metal, and for this reason d^0 metals cannot polymerize alkenes by the Green mechanism.

$$\underset{\underset{\displaystyle \text{III}}{|||}}{\overset{Me_3P_{,,,|}}{\underset{Me_3P}{}}}\overset{H}{\underset{}{Ta}}\diagdown\diagup\underset{R}{H} \quad \xrightleftharpoons{\textit{\alpha-hydride elimination}} \quad \underset{Me_3P}{\overset{Me_3P}{}}\overset{V}{Ta}=\diagup\overset{H\ H}{\underset{R}{}}$$

$$[2+2]\ \Big\uparrow\ H_2C=CH_2$$

$$\underset{Me_3P}{\overset{Me_3P_{,,,|}}{}}\overset{III}{Ta}\diagdown\diagup\diagdown\underset{R}{H\ H} \quad \xleftarrow{\textit{reductive elimination}} \quad \underset{Me_3P}{\overset{Me_3P_{,,V|}}{}}\overset{H\ H}{Ta}\diagdown R$$

When alkenes higher than ethylene are polymerized, regio- and stereochemical issues arise. The regiochemical issue arises when the alkene $H_2C=CHR$ inserts into the M−R′ bond. Either a 1° alkyl metal complex (M−CH$_2$CHRR′) or a 2° alkyl metal complex (M−CHRCH$_2$R′) can form. The 1° alkyl complex is usually thermodynamically favored and is usually observed. To understand the stereochemical issue, consider the polymer polypropylene. The Me groups on the extended polymer backbone may point in the same direction (*isotactic*), in alternating directions (*syndiotactic*), or randomly (*atactic*). The different stereoisomers have different physical properties. The isotactic polymer is in highest demand. Catalysts with special stereochemical features can be used to produce one or the other kind of polymer.

isotactic   syndiotactic   atactic

Problem 6.5. When alkenes are subjected to polymerization conditions in the presence of a sufficiently high concentration of H$_2$ gas, alkene hydrogenation occurs instead of polymerization. Draw a reasonable mechanism for this transformation.

6.2.5 Cyclopropanation, Epoxidation, and Aziridination of Alkenes (Cu, Rh, Mn, Ti)

You may remember that the formation of carbenoids from diazo compounds such as EtO$_2$CCHN$_2$ is catalyzed by Cu(II) and Rh(II) complexes and that these carbenoids undergo [2 + 1] cycloadditions with alkenes as if they were singlet carbenes. Now that we have developed an understanding of organometallic reactivity, we can draw a more complete mechanism for the reactions of Rh and Cu carbenoids with alkenes. Coordination of the nucleophilic diazo C to Rh(II) and loss of N$_2$ gives a Rh(IV) alkylidene complex (Rh=CHCO$_2$Et). Complexes with M=E π bonds generally undergo [2 + 2] cycloadditions, and this complex is no different. A [2 + 2] cycloaddition with the alkene affords a rhodacyclobutane, and reductive elimination gives the product cyclopropane and regenerates Rh(II). Note that the stereochemistry of the alkene is preserved in both the [2 + 2] cycloaddition and the reductive elimination steps.

(a) coordination; (b) [2 + 2] cycloaddition; (c) reductive elimination.

A number of other one-atom transfer reactions are mechanistically similar to the Rh- and Cu-catalyzed cyclopropanations. For example, chiral Mn(II) salen complexes are widely used to catalyze the *Jacobsen* or *Jacobsen–Katsuki* epoxidation of alkenes. The oxidant can be NaOCl (bleach), PhI=O, or another source of electrophilic O. Salen (salicylaldehyde ethylenediamine) is a dianionic, eight-electron, tetradentate ligand that resembles heme or porphyrin. A mechanism very similar to cyclopropanation can be drawn for the Jacobsen epoxidation. Addition of ClO⁻ (for example) to the (salen)Mn(II) complex gives an anionic Mn(II) *ate* complex (one in which the metal has a formal negative charge), and displacement of Cl⁻ by a lone pair on Mn gives a Mn(IV) oxo complex (Mn=O). Metal oxides are known to undergo [2 + 2] cycloadditions with alkenes, so a [2 + 2] cycloaddition of the alkene to the oxo complex to give a Mn(IV) metallaoxetane could occur next. Reductive elimination would give the epoxide and regenerate the catalyst.

(a) coordination;
(b) [2 + 2] cycloaddition;
(c) reductive elimination.

Ugly facts, unfortunately, sometimes invalidate a beautiful mechanism. The Jacobsen epoxidation sometimes proceeds with loss of configurational purity of acyclic alkenes. This feature of the reaction can be explained by invoking radicals. Homolysis of the Mn–C bond in the manganaoxetane intermediate would give a Mn(III) 1,4-diradical complex, and attack of the alkyl radical on O with displacement of Mn(II) would give the epoxide and regenerate the catalyst.

There still exists a problem with the mechanism: The square planar geometry of the salen ligand is unlikely to allow both a Mn–O and a Mn–C bond on the same face of the complex. An alternative mechanism that does not involve a Mn–C bond can be proposed. Electron transfer from the alkene to the Mn(IV) oxo complex might give a Mn(III) radical anion and an organic radical cation. Addition of the oxide to the cation would then give the same Mn(III) 1,4-diradical seen earlier. Whether or not a manganaoxetane intermediate actually occurs in the Mn-catalyzed epoxidation of alkenes is very controversial.

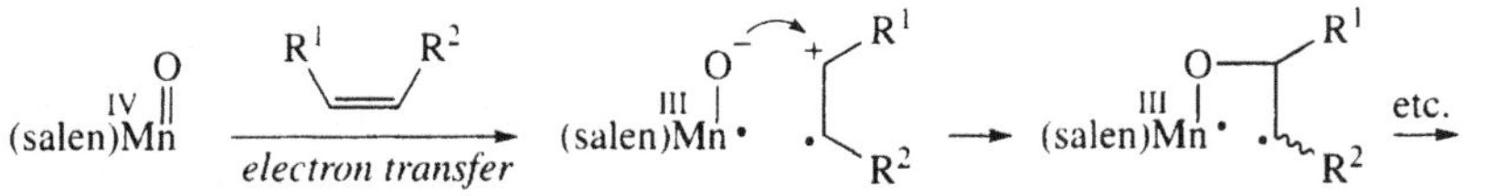

The O atom source can be replaced with a source of NR such as PhI=NTs, and in this case, an aziridination reaction is catalyzed. The catalyst is usually a Cu(II) complex. When the complex is chiral, asymmetric aziridination can be achieved.

Problem 6.6. Draw a reasonable mechanism for the following reaction. You may find it useful to draw PhI=NTs in an alternative resonance form.

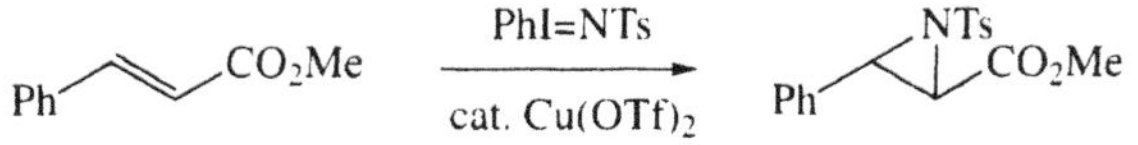

The *Sharpless epoxidation* proceeds by a completely different mechanism from these reactions. This asymmetric O-transfer reaction uses catalytic amounts of Ti(O-i-Pr)$_4$ and (+)- or (−)-diethyl tartrate to catalyze the epoxidation of allylic alcohols by t-BuOOH. The enantioselectivities are usually very good.

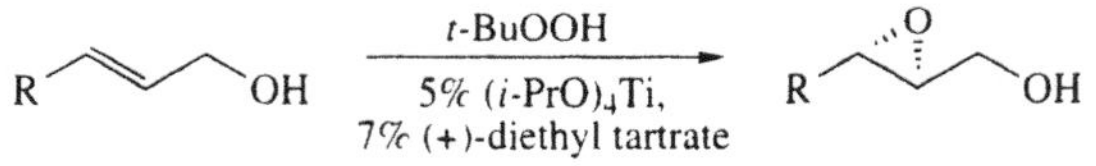

The metal acts essentially as a Lewis acid in this reaction. The terminal O atom of t-BuOOH is made more electrophilic by coordination of both its O atoms to a tartrate–allylic alcohol–Ti complex, and the terminal O atom is then transferred selectively to one enantioface of the nucleophilic C=C π bond. No change in Ti oxidation state or ligand count occurs. After the O atom transfer, a series of ligand substitutions regenerates the active species.

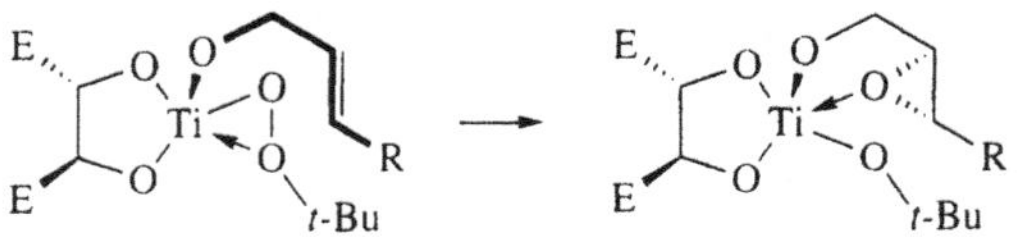

6.2.6   Dihydroxylation and Aminohydroxylation of Alkenes (Os)

When an alkene is treated with OsO$_4$, an *osmate ester* containing two new C–O bonds is formed. The osmate ester is hydrolyzed, usually with aqueous NaHSO$_3$, to give a 1,2-diol. The overall *dihydroxylation* reaction is stereospecifically syn.

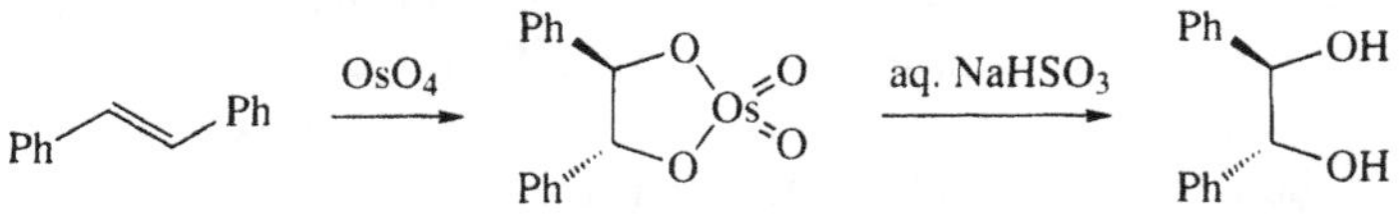

There are two major contenders for the mechanism of this reaction. One is a one-step, [3 + 2] cycloaddition between the alkene and OsO_4 to give the osmate ester. The other is a two-step mechanism: the alkene and OsO_4 undergo a [2 + 2] cycloaddition to give an osmaoxetane, and then the osmaoxetane undergoes α-insertion to give the osmate ester. Which mechanism is in fact operative remains controversial. In any case, the metal is reduced from Os(VIII) to Os(VI) upon osmylation. The hydrolysis of the osmate ester proceeds by simple ligand substitution reactions. The $NaHSO_3$, a mild reducing agent, facilitates ligand substitution by reducing the Os further.

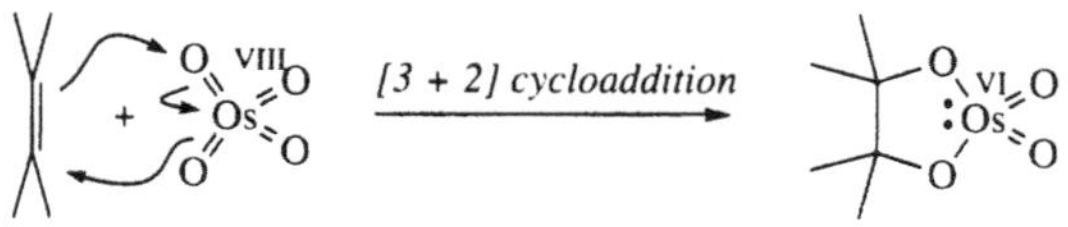

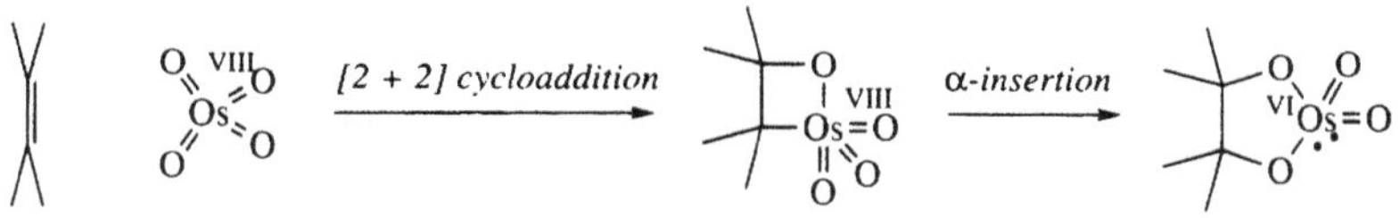

OsO_4 can catalyze the dihydroxylation of alkenes in the presence of a stoichiometric oxidant and water. The catalytic reaction is valuable because OsO_4 is very expensive and extremely toxic. The catalytic version also allows one to use the nonvolatile Os(VI) salt $K_2OsO_2(OH)_4$ in place of the volatile Os(VIII) complex OsO_4.

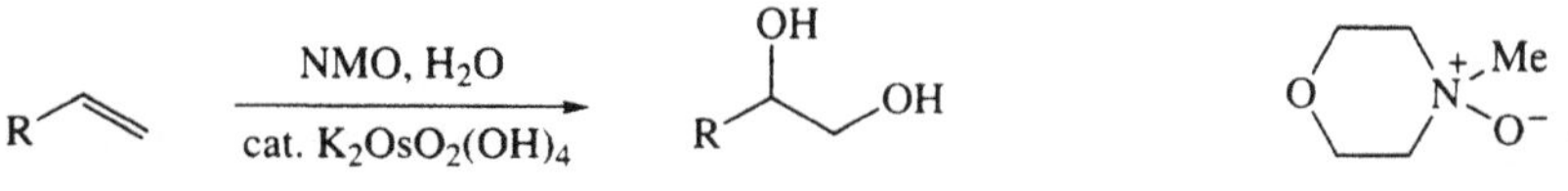

NMO

The mechanism of the catalytic osmylation is as follows. $K_2OsO_2(OH)_4$ is in equilibrium with $OsO_2(OH)_2$, and the latter complex is converted to OsO_4 by the stoichiometric oxidant. Addition of OsO_4 to the alkene by one of the mechanisms discussed above then gives an osmate(VI) ester, and water hydrolyzes the osmate ester in a ligand-substitution process to regenerate $OsO_2(OH)_2$. Stoichiometric oxidants include amine oxides R_3NO such as N-methylmorpholine N-oxide (NMO) and transition-metal salts such as potassium ferricyanide, $K_3Fe(CN)_6$. When the stoichiometric oxidant is $NaIO_4$, the diol is subsequently cleaved to two carbonyl compounds in a process that is an alternative to ozonolysis.

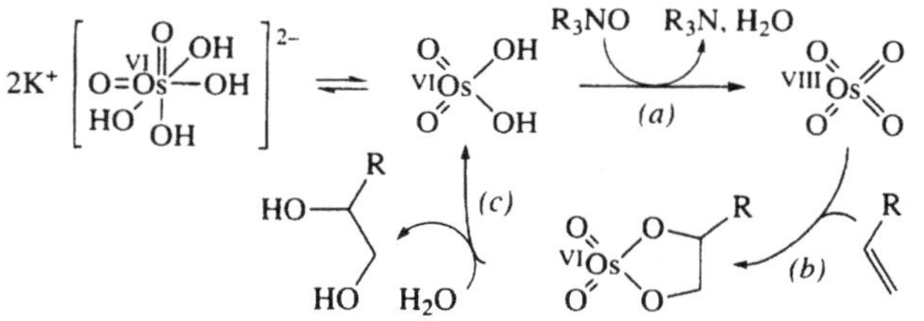

(a) oxidation;
(b) [3 + 2] cycloaddition;
(c) ligand substitution.

Problem 6.7. Propose a mechanism by which the oxidation of $OsO_2(OH)_2$ to OsO_4 by R_3NO may occur. *Hint:* The OsO_4 product contains the O atom derived from the amine oxide.

OsO_4-mediated dihydroxylation is greatly accelerated by amines. When an alkene is combined with a catalytic amount of a chiral amine, a catalytic amount of OsO_4 or $K_2OsO_2(OH)_4$, and stoichiometric amounts of an oxidant [usually NMO or $K_3Fe(CN)_6$] and H_2O, catalytic asymmetric or *Sharpless dihydroxylation* occurs. The chiral amine–OsO_4 complex adds preferentially to one face of the alkene over the other, but the mechanism of the dihydroxylation reaction is not fundamentally changed. The amine is usually a derivative of dihydroquinine (DHQ) or dihydroquinidine (DHQD), and exotic abbreviations such as $(DHQD)_2PHAL$ are commonly used for these compounds. The mixture of reagents required for Sharpless dihydroxylation is commercially available as AD-mix-α or AD-mix-β.

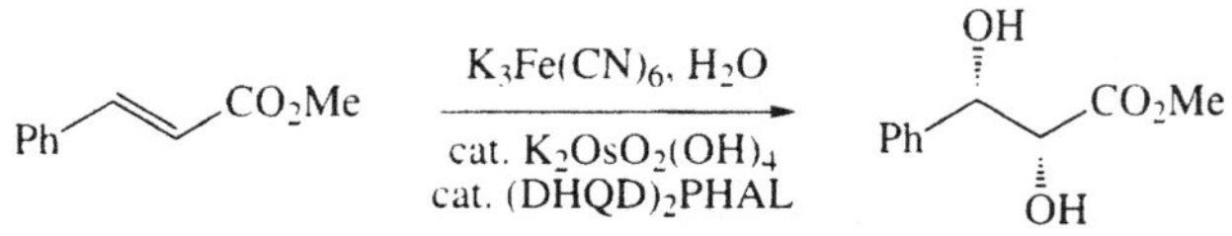

Chiral amines and $K_2OsO_2(OH)_4$ are also used to catalyze the *Sharpless asymmetric aminohydroxylation*. The stoichiometric oxidant in aminohydroxylation is a deprotonated *N*-haloamide, whose mechanistic behavior is very similar to NMO. The reaction proceeds by a mechanism essentially identical to that of Sharpless dihydroxylation.

Problem 6.8. Draw a mechanism for the following Sharpless aminohydroxylation:

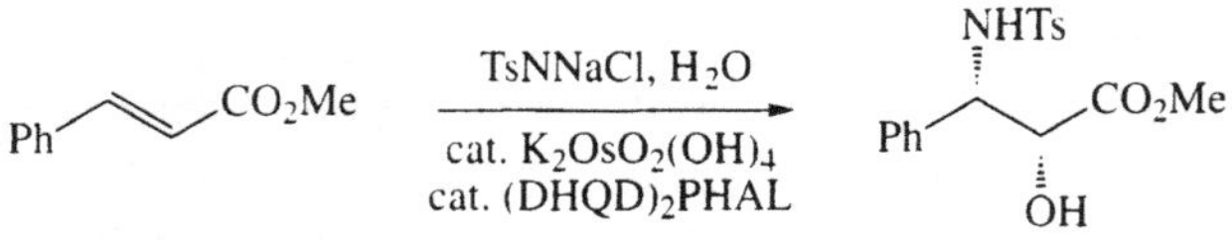

$KMnO_4$ can also dihydroxylate alkenes, particularly electron-poor ones. More electron-rich alkenes may be oxidized further to the dicarbonyl compound. $KMnO_4$ can also oxidize alkylbenzenes to the arylcarboxylic acids. The mechanism of the later reaction remains quite obscure.

6.2.7 Nucleophilic Addition to Alkenes and Alkynes (Hg, Pd)

You may remember that the acid-catalyzed addition of H_2O to alkenes is usually a difficult reaction to carry out because of the side reactions that can occur under the strongly acidic conditions required for the reaction. However, in the presence of Hg(II) salts such as $HgCl_2$, nucleophiles such as H_2O add readily to alkenes to give organomercury compounds. The C−Hg bond is readily converted to a C−H bond by the addition of NaBH₄.

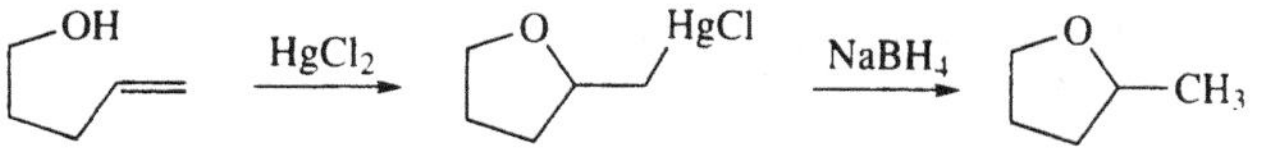

The alcohol is nucleophilic, so the alkene must be made electrophilic. The HgCl₂ first coordinates to the alkene to form a π complex. The coordination of the Hg(II) makes the alkene electrophilic. (In principle one could draw a metallacyclopropane resonance structure for the π complex, but in this resonance structure, Hg would be in the (IV) oxidation state, and Hg(IV) is a high-energy oxidation state.) The coordination weakens the C=C π bond and makes the carbon atoms electrophilic. The nucleophile then attacks one of the carbon atoms of the π complex, and the electrons from the C=C π bond move to form a σ bond between Hg(II) and the other C to give a stable, isolable organomercury compound.

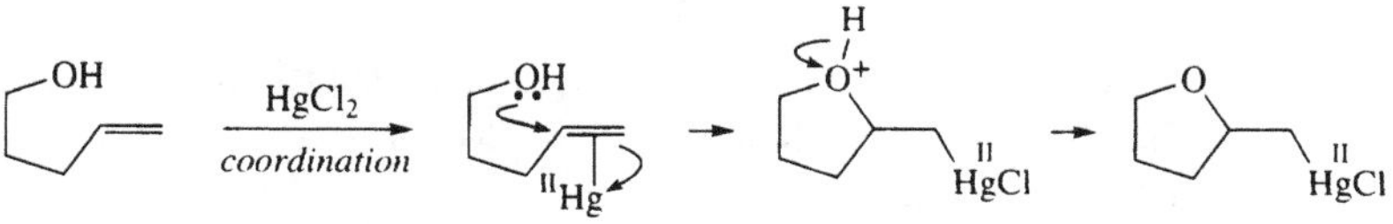

An alternative mechanism might begin with formation of an O−Hg bond. Insertion of the C=C π bond could then follow.

The second part of the reaction proceeds by a free-radical chain mechanism. First, the NaBH₄ converts RHgCl to RHgH by nucleophilic substitution. RHgH then undergoes free-radical decomposition to give RH and Hg(0). (Yes, a little ball of mercury appears at the bottom of the flask at the end of the reaction!)

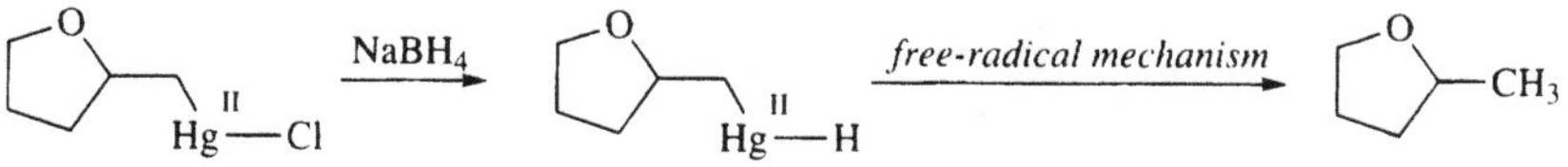

The decomposition of RHgH is initiated by R−Hg σ-bond homolysis. The alkyl radical thus obtained (R·) abstracts H· from H−Hg−R to give R−Hg(I), and homolysis of the R−Hg(I) bond gives Hg(0) and the chain-carrying species. Although in the particular example shown, the alkyl radical R· merely abstracts H· from the H−Hg bond, in other substrates R· may undergo other typical free-radical reactions like intramolecular addition to a π bond.

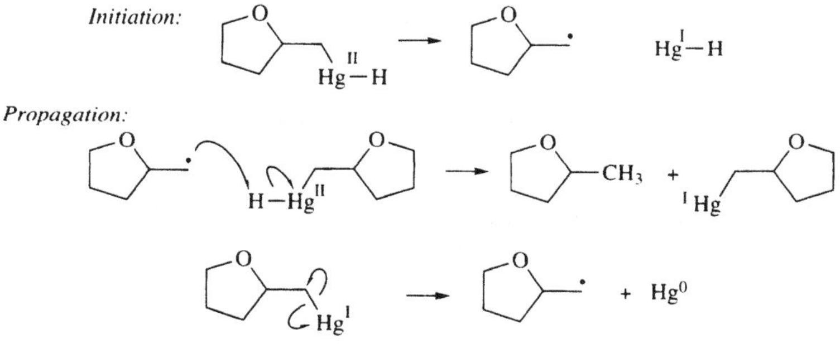

Initiation:

Propagation:

Mercury-mediated reactions of alkenes have largely been superseded by other, more efficient, less toxic procedures.

Problem 6.9. Draw a reasonable mechanism for the following reaction.

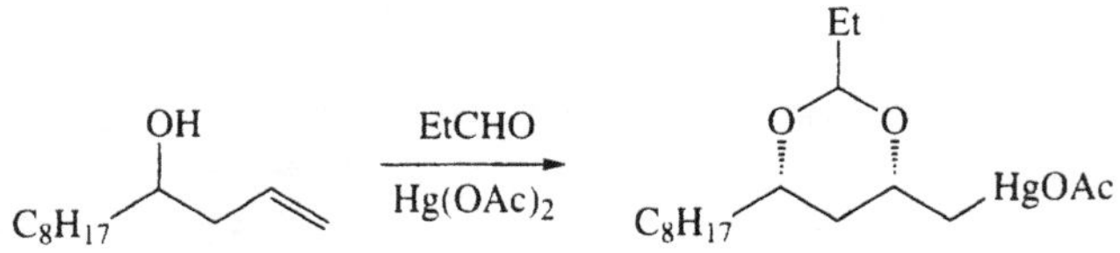

The addition of H_2O to alkynes is catalyzed by Hg(II) under acidic conditions. (The uncatalyzed reaction is extremely difficult to carry out due to the instability of alkenyl cations.)

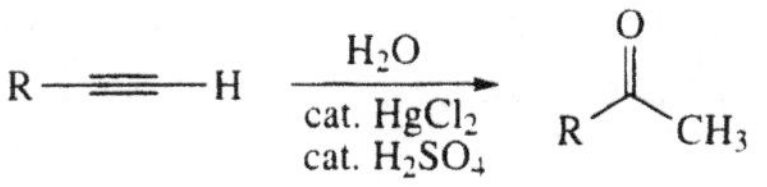

The mechanism begins the same way as the Hg-mediated nucleophilic addition to alkenes. In the first step, an electrophilic π complex forms between the alkyne and Hg(II). Water attacks one of the C atoms of the π complex in Markovnikov fashion to give a 2-hydroxy-1-alkenylmercury(II) compound, an enol, which is protonated to give a carbocation. Fragmentative loss of Hg(II) then occurs to give a metal-free enol, which tautomerizes to give the ketone product.

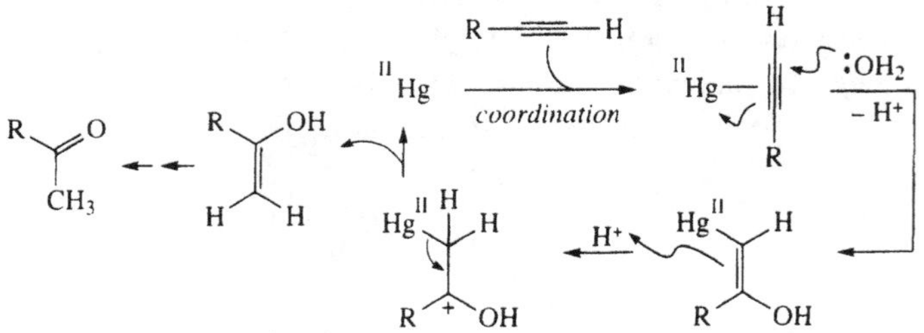

Palladium salts also promote the addition of nucleophiles to alkenes and alkynes. The Pd-catalyzed additions of nucleophiles to alkynes, which is useful for intramolecular cyclizations such as the isomerization of 2-alkynylphenols to benzofurans, proceeds by exactly the same mechanism as does the Hg-catalyzed reaction. However, the Pd-catalyzed additions of nucleophiles to alkenes takes the course of *substitution* rather than *addition* because alkylpalladium complexes are unstable toward β-hydride elimination. The Pd-catalyzed nucleophilic substitutions of alkenes are discussed later in this chapter (Section 6.3.6).

Problem 6.10. Draw a mechanism for the Pd-catalyzed cyclization shown.

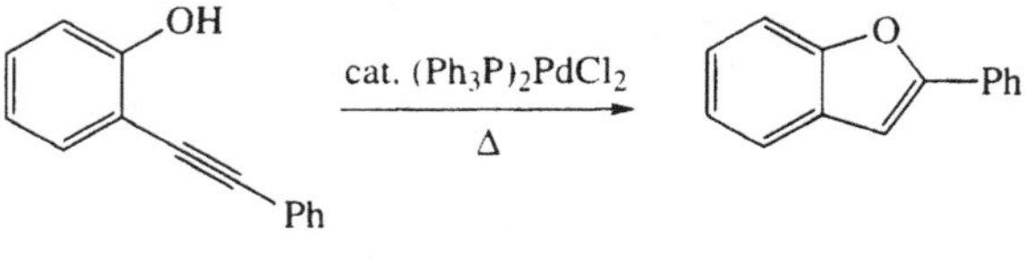

6.2.8   Conjugate Addition Reactions (Cu)

Compounds containing a C–Cu bond undergo conjugate addition to α,β-unsaturated carbonyl compounds, especially ketones. A wide variety of Cu-containing reagents undergo the reaction, including RCu, R_2CuLi, $R_2Cu(CN)Li_2$, and others. Different types of reagents are useful for different reactions; careful optimization of conditions is often required.

The mechanism by which Cu compounds undergo conjugate additions is highly controversial. The mechanism almost certainly changes based on the stoichiometry of the Cu species, the presence of other Cu-bound ligands, the counterion, the solvent, additives, and, of course, the substrate. Electron transfer steps have been implicated. Chemists' understanding of the mechanism is further complicated by a lack of information about the true nature of the Cu species. A very simple picture of the conjugate addition of Me_2CuLi to an enone has the 16-electron Cu species coordinate to the alkene to make an 18-electron species. Migratory insertion of the alkene into the Me–Cu bond and dissociation of neutral MeCu then give the product enolate. The enolate is sometimes trapped in situ by Me_3SiCl (which probably participates in the reaction mechanism) or an alkyl halide. The MeCu by-product does not participate in further reactions.

(a) coordination; (b) insertion; (c) dissociation.

In the presence of a catalytic amount of a Cu(I) salt, Grignard reagents add exclusively to the β-carbon of α,β-unsaturated carbonyl compounds. The reaction proceeds via transmetallation of the Grignard reagent (RMgX) with CuX to give either RCu or R_2Cu^-. The alkylcopper compound undergoes conjugate addition to the enone, as discussed above. Transmetallation of the Cu enolate with RMgX gives the Mg enolate and completes the catalytic cycle.

Problem 6.11. Propose a mechanism for the following Cu-catalyzed conjugate addition.

6.2.9   Reductive Coupling Reactions (Ti, Zr)

The group 4 metals, especially Ti and Zr, are most stable in their (IV) oxidation states. In their (II) oxidation states they act as reducing agents, but in a very specific way. They use the two excess electrons to create a new C–C bond between two C=X or C≡X species to give a new X–C–C–X or X=C–C=X compound. (X may be N, O, or C.)

The reductive coupling of two carbonyl compounds using reduced Ti reagents is called *McMurry coupling*. The products may be 1,2-diols or alkenes.

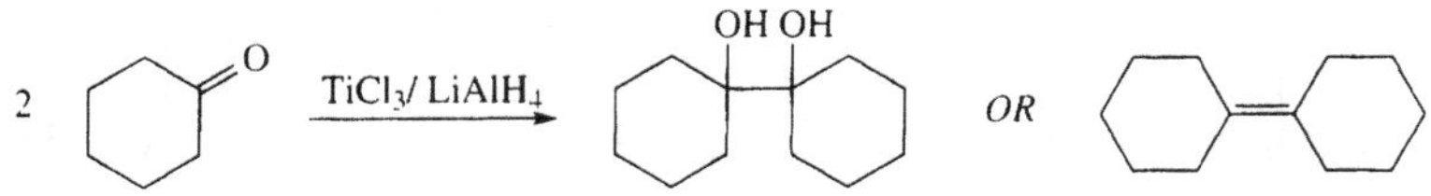

Many Ti reagents (e.g., TiCl$_3$/LiAlH$_4$, TiCl$_3$/Zn, etc.) have been used for this reaction. The nature of the active species in McMurry couplings is obscure, but it definitely involves Ti in a lower oxidation state, at least Ti(II) and possibly lower. The mechanism is also obscure. One can imagine that the reduced Ti(II) forms a π complex with one carbonyl compound to give a Ti(IV) titanaoxirane. Coordination and insertion of a second carbonyl gives the Ti complex of a 1,2-diol. If a second Ti(II) species comes along and coordinates to one of the O atoms, further reduction of the 1,2-diol can take place to give an alkene and two Ti(IV) species.

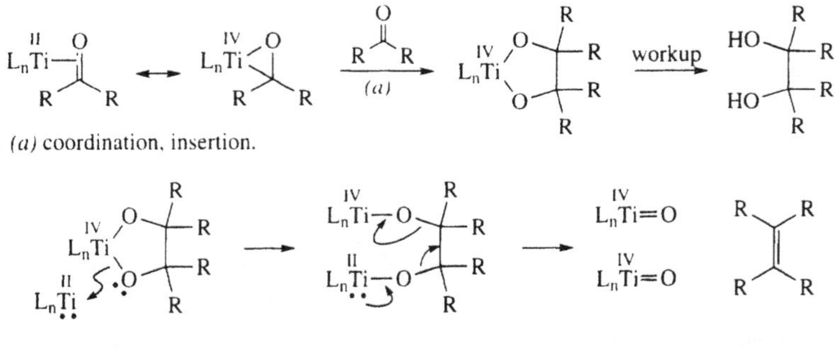

(a) coordination, insertion.

In another mechanism for McMurry coupling, a Ti(IV) ketyl forms from Ti(III) and the carbonyl. Radical–radical combination then gives the 1,2-diol in a pinacol coupling (Chapter 5).

Reductive coupling of alkynes or alkenes may be carried out with Cp$_2$ZrCl$_2$, Ti(O-i-Pr)$_4$, or other Ti(IV) and Zr(IV) complexes. The reaction is often carried out in an intramolecular fashion to achieve better control of regiochemistry. The diyne or other polyunsaturated compound is added to a mixture of the Zr(IV) or Ti(IV) complex and a reducing agent such as BuLi. After formation of the intermediate metallacycle, the C–M bonds are cleaved by treatment with an electrophile such as H$^+$ or I$_2$ to give the organic product.

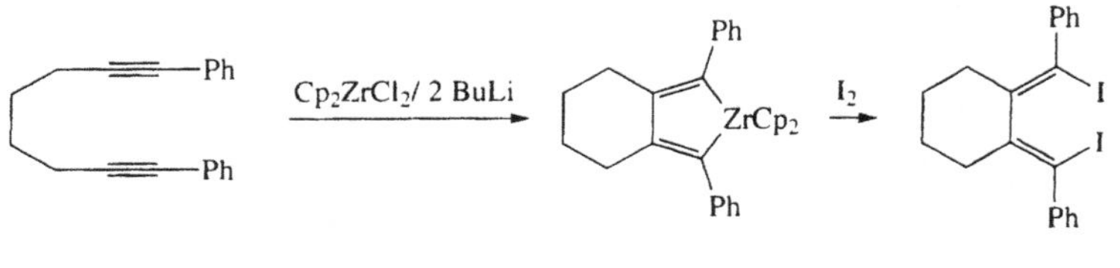

The reductive coupling requires that the Zr(IV) be reduced to Zr(II). Metals such as Mg may be used to reduce Zr(IV) to Zr(II) by electron transfer, but the most widely used method today involves the addition of two equivalents of BuLi to Cp_2ZrCl_2. Transmetallation to give Cp_2ZrBu_2 is followed by β-hydride abstraction to give the Zr(IV) metallacyclopropane, which can also be described as a Zr(II)–butene complex.

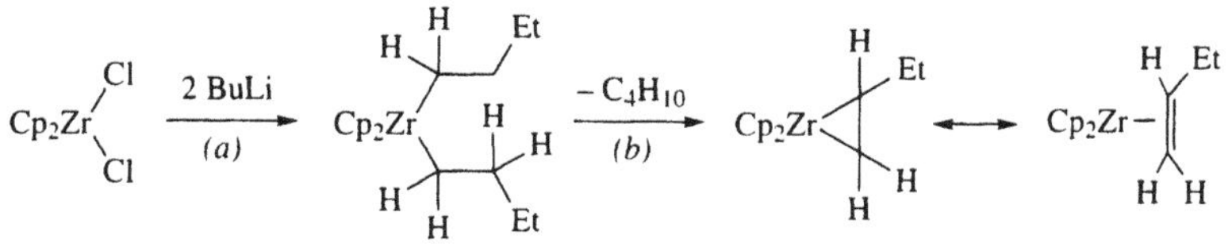

(a) transmetallation; (b) β-hydride abstraction.

Next, 1-butene is displaced from Zr(II) by an alkyne to give a new π complex; then, the other alkyne coordinates to the Zr(IV) metallacyclopropene and inserts into the C–Zr bond to give a Zr(IV) metallacyclopentadiene.

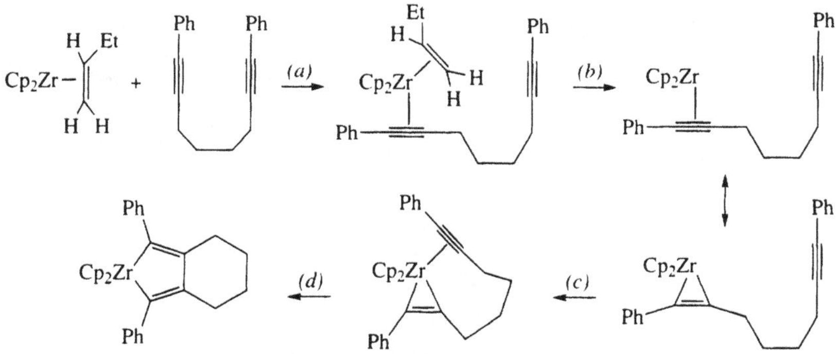

(a) association; (b) dissociation; (c) association; (d) insertion.

When an *enyne* is reductively coupled in an intramolecular fashion with Zr or Ti, the metallacyclopentene that is obtained initially may be carbonylated to the corresponding bicyclic cyclopentenone with CO. The reaction proceeds by insertion of CO and reductive elimination. In Ti-promoted cyclization, in fact, under certain conditions the Ti(II) fragment produced by reductive elimination can promote reductive coupling of another enyne, providing a basis for a variant of this reaction that is catalytic in Ti.

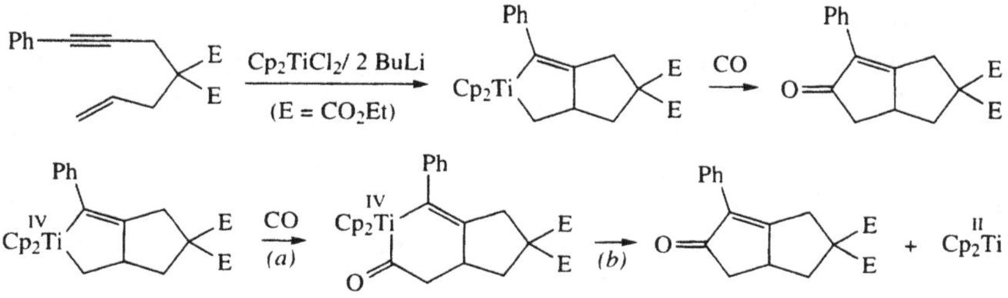

(a) coordination, insertion; (b) reductive elimination.

The π complex that is the common intermediate in all these reactions can be generated in other ways. For example, Cp_2ZrPh_2 undergoes β-hydride abstraction upon heating to generate a benzyne complex of $Cp_2Zr(II)$. The complex is highly reactive, and it can be trapped by a π compound such as an alkyne, alkene, nitrile, or carbonyl to give a five-membered metallacycle.

(a) β-hydride abstraction; (b) coordination, insertion.

In fact, a wide variety of π complexes $Cp_2Zr(X=Y)$, including complexes of alkynes, cycloalkynes, arynes, imines, and thioaldehydes, can be generated by β-hydride abstraction from complexes of the type $Cp_2Zr(Me)X-Y-H$. These complexes can in turn be made by hydrozirconation of $X=Y$ with $Cp_2Zr(H)Cl$ followed by addition of MeLi, or by addition of ^-X-YH to $Cp_2Zr(Me)Cl$.

(a) transmetallation; (b) β-hydride abstraction; (c) coordination, insertion.

The *Kulinkovich cyclopropanation*, in which an ester and two equivalents of a Grignard reagent are combined to make a cyclopropanol in the presence of a *catalytic* amount of Cp_2TiCl_2, is a useful variant of reductive coupling reactions. A cyclopropylamine can be obtained instead if a tertiary amide is substituted for the ester.

The catalytic cycle is entered by the addition of two equivalents of Grignard reagent to the Ti catalyst to make a dialkyltitanium compound, which undergoes β-hydride abstraction to give a titanacyclopropane. The C=O π bond of the ester inserts into this compound to give a titanaoxolane that is also essentially a hemiacetal. A β-alkoxide elimination gives an intermediate that has a nucleophilic Ti-C bond poised close to the ketone. Insertion of the C=O π bond into the Ti-C bond gives the cyclopropanol, and ligand substitution completes the catalytic cycle. The oxidation state of the Ti remains at (IV) throughout the reaction. Note that the reaction of the Grignard reagent with the Ti complex is faster than addition of the Grignard reagent to the ester(!).

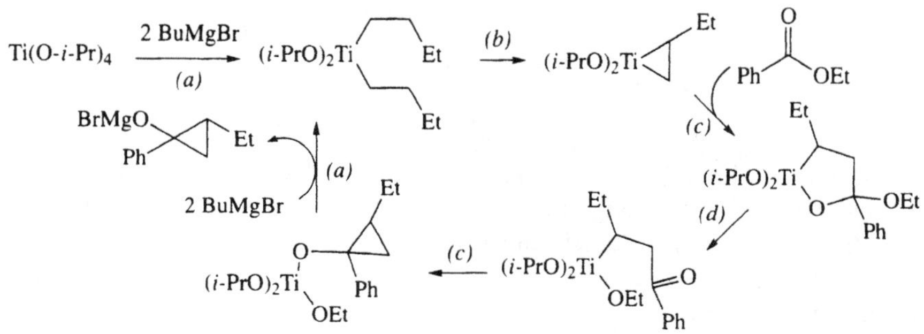

(a) ligand substitution; *(b)* β-hydride abstraction; *(c)* insertion; *(d)* β-alkoxide elimination

The conventional Kulinkovich cyclopropanation uses two equivalents of a Grignard reagent, but only one equivalent is incorporated into the product; the other is lost as alkane. This aspect of the reaction is not a problem when the Grignard reagent is inexpensive, but when the Grignard reagent is precious, its waste should be avoided. One of the intermediates in the Kulinkovich reaction is a titanacyclopropane, a compound that can also be described as an alkene–titanium complex. When an external alkene is added to the reaction mixture, ligand exchange can generate a new alkene–titanium complex into which the ester can insert. Thus, two equivalents of an inexpensive cheap Grignard reagent can be sacrificed to couple a precious alkene to the ester or amide. This reaction can also be carried out in intramolecular fashion.

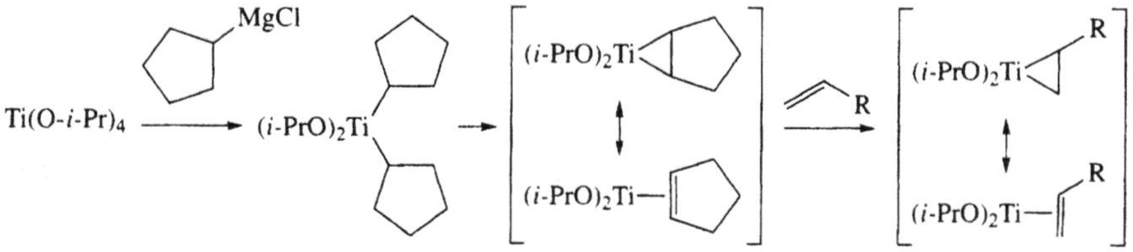

Problem 6.12. Write a reasonable mechanism for the following Kulinkovich cyclopropanation:

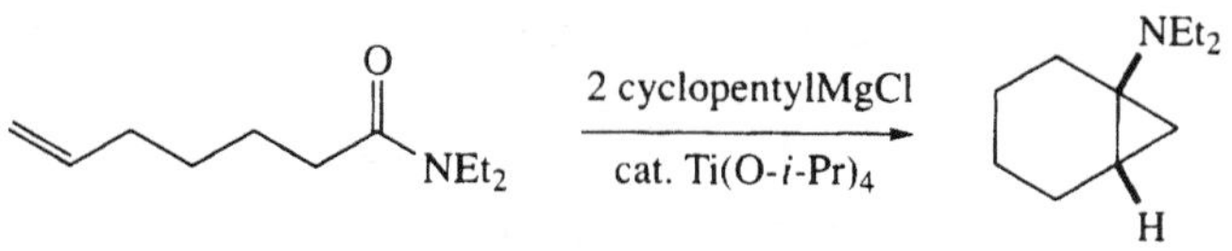

6.2.10   Pauson–Khand Reaction (Co)

The *Pauson–Khand reaction* combines an alkyne, an alkene, and CO to give a cyclopentenone. This highly convergent reaction is mediated by the complex $Co_2(CO)_8$, $(CO)_4Co–Co(CO)_4$, which contains an unusual metal–metal bond. Each Co atom in $Co_2(CO)_8$ has its full complement of 18 electrons, and each is in the Co(0) oxidation state.

$$R \equiv\!\!\!\equiv R \xrightarrow{\text{Co}_2(\text{CO})_8 \quad \text{H}_2\text{C}=\text{CH}_2}$$

The Pauson–Khand reaction gives the same product as the group 4 metal-mediated reductive coupling and carbonylation, and both reactions proceed by essentially the same mechanism: formation of an alkyne–metal π complex, insertion of an alkene, insertion of CO, and reductive elimination. Some details differ, however. When an alkyne is added to $\text{Co}_2(\text{CO})_8$, CO evolves, and an isolable, chromatographable alkyne–$\text{Co}_2(\text{CO})_6$ complex is obtained. This "butterfly" complex contains four Co(II)–C bonds, and the Co–Co bond is retained. The formation of the alkyne–$\text{Co}_2(\text{CO})_6$ complex involves the formation of an ordinary π complex of the alkyne with one Co(0) center, with displacement of CO. The π complex can be written in its Co(II) cobaltacyclopropene resonance structure. The π bond of the cobaltacyclopropene is then used to form a π complex to the other Co center with displacement of another equivalent of CO. This second π complex can also be written in its cobaltacyclopropene resonance structure. The alkyne–$\text{Co}_2(\text{CO})_6$ complex has two 18-electron Co(II) centers.

The alkyne–$\text{Co}_2(\text{CO})_6$ complex is now combined with an alkene such as $\text{CH}_2=\text{CH}_2$. Initially, nothing happens, because the alkyne–$\text{Co}_2(\text{CO})_6$ complex is a relatively inert 18-electron complex. Heat, light, or addition of N-methylmorpholine N-oxide (NMO) causes one CO ligand to leave to generate a 16-electron Co complex. Coordination of the alkene to Co then occurs. Migratory insertion of the alkene into one of the Co–C bonds occurs, followed by migratory insertion of CO. Reductive elimination gives a π complex of the cyclopentenone with $\text{Co}_2(\text{CO})_4$, and enone dissociation occurs to give the cyclopentenone product.

(a) dissociation; (b) association; (c) insertion; (d) reductive elimination; (e) dissociation.

Problem 6.13. NMO oxidizes one CO ligand of the alkyne–$Co_2(CO)_8$ complex to CO_2 and gives an alkyne–$Co_2(CO)_7$ complex. Write a mechanism for this transformation.

The Pauson–Khand reaction is especially useful for the intramolecular cyclization of 1,6-enynes to give bicyclo[3.3.0]octenones. The intermolecular reaction has problems of regioselectivity when unsymmetrical alkynes or alkenes are used, although some substrates give especially good selectivity for various reasons.

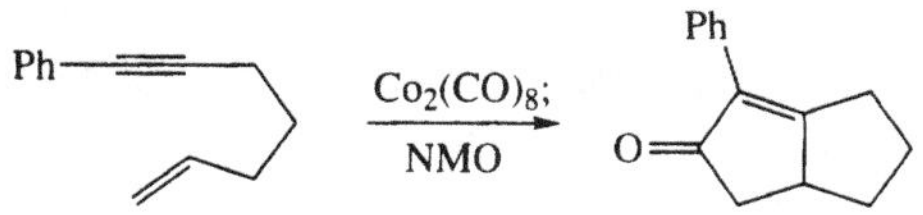

Other metal complexes have been found to promote the Pauson–Khand reaction, including $Mo(CO)_6$, $Fe(CO)_5$, $W(CO)_6$, and $Cp_2Ti(CO)_2$, and variants of the Pauson–Khand reaction that use catalytic amounts of the metal have also been developed, but stoichiometric $Co_2(CO)_8$ remains the most widely used method for effecting this transformation.

6.2.11   Dötz Reaction (Cr)

In the *Dötz reaction*, an unsaturated chromium carbene complex is combined with an alkyne to give a substituted phenol. The unsaturated group may be an alkenyl, cycloalkenyl, or aryl group. There is almost always a methoxy or other alkoxy group attached to the carbene.

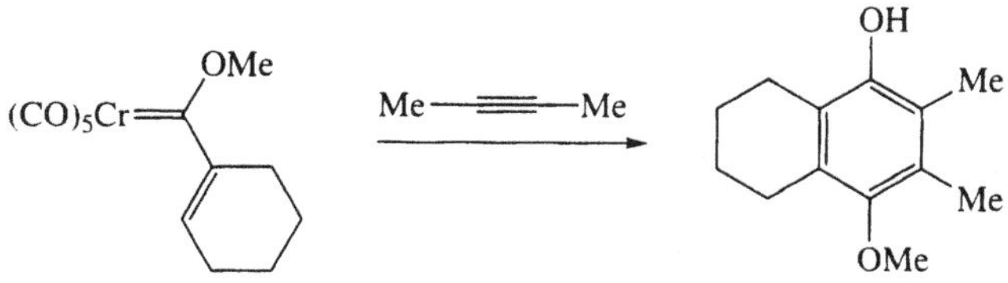

By numbering the atoms, we see that the alkylidene portion of the Cr complex contributes three C atoms to the new aromatic ring, and the alkyne contributes two. The sixth C atom with its associated O must come from a CO ligand. Make: C2–C7, C4–C5, C5–C8. Break: Cr–C2.

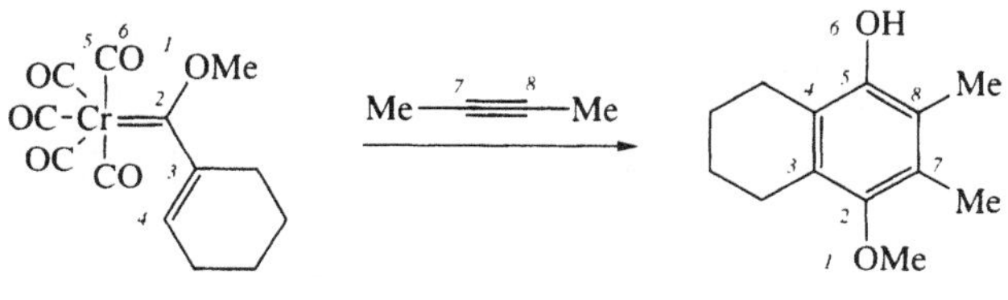

Compounds with multiple M–C bonds are prone to undergo [2 + 2] cycloadditions. Such a reaction in this system makes the C2–C7 bond and generates a metallacyclobutene. In this system, the [2 + 2] cycloaddition is preceded by loss of CO from the unreactive, 18-electron carbene complex and coordination of the alkyne.

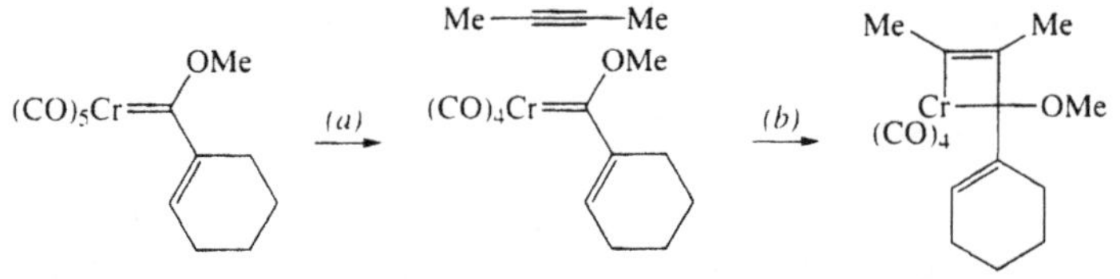

(a) dissociation; (b) coordination, [2 + 2] cycloaddition

Three reasonable possibilities can be formulated for incorporation of the CO group into the ring and formation of the two remaining C–C bonds. For example, the metallacyclobutene complex can undergo sequential electrocyclic ring opening and ring closing to give a metallacyclohexadiene complex. CO insertion, reductive elimination, and tautomerization then give the observed product.

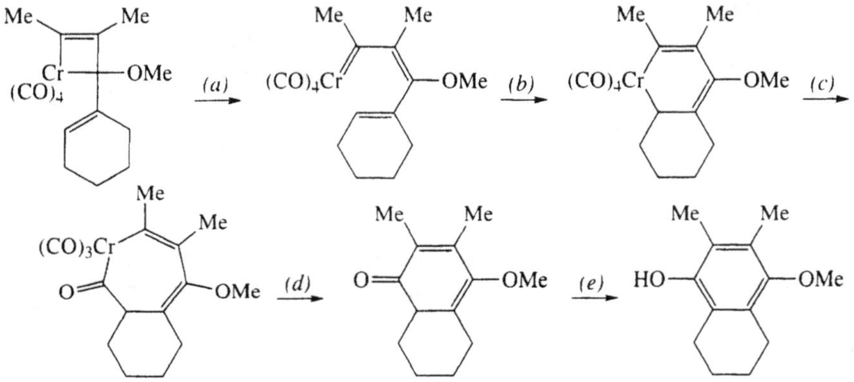

(a) electrocyclic ring opening; (b) electrocyclic ring closing; (c) insertion;
(d) reductive elimination; (e) tautomerization (two steps).

Alternatively, migratory insertion of a CO group of the metallacyclobutene into the Cr–C(sp^3) bond affords a metallacyclopentenone, which can undergo reductive elimination to give a cyclobutenone. Sequential electrocyclic ring opening, ring closing, and tautomerization gives the product.

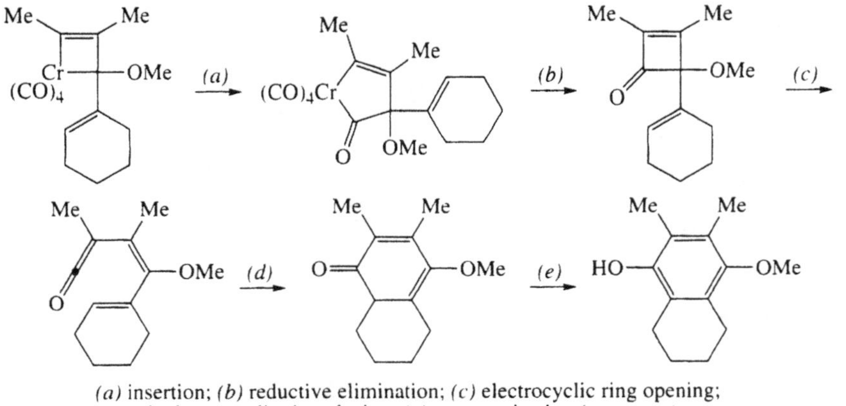

(a) insertion; (b) reductive elimination; (c) electrocyclic ring opening;
(d) electrocyclic ring closing; (e) tautomerization (two steps).

Finally, the dienylcarbene complex formed upon electrocyclic ring opening of the metallacyclobutene can undergo migratory insertion of a CO group into the Cr=C π bond (leaving the Cr–C σ bond unchanged) to give a chromacyclopropanone. A dienylketene–Cr(CO)$_3$ resonance structure can be drawn for this compound. Loss of Cr(CO)$_3$, electrocyclic ring closing, and tautomerization provide the phenol.

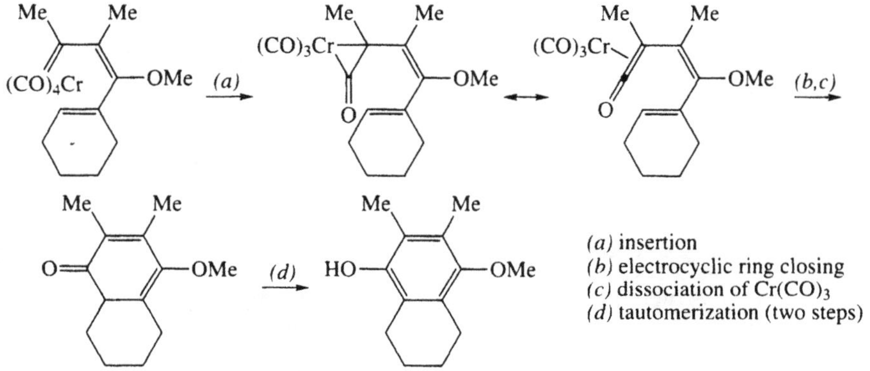

(a) insertion
(b) electrocyclic ring closing
(c) dissociation of Cr(CO)$_3$
(d) tautomerization (two steps)

When the Dötz reaction is executed in EtOH, products derived from addition of EtOH to a dienylketene are obtained, suggesting that either of the latter two mechanisms is correct.

The Cr-containing starting material for the Dötz reaction is usually prepared from Cr(CO)$_6$. The C atoms in Cr(CO)$_6$ are electrophilic, as is best illustrated by the metallaketene resonance structure of this compound. Addition of an unsaturated organolithium reagent to Cr(CO)$_6$ affords a chromaenolate, which is O-methylated (Meerwein's reagent (Me$_3$O$^+$ BF$_4$$^-$) is often used) to afford the carbene complex.

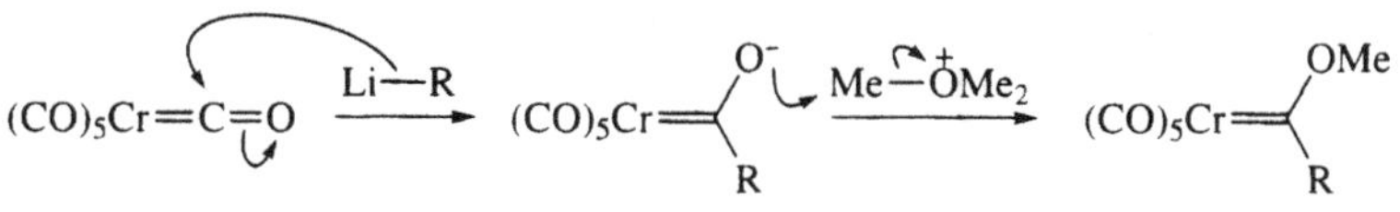

The carbene complex (CO)$_5$Cr=C(CH$_3$)OMe, prepared from Cr(CO)$_6$ and MeLi, can also be used to prepare Cr carbene complexes. The five CO groups attached to Cr in this compound exert such a strong electron-withdrawing effect on the Cr=C π bond that the latter becomes polarized toward Cr. In fact, the Cr(CO)$_5$ fragment is so electron-attracting that the CH$_3$ group becomes as acidic as a methyl ketone! Addition of an aldehyde and a Lewis acid causes an aldol reaction to occur, thereby producing an unsaturated Cr carbene complex.

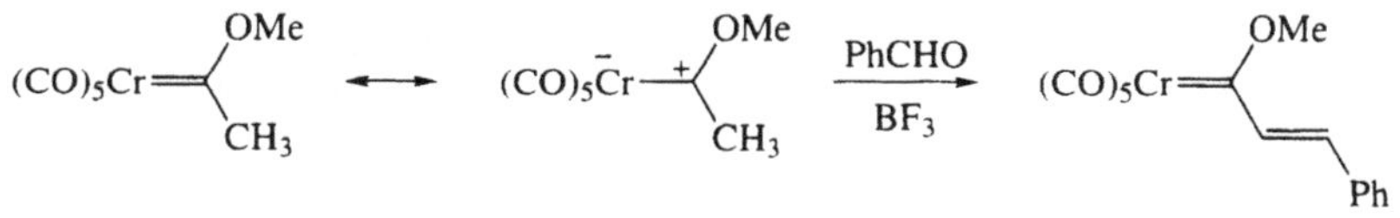

Problem 6.14. Draw a mechanism for the BF$_3$-promoted aldol reaction of (CO)$_5$Cr=C(CH$_3$)OMe and PhCHO.

Chromium carbene complexes undergo other reactions, too. The mechanisms of these reactions consist of steps ([2 + 2] cycloadditions, electrocyclic reactions) that are similar to those seen in the mechanism of the Dötz reaction.

Problem 6.15. Draw a mechanism for the following reaction:

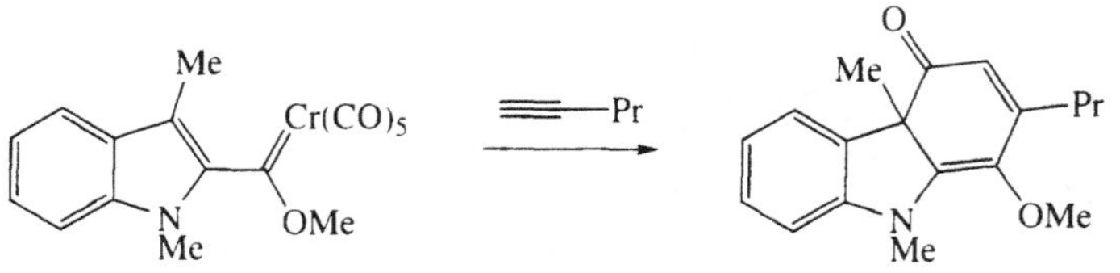

6.2.12   Metal-Catalyzed Cycloaddition and Cyclotrimerization (Co, Ni, Rh)

Certain late-transition-metal complexes have the ability to catalyze [4 + 2], [4 + 4], and [5 + 2] cycloadditions. Despite their appearance, these reactions are not concerted cycloadditions. For example, Ni(0) catalyzes intramolecular [4 + 4] cycloadditions.

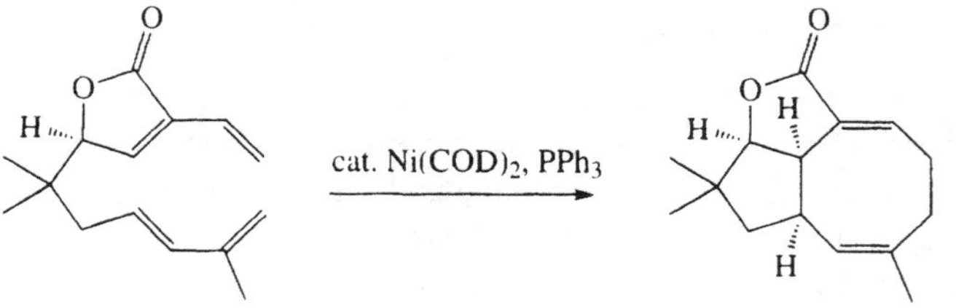

The mechanism probably begins by coordination of Ni(0) to the diene to give a Ni(0)–diene complex, which can also be described as a Ni(II) nickelacyclopentene. Insertion of a C=C π bond into the Ni–C bond, allylic rearrangement of the Ni–C bond, and reductive elimination completes the mechanism. Similar mechanisms can be drawn for Ni- and Rh-catalyzed [4 + 2] cycloadditions.

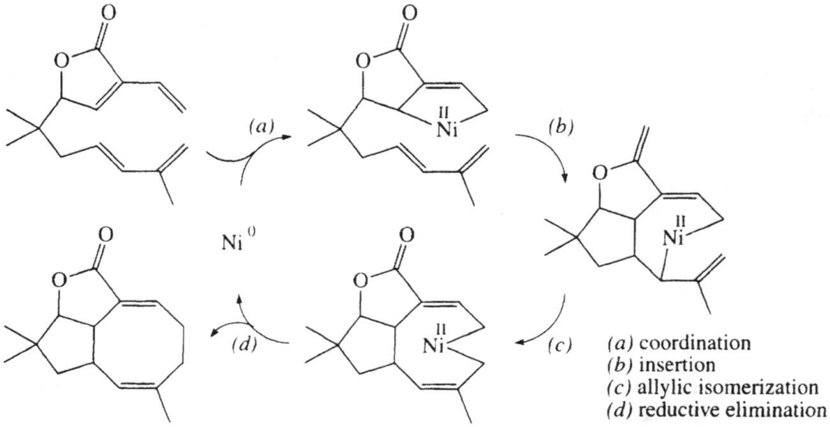

(a) coordination
(b) insertion
(c) allylic isomerization
(d) reductive elimination

Problem 6.16. Draw a reasonable mechanism for the following Ni-catalyzed [4 + 2] cycloaddition:

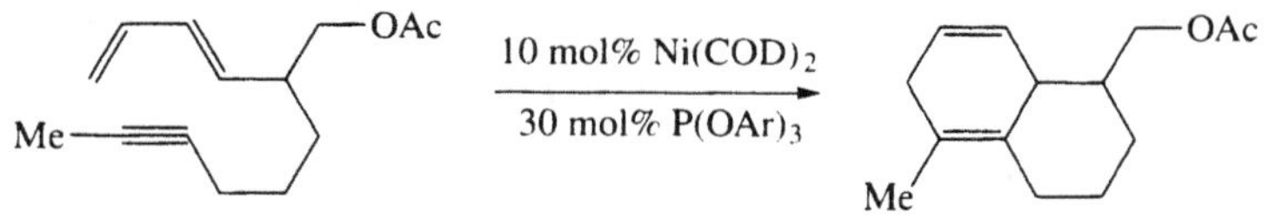

The Rh-catalyzed [5 + 2] cycloaddition combines a vinylcyclopropane and an alkyne (or an alkene or allene) to give a cycloheptadiene.

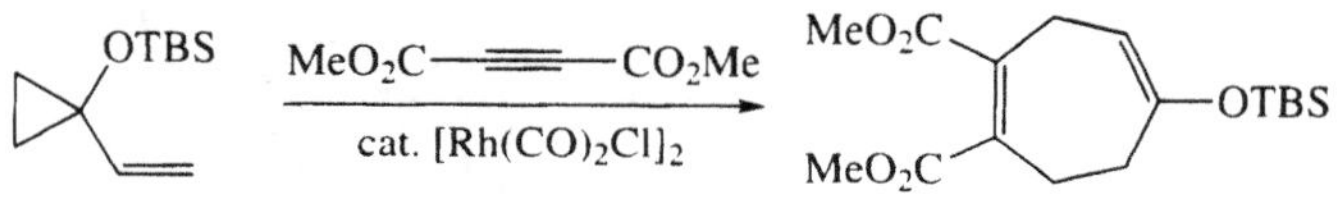

One reasonable mechanism for this reaction begins with coordination of Rh(I) to the alkyne to give a Rh(III) rhodacyclopropene. Reductive coupling of the π bond of the vinylcyclopropane with the alkyne gives a rhodacyclopentene, but this compound is also a cyclopropylmethylrhodium complex, and ring strain drives a very rapid homoallylic rearrangement. Reductive elimination then affords the cycloheptadiene.

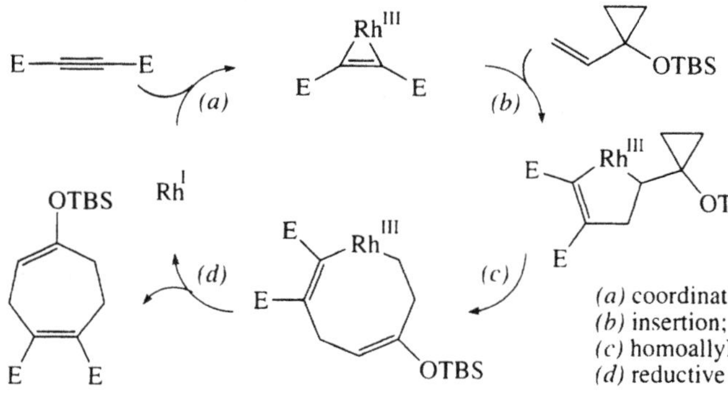

(a) coordination;
(b) insertion;
(c) homoallylic rearrangement;
(d) reductive elimination.

Problem 6.17. An alternative mechanism for the [5 + 2] cycloaddition begins with coordination of the Rh(I) to the vinyl group and puts the homoallylic rearrangement *before* coupling to the alkyne. Draw this mechanism.

The Rh-catalyzed [5 + 2] and [4 + 2] cycloadditions can be combined in a tandem process. A tandem reaction is one in which the product of one step is the starting material for the next.

Problem 6.18. Draw a reasonable mechanism for the following reaction:

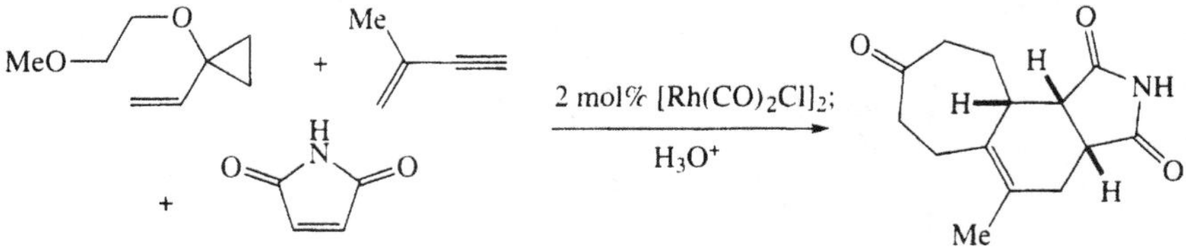

Various Ni and Co complexes catalyze cyclotrimerization reactions of alkynes to give arenes. These reactions are formally [2 + 2 + 2] cycloadditions. The cyclotrimerization of PhC≡CPh using CpCo(CO)$_2$ as catalyst is illustrative.

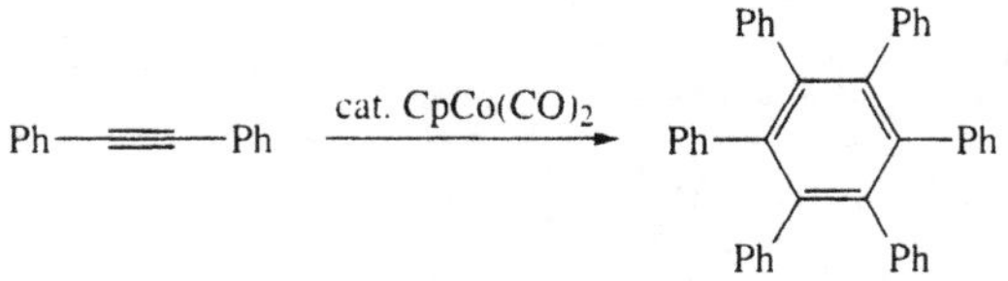

The 18-electron complex CpCo(CO)$_2$ must lose at least one CO ligand before it is able to coordinate to PhC≡CPh, another two-electron donor. One can draw two resonance structures for this complex, one of which is a cobaltacyclopropene, with Co(III). Loss of a second CO ligand (before or after coordination of PhC≡CPh) and coordination of another equivalent of PhC≡CPh is followed by an insertion reaction to give a cobaltacyclopentadiene. Coordination and insertion of another equivalent of PhC≡CPh gives a cobaltacycloheptatriene, and reductive elimination gives the product and the 14-electron complex CpCo(I), which recoordinates PhC≡CPh to reenter the catalytic cycle.

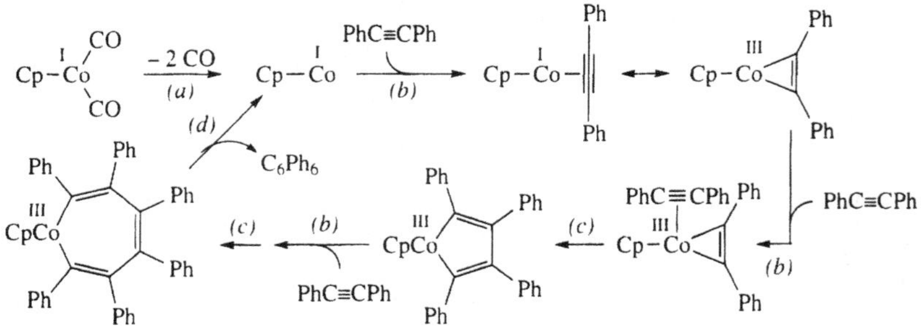

(a) dissociation; (b) coordination; (c) insertion; (d) reductive elimination.

The end of the catalytic cycle can be drawn differently. The cobaltacyclopentadiene is particularly well arranged to undergo a Diels–Alder reaction with another equivalent of PhC≡CPh. A [4 + 1] retro-cycloaddition then gives the product C$_6$Ph$_6$ and regenerates the catalyst CpCo(I).

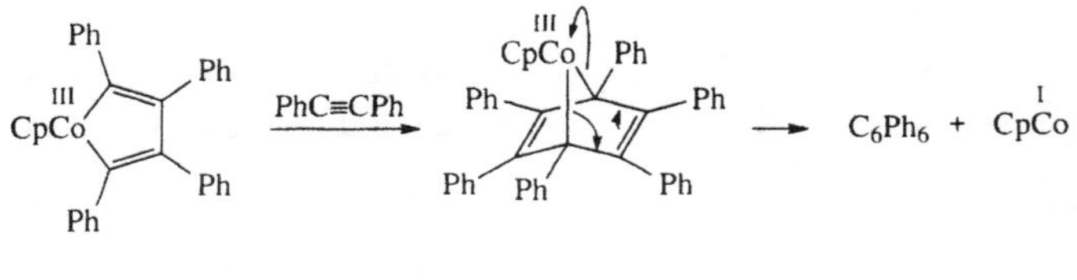

The cyclotrimerization reaction has been used to prepare some remarkable compounds. Bis(trimethylsilyl)acetylene ($Me_3SiC{\equiv}CSiMe_3$) does not undergo cyclotrimerization with itself due to steric hindrance, but it does cyclotrimerize with other alkynes. Slow addition of hexaethynylbenzene to $CpCo(CO)_2$ in $Me_3SiC{\equiv}CSiMe_3$ gives a triply cyclotrimerized product. The product shows bond alternation in the central ring due to its desire to avoid antiaromatic cyclobutadiene rings around its perimeter.

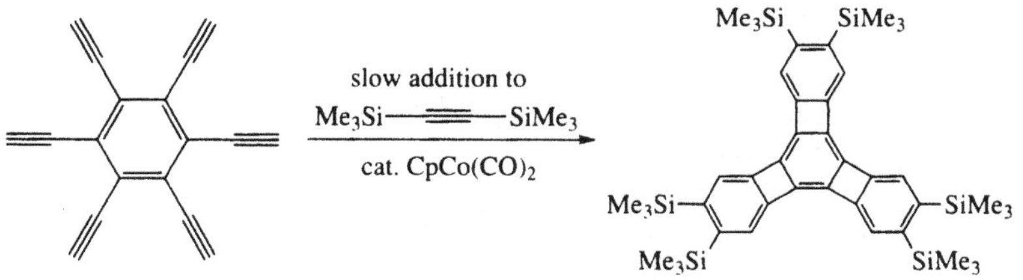

6.3 Substitution Reactions

6.3.1 Hydrogenolysis (Pd)

Hydrogenolysis catalyzed by Pd/C is widely used to convert benzylic ethers $ArCH_2$–OR to $ArCH_3$ and ROH. The reaction is often accelerated by acid. The simplest possibility for a catalytic cycle is oxidative addition of both H_2 and Bn–OR to Pd(0) to give a Pd(IV) complex, followed by reductive elimination of both Bn–H and H–OR to regenerate Pd(0). This mechanism seems very unlikely because the Pd(IV) oxidation state is high in energy.

(a) oxidative addition;
(b) reductive elimination.

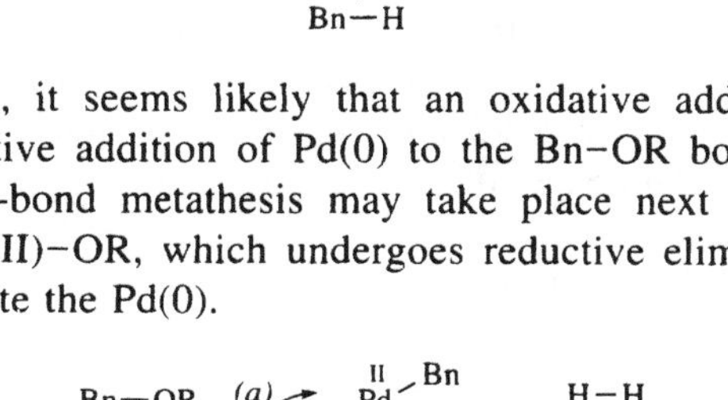

Still, it seems likely that an oxidative addition must be the first step. Oxidative addition of Pd(0) to the Bn–OR bond may give Bn–Pd(II)–OR. Sigma-bond metathesis may take place next to give Bn–H (toluene) and H–Pd(II)–OR, which undergoes reductive elimination to give ROH and regenerate the Pd(0).

(a) oxidative addition;
(b) σ-bond metathesis;
(c) reductive elimination.

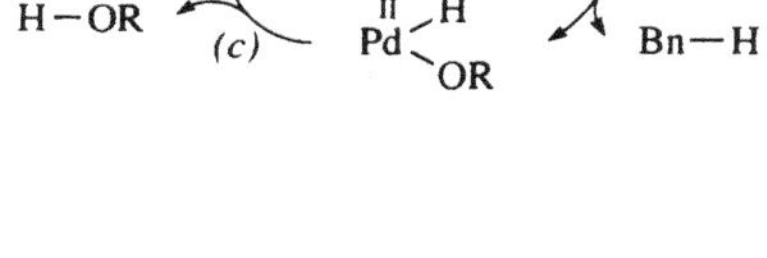

Another catalytic cycle can be drawn. The first step, oxidative addition of Pd(0) to Bn−OR, is the same as before. The Bn−Pd(II)−OR complex might then form a σ-bond complex with H_2, with the H−H σ bond acting as a two-electron donor. The H atoms are made electrophilic upon coordination. Intramolecular deprotonation of the σ complex by ¯OR would then give ROH and Bn−Pd(II)−H, and the latter would undergo reductive elimination to complete the catalytic cycle.

(a) oxidative addition;
(b) coordination;
(c) deprotonation;
(d) reductive elimination.

A third possibility begins with oxidative addition of H_2 to Pd(0) to give a H−Pd(II)−H complex. Deprotonation gives an anionic [Pd(0)−H]¯ complex. Simple S_N2 displacement of ROH from BnO(H)R by the anionic [Pd(0)−H]¯ complex then gives Bn−Pd(II)−H and ROH. Finally, reductive elimination of toluene regenerates Pd(0). Although it seems that the acceleration of hydrogenolyses by acid is not consistent with an anionic palladium intermediate, it is possible that the S_N2 step, not the deprotonation of Pd, is rate-limiting, and this step is likely to be promoted by acid.

(a) oxidative addition;
(b) deprotonation;
(c) reductive elimination.

Finally, a different sort of catalytic cycle can be drawn. Oxidative addition of H_2 to Pd gives a H−Pd(II)−H complex, and one of the π bonds of the benzyl ether undergoes insertion into the Pd(II)−H bond to give a cyclohexadienyl−Pd(II)−H complex. Isomerization to put the Pd on the ipso C is followed by a β-alkoxide elimination to give isotoluene and H−Pd(II)−OR, which undergoes reductive elimination to give ROH and regenerate the catalyst. The isotoluene undergoes isomerization to toluene under the reaction conditions, possibly catalyzed by Pd(0) via insertion and β-hydride elimination. If this mechanism were correct, one would expect that Pd would reduce aromatic rings under the conditions required for hydrogenolysis, which it does not. Nevertheless, this mechanism is accepted by some.

(a) oxidative addition; (b) insertion; (c) β-alkoxy elimination;
(d) β-hydride elimination; (e) reductive elimination.

6.3.2 Carbonylation of Alkyl Halides (Pd, Rh)

A Pd(0) catalyst such as $(Ph_3P)_4Pd$ catalyzes the *alkoxycarbonylation* of an organic halide (RX) to an ester (RCO_2Me) in basic MeOH under a CO atmosphere. (Any alcohol can be used.) *The first step in any late metal-catalyzed substitution reaction of a C–halogen bond is almost always oxidative addition.* Oxidative addition of R–X to Pd(0) gives R–Pd(II)–X, and coordination and insertion of CO gives RCO–Pd(II)–X. The acylpalladium complex can be regarded as a complicated acyl "halide." The Pd–C bond can be replaced by a MeO–C bond by an addition–elimination reaction. Loss of X^- from the Pd(0) complex regenerates the catalyst. The active catalytic species is often designated as L_nPd, as ligands may be associating and dissociating with the catalyst throughout the course of the reaction. The active species is probably $(Ph_3P)_2Pd$, derived from $(Ph_3P)_4Pd$ by loss of $2\ Ph_3P$, but it may be $(Ph_3P)_3Pd$.

(a) oxidative addition; *(b)* coordination; *(c)* insertion; *(d)* dissociation.

Other Pd ligands besides Ph_3P, usually di- or triarylphosphines like $Ph_2PCH_2CH_2PPh_2$ (diphos or dppe), 1,1'-bis(diphenylphosphino)ferrocene (dppf), $(o\text{-tolyl})_3P$, trifurylphosphine, and the like, are also used. Ph_3As is sometimes used, too. The ligands serve largely to keep the Pd(0) in solution. Experimentation is required to determine which ligands are best for any particular transformation.

Sometimes, a Pd(II) compound such as $(Ph_3P)_2PdCl_2$ or $Pd(OAc)_2$ is added to the reaction mixture as catalyst. The Pd(II) must be reduced to Pd(0) before the catalytic cycle can proceed. Et_3N can act as the reducing agent. Coordination to Pd(II) and β-hydride elimination gives a Pd(II) hydride, which is deprotonated to give Pd(0). Triarylphosphines can also reduce Pd(II) to Pd(0) by electron transfer.

(a) ligand substitution; *(b)* β-hydride elimination; *(c)* deprotonation, dissociation.

The Pd-catalyzed carbonylation of alkyl halides is an extraordinarily useful reaction. The "classic" way to accomplish this transformation would be to convert RX to the Grignard reagent, add CO_2, and esterify the carboxylic acid. Many functional groups cannot survive the strongly reducing and basic Grignard conditions. The Pd-catalyzed reaction proceeds at room temperature or slightly higher temperatures and requires only a weak base like Et_3N. The reaction proceeds best with X = I or Br. It also proceeds well with pseudo-halides; X = OTf is widely used, as it is easily prepared from the corresponding ketone. An important limitation to the reaction, however, is that R must almost always be C(sp²). The only C(sp³) halides that can undergo the reaction are those lacking β-hydrogens, such as methyl, benzyl, and neopentyl halides (Me₃CCH₂X). Alkyl halides that have β-hydrogens undergo β-hydride elimination after the oxidative addition step, affording only the product derived from elimination of HX.

Problem 6.19. Draw a reasonable mechanism for the following alkoxycarbonylation reaction.

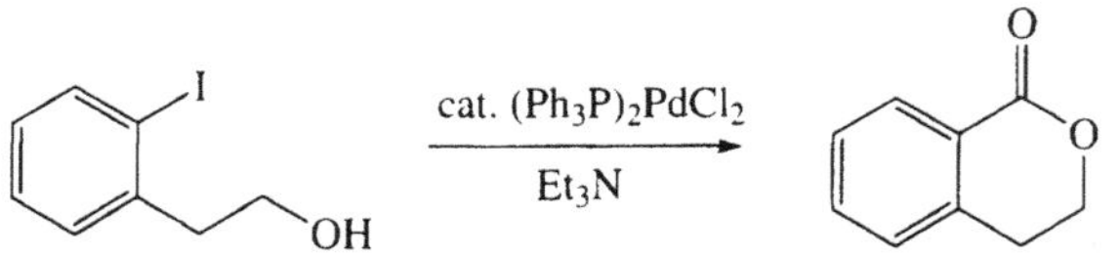

The *Monsanto process*, one of the most successful industrial homogeneous catalytic processes, uses a Rh complex and catalytic HI to carbonylate MeOH to MeCO₂H. A Rh precatalyst (almost any Rh complex will do) is converted into Rh(CO)₂I₂⁻, the active catalyst, under the reaction conditions. The mechanism of the reaction involves three steps. In the first step, MeOH and HI are converted to MeI and H₂O by an S_N2 mechanism. In the second step, MeI and CO are converted to MeCOI under Rh catalysis. In the third step, H₂O (generated in the first step) hydrolyzes MeCOI to afford MeCO₂H and to regenerate HI.

$$MeOH + HI \longrightarrow MeI + H_2O$$

$$MeI + CO \xrightarrow{\text{cat. Rh}} MeCOI$$

$$MeCOI + H_2O \longrightarrow MeCO_2H + HI$$

The catalytic cycle of the second step proceeds by oxidative addition of MeI to Rh(I) to give a MeRh(III) complex, insertion of CO to give the (MeCO)Rh(III) complex, coordination of more CO, and reductive elimination of MeCOI to regenerate the Rh(I) complex.

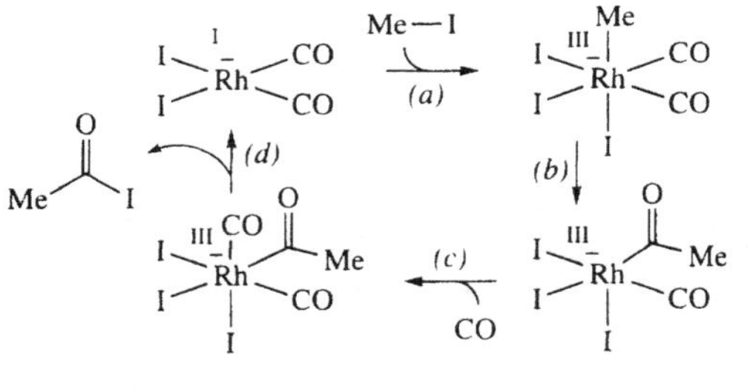

(a) oxidative addition;
(b) insertion;
(c) coordination;
(d) reductive elimination.

6.3.3 Heck Reaction (Pd)

In the *Heck reaction*, an aryl or vinyl halide (R–X) and an alkene ($H_2C=CHR'$) are converted to a more highly substituted alkene (R–CH=CHR') under Pd catalysis. Base is used to neutralize the by-product (HX). The Heck reaction can be carried out in intra- or intermolecular fashion. In intermolecular reactions, the reaction proceeds best when the alkene is electrophilic. In intramolecular reactions, more highly substituted alkenes can be used.

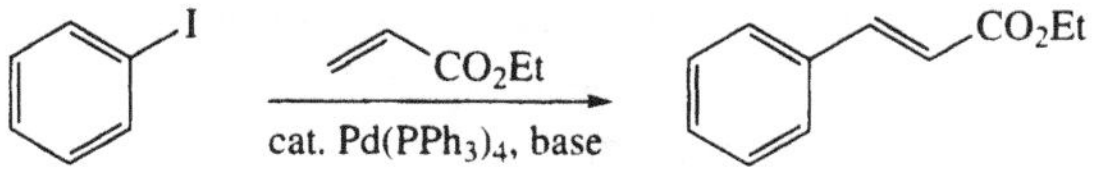

The C–I bond is broken, and a new C–C bond is formed. The first step, as always, is oxidative addition of Pd(0) to the C–I bond to give a Pd(II)–Ph complex. Insertion of the alkene into the Pd(II)–Ph bond now takes place to give the new C–C bond. A β-hydride elimination gives the product and H–Pd(II)–I, which is deprotonated by base to regenerate Pd(0).

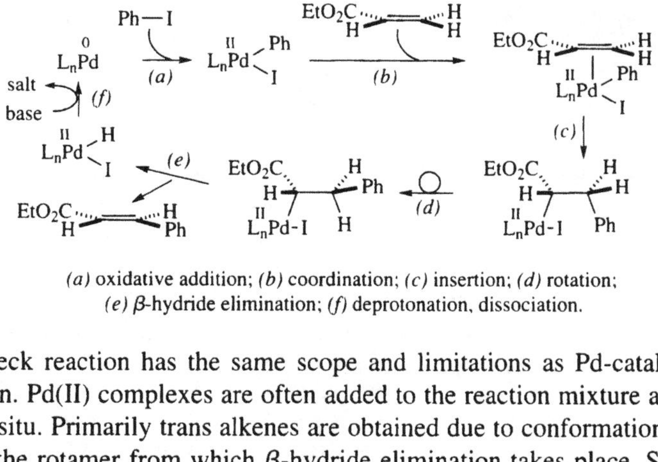

(a) oxidative addition; *(b)* coordination; *(c)* insertion; *(d)* rotation;
(e) β-hydride elimination; *(f)* deprotonation, dissociation.

The Heck reaction has the same scope and limitations as Pd-catalyzed carbonylation. Pd(II) complexes are often added to the reaction mixture and are reduced in situ. Primarily trans alkenes are obtained due to conformational preferences in the rotamer from which β-hydride elimination takes place. Sometimes the β-hydride elimination occurs away from the new C–C bond, and in this case a new stereocenter is formed. If the starting material is prochiral, the reaction can be made asymmetric by the use of chiral phosphine ligands like BINAP (2,2'-bis(diphenylphosphino)-1,1'-binaphthyl).

Problem 6.20. Draw a reasonable mechanism for the following Heck reaction.

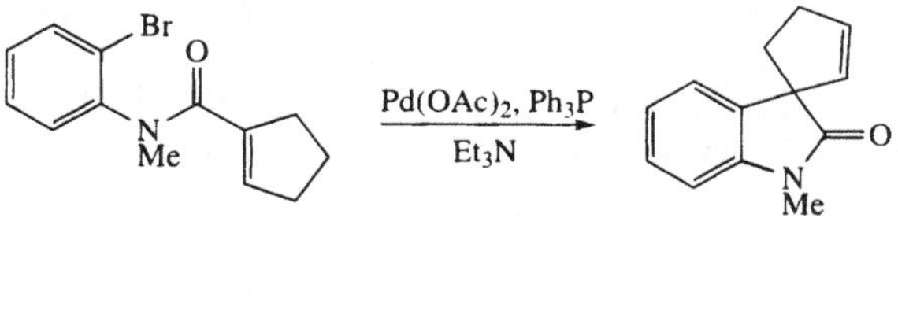

6.3.4 Coupling Reactions Between Nucleophiles and C(sp²)–X: Kumada, Stille, Suzuki, Negishi, Buchwald–Hartwig, Sonogashira, and Ullmann Reactions (Ni, Pd, Cu)

Aryl and vinyl halides can undergo substitution with nucleophiles by one of three mechanisms: addition–elimination, $S_{RN}1$, or elimination–addition (Chapter 2). The addition–elimination mechanism requires that C be electrophilic, which in the case of an aryl halide requires strongly electron withdrawing groups such as NO_2 groups on the ring. The $S_{RN}1$ reaction requires light or a nucleophile that can stabilize a radical, and the leaving group must be Br, I, or aromatic. Elimination–addition requires strongly basic conditions. Many aryl halide–nucleophile pairs fulfill none of these conditions, and consequently this transformation was for a long time one of the most difficult ones to accomplish.

In the mid-1970s, it was discovered that Ni complexes catalyzed the substitution of aryl halides with Grignard reagents at room temperature. The Ni catalysts were mostly phosphine complexes of $NiCl_2$ (e.g., $(Ph_3P)_2NiCl_2$), although other phosphine complexes sometimes gave better results. Alkyl (1°, 2°, or 3°), aryl, or alkenyl Grignard reagents could be used.

$$\text{Ph–Cl} \xrightarrow[\text{cat. }(Ph_3P)_2NiCl_2]{\text{BuMgBr}} \text{Ph–Bu}$$

The mechanism of the *Kumada coupling* was originally proposed to be as follows. The original Ni(II) complex underwent two transmetallation reactions with BuMgBr to give L_2NiBu_2. Reductive elimination (or β-hydride elimination followed by reductive elimination) gave an $L_2Ni(0)$ complex, the active catalytic species. The catalytic cycle involved oxidative addition of Ar–Cl to Ni(0) to give a Ni(II) complex, transmetallation to give Ar–Ni(II)–Bu, and reductive elimination to give Ar–Bu and regenerate the Ni(0) complex.

$$Ph_3P{-}Ni(II){-}Cl,\ Cl \xrightarrow[(a)]{\text{BuMgBr}} Ph_3P{-}Ni(II){-}Bu,\ Bu \xrightarrow{(b)} Ph_3P{-}Ni(0)$$

(a) transmetallation; (b) reductive elimination; (c) oxidative addition.

This mechanism was perfectly reasonable, but experimental evidence soon suggested a different mechanism was operative. The reaction was sensitive to O_2 and the presence of radical inhibitors, and it was found to have an induction period. These results suggested that odd-electron species were involved, and it was proposed that the reaction actually involved a Ni(I)/Ni(III) couple, not the Ni(0)/Ni(II) couple originally proposed. The active Ni(I) catalyst could be formed from the L_2NiCl_2 starting material by electron transfer from the Grignard reagent.

$$Ph_3P \underset{Ph_3P}{\overset{\text{II}}{\diagdown}} Ni \diagup^{Cl}_{Cl} \quad \xrightarrow[\text{(from Bu–MgBr)}]{\textit{electron transfer}} \quad \left[Ph_3P \underset{Ph_3P}{\overset{\text{I}}{\diagdown}} Ni \diagup^{Cl}_{Cl} \right]^{-} \quad \xrightarrow{\textit{dissociation}} \quad Ph_3P \underset{Ph_3P}{\overset{\text{I}}{\diagdown}} Ni{-}Cl \quad \longrightarrow \text{ etc.}$$

After the discovery of the Kumada reaction, intensive efforts were made to discover other group 10 metal-catalyzed reactions. These efforts paid off tremendously with the development of some of the most widely used C–C bond-forming reactions in organic synthetic methodology, including the *Stille* (pronounced "still-ie") *coupling*, the *Suzuki coupling*, the *Negishi coupling*, and related reactions. In these reactions, an aryl or vinyl halide or pseudohalide undergoes a Pd-catalyzed substitution reaction with a "nucleophilic" alkylmetal compound such as R–SnR′$_3$ (Stille coupling), R–B(OH)$_2$ (Suzuki coupling), R–ZnCl (Negishi coupling), or another. The catalytic cycle, shown for the Stille coupling, involves oxidative addition to the Ar–X bond, transmetallation, and reductive elimination. The catalytic cycle is exactly the same for the Suzuki, Negishi, and other alkylmetal couplings.

(a) oxidative addition;
(b) transmetallation;
(c) reductive elimination.

The catalyst that is added to the reaction mixture may be either a Pd(0) species like (Ph$_3$P)$_4$Pd or Pd$_2$(dba)$_3$ (dba = di̠benzylideneac̠etone) or a Pd(II) species like (Ph$_3$P)$_2$PdCl$_2$ or Pd(OAc)$_2$/2 AsPh$_3$. When the added catalyst is Pd(II), it is reduced to Pd(0) before the catalytic cycle begins. The reduction probably proceeds by two transmetallations from the nucleophile and a reductive elimination.

Problem 6.21. Draw a mechanism for the following Stille coupling.

papuamine

The reactions have a very wide scope. Acyl halides serve as substrates for Stille couplings in addition to the usual aryl and alkenyl halides; however, most alkyl halides cannot be used as substrates. Like Pd-catalyzed carbonylations, reactions proceed most quickly when X = I and rather slowly when X = Cl, although especially bulky phosphine ligands such as *t*-Bu$_3$P allow even aryl chlorides to undergo coupling at room temperature. Again, triflates are also widely used, especially alkenyl triflates, which are easily prepared from the ketones. The nucleophile may be C(sp), $C(sp^2)$, or $C(sp^3)$.

Metal amides and alkoxides are also coupled to $C(sp^2)-X$ electrophiles under Pd catalysis in *Buchwald–Hartwig amination* or *etherification*. The use of amines or alcohols in Pd-catalyzed coupling might seem like an obvious extension from carbon nucleophiles, but, in fact, the Buchwald–Hartwig reactions were discovered much later. Especially bulky phosphine ligands such as t-Bu$_3$P are required for these reactions to proceed. The ligands are thought to enforce a lower coordination number of the Pd catalyst, making a more active catalyst. In broad terms, however, the mechanism of Buchwald–Hartwig amination is the same as the Stille or Suzuki coupling: oxidative addition, transmetallation, and reductive elimination.

Problem 6.22. Draw a mechanism for the following Buchwald–Hartwig amination.

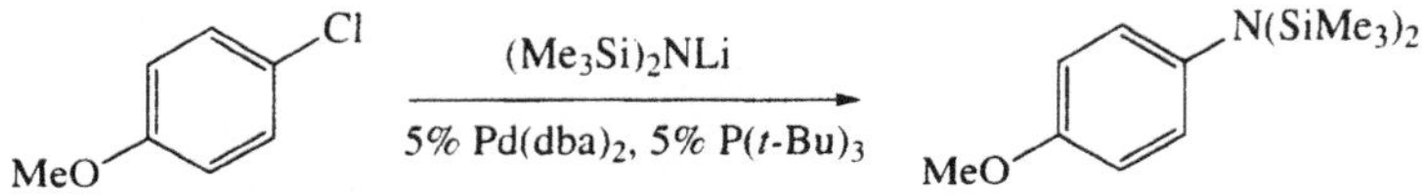

Many compounds containing metal–metal bonds undergo "Stille" and "Suzuki" couplings, too. Two of the most popular reagents for this purpose are hexamethyldistannane (Me$_3$Sn–SnMe$_3$) and pinacoldiborane [(pin)B–B(pin), where pin = 1,1,2,2-tetramethylethylene-1,2-dioxy]. In fact, even (pin)BH itself will convert aryl halides to arylboronates when an amine is present to neutralize the HX by-product. In this case, the mechanism by which the B–H bond is converted to a B–Pd bond is somewhat murky.

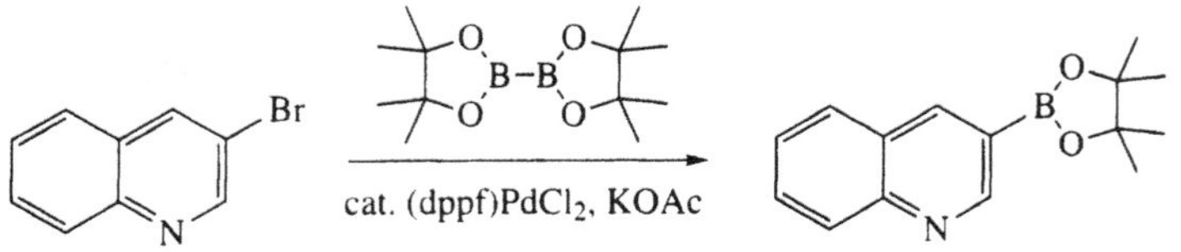

The *carbonylative Stille coupling* is another extremely useful reaction. When the Stille coupling of Ar–X and R–SnR$'_3$ is carried out under an atmosphere of CO, the product is the ketone (ArCOR). An insertion of CO intervenes between oxidative addition and reductive elimination.

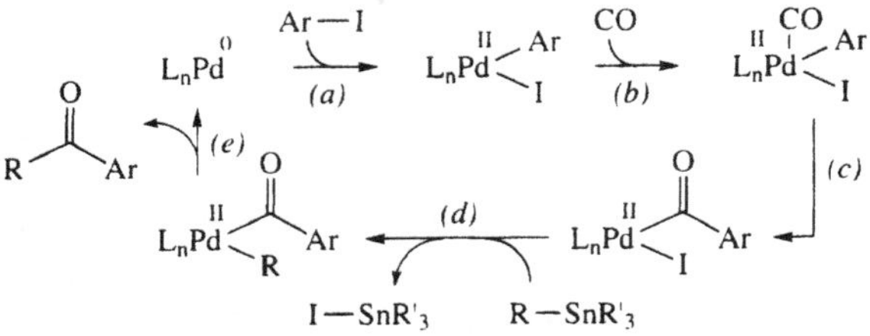

(a) oxidative addition;
(b) coordination;
(c) insertion;
(d) transmetallation;
(e) reductive elimination.

Problem 6.23. Draw a mechanism for the following carbonylative Stille coupling.

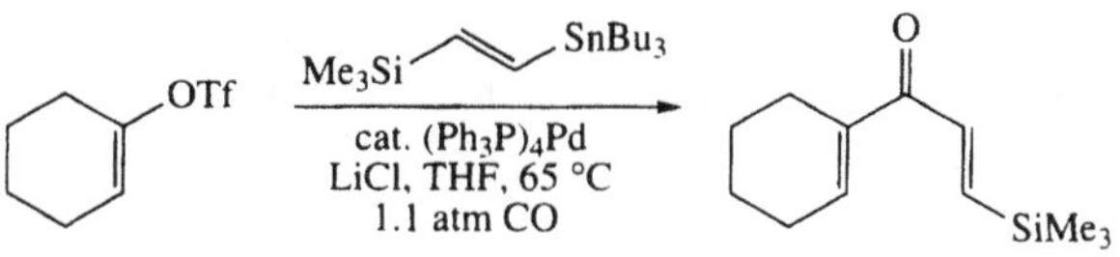

Terminal alkynes (RC≡CH) also undergo Pd-catalyzed *Sonogashira coupling* to alkyl halides (R′X) in the presence of base and a subcatalytic amount of CuI to give terminal alkynes (RC≡CR′).

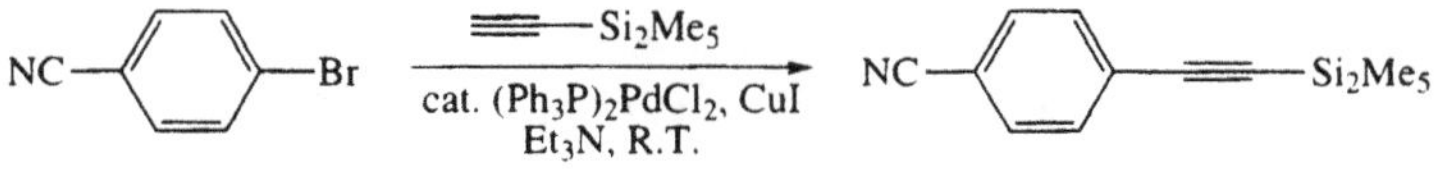

One can imagine that the Sonogashira coupling proceeds by a catalytic cycle very similar to the Stille coupling. The transmetallation step of the Stille coupling is replaced with a ligand substitution reaction, in which a deprotonated alkyne displaces X⁻ from the Pd(II) complex. The alkyne may be deprotonated by the Et₃N that is present in the reaction mixture.

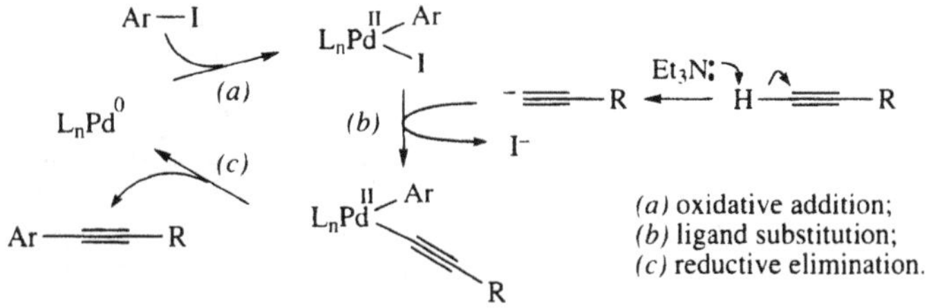

(a) oxidative addition;
(b) ligand substitution;
(c) reductive elimination.

The problem with this scenario is that alkynes are more acidic than most hydrocarbons (pK_a ≈ 25), but they are not sufficiently acidic to be deprotonated by amines (pK_b ≈ 10). The answer to this problem is found in the CuI that cocatalyzes the reaction. CuI lowers the temperature required for the Sonogashira coupling from >100 °C to room temperature. The CuI may convert the alkyne (RC≡CH) to a copper(I) acetylide (RC≡C–Cu), a species that can undergo transmetallation with Pd(II). Of course, now the question is, How is RC≡C–H converted to RC≡C–Cu? The alkyne may form a π complex with Cu, and this complex may be deprotonated (E2-like elimination) to give the Cu acetylide, which can transmetallate with Pd.

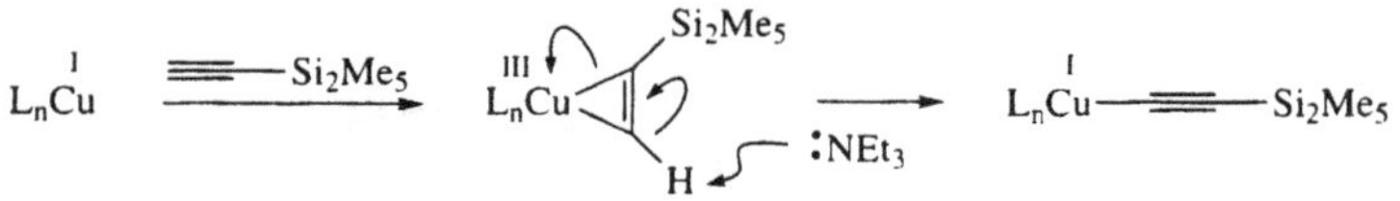

Copper(I) salts such as CuCN and ROCu undergo aromatic substitution reactions very readily with ordinary aryl halides. The mechanism has not been established with certainty. One reasonable possibility is an $S_{RN}1$ mechanism. Another reasonable possibility involves oxidative addition of Ar–X to N≡C–Cu(I) to give a Cu(III) complex, followed by reductive elimination of Ar–CN to give CuX.

$$
\begin{array}{c}
\text{MeO}-\!\!\!\bigcirc\!\!\!-\text{Br} \xrightarrow{\text{CuCN}} \text{MeO}-\!\!\!\bigcirc\!\!\!-\text{CN} + \text{CuBr}
\end{array}
$$

$$
\text{NC}-\text{Cu}^{I} \xrightarrow[\text{oxidative addition}]{\text{Ar}-\text{Br}} \underset{\overset{|}{\underset{Br}{}}}{\text{NC}-\overset{Ar}{\underset{|}{\text{Cu}}}^{III}} \xrightarrow[\text{reductive elimination}]{\text{Ar}-\text{CN}} \text{Br}-\text{Cu}^{I}
$$

The only problem with the oxidative addition–reductive elimination mechanism is that Cu(III) is a relatively high energy species. The mechanism would be much more reasonable if the Cu cycled between Cu(0) and Cu(II) instead of Cu(I) and Cu(III). Such a mechanism can be proposed if an initial electron transfer from CuCN to CuCN to give [N≡C–Cu(0)]⁻ is supposed. Oxidative addition of Ar–Br to Cu(0)⁻ and reductive elimination of Ar–CN from Cu(II)⁻ affords the organic product and [Br–Cu(0)]⁻. Finally, ligand exchange with another equivalent of N≡C–Cu(I) regenerates [N≡C–Cu(0)]⁻.

$$
\text{NC}-\text{Cu}^{I} + \text{NC}-\text{Cu}^{I} \xrightarrow{\text{electron transfer}} \left[\text{NC}-\text{Cu}^{0}\right]^{-} + \left[\text{NC}-\text{Cu}^{II}\right]^{+}
$$

$$
\left[\text{NC}-\text{Cu}^{0}\right]^{-} \xrightarrow[\text{oxidative addition}]{\text{Ar}-\text{Br}} \left[\underset{\overset{|}{\underset{Br}{}}}{\text{NC}-\overset{Ar}{\underset{|}{\text{Cu}}}^{II}}\right]^{-} \xrightarrow[\text{reductive elimination}]{\text{Ar}-\text{CN}} \left[\text{Br}-\text{Cu}^{0}\right]^{-}
$$

ligand substitution of Br⁻ by CN⁻ from another equivalent of CuCN

Other copper nucleophiles such as R_2CuLi and $R_2Cu(CN)Li_2$ also undergo substitution reactions with aryl and alkenyl halides. Retention of double-bond geometry about alkenyl halides is observed. The mechanisms of the reactions of these other Cu(I) nucleophiles are likely very similar to the one shown for the reaction of CuCN and ArBr.

In the *Ullmann reaction*, Cu metal promotes the coupling of Ar–I to give Ar–Ar. The reaction mechanism almost certainly involves oxidative addition of Cu(0) to Ar–I to give Ar–Cu(II), then reduction by another equivalent of Cu(0) to give Ar–Cu(I). This species can then react with ArI by the same mechanism that has just been written for other Cu(I) salts.

6.3.5 Allylic Substitution (Pd)

A nucleophile such as $(EtO_2C)_2\overset{-}{C}Me$ can undergo a substitution reaction with an allylic halide by an S_N2 mechanism. Allylic carbonates and allylic acetates do not normally undergo this reaction because $ROCO_2^-$ and AcO^- are not sufficiently good leaving groups. In the presence of a Pd catalyst, though, allylic carbonates and acetates are substrates for the substitution reaction. Significantly, the Pd-catalyzed reaction occurs with *retention* of stereochemistry, not the inversion of stereochemistry observed in S_N2 substitution reactions.

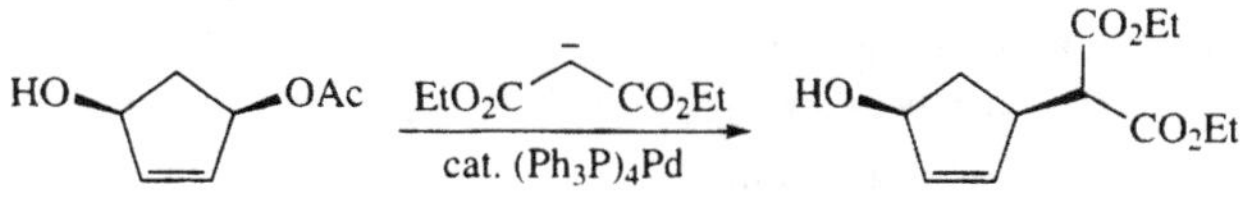

Whenever you see retention of stereochemistry, you should think "double inversion," and in fact double inversion occurs in this reaction. The Pd(0) complex acts as a nucleophile toward the allylic carbonate or acetate, displacing $MeOCO_2^-$ or AcO^- by backside attack and giving an allylpalladium(II) complex. The nucleophile then attacks the allylpalladium(II) complex, displacing Pd by backside attack to give the product and regenerate Pd(0). The regiochemistry of attack (S_N2 or S_N2') is dependent on the structure of the substrate.

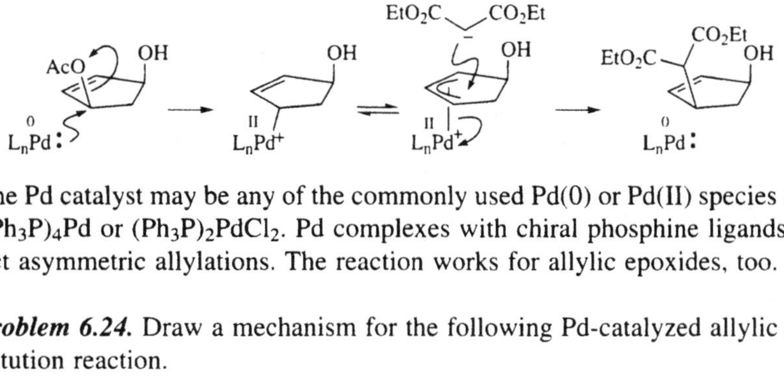

The Pd catalyst may be any of the commonly used Pd(0) or Pd(II) species such as $(Ph_3P)_4Pd$ or $(Ph_3P)_2PdCl_2$. Pd complexes with chiral phosphine ligands can effect asymmetric allylations. The reaction works for allylic epoxides, too.

Problem 6.24. Draw a mechanism for the following Pd-catalyzed allylic substitution reaction.

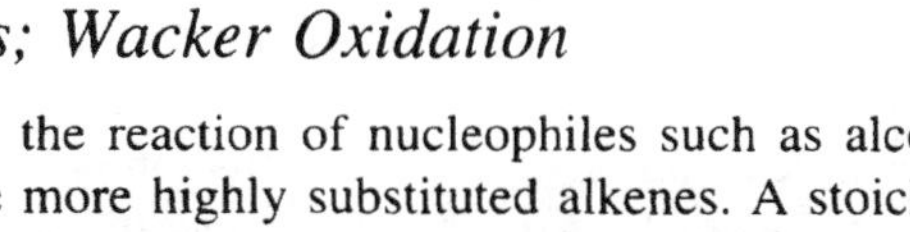

6.3.6 Palladium-Catalyzed Nucleophilic Substitution of Alkenes; Wacker Oxidation

Pd(II) salts catalyze the reaction of nucleophiles such as alcohols and amines with alkenes to give more highly substituted alkenes. A stoichiometric amount of an oxidant such as benzoquinone, CuCl, or O_2 (with or without catalytic Cu salts) is required for the reaction to be catalytic in Pd.

The mechanisms of these reactions begin the same way as Hg-mediated nucleophilic addition to alkenes. The alkene coordinates directly to Pd(II) to form an electrophilic π complex. The nucleophile attacks one of the carbons of the π complex, and the electrons from the C=C π bond move to form a σ bond between Pd(II) and the other carbon to give an alkylpalladium(II) compound. The alkylpalladium(II) compound then undergoes β-hydride elimination to give the observed product and a palladium(II) hydride. Loss of H^+ from the palladium(II) hydride gives Pd(0), which is oxidized back to Pd(II) by the stoichiometric oxidant. The Pd(II)−H sometimes catalyzes the migration of the π bond by a series of insertions and β-hydride eliminations.

The addition of nucleophiles to alkenes is *mediated* by Hg(II) salts and *catalyzed* by Pd(II) salts. The difference between the two reactions is the fate of the alkylmetal(II) intermediate obtained after addition of the nucleophile to the π complex. The alkylmercury(II) intermediate is stable and isolable, whereas the alkylpalladium(II) intermediate undergoes rapid β-hydride elimination.

The addition of nucleophiles to *alkynes* is catalyzed by both Hg(II) and Pd(II) salts. The intermediate alkenylmetal(II) complex is stable to β-hydride elimination, but the C−metal bond can be replaced by a C−H bond by a protonation–fragmentation mechanism.

The *Wacker oxidation* (pronounced "vocker") is used industrially to convert ethylene and O_2 into acetaldehyde. The Wacker oxidation is catalyzed by $PdCl_2$ and $CuCl_2$ and requires H_2O as solvent. The O atom in the product comes from the *water*, not the O_2.

$$H_2C=CH_2 \ + \ 1/2 \, O_2 \ \xrightarrow[\text{H}_2\text{O}]{\text{cat. PdCl}_2, \text{CuCl}_2} \ CH_3CHO$$

The mechanism of the Wacker oxidation is simply another Pd-catalyzed nucleophilic substitution of an alkene, with H_2O as the nucleophile. H_2O adds to a Pd(II) complex of ethylene, and β-hydride elimination occurs to give a π complex of the enol of acetaldehyde. After rotation about the Pd(II)−alkene σ bond, the alkene reinserts into the Pd(II)−H bond to give a new Pd(II)−alkyl. This complex undergoes β-hydride elimination one more time to give acetaldehyde itself and Pd(II)−H. Deprotonation of the Pd(II) complex converts it to Pd(0), and oxidation of Pd(0) by air (see below) brings it back to Pd(II).

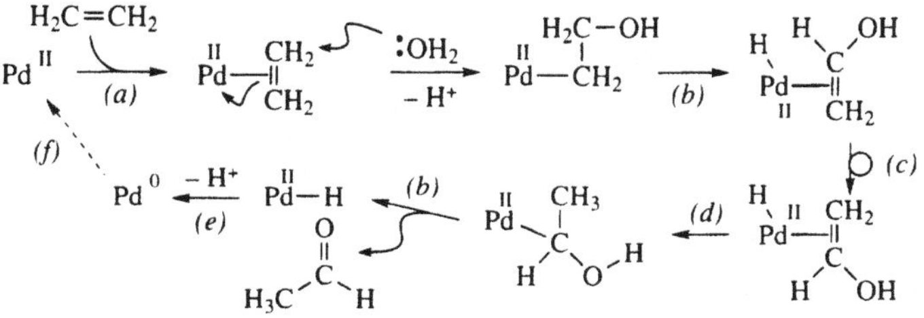

(a) coordination; (b) β-hydride elimination; (c) rotation;
(d) insertion; (e) deprotonation; (f) oxidation.

After the first β-hydride elimination, it is reasonable simply to allow the enol to dissociate from Pd and then undergo acid-catalyzed tautomerization.

For the reaction to be catalytic in Pd, the Pd(0) has to be reoxidized back to Pd(II). Two equivalents of $CuCl_2$ convert Pd(0) to Pd(II) and produce 2 CuCl. The Cu(I) is then reoxidized back to Cu(II) by O_2.

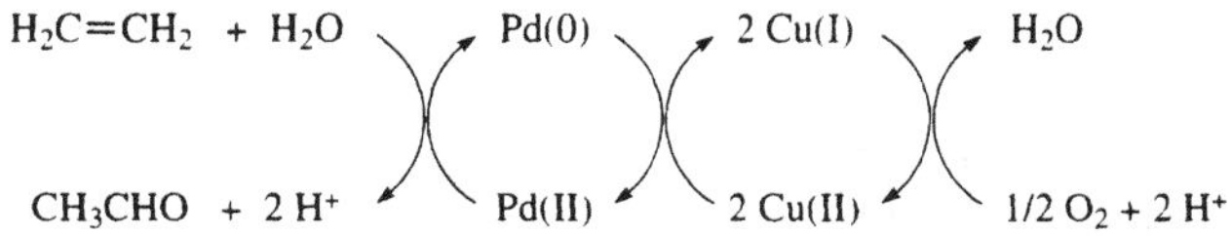

The Wacker oxidation works well for terminal alkenes, too. The products are methyl ketones, not aldehydes, as expected from Markovnikov attack of H_2O on the Pd–alkene π complex.

Problem 6.25. Draw mechanisms for the following Pd-catalyzed nucleophilic substitutions.

(a)

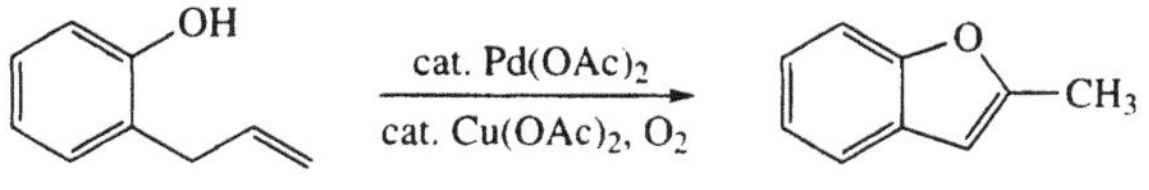

(b)

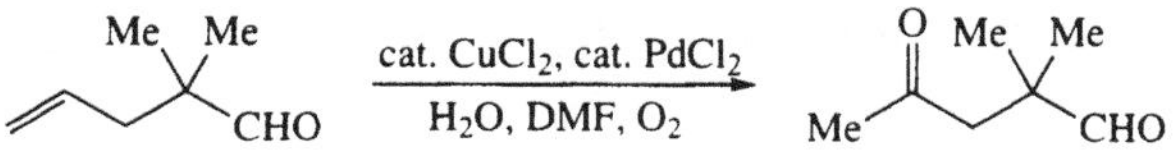

6.3.7   Tebbe Reaction (Ti)

The *Tebbe reaction* is an early-metal organometallic version of the Wittig reaction (Chapter 4). A carbonyl compound is converted to the corresponding methylene compound by treatment with the Tebbe reagent, which is a complex of $Cp_2Ti=CH_2$ with Me_2AlCl. The Tebbe reagent is prepared by combining Cp_2TiCl_2 with two equivalents of Me_3Al. Two transmetallations give Cp_2TiMe_2, which undergoes α-hydride abstraction to give $Cp_2Ti=CH_2$. Coordination to $ClAlMe_2$ gives the Tebbe reagent.

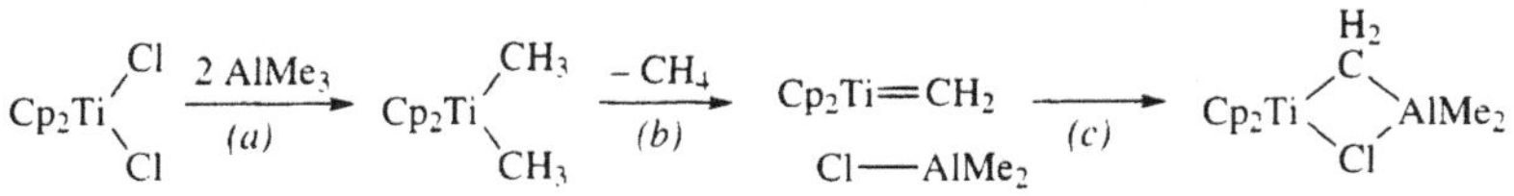

(a) transmetallation; (b) α-hydride abstraction; (c) coordination.

The presumed mechanism of action of the Tebbe reagent is very simple: after dissociation of Me$_2$AlCl, [2 + 2] cycloaddition of Cp$_2$Ti=CH$_2$ and R$_2$C=O is followed by [2 + 2] retro-cycloaddition to give the product. Unlike the conventional Wittig reaction, the Tebbe reaction works well with esters.

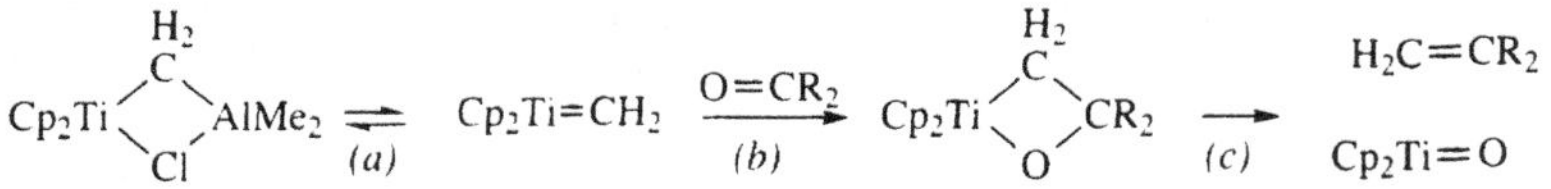

(a) dissociation; (b) [2 + 2] cycloaddition; (c) [2 + 2] retro-cycloaddition.

In the *Petasis reaction*, Cp$_2$TiMe$_2$ itself is used to carry out Tebbe reactions. Cp$_2$TiMe$_2$ is much easier to prepare (from Cp$_2$TiCl$_2$ and MeLi) and handle than is the Tebbe reagent. Upon heating, it undergoes α-hydride abstraction to give Cp$_2$Ti=CH$_2$. More highly substituted Cp$_2$Ti(CH$_2$R)$_2$ complexes also convert carbonyl compounds to the corresponding alkenes. Unlike Wittig reagents, Tebbe reagents are limited to transferring alkylidene groups lacking β-hydrogens, such as methylidene and benzylidene. Cp$_2$TiR$_2$ compounds bearing alkyl groups that have β-hydrogens, such as Cp$_2$TiBu$_2$, undergo β-hydride abstraction much more readily to give titanacyclopropanes than they undergo α-hydride abstraction.

Problem 6.26. Draw a reasonable mechanism for the following Petasis reaction.

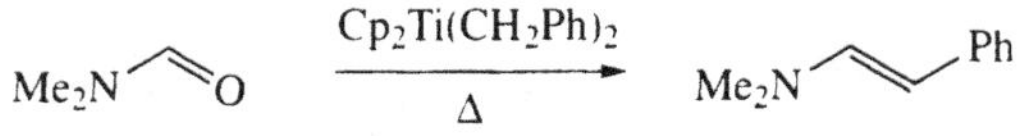

6.3.8   Propargyl Substitution in Cobalt–Alkyne Complexes

Alkyne–Co$_2$(CO)$_6$ complexes have uses in organic chemistry other than the Pauson–Khand reaction. When the complex is formed from a propargyl alcohol or ether, the C–O bond is especially prone to undergo S$_N$1 substitution reactions, as the cation is hyperconjugatively well stabilized by the neighboring high-energy Co–C bonds. The nucleophile is delivered exclusively to the propargylic C. Substitution reactions at Co-free propargylic C atoms are often plagued by the allenic products derived from addition of the nucleophile to the distal alkyne C.

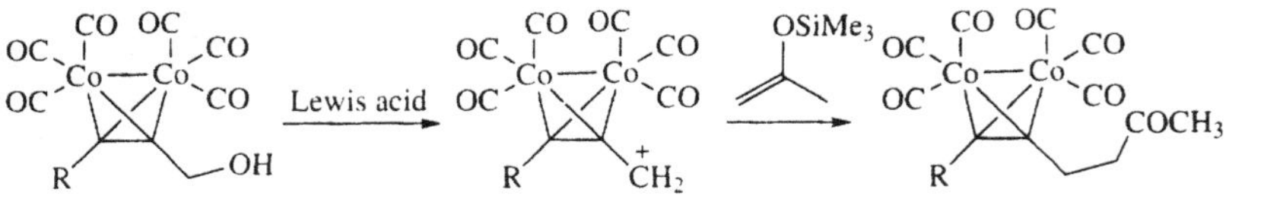

The complex of an alkyne with $Co_2(CO)_6$ can also be used to protect an alkyne in the presence of an alkene (e.g., in hydroborations). The $Co_2(CO)_6$ can then be removed from the alkyne by oxidation with Fe(III) ion.

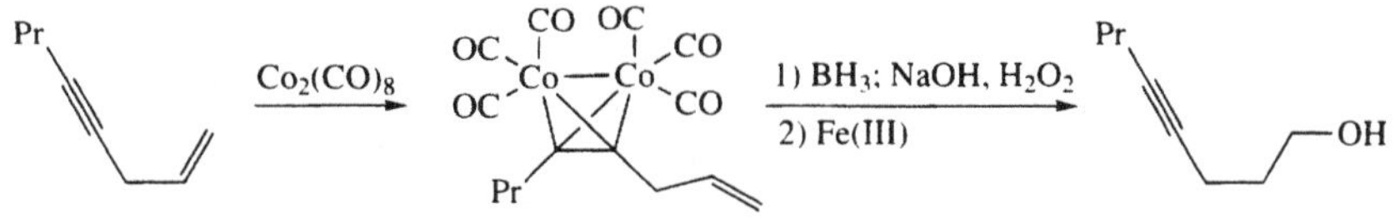

6.4   Rearrangement Reactions

6.4.1   Alkene Isomerization (Rh)

Wilkinson's catalyst catalyzes the isomerization of an alkene to its thermodynamically most stable isomer. Isomerization of allyl ethers in this manner gives enol ethers, which can be hydrolyzed to give the free alcohol and a carbonyl compound.

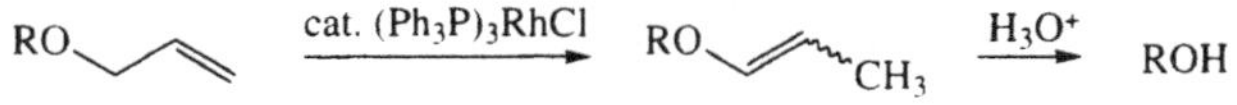

Other double-bond isomerizations (e.g., those that occur in the course of Pd-catalyzed hydrogenolyses or hydrogenations) proceed by insertion of the alkene into a M–H bond followed by β-hydride elimination. Wilkinson's catalyst, though, lacks a Rh–H bond into which an alkene can insert. The reaction may proceed by oxidative addition to an allylic C–H bond, then reductive elimination at the other end of the allylic system.

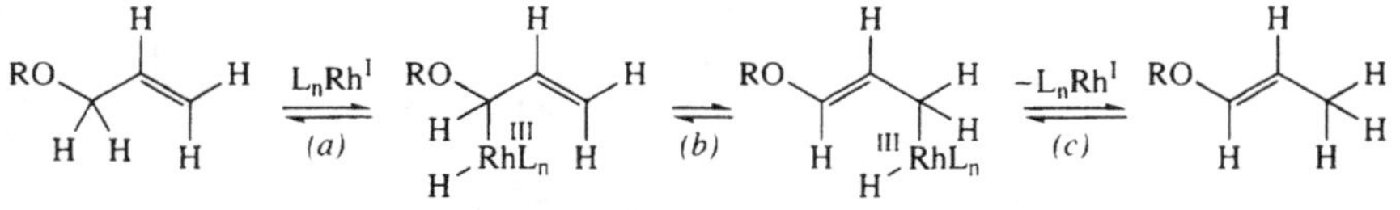

(a) oxidative addition; (b) allylic transposition; (c) reductive elimination.

6.4.2   Olefin and Alkyne Metathesis (Ru, W, Mo, Ti)

In the olefin metathesis reaction, the alkylidene fragments of two alkenes ($R^1R^2C=CR^3R^4$) are swapped to give $R^1R^2C=CR^1R^2$ and $R^3R^4C=CR^3R^4$.

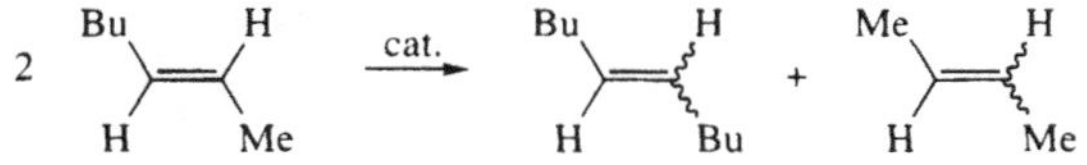

This remarkable reaction is catalyzed by early and middle metals, primarily complexes of Ti, Mo, W, and Ru. All olefin metathesis catalysts either have a M=C π bond or are converted into compounds that have a M=C π bond under the reaction conditions. Some of the most widely used homogeneous catalysts are shown, but many other compounds have been shown to catalyze these reactions.

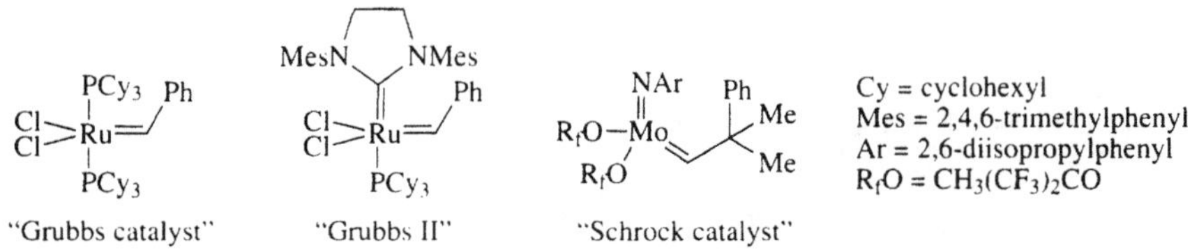

Cy = cyclohexyl
Mes = 2,4,6-trimethylphenyl
Ar = 2,6-diisopropylphenyl
$R_fO = CH_3(CF_3)_2CO$

"Grubbs catalyst"   "Grubbs II"   "Schrock catalyst"

The olefin metathesis reaction proceeds by a series of [2 + 2] cycloadditions and retro-cycloadditions. No change in metal oxidation state occurs in the course of the reaction. The reaction sets up an equilibrium between all the possible alkenes, but it is possible to drive the equilibrium in one direction or the other, for example, by removal of gaseous ethylene. The two Grubbs catalysts undergo dissociation of a Cy_3P ligand before the catalytic cycle begins so that coordination of the alkene to the metal may precede each [2 + 2] cycloaddition.

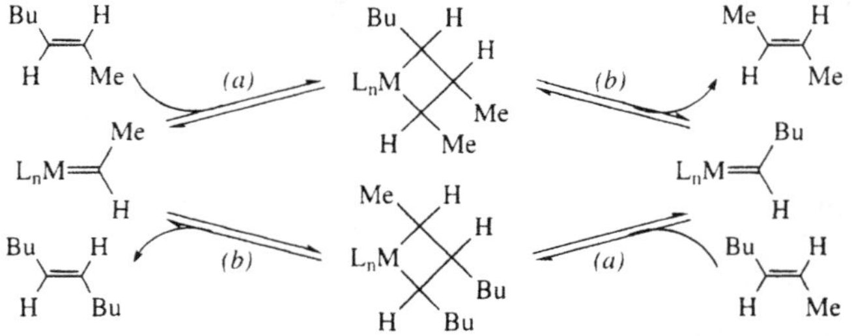

(a) [2 + 2] cycloaddition; *(b)* [2 + 2] retro-cycloaddition.

The olefin metathesis reaction has been known for a long time, but the first catalysts were heterogeneous, poorly characterized, and intolerant of functional groups, so the reaction was originally applied only to very simple olefins of industrial interest. More recently, homogeneous, functional-group-tolerant catalysts that work under mild conditions have been developed, causing applications to complex molecule synthesis to proliferate. One of the most widely used variations of olefin metathesis is the *ring-closing metathesis* reaction (RCM), in which a diene is allowed to undergo intramolecular olefin metathesis to give a cyclic alkene and ethylene. The evaporation of the gaseous ethylene drives the equilibrium forward.

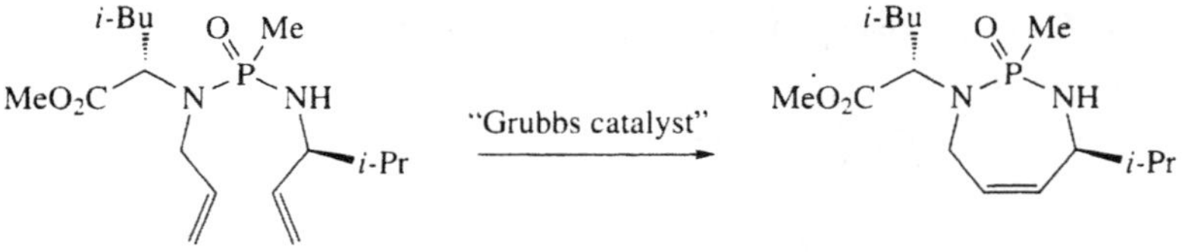

"Grubbs catalyst"

Ring-closing metathesis has quickly been adopted as a premier method for preparing compounds with very large rings. However, substrates that would lead to strained rings preferentially undergo polymerization in favor of cyclization when subject to metathesis catalysts.

Problem 6.27. Draw a mechanism for the following ring-closing metathesis reaction.

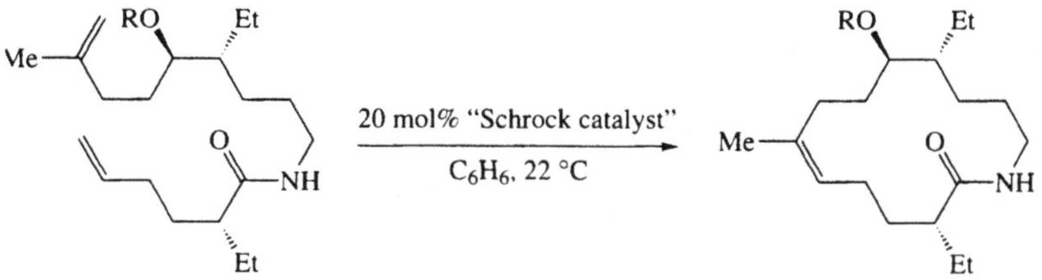

There are other variations of olefin metathesis. In the *cross-metathesis* reaction (CM), two olefins are coupled. This reaction gives good yields and nonstatistical mixtures if the two alkenes have electronically different properties.

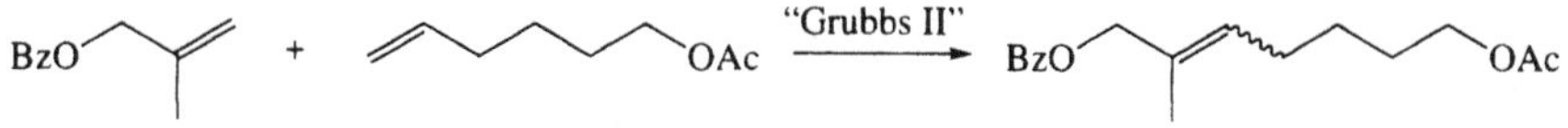

In *ring-opening methathesis polymerization* (ROMP), a strained alkene such as norbornene is allowed to undergo metathesis with itself to give a polymer.

Problem 6.28. Draw a mechanism for the following ring-opening metathesis reaction.

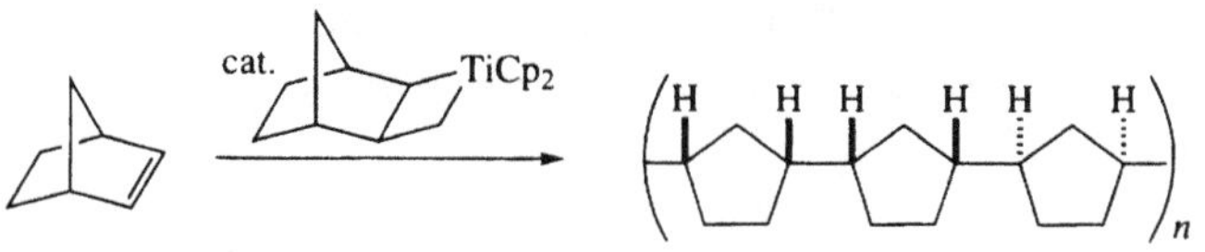

Even more recently, alkyne metathesis catalysts have been developed and applied to organic synthesis. The catalysts for these reactions either have a M≡C triple bond or are converted into compounds that have a M≡C triple bond under the reaction conditions.

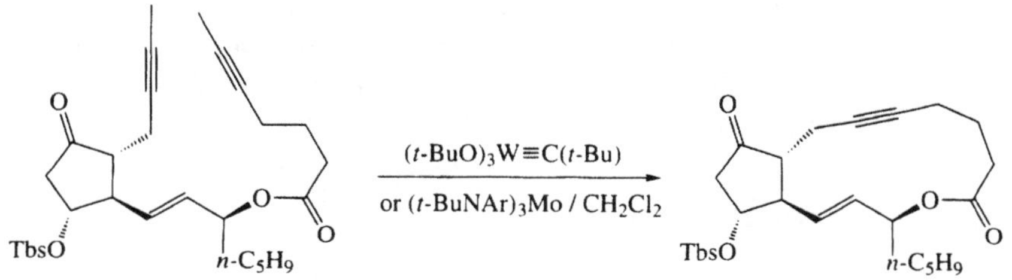

The mechanism of alkyne metathesis also consists of a series of [2 + 2] and [2 + 2] retro-cycloadditions. The key intermediate, a metallacyclobutadiene, appears to be antiaromatic, but apparently the metal uses a d orbital to form the W=C π bond so that the ring is aromatic.

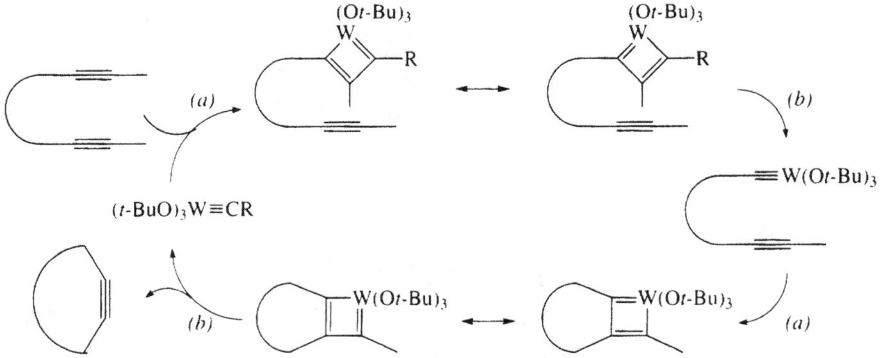

(a) coordination, [2 + 2] cycloaddition; *(b)* [2 + 2] retro-cycloaddition.

6.5   Elimination Reactions

6.5.1   Oxidation of Alcohols (Cr, Ru)

Cr(VI) reagents are widely used to oxidize alcohols to aldehydes/ketones or carboxylic acids. Commonly used reagents include CrO_3, Jones reagent, PDC (pyridinium dichromate), and PCC (pyridinium chlorochromate). Alcohols coordinate to Cr(VI) reagents and undergo a β-hydride elimination to give the carbonyl compound and Cr(VI)−H. The hydride is deprotonated to give a Cr(IV) species, which may undergo further redox chemistry. Harsher reagents such as CrO_3 are able to oxidize the aldehyde further to the acid via the aldehyde hydrate.

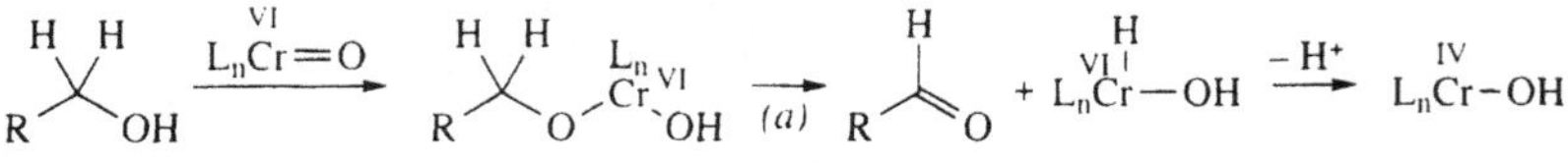

(a) β-hydride elimination.

Practical Cr-*catalyzed* oxidations of alcohols have not been adopted widely, but Pr_4N^+ RuO_4^- (TPAP, tetrapropylammonium perruthenate) catalyzes the oxidation of alcohols to aldehydes by a stoichiometric oxidant such as NMO, H_2O_2, or O_2 itself. The Ru(VII) complex oxidizes alcohols by the same mechanism described earlier for stoichiometric Cr species. The stoichiometric oxidant then reoxidizes the Ru(V) by-product back to Ru(VII).

6.5.2   Decarbonylation of Aldehydes (Rh)

Wilkinson's catalyst mediates the decarbonylation of aldehydes (RCHO → RH + CO). It's quite a remarkable reaction, yet the reaction mechanism is quite simple. Nucleophilic addition of Rh(I) to the aldehyde is followed by a 1,2-hydride shift from C to Rh to give an acyl–Rh(III) complex, the product of an overall oxidative addition to the aldehyde C−H bond. Elimination of CO then gives the alkyl–Rh(III)–H complex, which undergoes reductive elimination to give the alkane and Rh(I). Unfortunately, the reaction is stoichiometric (not catalytic) in Rh, as the product complex $(Ph_3P)_2Rh(CO)Cl$ is inert toward oxidative addition. The electron-withdrawing CO group makes the Rh complex too electron-poor. Because Rh is so expensive, the reaction is useful only when the product is very valuable or when the reaction is carried out on very small amounts of material.

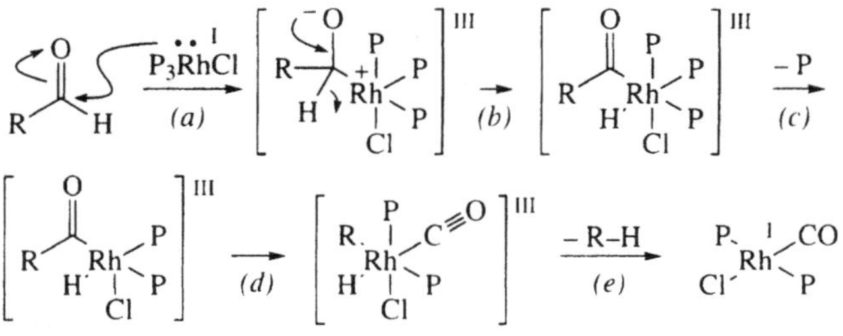

P = Ph₃P

(a) nucleophilic addition; (b) 1,2-hydride shift;
(c) dissociation; (d) elimination; (e) reductive
elimination.

6.4 Summary

Metals catalyze a diverse set of reactions, and it is difficult to make generalizations, but some basic principles can be kept in mind.

• Any reaction involving a metal should be one of the typical reactions that have been discussed.

• Ligand association, dissociation, and substitution processes are facile, so the exact number of ligands on a metal center is usually not a major concern when mechanisms and catalytic cycles involving transition metals are drawn.

• The key step of reactions that involve addition of X–Y across a π bond is *insertion of the π bond into an M–X bond*. The M–X bond is often generated by oxidative addition of M across an X–Y bond.

• Substitution reactions of organic halides proceed by oxidative addition to the C–X bond.

• d^0 Metals are especially prone to undergo σ-bond metathesis reactions.

• Compounds with M=X bonds tend to undergo [2 + 2] cycloadditions, and metallacyclobutanes tend to undergo [2 + 2] retro-cycloadditions.

• Third-row transition metals tend to undergo α-insertion and α-elimination reactions.

• Palladium is used in a wide variety of popular reactions with diverse mechanistic pathways. How do you know how to start when facing a Pd-catalyzed reaction?

• When a leaving group (usually attached to C(sp²) or C(sp), allylic, or benzylic) undergoes substitution by a nucleophile, the first step of the catalytic cycle is usually oxidative addition of Pd(0) to the C–X bond. (In hydrogenolysis, though, the oxidative addition step is preceded in the catalytic cycle by generation of the H–Pd(0)$^-$ intermediate.)

• When *nonpolar* X–Y is added across a π bond, the first step of the catalytic cycle is usually oxidative addition of Pd(0) across the X–Y σ bond. However, when a *nucleophile* adds to a π bond, a Pd(II) complex of the π bond may be the first intermediate.

• The catalytic cycles of all Pd-catalyzed reactions other than nucleophilic substitutions at alkenes begin with Pd(0), even when a Pd(II) complex is added to the reaction mixture.

No matter what the metal and what the transformation, remember to label the atoms, make a list of bonds to make and break, and obey Grossman's rule!

PROBLEMS

1. Palladium complexes are versatile and popular catalysts, catalyzing a wide variety of transformations in good yield under mild conditions. Draw mechanisms for the following Pd-catalyzed reactions.

(a)

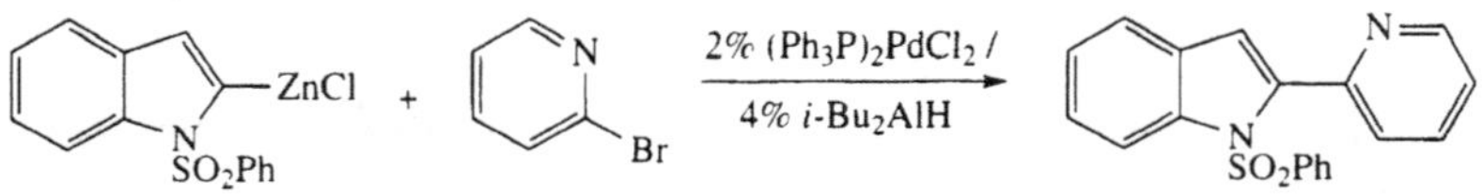

indole–ZnCl, N–SO$_2$Ph + pyridine–Br → 2% (Ph$_3$P)$_2$PdCl$_2$ / 4% *i*-Bu$_2$AlH → product (2-pyridyl indole, N–SO$_2$Ph)

(b) chiral diphosphinite = (*S*)-2,2'-bis(diphenylphosphinoxy)-1,1'-binaphthyl.

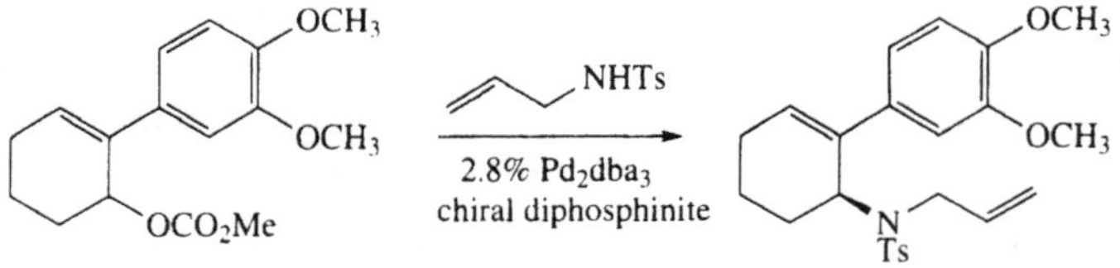

starting material (OCH$_3$, OCH$_3$, OCO$_2$Me) + NHTs → 2.8% Pd$_2$dba$_3$ / chiral diphosphinite → product (OCH$_3$, OCH$_3$, N–Ts)

(c)

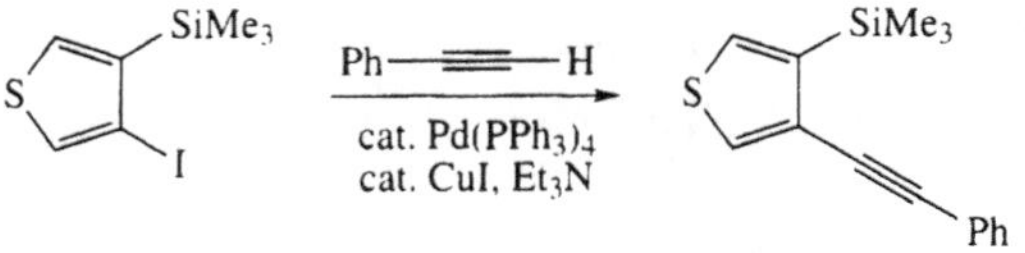

thiophene (SiMe$_3$, I) + Ph≡H → cat. Pd(PPh$_3$)$_4$ / cat. CuI, Et$_3$N → product (SiMe$_3$, ≡Ph)

(d)

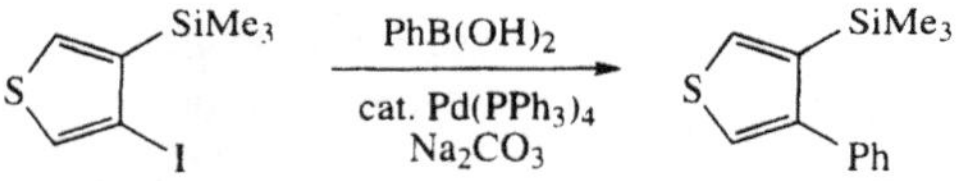

thiophene (SiMe$_3$, I) + PhB(OH)$_2$ → cat. Pd(PPh$_3$)$_4$ / Na$_2$CO$_3$ → product (SiMe$_3$, Ph)

(e)

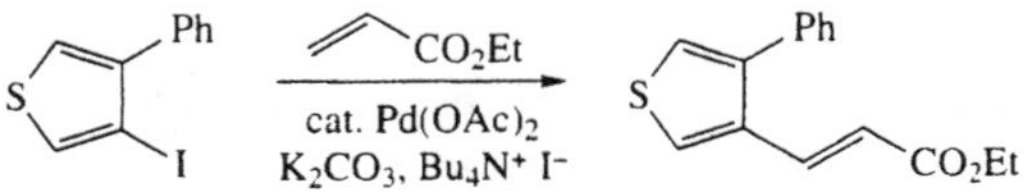

thiophene (Ph, I) + ≡CO$_2$Et → cat. Pd(OAc)$_2$ / K$_2$CO$_3$, Bu$_4$N$^+$ I$^-$ → product (Ph, CH=CH–CO$_2$Et)

(f)

aziridine (R, H, N–SO$_2$Ar, vinyl) ⇌ 5 mol% Pd(PPh$_3$)$_4$ ⇌ product (R, H, N–SO$_2$Ar, vinyl)

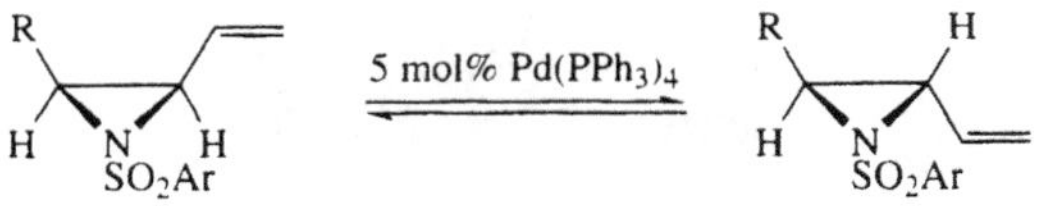

(g) dppf = 1,1'-bis(diphenylphosphino)ferrocene.

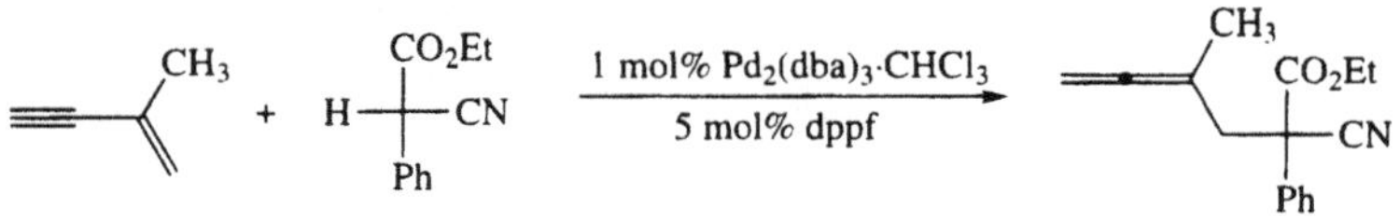

(h)

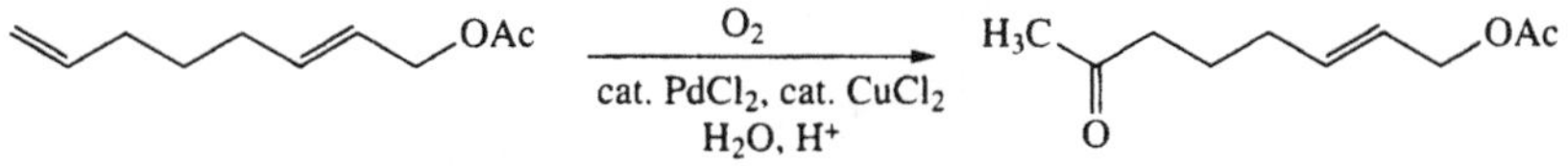

(i)

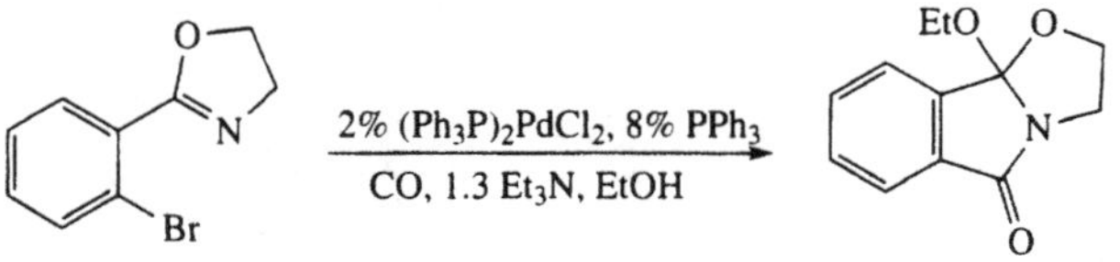

(j)

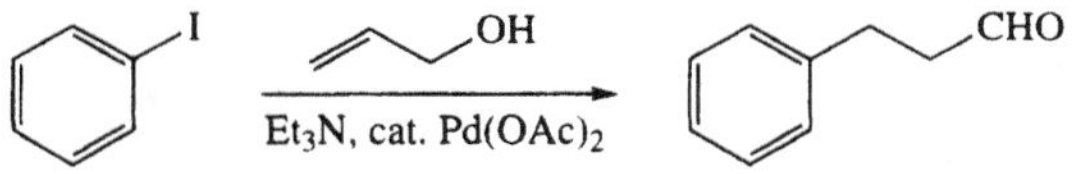

(k) One could draw a mechanism for this reaction that did not require Pd, but
     in fact both bond-forming steps are Pd-catalyzed.

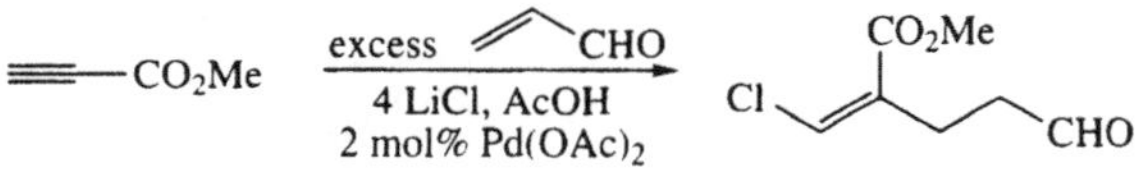

(l) DIBAL = i-Bu₂AlH.

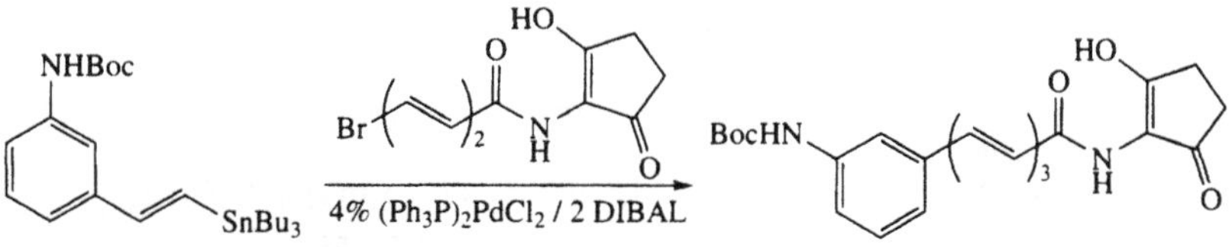

(m)

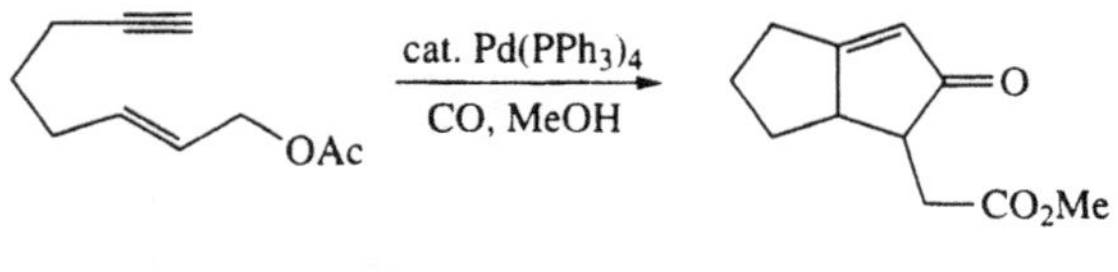

(n)

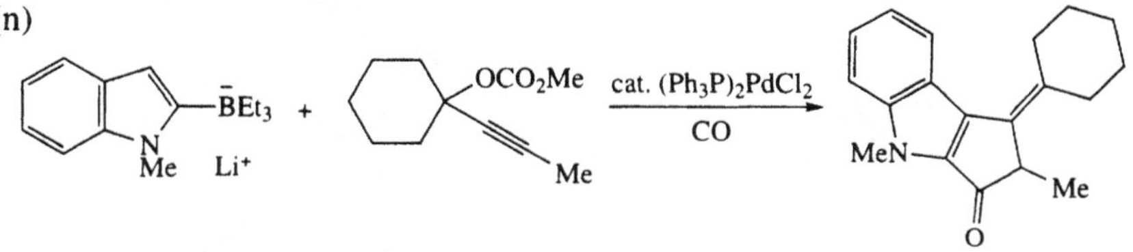

(o) This reaction is Pt-catalyzed, but its mechanism is very similar to many
     Pd-catalyzed reactions. *Note*: The active catalyst is a Pt(II) species,
     whereas H₂PtCl₆ is a Pt(IV) species (PtCl₄·2HCl).

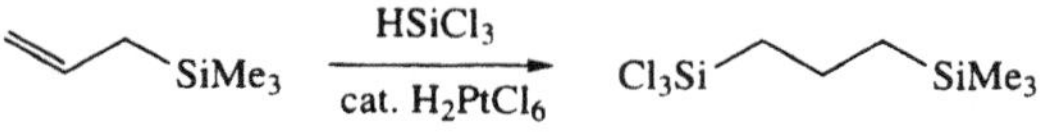

(p)

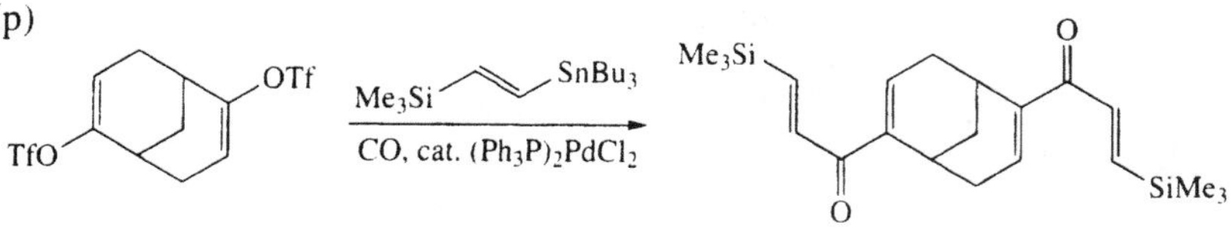

(q)

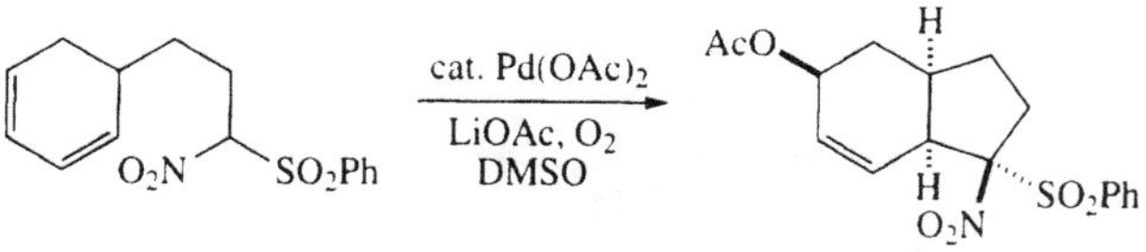

(r)

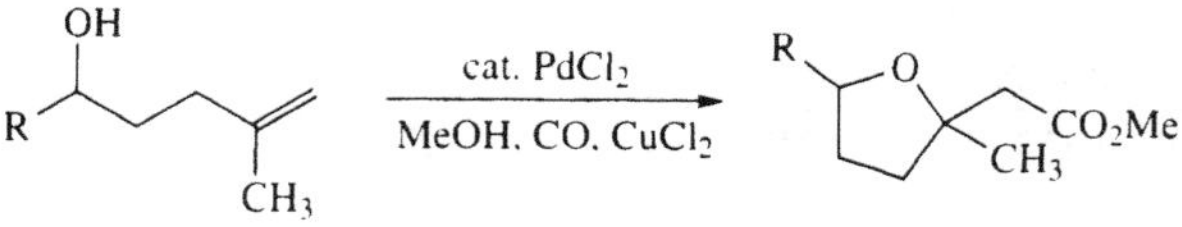

(s)

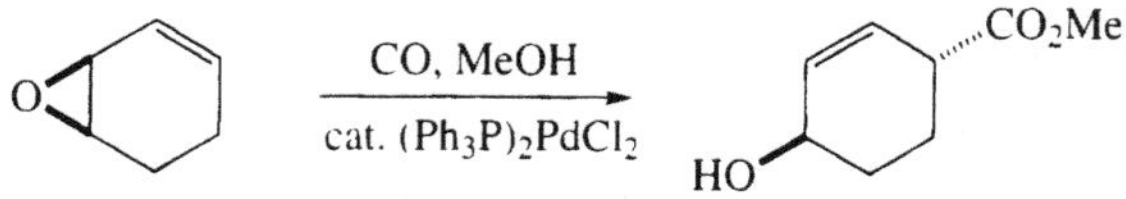

(t)

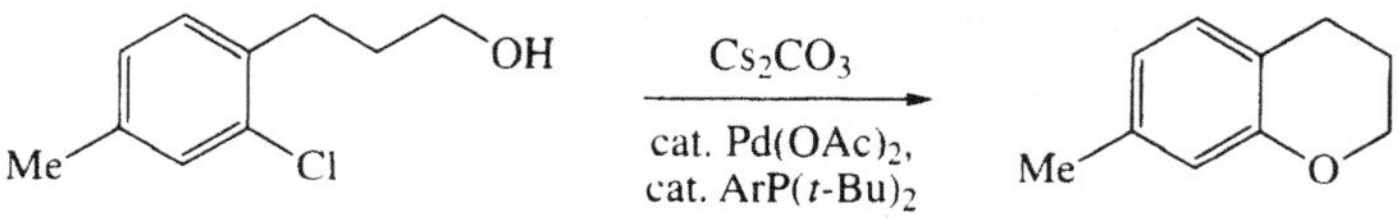

(u)

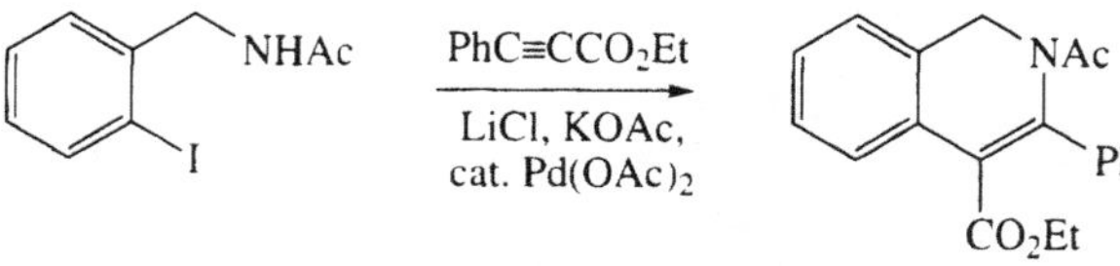

2. Many transition metals other than Pd also catalyze organic reactions. Draw mechanisms for the following metal-catalyzed reactions.

(a)

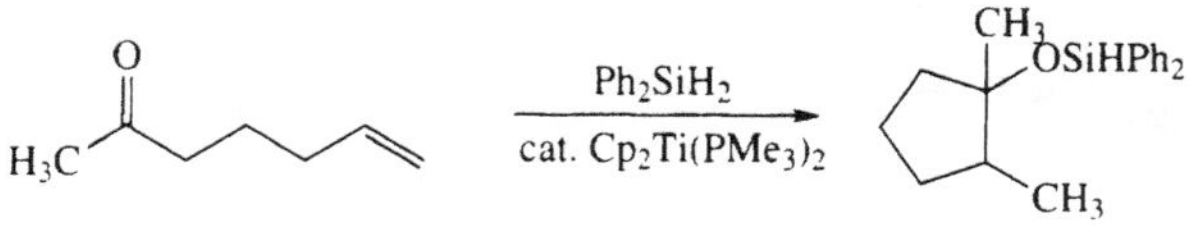

(b) Ar = 2,6-dimethylphenyl.

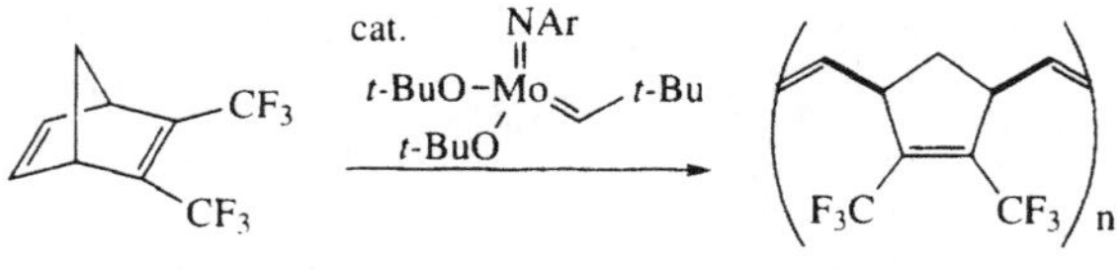

(c) BINAP = 2,2'-bis(diphenylphosphino)-1,1'-binaphthyl.

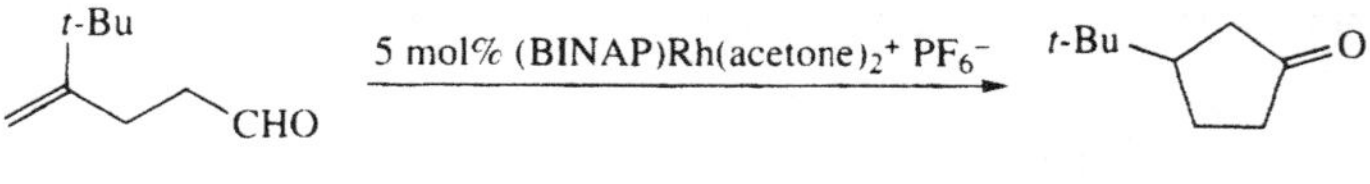

(d) BuLi is added to the Rh complex, *then* the substrate is added.

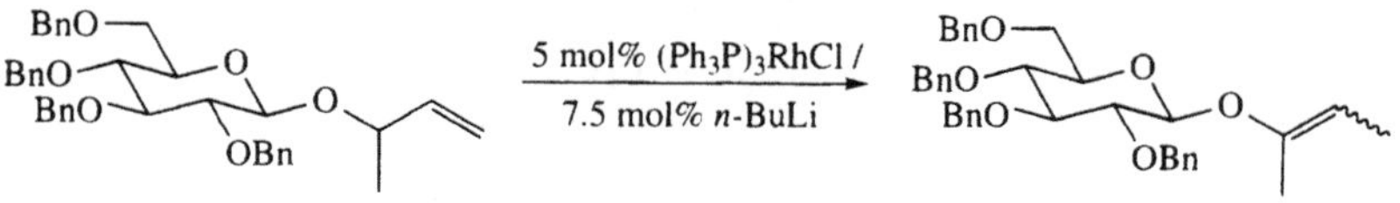

$$\xrightarrow[\text{7.5 mol\% } n\text{-BuLi}]{\text{5 mol\% (Ph}_3\text{P)}_3\text{RhCl}/}$$

(e) Fcm = ferrocenylmethyl, a protecting group for N. Cy = cyclohexyl.

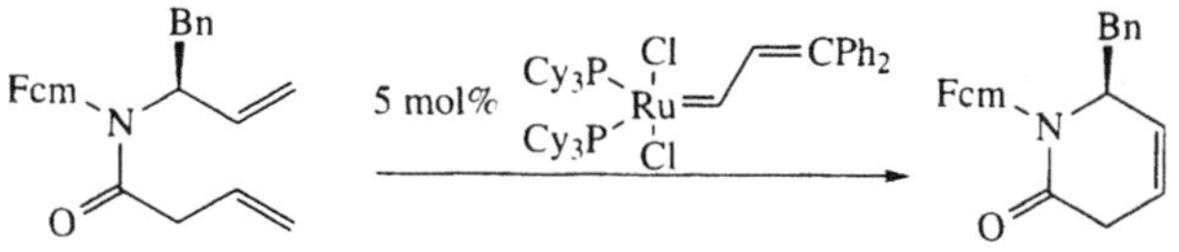

5 mol%

(f)

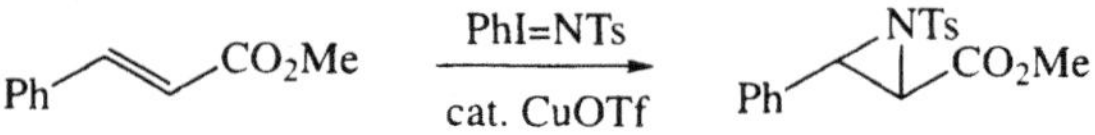

$$\xrightarrow[\text{cat. CuOTf}]{\text{PhI=NTs}}$$

(g) R = n-hexyl; cod = 1,5-cyclooctadiene, a weak four-electron donor ligand.

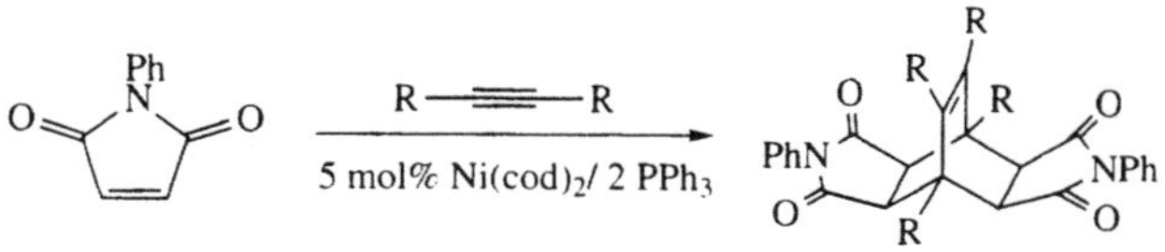

$$\xrightarrow[\text{5 mol\% Ni(cod)}_2/\text{ 2 PPh}_3]{\text{R}\!\equiv\!\equiv\!\text{R}}$$

(h) Cp* = pentamethylcyclopentadienyl, $C_5Me_5^-$.

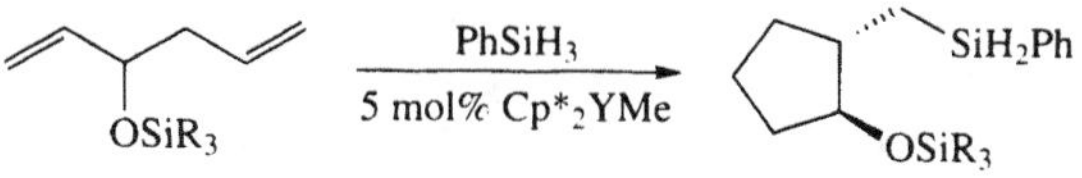

$$\xrightarrow[\text{5 mol\% Cp*}_2\text{YMe}]{\text{PhSiH}_3}$$

(i) acac = acetylacetonate (MeC(O)CH=C(O$^-$)Me).

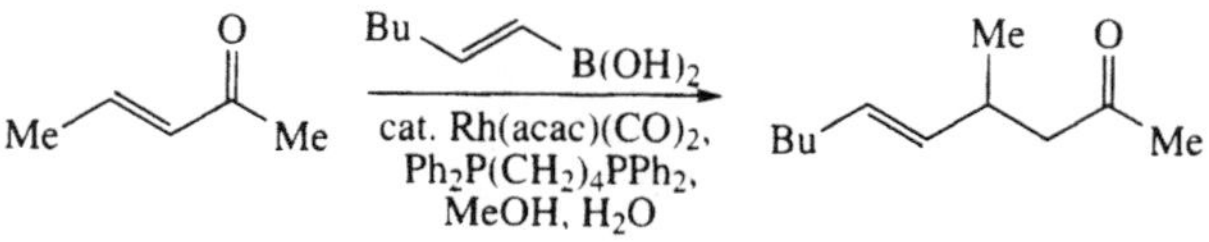

$$\xrightarrow[\substack{\text{cat. Rh(acac)(CO)}_2,\\ \text{Ph}_2\text{P(CH}_2)_4\text{PPh}_2,\\ \text{MeOH, H}_2\text{O}}]{\text{Bu} \diagup\!\!\diagdown \text{B(OH)}_2}$$

(j)

$$\xrightarrow{\text{"Grubbs catalyst"}}$$

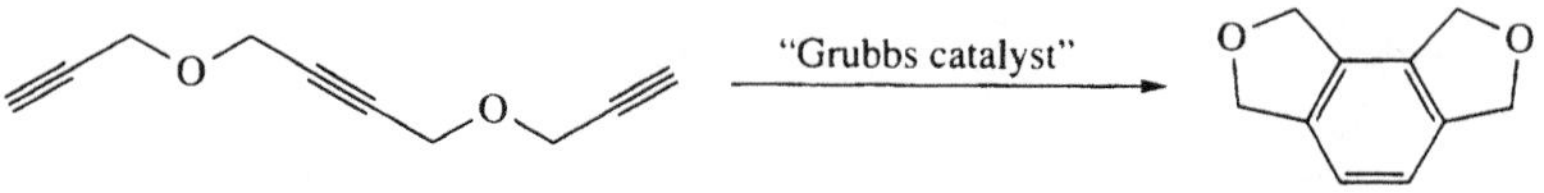

(k) The role of the AgOTf is simply to ionize the Rh–Cl bond and to make the Rh a more active catalyst.

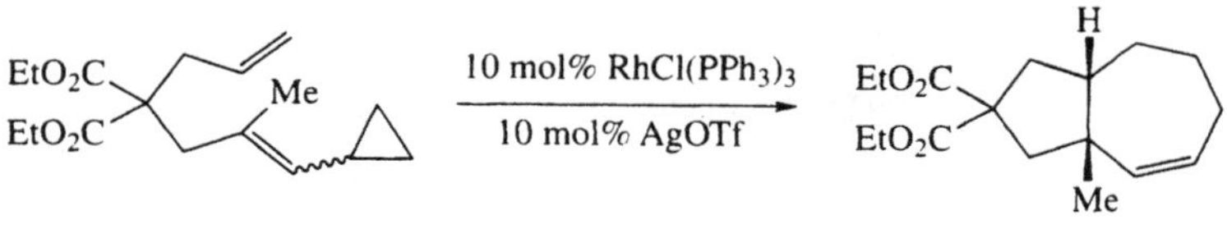

$$\xrightarrow[\text{10 mol\% AgOTf}]{\text{10 mol\% RhCl(PPh}_3)_3}$$

(l)   acac = acetylacetonate [MeC(O)CH=C(O⁻)Me].

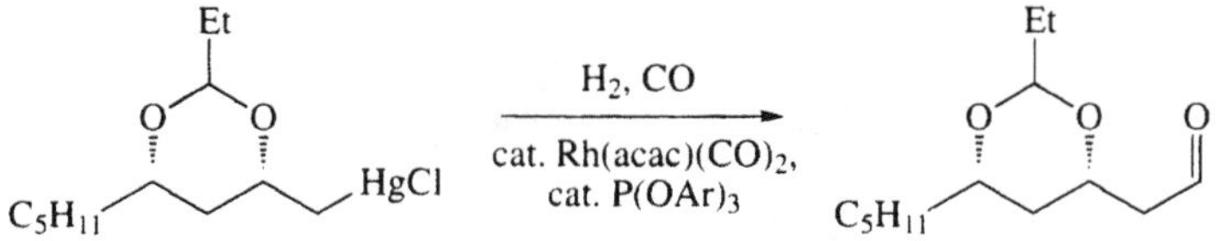

3. Metal-*mediated* reactions are not as popular as metal-*catalyzed* ones, but they are still indispensable for some transformations. Draw mechanisms for the following metal-mediated reactions.

(a)

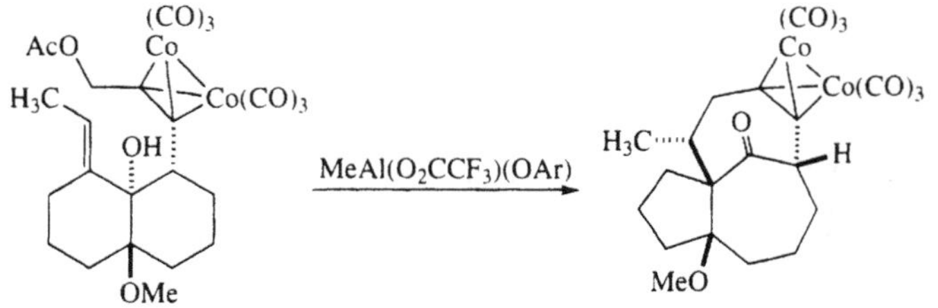

(b)

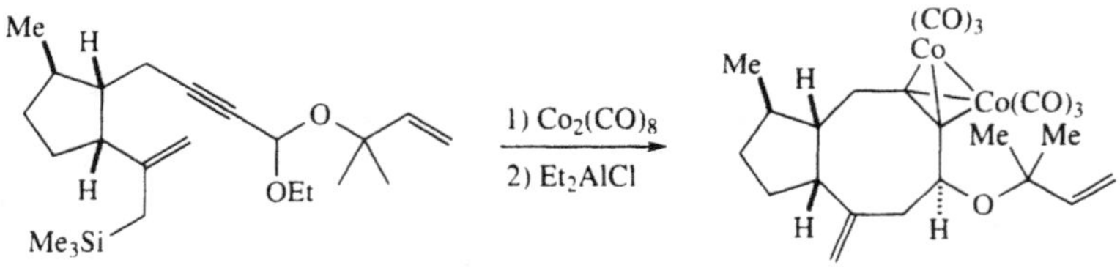

(c)

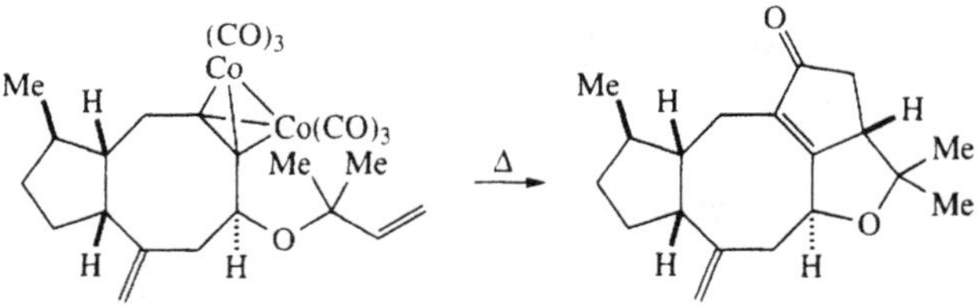

(d) The Grignard reagent reacts with the Ti complex *first*.

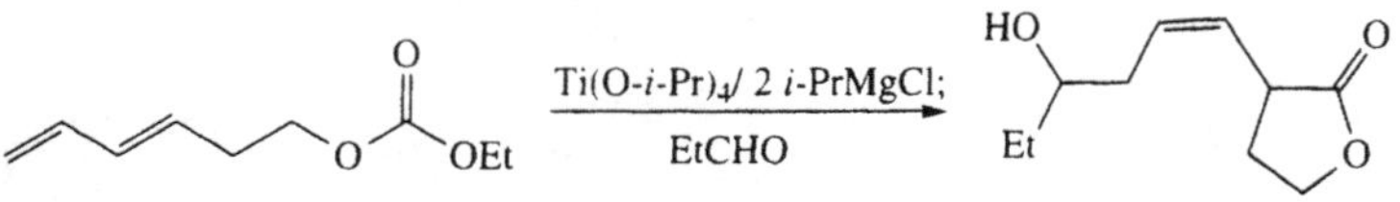

(e)

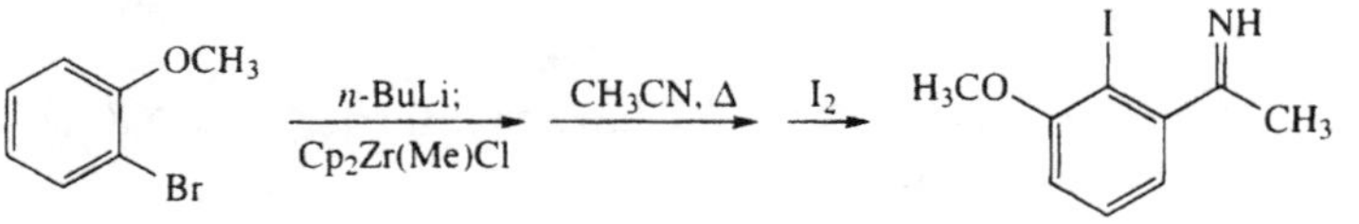

(f)

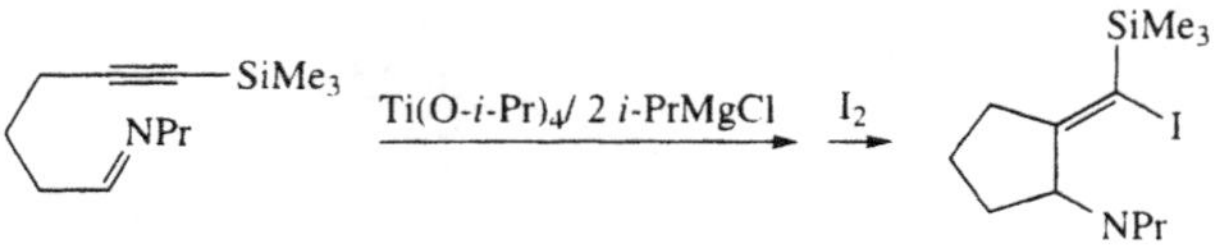

(g) CuBr·SMe$_2$ is simply a more soluble form of CuBr.

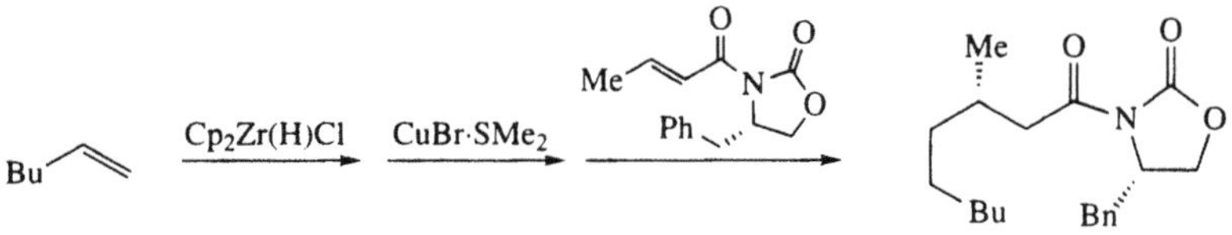

(h)

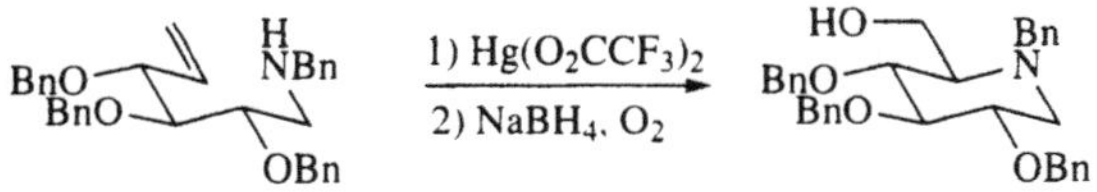

(i)

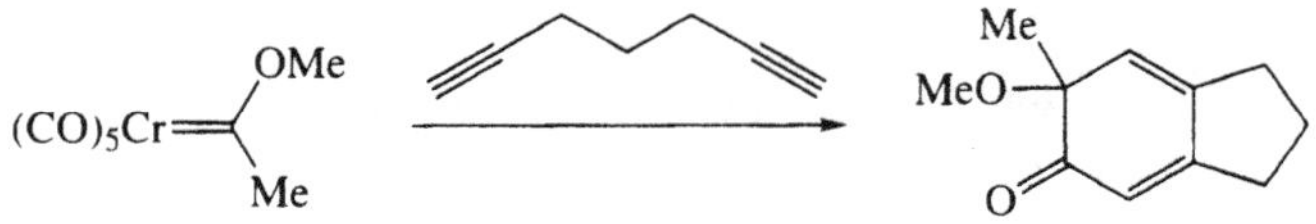

(j)

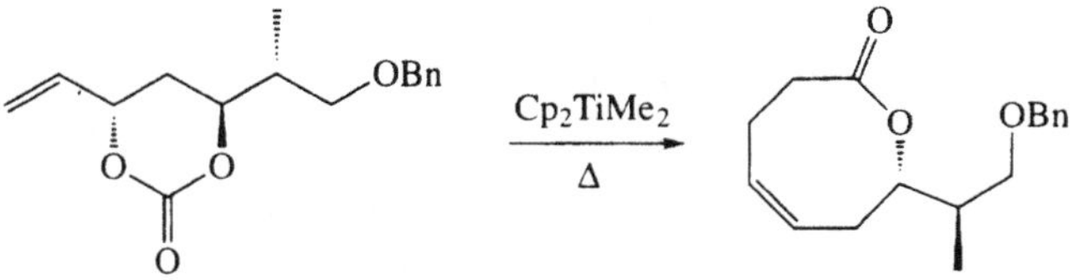

4. The order of reactivity of aryl or alkenyl halides in Pd-catalyzed cross-couplings is I > Br > Cl. The preparation of enediynes by a Sonogashira coupling, however, is usually carried out with *cis*-dichloroethylene as the substrate; yields are considerably poorer when *cis*-dibromoethylene is used instead. Why? (*Hint*: The key intermediate may be more prone to undergo what side reaction when the dibromide or diiodide is used?)

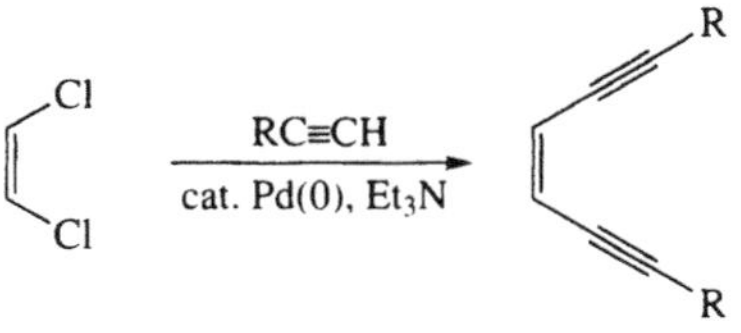

7

Mixed-Mechanism Problems

In Chapters 2 through 6 you learned how to draw polar basic, polar acidic, pericyclic, free-radical, and transition-metal-mediated and -catalyzed mechanisms. The reactions in the following problems may proceed by any of these mechanisms. Before you solve each problem, then, you need to identify its mechanistic class. See Chapter 1 if you have forgotten how to do so.

1. Solve the mechanism problems (problems 3 and 4) at the end of Chapter 1.

2. The following sequence of reactions was reported recently as part of a synthesis of the natural product qinghaosu, the active ingredient in a number of Chinese folk medicines. Qinghaosu is an antimalarial drug, a property of increasing importance as new strains of malaria appear that are resistant to the drugs that have been used until now.

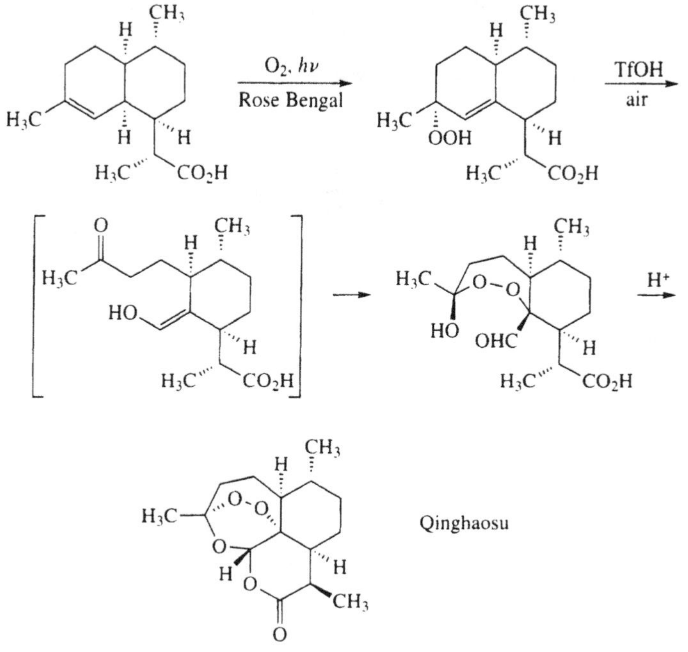

The compound in brackets has been shown to be an intermediate in the conversion of the hydroperoxide to the endoperoxide. Air is not required for its formation. In fact, it is isolable at low temperatures (-20 °C) when air is excluded. If it is then exposed to air, it is transformed to the endoperoxide.

Draw mechanisms for each of the steps in this sequence. Your mechanisms should take the preceding information into account.

3. The following questions are based on a total synthesis of isocomene, an angular triquinane natural product.

    (a) (i)   Draw a mechanism for the formation of **1.**

    (ii) Name any pericyclic reactions in your mechanism. Be as specific as possible.

    (iii) Explain why **1** is obtained diastereoselectively.

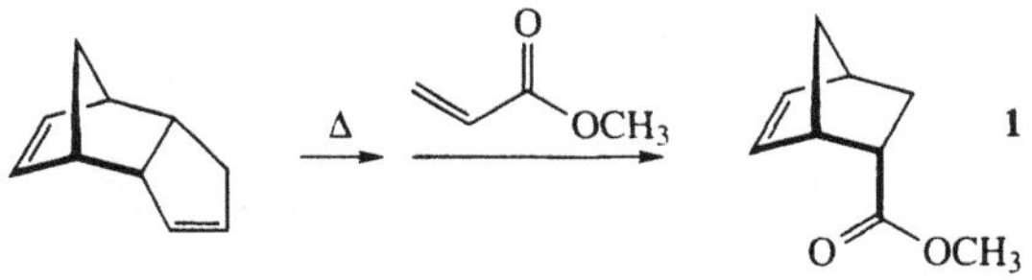

(b) Draw mechanisms for the conversion of **1** to **2** and the conversion of **2** to **3.**

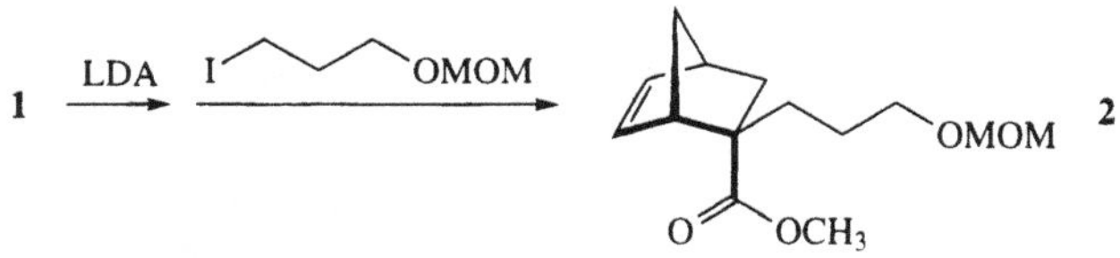

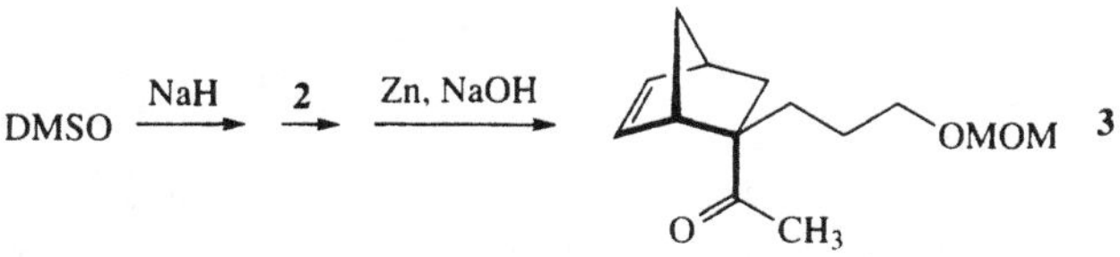

(c) (i)   The conversion of **3** to **4** proceeds by a pericyclic mechanism. Name it.

    (ii) Does this reaction proceed thermally or photochemically?

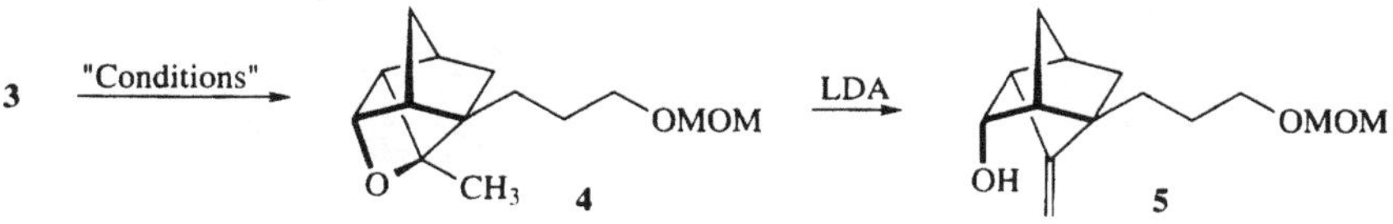

(d) Draw mechanisms for each transformation from **4** to **11** (isocomene).

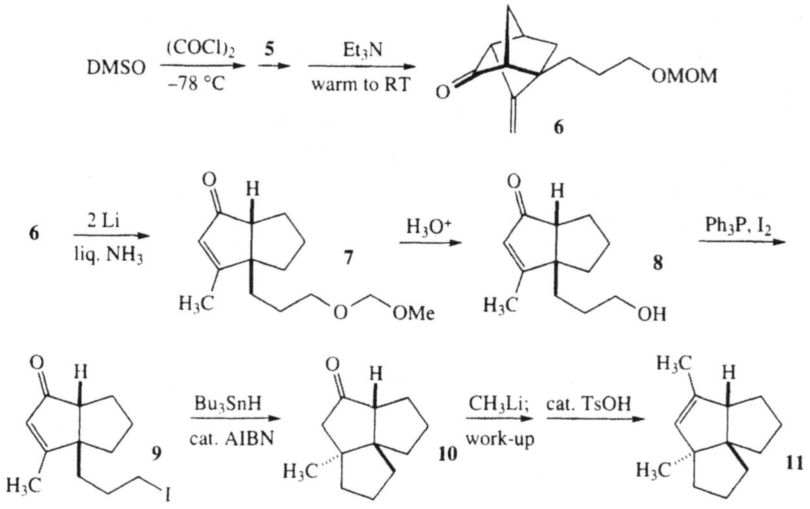

4. The technology used in the preparation of isocomene can be modified to carry out several other interesting transformations.

(a) Draw mechanisms for each transformation from **1** to **4**.

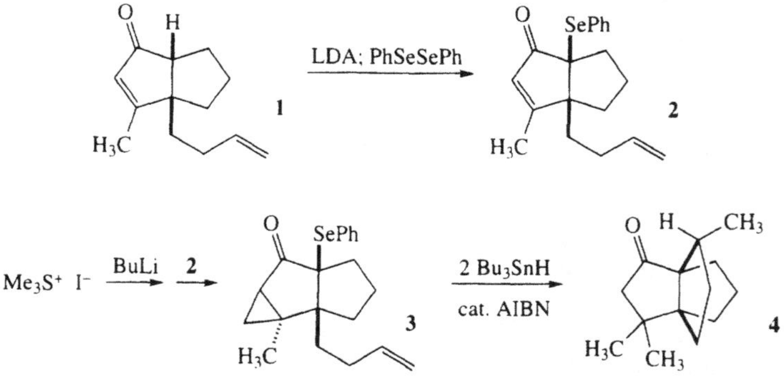

(b) When **5** is treated with LiAlH₄, an intermediate is obtained that retains the alcohol group. Al(O-*i*-Pr)₃ and acetone accomplish the oxidation of the alcohol to the corresponding ketone, which then spontaneously transforms into **6**. Draw mechanisms for the conversion of **5** to the intermediate alcohol and the conversion of the derived ketone to **6**.

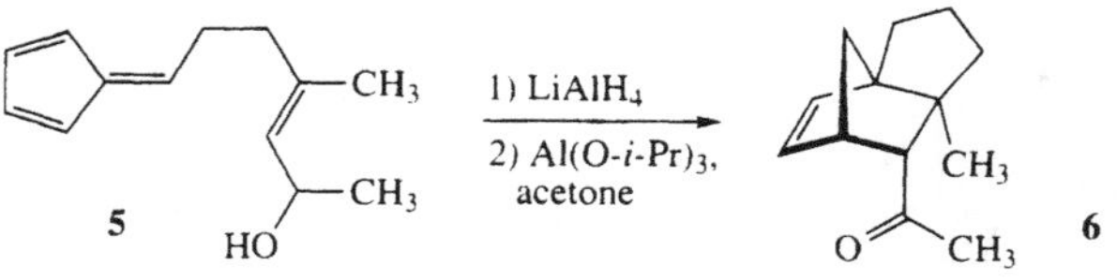

(c) Draw a mechanism for the conversion of **7** to **8**. LiDBB (lithium 4,4'-di-
t-butylbiphenylide) is a source of Li metal that is soluble in THF; you may
treat it as if it were Li metal.

5. Draw a reasonable mechanism for each step in the following synthetic sequence.

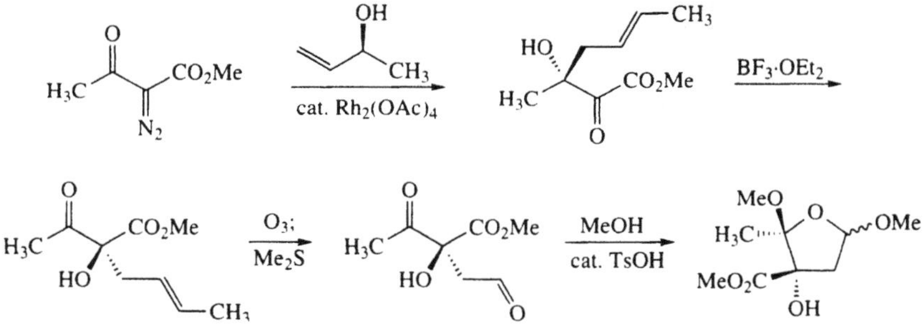

6. Two multistep mechanisms may be drawn for the following reaction. The two
mechanisms differ both in the order of bond formation and the nature of some
of the individual steps. Draw one or both of them.

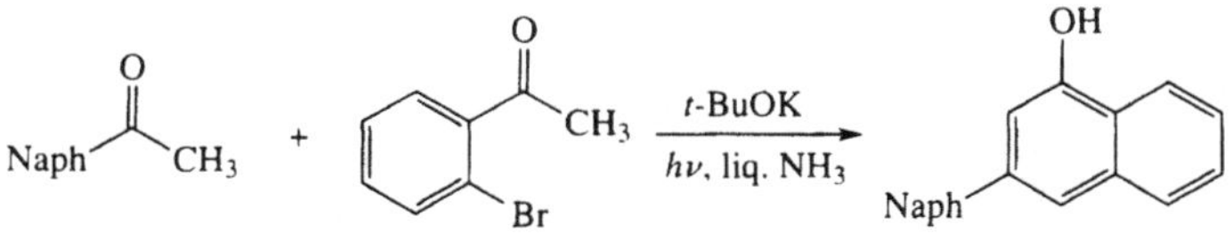

7. Draw a reasonable mechanism for the following reaction.

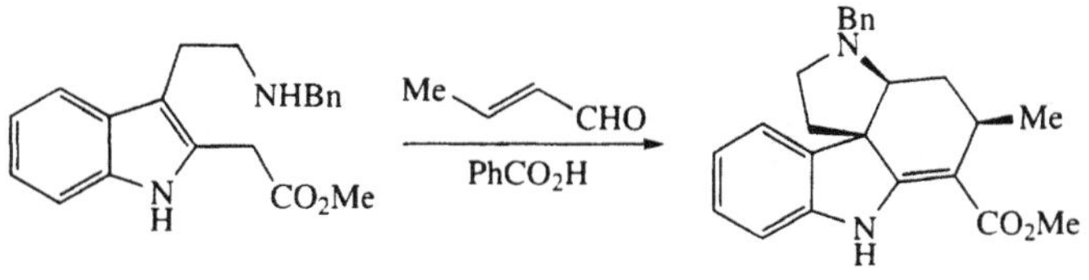

Hints:

(i) The order of bond formation is important.

(ii) One of the new bonds in the product is formed, then broken, and then re-
formed.

8. The following synthetic sequence recently appeared as the key part of a synthesis of some morphine analogs. Draw reasonable mechanisms for each step. Dbs = 2,6-dibenzosuberyl, a protecting group for N.

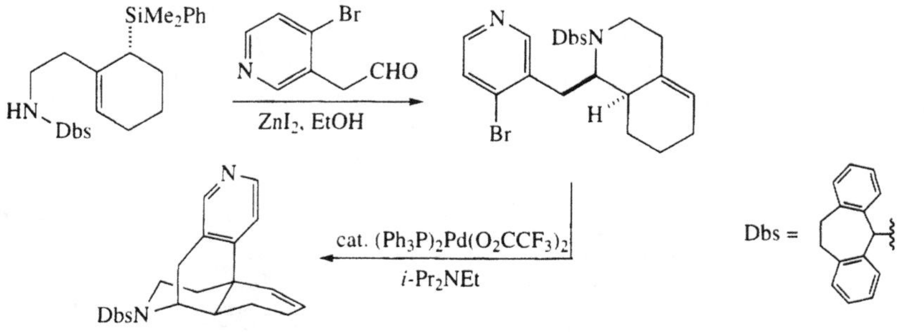

A Final Word

The purpose of this book has been to teach you how to draw a reasonable organic mechanism for almost any organic reaction you encounter. Sometimes, though, you may be unsure whether your mechanism is reasonable, sometimes more than one reasonable mechanism can be drawn, and sometimes what at first seems to be a reasonable mechanism may seem less reasonable when further information has been gathered. In these cases, you may want to go to the literature to see what is already known about the mechanism of the reaction.

Many sources discuss the mechanisms of particular organic reactions in greater detail than has been possible in this book. Smith and March's *Advanced Organic Chemistry*, 5th ed. (New York: Wiley, 2001) is an indispensable reference for the synthetic organic chemist, both for its compendium of synthetic procedures, its detailed discussion of the mechanisms of many of these reactions, and its huge number of references to the primary and secondary literature. The *Organic Reactions* series features discussions of the mechanisms of many widely used reactions. The encyclopedias *Comprehensive Organic Synthesis* (Elmsford, NY: Pergamon, 1991), *Comprehensive Organometallic Chemistry II* (Pergamon, 1995), and *Comprehensive Organic Functional Group Transformations* (Pergamon, 1997) are also good places in which to look for the mechanisms of well-known reactions. The scientific publisher Thieme has compiled a database of approximately 12,000 English-language review articles of interest to synthetic organic chemists; you can download it for free from http://www.chem.leeds.ac.uk/srev/srev.htm. This database includes reviews on almost every imaginable subject in organic synthesis. Reviews are sometimes the only places in the journal literature where reaction mechanisms are discussed in any detail.

Finally, the canon of knowledge presented in this text and the preceding references has been developed over a long period of time from difficult, detailed experimental investigations of the effects of concentration, solvent, isotopic substitution, substrate structure, and other variables on the rates and yields of reactions. In fact, writing a reasonable mechanism from one's own knowledge is a piece of cake compared with the work required to verify it experimentally. The experimental methods used to determine reaction mechanisms are discussed in great detail in several textbooks, including Lowry and Richardson's *Mechanism and Theory in Organic Chemistry*, 3rd ed. (New York: Addison Wesley, 1987), Carey and Sundberg's *Advanced Organic Chemistry, Part A*, 4th ed. (New York: Plenum, 2000), and Carroll's *Perspectives on Structure and Mechanism in Organic Chemistry* (Monterey, CA: Brooks/Cole, 1998). These books and others provide a more in-depth look at the very complex field of organic reaction mechanisms.

Index

Note: t = table.

persistent, 227
stability, 224–27
Free-radical reactions, 26, 212, 255
  addition and fragmentation, 225–26
  mechanisms for, 38–39
  solvents of choice, 235
  substitution, 238–39
  typical, 232–38
Friedel–Crafts reactions, 126
Frontier molecular orbital (FMO)
    theory
  and butadiene electrocyclic ring
    closing, 164
  and Diels–Alder reactions, 173–76,
    183–84
  and sigmatropic rearrangements,
    200–3
Furan, 15

Galvinoxyl, 227
Germanes, 326
Gilman reagents, 270
Glucals, 175
Glycosylation reaction, 119
Green mechanism, 289
Grignard reactions
  dibromoethane in, 83
  metal insertion reaction, 72–73
Grignard reagents
  addition to carbonyl compounds, 59
  conjugate addition reactions of, 297
  in imine production, 66
  reaction with esters, 72–73
Grossman's rule
  and carbocations, 105
  and conventions of drawing
    structures, 1–3
  and drawing mechanisms, 23–24
Ground state, of resonance structures,
    8–9

H shifts, [1,3] thermal, 202–3
β-Halide eliminations, 279
Halocyclopropanes, 168–69
Halogenation, of alkanes, 239–40
Halogen-metal exchange
  C(sp^2)–X σ bonds, 69–80
  metal insertion, 78–80, 83–84
Halonium ions, 123–24

Hammond postulate
  and carbocations, 106
  and Markovnikov's rule, 123
  on TSs, 22
Heck reaction, 313–14
Hemiacetals, 59–60, 133, 136t
Hemiaminals, 61
Hetero-ene reaction, 152
Heterolytic bond strength, 235
Hexachlorocyclopentadiene, 174
1,3,5-Hexatrienes
  cyclohexadiene equilibrium with, 157
  electrocyclic ring closings, 165
  MOs of, 156
Highest occupied molecular orbitals
    (HOMOs), 155
Hillman–Baylis reaction, 100
Hofmann rearrangement, 90
Hofmann–Loeffler–Freytag reaction, 243
HOMO (Highest occupied molecular
    orbitals), 155
Homogenesis, 236
Homolytic bond strengths, 214, 235
Hunsdiecker reaction, 252
Hybrid orbitals, 12–13
Hybridization
  molecular shape, 9–13
  sp, 12
  sp^2, 12
  sp^3, 12
  and stability of carbocations, 106, 108
Hydrates, 60
Hydrazone, 60
α-Hydride abstraction, 321–22
β-Hydride abstraction, 298, 301–2
α-Hydride eliminations, 289
β-Hydride eliminations, 279, 287–88
1,2-Hydride shifts, 114–16
1,5-Hydride shifts, 114
Hydroboration, 285, 292–93
Hydrochlorofluorocarbons (HCFCs), 265
Hydroformylation, 285–86
Hydrogen
  addition across π bonds, 254–55
  near reaction centers, 2
  tracking, 1
Hydrogenation, metal-catalyzed, 285,
    293–94
Hydrogenolysis, 309

Weinreb amides, 73
Wilkinson's catalyst, 284, 323
Wittig reactions, 180, 188
Wittig rearrangement, 199–200, 261
Wolff rearrangement, 87, 89, 179
Woodward–Hoffmann rules, 155
    for cycloadditions, 189–90
    for electrocyclic reactions, 165–67
    for pericyclic reactions, 215

    for sigmatropic rearrangements,
       204–5, 262
    and the Wittig reaction, 188
o-Xylylenes, 157, 171

Ziegler–Natta catalysts, 288–89
Zinc, 69–80
Zirconium (Zr), 292–93, 297–98
Zirconocene, 167

CPSIA information can be obtained
at www.ICGtesting.com
Printed in the USA
LVHW05s2315040818
585985LV00008B/106/P